Chemical and Materials Engineering

Chemical and Materials Engineering

Edited by **Sally Renwick**

Willford Press
New York

Published by Willford Press,
118-35 Queens Blvd., Suite 400,
Forest Hills, NY 11375, USA
www.willfordpress.com

Chemical and Materials Engineering
Edited by Sally Renwick

International Standard Book Number: 978-1-68285-321-4 (Hardback)

Printed in China.

Contents

	Preface	IX
Chapter 1	**Application of statistical technique to the analysis of passivation of Al - Zn alloy systems in brine** E. Chinedu Ekuma, E. Ndubuisi Idenyi and O. Israel Owate	1
Chapter 2	**Corrosion inhibition and adsorption behavior of methionine on Mg- Al-Zn alloy** Nadia H. Helal	8
Chapter 3	**Optimization of Li-Zn ferrite synthesis parameters using the experimental design method** M. Kavanloui, B. Hashemi and E. Nourafkan	19
Chapter 4	**The estimation of the cooling tower height by modeling the water and air contact situation in cooling tower falling film** Mortaza Gholizadeh and Mosen Momayyeza	27
Chapter 5	**Design and simulation reconfigurable liquid crystal patch antennas on foam substrate** S. Missaoui, A. Gharbi and M. Kaddour	34
Chapter 6	**Algorithm and computational complexity of biochemistry** Lutvo Kurić	41
Chapter 7	**$CuO-CeO_2$ catalytic systems destined for CO removal synthesized by means of the Pechini method: An evaluation of the structures obtained** L. S. Neiva, H. M. C. Andrade and L. Gama	48
Chapter 8	**Equilibrium modeling for a downdraft biomass gasifier for cotton stalks biomass in comparison with experimental data** Koroneos C. and Lykidou S.	55
Chapter 9	**Fracture behavior of cryogenically solidified aluminumalloy reinforced with Nano-ZrO_2 metal matrix composites (CNMMCs)** Joel Hemanth	63
Chapter 10	**Investigation of turbulence mixed convection in air filled enclosures** Hamid Reza Goshayeshi and Mohammad Reza Safaiy	75
Chapter 11	**Short review on the crystallization behavior of PET/clay nanocomposites** Ranjana Sharma, Himani Joshi and Purnima Jain	84

Chapter 12 **A kinetic and mechanistic study on the oxidation of arginine and lysine by hexacyanoferrate (III) catalysed by iridium (III) in aqueous alkaline medium** 89
Anjali Goel and Ruchi Sharma

Chapter 13 **Tween 60 – amido black 10B – ascorbic acid system: Studies of photogalvanic effect and solar energy conversion** 95
K. R. Genwa and C. P. Sagar

Chapter 14 **Factors controlling the structural properties of carbon nitride films deposited by electrochemical method** 104
M. A. Moharram, F. H. Abd El-kader, Kh. I. Grais, M. G. Khafagi and Fathia Mamdouh

Chapter 15 **Emissions control catalysts: Links between Fecralloy preoxidation and coating adherence** 115
Stephen A. Adegbite

Chapter 16 **Modified exponentially weighted moving average (EWMA) control chart for an analytical process data** 121
Alpaben K. Patel and Jyoti Divecha

Chapter 17 **Enhanced photoluminescence from ZnO/ZnS core-shell structure** 130
Sarmila Dutta, Shibabrata Basak and Pijus Kanti Samanta

Chapter 18 **Chlorine adsorption kinetics of activated carbon from selected local raw materials** 135
Yusufu, Mohammed I., Ariahu, C. C., Nkpa, N. N. and Igbabul, B. D.

Chapter 19 **Synthesis and characterization of (Thallium-Tin) doped Bismuth based superconducting materials** 142
Rashid Mahmood and Muhammad Javed Iqbal

Chapter 20 **Kinetic and thermodynamic studies on the adsorption behavior of Rhodamine B dye onto animal bone meal** 150
Mohammadine El Haddad

Chapter 21 **Autocatalysed oxidation of etophylline by permanganate in aqueous sulphuric acid medium- kinetics and mechanistic study** 157
H. V. Rajeshwari, K. S. Byadagi, S. T. Nandibewoor and S. A. Chimatadar

Chapter 22 **Removal of ammonium and phosphates from aqueous solutions by activated and modified Bulgarian clinoptilolite** 171
Nadejda Taneva

Chapter 23 **Electrochemical behaviour of some copper alloys in sodium chloride solutions containing different inorganic additives** 178
H. E. El-Feky, N. H. Helal and M. R. Negem

Chapter 24 **Purification of an industrial aluminum alloy by melt stirring during ohno continuous casting process** 193
Simbarashe Fashu, Li Jianguo and Hu Qiao Dan

Chapter 25 **Preparation and characterization of exfoliated nanocomposite of polyvinyl acetate and montmorillonite** 202
Hui-Wang Cui and Guan-Ben Du

Chapter 26 **Biosorption of copper(II) and lead(II) ions from aqueous solutions by modified loquat (*Eriobotrya japonica*) leaves (MLL)** 209
Akl M. Awwad and Nidà M. Salem

Chapter 27 **Characterization of stainless steel SS-1672** 220
Farah Noori and Solveig Melin

Chapter 28 **Comparison of unconstrained search methods for correlation of binary vapor liquid equilibrium (VLE) by five variants of Wilson model** 233
A. Aravind Kumar and A. Ravi Prasad

Chapter 29 **Solar photocatalytic degradation of phenol using nanosized ZnO and α-e_2O_3** 240
Hamza A., Fatuase J.T., Waziri, S. M. and Ajayi, O. A.

Chapter 30 **Time dependent mathematical model of air pollutants emitted from time-dependent elevated line source into a stable atmospheric boundary layer** 246
Moka Shekhu and C. Sulochana

Chapter 31 **Inter-provincial cooperative control of air pollution in China** 259
Jian Xue and Changmin Li

Chapter 32 **Extreme pressure property of Carbon Nano Tubes (CNT) based nanolubricant** 266
Shubrajit Bhaumik, S.Prabhu, VarunKarpurapu and Nihit Uppala

Permissions

List of Contributors

Preface

This book was inspired by the evolution of our times; to answer the curiosity of inquisitive minds. Many developments have occurred across the globe in the recent past which has transformed the progress in the field.

The fields of materials science and chemical engineering are closely associated. The advances in one field contribute to the progress of the other. Included in this book are significant topics like semiconductors, high temperature alloys, optimization of mixed materials, etc. The chapters included within consist of contributions made by international experts of chemical engineering and materials science. Students, researchers, experts and all associated with chemical engineering, materials science and allied fields will benefit alike from this book.

This book was developed from a mere concept to drafts to chapters and finally compiled together as a complete text to benefit the readers across all nations. To ensure the quality of the content we instilled two significant steps in our procedure. The first was to appoint an editorial team that would verify the data and statistics provided in the book and also select the most appropriate and valuable contributions from the plentiful contributions we received from authors worldwide. The next step was to appoint an expert of the topic as the Editor-in-Chief, who would head the project and finally make the necessary amendments and modifications to make the text reader-friendly. I was then commissioned to examine all the material to present the topics in the most comprehensible and productive format.

I would like to take this opportunity to thank all the contributing authors who were supportive enough to contribute their time and knowledge to this project. I also wish to convey my regards to my family who have been extremely supportive during the entire project.

Editor

Application of statistical technique to the analysis of passivation of Al - Zn alloy systems in brine

E. Chinedu Ekuma[1]*, E. Ndubuisi Idenyi[2] and O. Israel Owate[1]

[1]Department of Physics, University of Port Harcourt, P. M. B. 5323, Choba, Port Harcourt, Rivers State, Nigeria.
[2]Department of Industrial Physics, Ebonyi State University, P. M. B. 053, Abakaliki, Ebonyi State, Nigeria.

The high demand for light metals as engineering materials lies in their high strength - to - weight ratio with good corrosion resistance in diverse environments. Passivation phenomena in these materials especially Al - based binary alloys have well been documented. However, the present study which aims at using model equations to investigate the passivation trend in various compositions of duplex Al - Zn alloy systems subjected to different concentrations of brine is totally new. The comparative plots of the generated values from the model equations with experimental data show that it correlated well with experimental observations.

Key words: Model equations, correlation, passivation, Duplex alloy systems.

INTRODUCTION

The production of materials with high corrosion resistance and relatively low cost has been the focus of material scientists and engineers. This has largely been hampered by the composite nature of many materials even metals coupled with the complex nature of material degradation in diverse environments. Even at that, many positive results have been achieved in this regard (Kontkova et al., 1995; Feliu and Morcillo, 1993; Ekuma and Idenyi, 2006; Idenyi et al., 2006).

The intricate nature of the corrosion process may be better appreciated by recognizing that many variables are involved: environmental, electrochemical and metallurgical history of the composite etc (Ekuma and Idenyi, 2006) which is basically an electrolytic (Feliú and Morcillo, 1982; Rozenfeld, 1972) process. The electrolyte is a scale film of moisture (normally few monolayers), or an aqueous film (a thickness of hundreds of microns) when metals or composites show up perceptibly damping. Among the external factors that are determinant on the corrosion rate definition especially in aqueous environment, it is possible to mention: 1) the lifetime of the electrolytic film on the metal surface, 2) the chemical composition of environment and 3) the ambient temperature. These factors are important in corrosion measurement and general corrosion design.

The weight difference per unit area (scale film thickness) can be a good passivation monitoring parameter especially with regards to the statistical nature of passivation rate processes governed by a number of variables, though its widespread application have been hampered largely because, the use of statistics requires specialist knowledge and no reference standards exist (Ekuma et al., 2007).

However, for a passivated film to be effective, it must provide a protective barrier that keeps the corrosion current on the metal-environment interface at relatively low enough rate so that the extent of corrosion damage is reduced. An effective film is one that resists the breakdown of the passive film and if abraded, repassivates immediately (Frankenthal and Kruger, 1978); hence, the conditions that promote corrosion of aluminum and its alloys, must be those that continuously abrade the film mechanically or promote reactions that locally degrade this protective oxide film by minimizing the availability of oxygen to rebuild it (Ekuma et al., 2007).

The scale film thickness is unavoidably a good index for passivation characterization especially the statistical nature of passivation phenomenon. It clearly gives an insight into materials behaviour in a passivating media. The formation of thin protective film at metal interfaces, which could be successfully used for corrosion control, has long been the area of research for corrosion experts.

*Corresponding author. E -mail: panaceamee@yahoo.com.

The layers are formed either during the adsorption of corrosion inhibitors or in reactions of the cations of the metal to be protected with some components of the medium (Ekuma et al., 2008). The latter case involves not only nanosized layers formed by corrosion inhibitors but also thick conversion (oxide, phosphate, etc) coatings (Kuznetsov, 2006).

Also, the scale film deposition rate has imbedded in it the vital property of the rate at which adsorption of corrosion products occurs on the surface of any passivating material subjected to simulated corrosive environments and the stability of metals or alloys in such environments will depend on the protective properties of organic or inorganic films, as well as on the layer of corrosion products. Some film characteristics, such as chemical composition, conductivity, adhesion, solubility, hygroscopicity and morphology, determine its ability as controlling barrier to different kinds of attack and corrosion rates (Stratmann et al., 1983). On the other hand the stated characteristics depend, in turn, on chemical composition and metallurgical history of the metal, on physicochemical properties of coating and on environmental variables such as atmospheric conditions, type and amount of pollutants, wet-dry cycles, etc (Stratmann et al., 1983; Stratmann and Streckel, 1990).

The use of statistical models for predicting the passivation (by extension, corrosion) characteristics of metals and composite materials in diverse environments have been predicted for decades but its application has largely been hampered due to the special skill and above all no reference standard that is involved in the analysis. Deterministic and statistical models have both been formulated for better perceptive of the environment. Deterministic models are based on the basic mathematical descriptions of the environmental processes, in which effects are generated by causes while the statistical models are based on semi-empirical statistical relations between available data and measurements that do not necessarily reveal any relation between cause and effect, but reveals the underlying correlation between sets of input data (predictors) and targets (predictands). Example of the deterministic model is Euler models (Ekuma and Idenyi, 2007) and that for statistical models are time series analysis (Hsu, 1992) and artificial neural networks (Abdu-Wahab and Al-Alawi, 2002).

Aluminium based alloys have of recent found diverse usage in engineering industry. This can be attributed to its low density (2.71kg/cm^3); low emission, hence its frequent use as heat radiators; does not become brittle at low temperatures (strength and ductility slightly increases) like most bcc lattice materials (e.g. steel); high coefficient of linear expansivity in comparison with other metals (for instance, it is twice that of steel); specific strength with light weight as engineering materials for transportation which greatly reduces fuel consumption; good conductors of heat and electricity etc. These out lined properties of aluminium may be greatly enhanced by cold working and by alloying for instance with zinc (Callister, 1997).

In this paper, the statistical approach will be adopted to analysis the passivation trend of various alloys of aluminium - zinc alloy: Al - 2.0% Zn; Al - 2.5% Zn and Al - 3.5% Zn subjected to varying concentrations of brine: 0.1; 0.25; 0.5 and 1.0 M which are simulated from 58.5% pure assay of NaCl using standard procedure. This range of brine adequately accommodates all the positively saline environments habouring Al - based alloys as it contains the aggressive ion, chloride, which is of practical importance because, it is present in large amount both in sea water, road salts and some soils and in lower concentration in other natural sources. The aim however, will be to formulate model equations based on logarithmic regression analysis for the investigation of passivation trend in these various Al - Zn alloys. The validity of the developed model equations will be checked by plotting comparative plots of the modeled values with experimental data. It is however envisaged that the present study will aid greatly in gaining more insight into the dynamics of passivation in Al - based alloys and even in other passivating engineering materials.

DEVELOPMENT OF THE MODEL EQUATION

The rate of material degradation in any environment is better evaluated in terms of the corrosion penetration rate (CPR) which by extension gives the passivation rate (PR). This is obviously due to the fact that it gives reliable information to corrosion experts and is given mathematically as:

$$CPR = PR = \frac{k\Delta w}{\rho A t} \qquad (1)$$

Where Δw is weight difference after exposure time t, ρ and A are density and exposed specimen area respectively and k is a constant whose magnitude depends on the system of units used. For instance, when k = 87.6, CPR is in mm/yr and Δw, t, ρ and A are expressed in mg, hrs, g/cm^3 and cm^2 respectively.

From equation (1), we can have that:

$$InPR = \text{In}\left(\frac{k\Delta W}{\rho A}\right) - \text{Int} \qquad (2)$$

It is straightforward to show from equation (2) noting that $\Delta W, \rho, A$ and k are all constants that:

$$PR = \eta - \lambda \text{Int} \qquad (3)$$

Where η is a constant $= \frac{k\Delta w}{\rho A}$ ($k, \Delta w, \rho$ and A are all constants) and λ is an empirical constant which is model fitted.

Equation (3) is the compact form of the model equation used in this analysis that is in tandem with the observations made from the passivation rate profile of experimental values which utmostly forms the basis of the adoption of logarithmic model for the present study.

EXPERIMENTAL PROCEDURE

The data of Ekuma (2006), a destructive testing method which adopted the weight loss technique is used in this analysis. After the initial treatment of the samples (casting and machining to sizeable coupons; polishing of the surface according to ASTM standards), 72 test coupons were divided into three groups (of six each) with each of the alloy containing varying fractions of Zn in each of the four (4) simulated concentrations of brine. A set of coupon was withdrawn 24 hourly, washed with distilled water, cleaned with acetone and dried in an open air. The final weight of each of the test coupon was determined using the digital analytic weighing balance as to enable calculation of the passivation rate from the weight difference. In all, the samples were allowed to stand for 144 h (6-days) with a set of test coupons withdrawn 24 hourly for corrosion rate characterization.

RESULTS

Table 1 is the table of the modeled equations, coefficient of correlation (R - value) and coefficient of determination (R^2 - values) while Figures 1 - 12 show the comparative plots of passivation rate as a function of time for the experimental data and model data for the various samples: Al - 2.0% Zn; Al - 2.5% Zn and Al - 3.5% Zn in various concentration media (0.1; 0.25; 0.5 and 1.0 M of brine environment).

ANALYSIS OF RESULT

The statistical models were developed from the regression analysis of the experimental data based on the fore - knowledge of the passivation trend of the studied binary alloy system which is basically logarithmic (or exponential) in nature (Equation 2).

From the comparison plots (Figures 1 - 12) and Table 1, it can be verified that the observed experimental values correlated well with the modeled values. This is clearly significant from the high correlation coefficient values (R - values) of all the binary alloy systems being studied. For the entire composite, it can be seen that the correlation coefficient is in the range of $0.92 \leq R \leq 0.99$ (Table 1) which depicts that $R \approx 1$ hence, nearly perfect positive correlation. In order to measure the total variation in passivation rate that has been adequately accounted for by the variation in exposure time, we measured the coefficient of determination of the various duplex alloy systems being studied. It can be inferred from the table that the range of the coefficient of determination which is $0.82 \leq R^2 \leq 0.98$. The conclusion that can be drawn from this observation is that the model fit developed from the

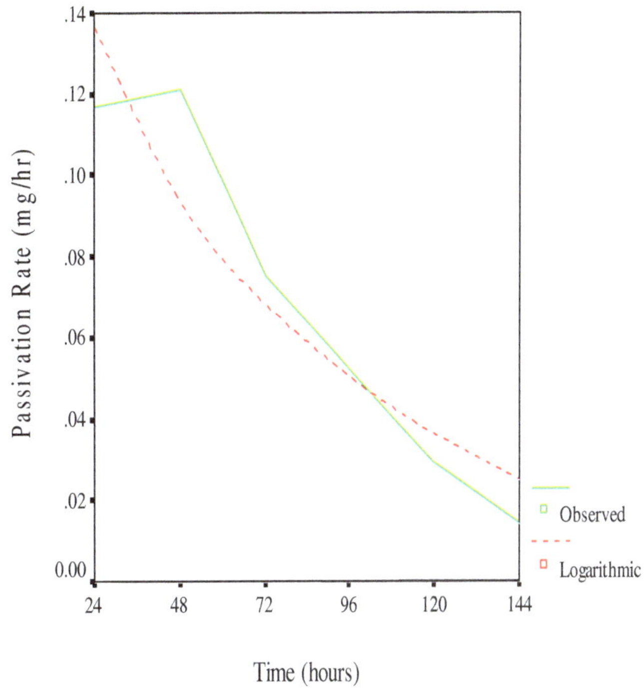

Figure 1. The graph of passivation rate (mg/h) for the experimental data and model data for Al - 2.0% Zn in 0.1 M NaCl media concentrations.

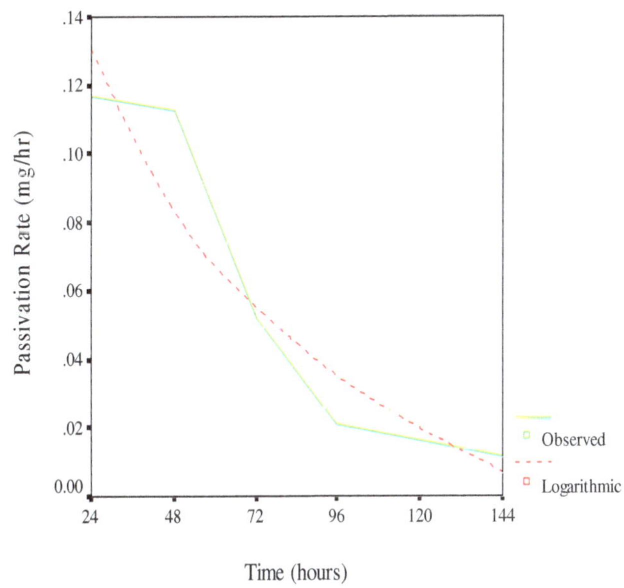

Figure 2. The graph of passivation rate (mg/h) for the experimental data and model data for Al - 2.0% Zn in 0.25 M NaCl media concentrations.

model equations is accurate and will be a good predicator of the passivation behaviour of the alloy of aluminium (in

Table 1. The modeled passivation parameters for the various Al - Zn alloys in different brine environment.

Media concentration (M)	Coefficient of correlation (R - value)	Coefficient of determination (R^2 - value)	Model equation
	Al - 2.0% Zn in various concentrations of NaCl		
0.1	0.92824	0.86163	$PR = 0.334309 - 0.062312 \ln t$
0.25	0.94223	0.88780	$PR = 0.349456 - 0.068950 \ln t$
0.5	0.95754	0.91689	$PR = 0.163375 - 0.031833 \ln t$
1.0	0.93007	0.86502	$PR = 0.663981 - 0.119952 \ln t$
	Al - 2.5% Zn in various concentrations of NaCl		
0.1	0.95968	0.92098	$PR = 0.390495 - 0.063541 \ln t$
0.25	0.92365	0.85312	$PR = 0.607812 - 0.103443 \ln t$
0.5	0.97318	0.94708	$PR = 0.150609 - 0.025135 \ln t$
1.0	0.94612	0.89514	$PR = 0.073447 - 0.008324 \ln t$
	Al - 3.5% Zn in various concentrations of NaCl		
0.1	0.94362	0.89043	$PR = 0.404876 - 0.078864 \ln t$
0.25	0.90783	0.82415	$PR = 0.305115 - 0.053806 \ln t$
0.5	0.97666	0.95386	$PR = 0.227076 - 0.034941 \ln t$
1.0	0.99213	0.98433	$PR = 0.262394 - 0.047547 \ln t$

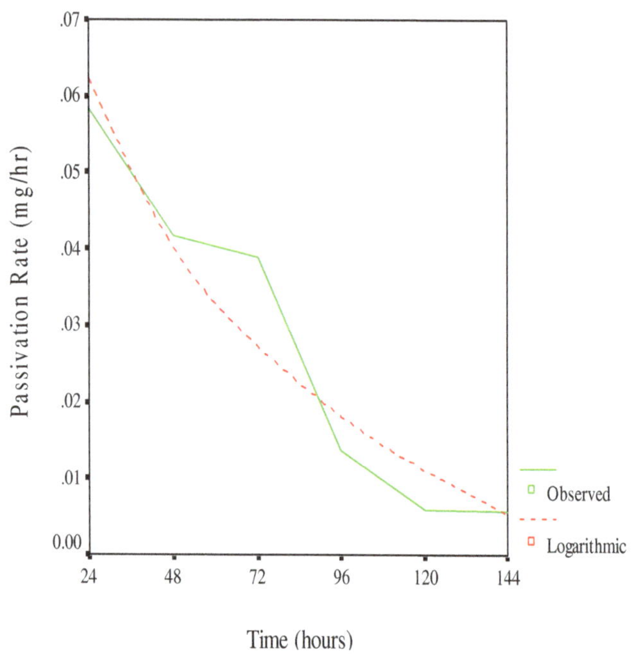

Figure 3. The graph of passivation rate (mg/h) for the experimental data and model data for Al - 2.0% Zn in 0.5 M NaCl media concentrations.

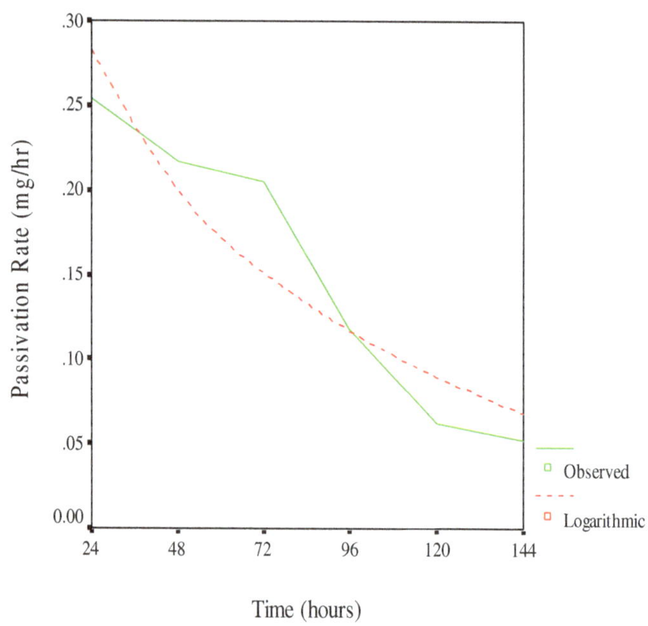

Figure 4. The graph of passivation rate (mg/h) for the experimental data and model data for Al -2.0% Zn in 1.0 M NaCl media concentrations.

this case Al - Zn) in the brine environments being studied. It further points out that approximately over 82.42 to 98.43% of the total variation in passivation rate is accounted for by corresponding variation in the exposure time while only approximately 17.58 to 1.57% are due to the influence of other factors that are not incorporated into the model equations. This model no doubt provides a

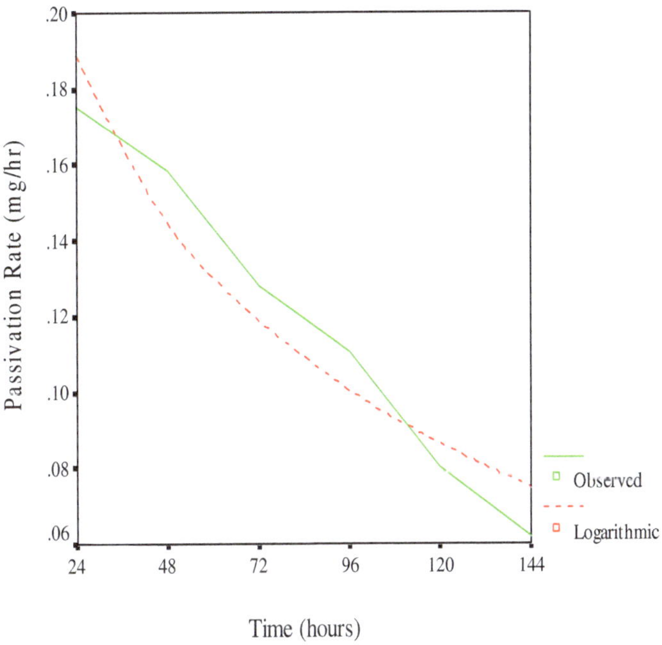

Figure 5. The graph of passivation rate (mg/h) for the experimental data and model data for Al - 2.5% Zn in 0.1 M NaCl media concentrations.

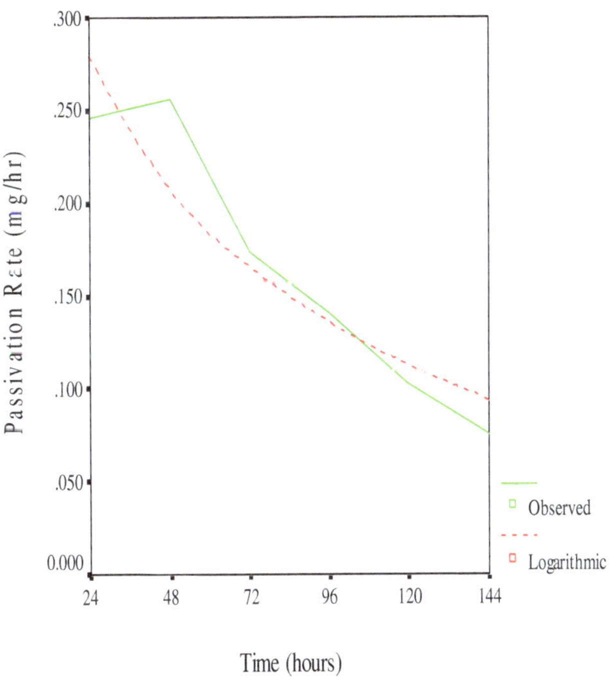

Figure 6. The graph of passivation rate (mg/h) for the experimental data and model data for Al - 2.5% Zn in 0.25 M NaCl media concentrations.

better insight into passivation trend in studied duplex aluminium alloy systems as it can be a good working tool for corrosion experts during corrosion designing of Al - Zn

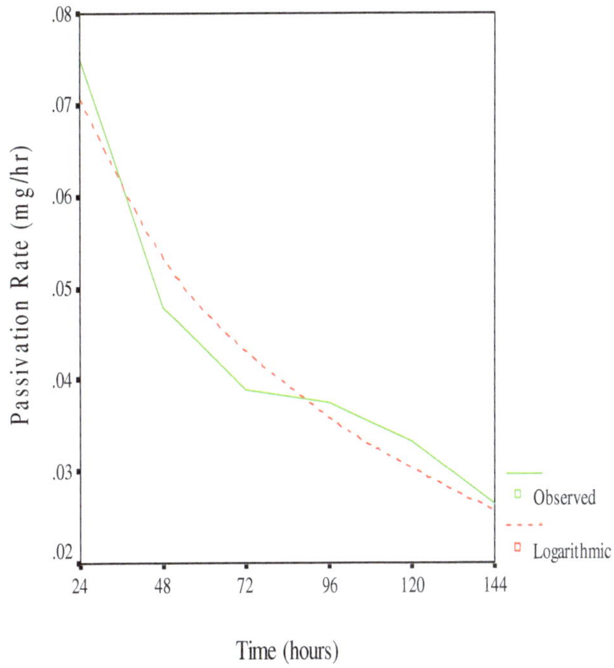

Figure 7. The graph of passivation rate (mg/h) for the experimental data and model data for Al - 2.5% Zn in 0.5 M NaCl media concentrations.

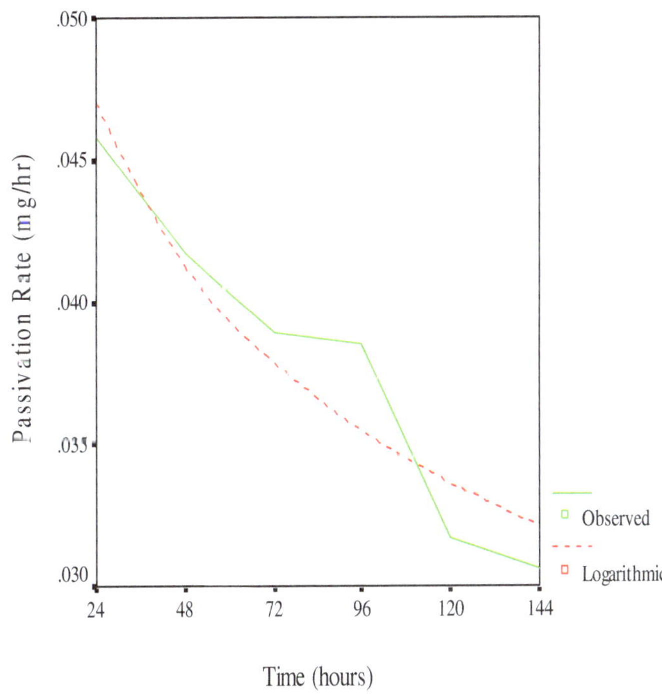

Figure 8. The graph of passivation rate (mg/h) for the experimental data and model data for Al - 2.5% Zn in 1.0 M NaCl media concentrations.

alloys (even other passivating light metallic alloys) in a positively saline environment.

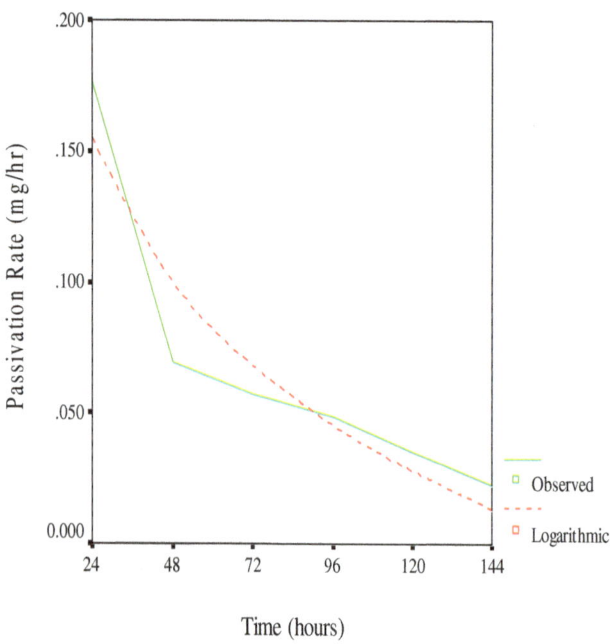

Figure 9. The graph of passivation rate (mg/h) for the experimental data and model data for Al - 3.5% Zn in 0.1M NaCl media concentrations.

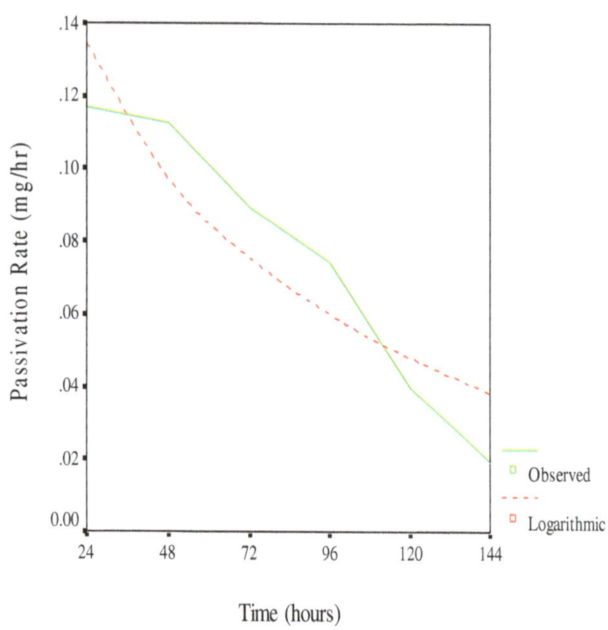

Figure 10. The graph of passivation rate (mg/h) for the experimental data and model data for Al - 3.5% Zn in 0.25 M NaCl media concentrations.

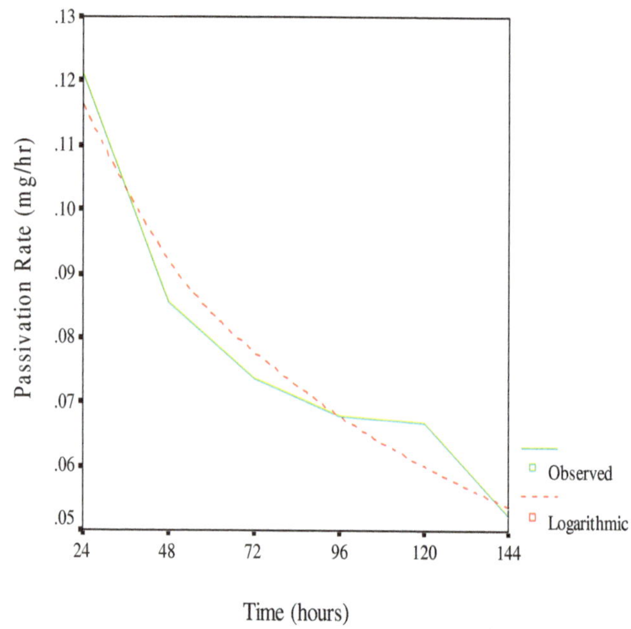

Figure 11. The graph of passivation rate (mg/h) for the experimental data and model data for Al - 3.5% Zn in 0.5 M NaCl media concentrations.

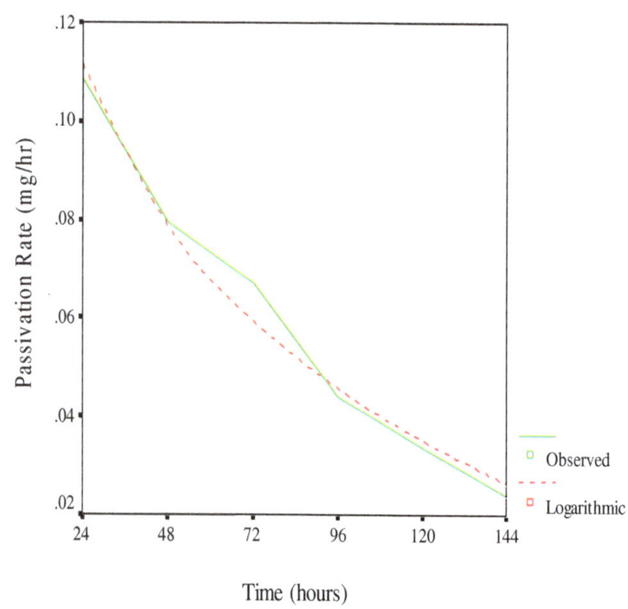

Figure 12. The graph of passivation rate (mg/h) for the experimental data and model data for Al - 3.5% Zn in 1.0 M NaCl media concentrations.

Conclusion

From the model Equations (1 - 12) developed and the statistical parameters obtained for each case (Table 1), it is evident that all the model equations are good predicators of the passivation characteristics of the studied binary alloy systems in the various media concentrations being considered. The high positive coefficient of correlation ($R \approx 1$) is consistent for all the cases and also, the high coefficient of determination for all the cases confirms this assertion. The authors wishes to state that the method used in this work is totally new and has not been used elsewhere except by the authors to their best of knowledge. Thus, we recommend that this work in itself is not conclusive and as such more elaborate work for other alloy compositions be carried out at expanded time scale to really determine the limit of validity of this technique.

ACKNOWLEDGEMENTS

To all our families for their patient. ECE wishes to thank Prof. (Emeritus) Alex, O.E. Animalu and Prof. C.M.I. Okoye for their encouragements during and after this work.

REFERENCES

Abdu-Wahab SA, Al-Alawi S (2002). Assessment and prediction of Tropospheric Ozone Concentration Levels Using Artificial Neural Networks, Environ. Modeling Software 17(6): 563-570.

Ekuma CE (2006). Effects of Zinc Addition on the Corrosion Susceptibility of Al – Alloys in Selected Media Concentration. B.Sc. Project, EBSU, Nigeria.

Ekuma CE, Idenyi NE (2006). The Inhibition Characteristic of Brine on the Corrosion Susceptibility of Al- Zn Alloy Systems. J. Appl. Sci. 6(8): 1751-1755.

Ekuma CE, Idenyi NE (2007). Statistical Analysis of the Influence of Environment on Prediction of Corrosion from its Parameters, Res. J. Phys. USA, 1(1): 27-34.

Ekuma CE, Idenyi NE, Onwu FK, Umahi AE (2008). The Influence of Media Concentrations on the Passivation Layer Characteristics of Al-Zn Alloys in Brine Environment, Asian J. Sci. Res. Pakistan 1(2): 113-121.

Ekuma CE, Idenyi NE, Umahi AE (2007). The Effects of Zinc Additions on the Corrosion Susceptibility of Aluminium Alloys in various Tetraoxosulphate (vi) Acid Environments". Inform. Technol. J. ANSINET, 7(2): 237-241.

Feliú S, Morcillo M (1982). Corrosión y protección de los metales en la atmósfera, Bellaterra.

Feliu S, Morcillo M (1993). The Prediction of Atmospheric Corrosion from Meteorological and Pollution Parameters- I. Long Term Forecast", Corros. Sci. 34(3): 403-414.

Frankenthal RP, Kruger J (1978). Passivity of Metals, The Electrochemical Society, Pennington NJ, USA.

Hsu KJ (1992). Time Series Analysis of Interdependence Among Air Pollutants, Atmospheric Environ. 26: (4)491-503.

Idenyi NE, Ekuma CE, Owate IO (2006). The Influence of Alloy Composition on the Passivation Layer Characteristics of Al-Zn Alloys Systems". Proceedings of Materials Science and Technology 2006 Meeting and Exhibition, Cincinnati, Ohio, USA (October 15th –19th).

Kontkova D, Boschek P, Kreislova K (1995). Effects of Acidification on Atmospheric Corrosion of Structural Metals in Europe", Water, Air Soil Pollut. 85: 2661-2666.

Kuznetsov YI (2006). Protection of Metals, Zashchita Metallov, 1: 42: 3-12.

Rozenfeld IL (1972). Atmospheric Corrosion of Metals. NACE, Houston, USA

Stratmann M, Bohnenkamp K, Engel WJ (1983). An electrochemical study of phase-transitions in rust layers", Corros. Sci. 23: 969-985.

Stratmann SG, Streckel H (1990). On the atmospheric corrosion of metals which are covered with thin electrolyte layers. II. Experimental results. Corros. Sci. 30: 697- 714.

2

Corrosion inhibition and adsorption behavior of methionine on Mg- Al-Zn alloy

Nadia H. Helal

Chemistry Department, Faculty of Science, Fayoum University – Fayoum, Egypt.

Corrosion inhibition behavior of Mg-Al-Zn alloy was investigated by different electrochemical techniques in neutral solutions containing chloride ions at different concentrations. The corrosion rate was calculated in the absence and presence of methionine as an environmentally safe corrosion inhibitor using polarization and impedance techniques. The results reveal that methionine inhibits the corrosion reaction by adsorption onto the metallic surface. Polarization results revealed that methionine acts as an anodic inhibitor.

Key words: Mg-Al-Zn alloy, corrosion, impedance, methionine, polarization.

INTRODUCTION

Because of the inherent superior properties, such as low density, outstandingly high strength to weight ratio and excellent electromagnetic shielding characteristic, Mg alloys are now being utilized increasingly as structural materials, especially in the fields where weight reduction is critical or particular technical requirements are needed, including automotive, aeronautic, electronic and military industries, etc. (Zhang et al., 2002; Gray and Luan, 2002; Ambat et al., 2000). But the inferior corrosion resistance primarily attributed to chemistry activity of Mg has been the most severe challenge to the application of Mg alloys to wider realm, especially to aggressive environments (Ambat et al., 2000; Zhang et al., 2002).

The corrosion resistance of magnesium alloys is generally inadequate and has limited their use. Formation of conversion coating is commonly applied to magnesium alloys in order to increase the corrosion resistance. Such processes vary from the widely used chromate treatment to rare earth metal salts dip treatment (Rudd et al., 2000; Yfantis et al., 2002). In recent years, much research has been carried out to search for more environmentally acceptable processes instead of chromate treatment in surface modification process of magnesium alloys.

Natural products extracted from plant sources (Oguzie et al., 2006; Morad et al., 2002), as well as some groups nontoxic organic compounds, which contain polar with nitrogen, oxygen, and/or sulfur in conjugated systems in their molecules (Moretti et al., 2004; Zhang et al., 2003; Oguzie et al., 2004), have been effectively used as inhibitors in many corrosion systems. The inhibiting action of such compounds is attributed, as a first stage, to the adsorption of the additives on the metallic surface. The roles of the individual functional groups in the inhibition mechanism of organic molecules have been widely reported, but the effect of having two different groups in one molecule, which may influence the corrosion process by different mechanisms, has not been properly documented and is an important consideration in this study. The test inhibitor for the present study is methionine [CH_3–S–CH_2–CH_2–CH(CO_2H)NH_2], an amino acid containing both the (–NH_2) and (–S–CH_3) groups in its molecule. Amino acids are attractive as corrosion inhibitors because they are relatively easy to produce with high purity at low cost and are soluble in aqueous media. Ashassi-Sorkhabi et al. (2005) reported the successful application of methionine as a corrosion inhibitor for aluminum in mixed acid solution and pointed out the role of the S atom in increasing the interaction of the molecule with the metal surface. Morad et al. (2002), in their study of the inhibiting effect of methionine on mild steel corrosion in phosphoric acid, observed that adsorption of the compound onto the metal surface obeyed the Frumkin isotherm with an adsorption free energy (ΔG^0_{ads}) value of 25.4 kJ mol^{-1}.

*Corresponding author. E-mail: nhm00@fayoum.edu.eg.

In the present work the effect of chloride anions on the dissolution of Mg-Al-Zn alloy was studied and methionine (MTI) as naturally occurring amino acid was used as an inhibitor for the corrosion of the alloy in aqueous 0.6 M NaCl at pH 7. The corrosion rate was calculated and the corrosion inhibition process was investigated. For these investigations conventional electrochemical techniques such as potentiodynamic polarization and electro-chemical impedance spectroscopy (EIS) were used.

EXPERIMENTAL

Materials

The chemical composition of the working electrodes is Mg (91 to 93%), Al (6 to 7%) and Zn (1 to 2%). The electrode consists of massive rods, mounted into glass tubes by two- component epoxy resin leaving a surface area of 0.5 cm^2 to contact the solution. The cell was a three- electrode all- glass cell, with a platinum counter electrode and saturated calomel reference electrode. Before each experiment, the working electrode was polished mechanically using successive grades of emery papers up to 2000 grit, rubbed with a smooth polishing cloth, then washed with triple distilled water and transferred quickly to the electrolytic cell.

Test medium

Electrochemical measurements were carried out in buffer solutions of pH 7, the solutions were prepared from analytical grade reagents and triple distilled water. The effect of chloride ion concentration was studied in solutions containing various amounts of NaCl (0.1 to 0.6 M) at pH 7. The corrosion inhibition by methonine (Market Harborough, U K), was carried out in solutions containing 0.6 M Cl-. The structure of methonine is given in Figure.1.

All experiments were conducted at room temperature (25 ± 1 °C). The test electrolytes were buffer solutions of pH 7 (127.8 ml 0.1 M) $COOKC_6H_4COOH$ + (122.2 ml 0.1M) NaOH and completed to 250 ml.)

Electrochemical tests

The elcotrochemical impedance investigations and polarization measurements were performed using the voltalab 10 PGZ 100 "All in-one" potentiostat /Galvanostat system. The potentials were measured against and referred to the saturated calomel electrode, SCE (E = 0.245 mV). All potentiodynamic polarization measurements were carried out using a scan rate of 10 mVs^{-1}.

RESULTS AND DISCUSSION

Potentiodynamic polarization measurements

Behavior of Mg-Al-Zn alloy in inhibitor- free aqueous solutions

The electrochemical behavior of Mg-Al-Zn alloy was investigated under polarization conditions, the linear polarization and Tafel extrapolation techniques were used. Unless otherwise stated the polarization experiments were carried out at a scan rate of 10 mV/s. Corrosion current density (i_{corr}), corrosion potential (E_{corr}) and polarization resistance (R_p) are often used to evaluate the protective property of the films. In addition, E_{corr}, and i_{corr} can be calculated from these experiments. Then, based on the approximate linear polarization at the E_{corr}, R_p values are determined by the following relationship.

$$R_p = \beta_a \beta_c / 2.3\, i_{corr} (\beta_a + \beta_c) \qquad \text{..................1}$$

The potentiodynamic polarization curves of Mg-Al-Zn alloy after holding the alloy at the open-circuit potential for 90 min in naturally aerated solutions of pH 7 in the chloride free and chloride containing solutions are presented in Figure 2. The corrosion parameter that is corrosion potential, E_{corr}, corrosion current density, i_{corr}, and the corrosion rate, of the alloy in chloride- free and chloride-containing naturally aerated stagnant aqueous solutions of pH 7 were calculated and presented in Table 1. No passivity was observed in any of the chloride containing solutions (Figure 2). At low chloride ion concentration, ($[Cl^-] \leq 0.2$ M), the cathodic reaction was more affected than the anodic reaction. At higher chloride ion concentration the rates of both cathodic and anodic reaction were accelerated. The interesting point of the anodic and cathodic curve (E_{corr}) is that they did not show a significant shift with increase of chloride ion concentration up to 0.4M, but for 0.5 and 0.6 M NaCl a large negative shift (towards active side) was observed.

Figure 3 shows the corrosion current density, i_{corr} and the polarization resistance, R_p as a function of chloride ion concentration for Mg alloy which showed a 10 fold increase in current density and decrease the polarization resistance for an increase in chloride ion concentration from 0.1 to 0.6 M. The increase in corrosion rate with increasing chloride ions concentration may be attributed to the participation of chloride ions in the dissolution reaction. Chloride ions are aggressive for both magnesium and aluminium. The adsorption of chloride ions on oxide covered magnesium surface transforms $Mg(OH)_2$ easily to soluble $MgCl_2$ (Ambat et al., 2000).

In agreement with the above observations, most of the polarization curves exhibited an increase in anodic current density with increase in chloride ion concentration (Figure 2). Cathodic curve also shifted to higher current density values with increase in chloride ion concentration. The shift of OCP to more negative values with increase in chloride ion (Table 1) concentration may be explained by the adsorption of these ions on the alloy surface at weak regions of the oxide film (Frumkin et al., 1961).

When unalloyed magnesium is exposed to the atmosphere or aqueous solutions, a grey oxide (mainly magnesium hydroxide, brucite) forms on its surface, which is stable in basic solutions (Shaw BA., 2003). Nevertheless, in presence of chloride anions, this surface

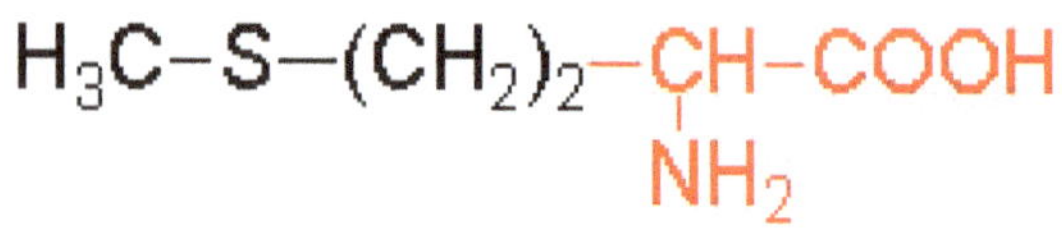

Figure 1. Structure of methionine.

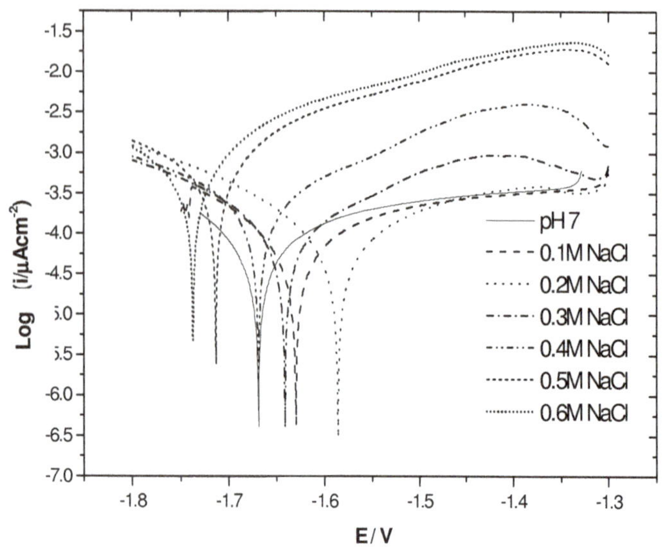

Figure 2. Potentiodynamic polarization curves of Mg-Al-Zn alloy after 90 min of electrode immersion in chloride- free and chloride - containing aqueous solutions at pH 7.

film breaks down and magnesium appears unprotected. Although corrosion mechanism of magnesium needs further investigation, it is generally reported that the following anodic and cathodic reactions take place:

$Mg \rightarrow Mg^{2+} + 2e^-$ Anodic reaction 2

$2H_2O + 2e \rightarrow 2OH^- + H_2$ Cathodic reaction 3

Firstly, magnesium dissolves and Mg^{2+}(aq) cations are produced possibly through intermediate steps involving monovalent magnesium ion (Makar et al., 1990; Hoey and Cohen, 1958). Secondly, magnesium dissolution is accompanied by hydrogen evolution, since magnesium in acidic and neutral solutions is well below the region of water stability. Finally, pH raises along with the cathodic reaction due to the formation of OH^-, which favours the formation of $Mg(OH)_2$(s) according to Pourbaix diagram (Van Muylder et al., 1966). Thus, the overall reaction could be expressed as

$Mg^{2+} + 2OH^- \rightarrow Mg(OH)_2$ (Corrosion product) 4

And the adsorption of chloride ions to oxide or hydroxide covered magnesium surface transforms $Mg(OH)_2$ to easily soluble $MgCl_2$ via the following reaction (Ambat et al., 2000)

$Mg(OH)_2 + 2Cl^- \rightarrow MgCl_2 + 2OH^-$ 5

Behavior of Mg-Al-Zn alloy in inhibitor containing aqueous solutions

Polarization experiments were undertaken specifically to separate the different effects of methionine, MTI, on the anodic and cathodic partial reactions of the corrosion process. Figure 4 illustrates typical anodic and cathodic polarization curves of *Mg-Al-Zn alloy* in 0.6 M NaCl with and without methionine. The corrosion parameter that is corrosion potential, E_{corr}, corrosion current density, i_{corr}, and the corrosion resistance, R_{corr}, of *Mg-Al-Zn alloy* electrode in methionine-free and methionine-containing naturally aerated stagnant neutral 0.6 M NaCl aqueous solutions were calculated from the potentiodynamic polarization curves and presented in Table 2.The corrosion potential, E_{corr} is shifted to the positive direction in the presence of methionine, which indicates inhibition of the anodic process of the corrosion reaction. This implies that MTI functions herein as an anodic type inhibitor.

The plots also show that MTI at the same time exerted a stimulating effect on the cathodic reaction. Also, MTI caused no pronounced change in the anodic and cathodic Tafel slopes, suggesting that the inhibitor adsorbs onto the metal surface and retards corrosion by blocking the active sites without altering the anodic and cathodic reaction mechanisms. It is clear from the potentiodynamic data that the presence of the amino acid in the solution decreases the corrosion rates, that is, the value of i_{corr} decreases. The inhibition efficiency, η, was calculated from the values of the corrosion current density without inhibitor, i_{corr}, and its value in the presence of the inhibitor, i_{corr}(inh), according to

$\eta = (i_{corr} - i_{corr}(inh)) / i_{corr} \times 100$ 6

Generally with increasing inhibitor concentration, the corrosion current density and corrosion rate decrease and polarization resistance increases (Table 2) from 0.5 to 5 mM of MTI but with the increase > 5 mM of MTI concentration the inhibition efficiency decreases. In neutral solution, MTI can be protonated at the amine group, even though the presence of $S–CH_3$ decreases the stability of the positive charge. The inhibitor could interact with the corroding alloy surface via the protonated amino groups or via the S atom in the aliphatic chain which can be adsorbed at anodic sites and retard Mg electrodissolution.

The sulfur containing amino acids can be adsorbed as bidentate ligands in which surface coordination is taking place through both the amino group (or carboxylic group) and the –S– moiety. The compound thus has the ability to influence both the anodic partial reactions, giving rise

Table 1. The corrosion parameter of Mg-Al-Zn alloy n buffer solution containing different concentration of chlorides.

Cl^- / M	E_{corr} / mV	i_{corr} / μAcm^{-2}	Rp / Ωcm^2	B_a / mV	B_c / mV	Corrosion rate/mmy^{-1}
0.0	-1668	44	410	149	-71	1.0
0.1	-1629	62	401	195	-130	1.4
0.2	-1585	72	400	203	-140	1.6
0.3	-1640	76	288	156	-119	1.7
0.4	-1668	153	137	131	-138	2.5
0.5	-1712	496	46	112	-191	9.2
0.6	-1737	516	46	116	-173	11.8

Table 2. The corrosion parameter of Mg-Al-Zn alloy in buffer of pH 7 and 0.6 M NaCl solution containing different concentration of methonine (MTI).

MTI concentration / mM L^{-1}	E_{corr} / mV	i_{corr} / μAcm^{-2}	Rp / Ωcm^2	η %
0.0	-1737	516	46	.....
0.5	-1716	450	40	13
1.0	-1700	155	80	70
5.0	-1697	124	93	75
10.0	-1708	370	55	28

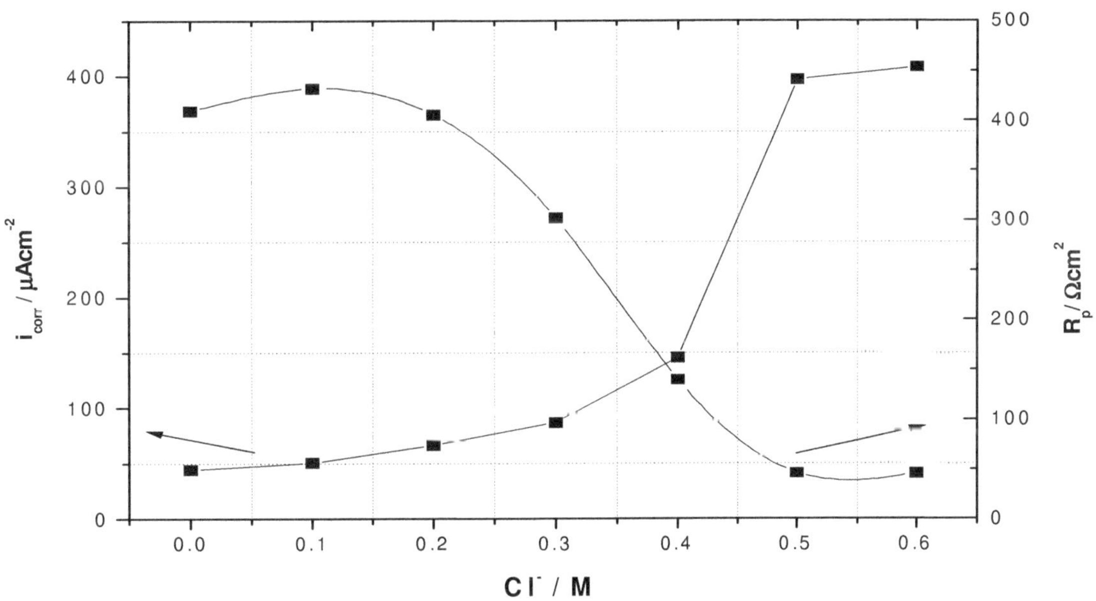

Figure 3. Variation of corrosion current density i_{corr} and polarization resistance of Mg-Al-Zn alloy in stagnant naturally aerated buffer solution (pH 7.0) containing different concentration of chloride at 25°C.

to the recorded anodic inhibition mechanism. The decreasing of the inhibition at high concentration (10 mM) may be due to some steric interaction from coordination of sulfur atom that the substituted methyl group may exert.

The electrochemical impedance measurements:

The results of the potentiodynamic polarization experiments were confirmed electrochemical impedance spectroscopy (EIS), where EIS is a powerful technique in

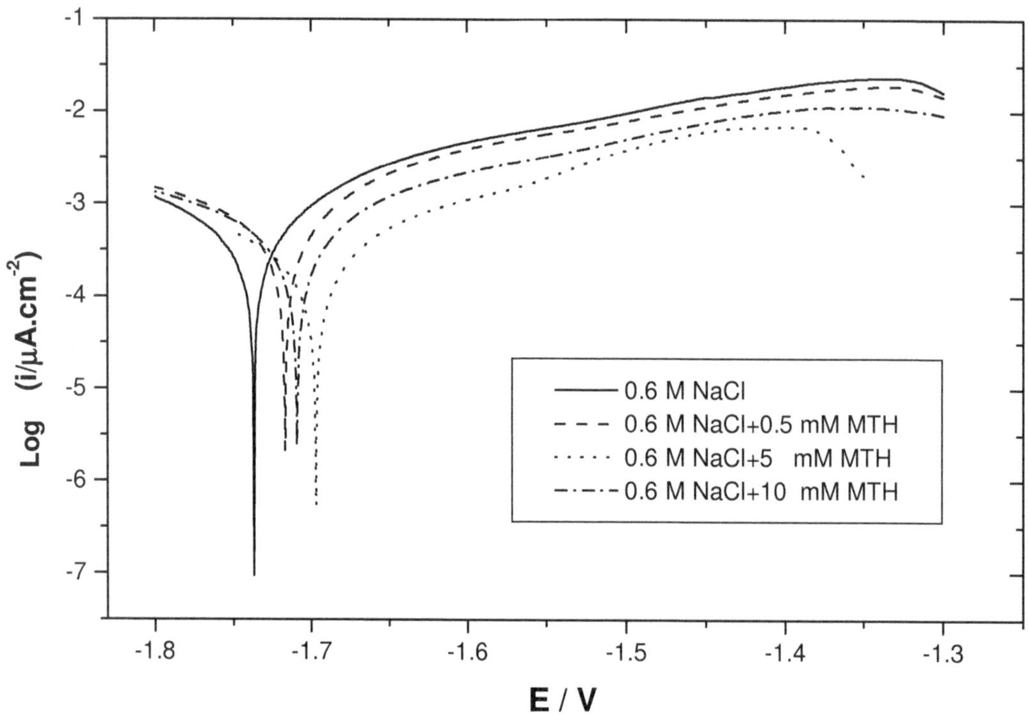

Figure 4. Potentiodynamic polarization curves of Mg-Al-Zn alloy after 90 min of electrode immersion in methonine free and methonine containing 0.6 M NaCl aqueous solutions at pH 7.

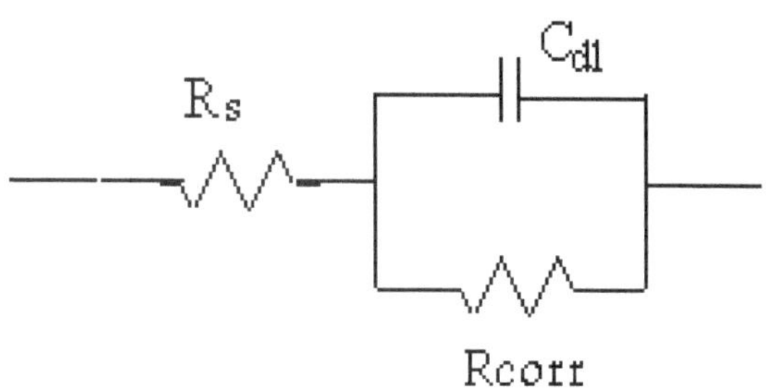

Figure 5. Equivalent circuit used in fitting impedance data.

studying corrosion mechanisms and adsorption phenomena (Macdonald et al., 1987). The experimental impedance results are simulated to pure electronic models that can verify the mechanistic models and enables the calculation of numerical values corresponding to the physical and/or chemical properties of the electrochemical system under investigation (Badawy et al., 1999; Hladky et al., 1980). The impedance data were analyzed using software provided with the electrochemical workstation, where the dispersion formula was used. A simple equivalent circuit model consisting of a parallel combination of a capacitance, C_{dl}, and resistor, R_{corr}, representing the corrosion resistance of the electrode, in series to a resistor, R_s, representing the solution resistance, is shown in Figure 5.

The electrode impedance, Z, for such a circuit is given by:

$$Z = R_s + [Rcorr / 1+ (2\pi f R_{corr} C_{dl})^{\alpha}] \qquad \text{...........7}$$

The effect of chloride ions concentration on the corrosion behavior of Mg-Al-Zn alloy in aqueous neutral solutions at pH 7 was investigated. The Bode plots for Mg-Al-Zn alloy in chloride-free and chloride-containing aqueous solutions of pH 7 are presented in Figure 6a. The diagrams show resistive regions at high and low frequencies and capacitive contribution at intermediate frequencies. The low frequency loop was related to the charge transfer resistance and the double-layer capacitance of the electrode, while the high frequency one was attributed to the presence of a partially protective surface layer of MgO or $Mg(OH)_2$ This feature can be observed in the Nyquist representation and the phase angle (Figure 6b).

As shown in Figure 6a, the impedance (|Z|) of Mg alloy is dependant on the concentration of the chloride ions. The increase in the concentration of Cl^- in the solution decreases the |Z| value, while the phase shift (Θ)max increases with a concomitant shift to higher frequencies (Figure 6b). This trend is most likely due to an increase in the surface film capacitance with an increase in the adsorbed amount of the Cl^- ions on the electrode surface. The impedance of Mg alloy show two well capacitive loops at low concentration (0.1 to 0.3 M NaCl) while at high concentration (>0.3 M) the second loop is depressed

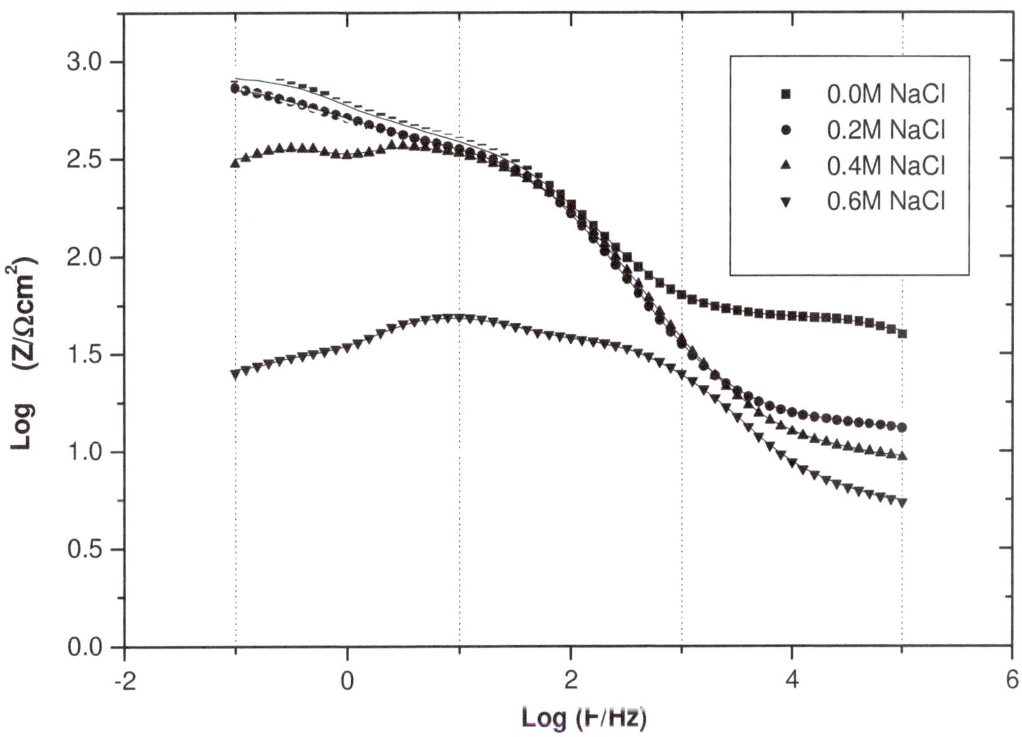

Figure 6a. Bode plots for Mg-Al-Zn alloy after 90 min of the electrode immersion in neutral aqueous solutions containing different concentration of chloride at 25°C.

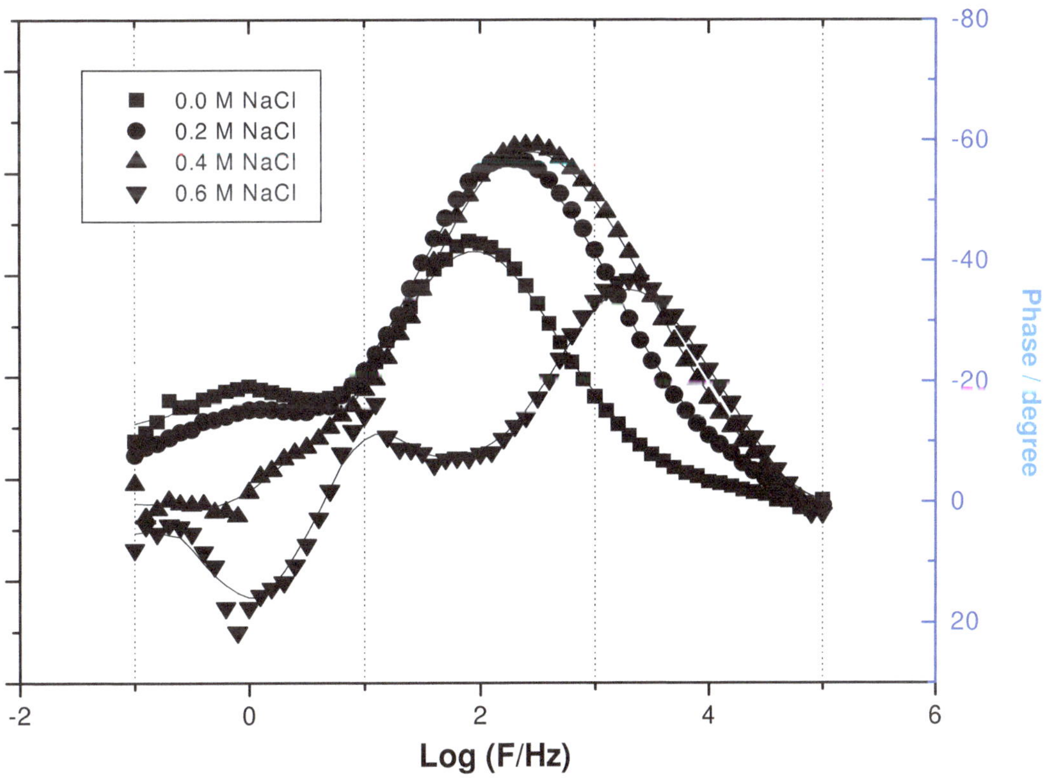

Figure 6b. Phase plots for Mg-Al-Zn alloy after 90 min of the electrode immersion in aqueous solutions containing different concentration of chloride at 25°C of pH 7.

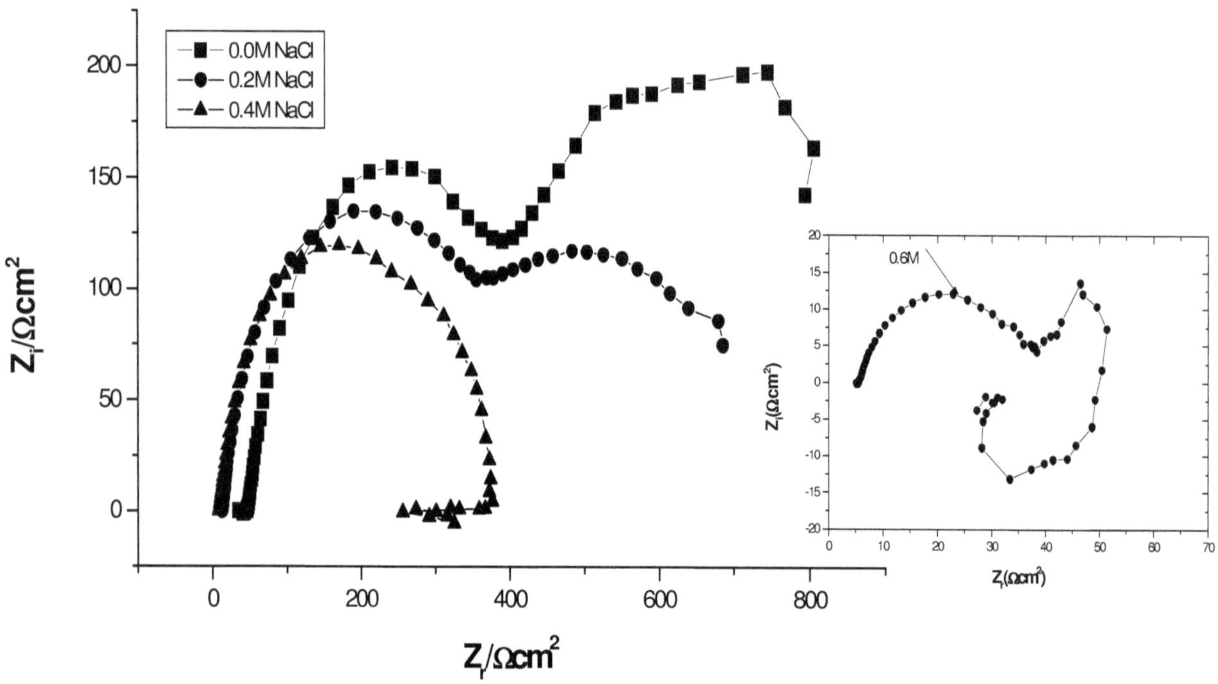

Figure 6c. Nyquist plots for Mg-Al-Zn alloy after 90 min of the electrode immersion in aqueous solutions containing different concentrations chloride at 25℃ of pH 7.

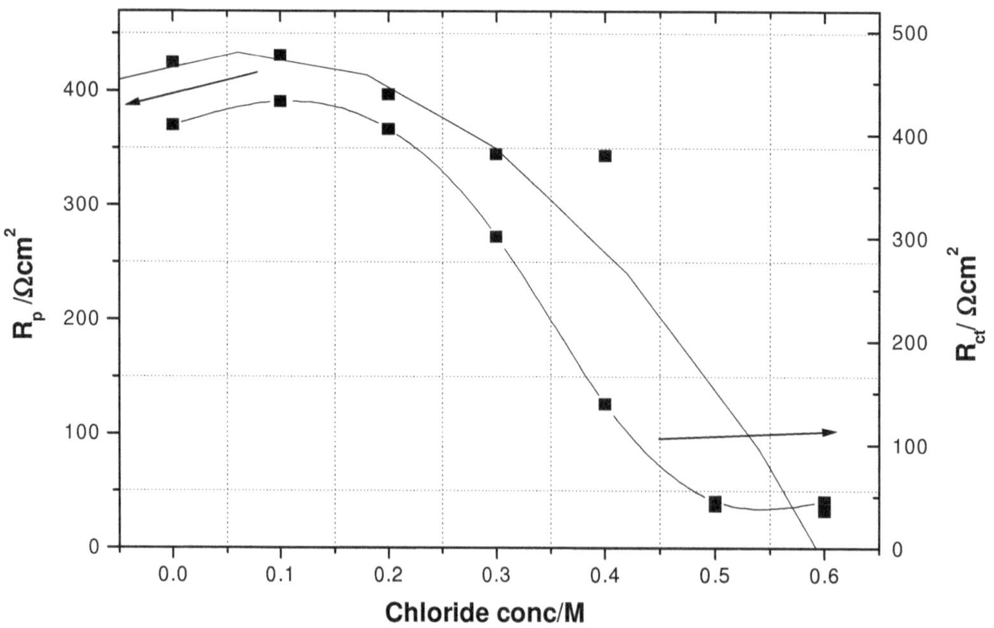

Figure 7. Variation of charge transfer resistance and polarization resistance of Mg-Al-Zn alloy in stagnant naturally aerated buffer solution (pH 7.0), containing different concentration of chloride at 25℃.

and shows a loop at higher frequency. This feature is associated to the presence of an inductor in an electrical equivalent circuit. This inductor is due to adsorption and desorption phenomena occurring on the surface of the sample (Kim and Kim, 2000; Greenblatt JH., 1956). The Bode representation of the impedance does not show a clear sharp maximum (Figure 6a), therefore, the presence of two time constants (associated to two capacitors) was assumed.

Figure 7 shows the calculated values for the charge

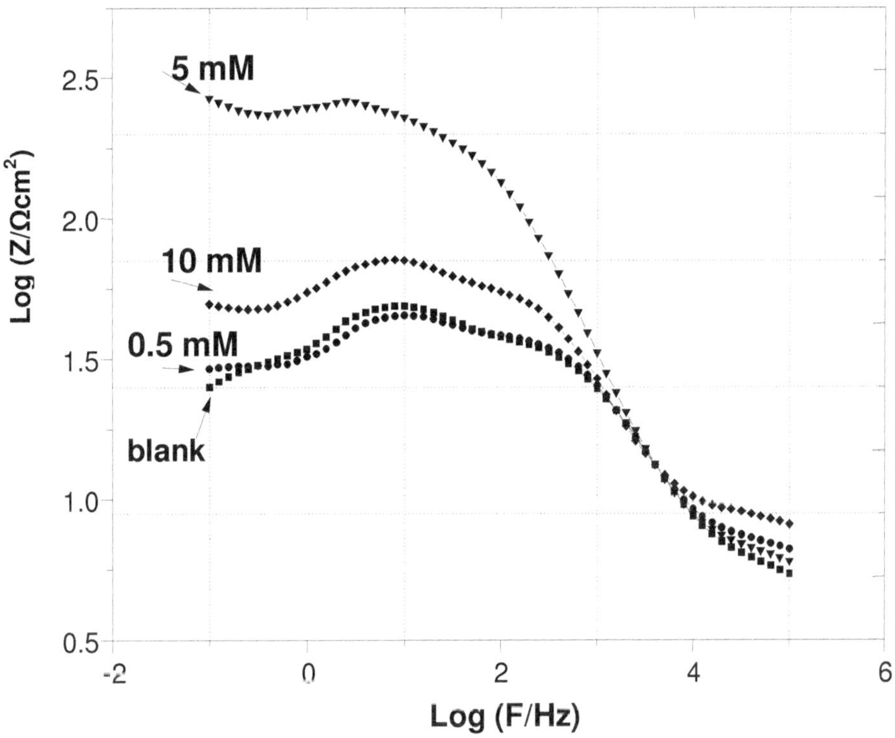

Figure 8a. Bode impedance plots of Mg-Al-Zn alloy after 90 min of electrode immersion in methonine free and methonine containing 0.6 M NaCl aqueous solutions at pH 7.

transfer resistance (R_{ct}) obtained from the impedance measurements and polarization resistance (R_p) which were obtained from polarization measurements with different concentration of chloride ions. The figure indicates that the resistance of Mg alloy to corrosion decreases with the increase of chloride ion concentration and the impedance measurements are inline with the polarization measurement.

Measurements were undertaken to assess the impedance parameters of the alloy surface/electrolyte interface in the presence of different concentrations of MTI. Representative examples of the impedance spectra are given in Figure 8, panels a, and b, which exemplify, respectively, the Bode, and phase angle plots obtained for Mg-Al-Zn alloy in 0.6 M NaCl solution in the absence and presence of 0.5 to 10 mM MTI. The spectra obtained with inhibitor consist of two depressed capacitive loops corresponding to two time constants. The high-frequency part of the impedance and phase angle describe the behavior of an inhomogeneous surface layer, whereas the low-frequency component depicts the kinetic response for the charge transfer reaction (Khaled and Hackerman, 2003). It is observed that increasing the concentration of MTI results in an increase in the impedance of the interface in (Figure 8a), and in the maximum phase angle (Figure 8b), which indicates inhibition of the corrosion process.

The values of the electrochemical impedance parameters derived from the Nyquist plots are given in Table 3. It is observed that introduction of MTI into the chloride containing solution leads to an increase in the charge transfer resistances, which become more pronounced as the MTI concentration is increased. This implies that an increase in MTI concentration, correspondingly, reduces the corrosion rate of the Mg-Al-Zn alloy by virtue of adsorption onto the metal/electrolyte interface, thereby protecting the alloy from the attack of the corrodent.

SEM techniques

Figures 9a, b and c present the macroscopic surface appearance of the corroded samples after 30 min immersion period in stagnant naturally aerated solution of pH 7 with and without chloride and mithionine. There was no obvious corrosion for the samples exposed to 0 M NaCl pH 7 solution indicating that the formed film provided resistance to corrosion initiation. The whole surface was homogenously covered by a layer of corrosion products, indicating that the corrosion was uniform. The amount of visible corrosion qualitatively increased in the presence of chloride ion at pH 7 (Figure 9b). Figure 9c shows the Mg alloy surface after immersion in 0.6M NaCl in the presence of methionine (5 mM). The addition of

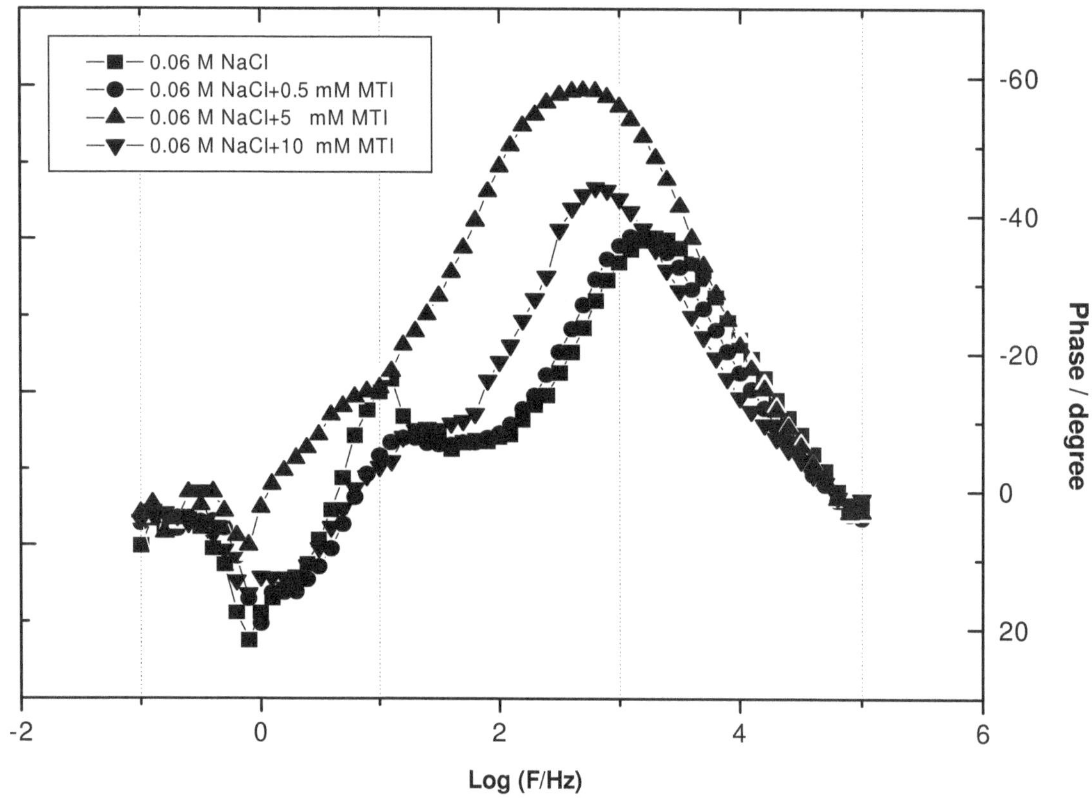

Figure 8b. Phase plots for Mg-Al-Zn alloy after 90 min of electrode immersion in 0.6 M NaCl solution containing different concentrations of methionine.

a

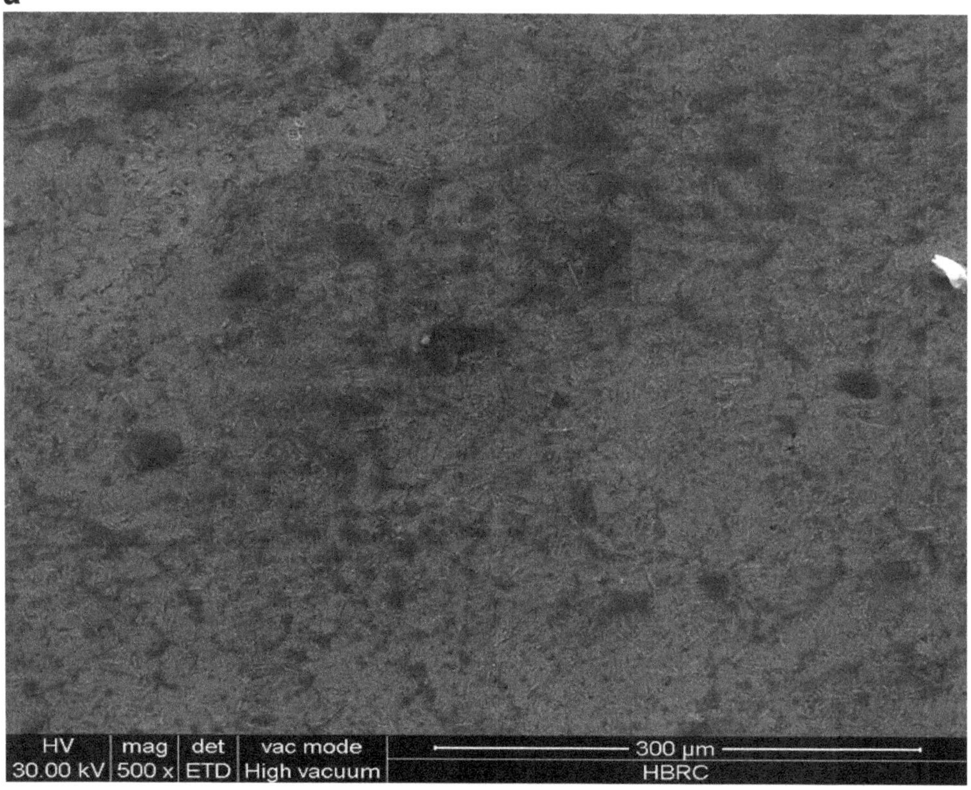

b

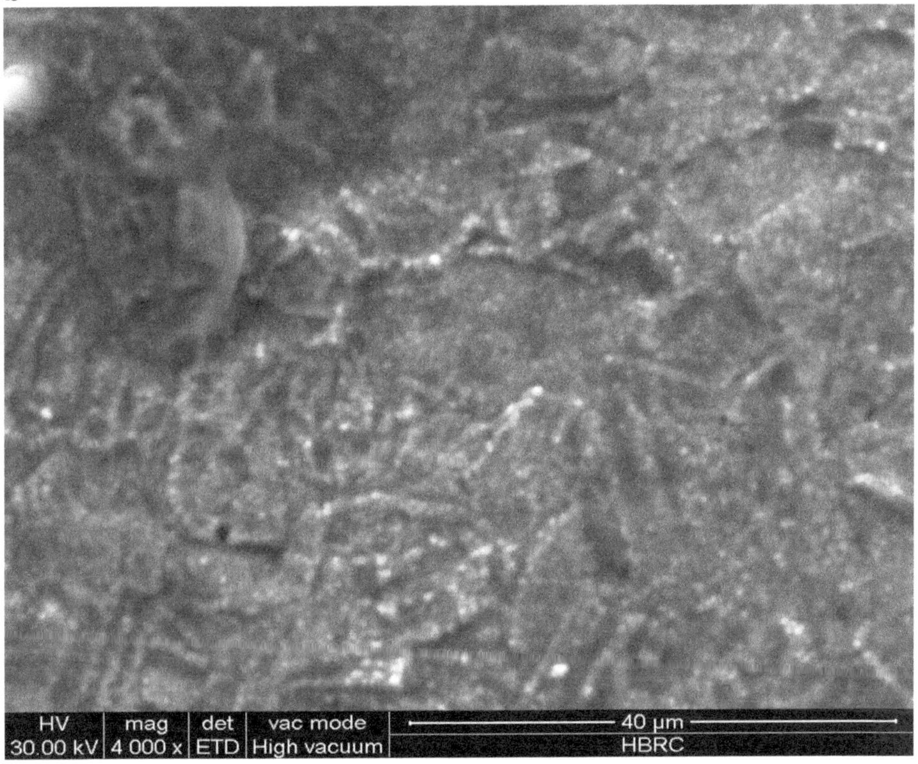

c

Figure 9. Surface of the corroded samples after immersion in pH 7 buffer solution containing. (a) No NaCl; (b) 0.6 M NaCl; (c) 0.6 M NaCl and 5 mM methionine.

Table 3. The electrochemical impedance parameters of Mg-Al-Zn alloy in buffer of pH 7 and 0.6 M NaCl solution containing different concentration of methonine (MTI).

MTI concentration / mM L^{-1}	R_s / Ωcm^2	R_{corr} / Ωcm^2	C_{dl} / μFcm^2
0.0	6	33	7.5
0.5	7	34	7.4
1.0	8	214	14.8
5.0	5	255	12.4
10.0	9	54	11.7

methionine increase the covered surface and decreases dissolution of magnesium.

Conclusion

The corrosion of Mg-Al-Zn alloy in solutions depends on the chloride ion concentration. The corrosion behavior is consistent with the current understanding that the corrosion behaviour of magnesium alloys is governed by a partially protective surface film, with the corrosion reactions occurring predominantly at the imperfections of the partially protective film. The implication is that the fraction of film-free surface increases with increasing chloride ion concentration. This is consistent with the known tendency of chloride ions to cause film breakdown.

Methionine functions as an inhibitor of Mg-Al-Zn alloy corrosion in chloride-containing aqueous solutions. An anodic inhibition mechanism is proposed. The corrosion inhibition efficiency increases with the concentration of the inhibitor and is related to the concentration and chemical structure of the amino acid.

REFERENCES

Ambat R, Aung NN, Zhou W (2000). Evaluation of microstructural effects on corrosion behaviour of AZ91D magnesium alloy. Corros. Sci., 42: 433.

Ashassi-Sorkhabi H, Ghasemi Z, Seifzadah D (2005). The inhibition effect of some amino acids towards the corrosion of aluminum in 1 M HCl + 1 M H_2SO_4 solution. Appl. Surf. Sci., pp. 249-408.

Badawy WA, Al–Kharafi FM, El-Azab AS (1999). Electrochemical behaviour and corrosion inhibition of Al, Al-6061 and Al–Cu in neutral aqueous solutions Corros. Sci., 41:709.

Frumkin A (1961). Transactions of the Symposium on `Electrode Process', E Yeager (Ed.) J. Wiley and Sons, New York, pp. 1.

Gray JE, Luan B (2002). Protective coatings on magnesium and its alloys — a critical review. J. Alloy. Comp., 336: 88.

Greenblatt JH (1956). A Mechanism for the Anodic Dissolution of Magnesium, J. Electrochem. Soc., 103: 539.

Hoey GR, Cohen M (1958). Corrosion of Anodically and Cathodically Polarized Magnesium in Aqueous Media, J. Electrochem. Soc., 105: 245.

Hladky K, Calow M, Dawson JL (1980). Corrosion rates from impedance measurements: an introduction. Br. Corros. J., 15: 20.

Kim JG, Koo SJ (2000). Effect of alloying elements on electrochemical properties of magnesium-based sacrificial anodes Corrosion. 56: 380.

Khaled KF, Hackerman N (2003). Investigation of the inhibitive effect of ortho-substituted anilines on corrosion of iron in 1 M HCl solutions Electrochim. Acta., 48: 2715.

Macdonald JR (1987). Impedance Spectroscopy, John Wiley and Sons, New York, pp. 1-26.

Makar GL, Kruger J (1990). Corrosion Studies of Rapidly Solidified Magnesium Alloys J. Electrochem. Soc., 137: 414.

Morad MS, El-Hagag A, Hermas A, Abdel AMS (2002).Effect of amino acids containing sulfur on the corrosion of mild steel in phosphoric acid solutions polluted with Cl^-, F^- and Fe3 ions behaviour near and at the corrosion potential. J. Chem. Technol. Biotechnol., 77: 486.

Moretti G, Guidi F, Grion G (2004). Tryptamine as a green iron corrosion inhibitor in 0.5 M deaerated sulphuric acid Corros. Sci., 46: 387.

Oguzie EE, Unaegbu C, Ogukwe CN, Okolue BN, Onuchukwu AI (2004). Inhibition of mild steel corrosion in sulphuric acid using indigo dye and synergistic halide additives Mater. Chem. Phys., 84: 363.

Oguzie EE (2006). Studies on the inhibitive effect of *Occimum viridis* extract on the acid corrosion of mild steel Mater. Chem. Phys., 99: 441.

Rudd AL, Breslin CB, Mansfeld F (2000). The corrosion protection afforded by rare earth conversion coatings applied to magnesium Corros. Sci., 42: 257.

Shaw BA (2003). Corrosion Resistance of Magnesium Alloys, in: ASM Handbook. Corrosion Fundamentals, Testing Protect, 13: 692.

Van Muylder J, Pourbaix M (1996). Magnesium in: M. Pourbaix (Ed.), Atlas of Electrochemical Equilibria in Aqueous Solution, Pergamon Press, Oxford. pp. 139–145.

Yfantis A, Paloumpa I, Schmei D, Yfantis D (2002). Novel corrosion-resistant films for Mg alloys, Surf. Coat. Technol., pp. 151- 400.

Zhang YJ, Yan CW, Wang FH, Lou HY, Cao CN (2002). "Study on the Environmentally Friendly Anodizing of AZ91D Magnesium Alloy." Surf. Coat. Tech., 161: 37.

Zhang YJ, Yan CW, Wang FH, Cao CN (2002). Study on the Environmentally Friendly Anodizing of AZ91D Magnesium Alloy Mater. Prot., 35: 4.

Zhang DQ, Gao LX, Zhou GD (2003). Synergistic effect of 2-mercapto benzimidazole and KI on copper corrosion inhibition in aerated sulfuric acid solution. J. Appl. Electrochem., 33: 361.

Optimization of Li-Zn ferrite synthesis parameters using the experimental design method

M. Kavanloui[1]*, B. Hashemi[1] and E. Nourafkan[2]

[1]Department of Materials Science and Engineering, School of Engineering, Shiraz University, Shiraz, Iran.
[2]Department of Chemical and Petroleum Engineering, School of Engineering, Shiraz University, Shiraz, Iran.

The conventional synthesis of ferrites involves mixing and sintering powder oxides. However, high sintering temperature deteriorates electrical properties of ferrites due to the formation of Fe^{2+} ions. Using sintering aid and decreasing particle size are easy and economical methods to decrease the sintering temperature and at the same time improve electrical properties of ferrites. In this research, the influence of sintering temperature, sintering aids B_2O_3 and nano SiO_2, milling time and the rotation speed of the mill on the density, dielectric constant and dielectric loss of Li-Zn ferrite were investigated. Response surface method (RSM) was used to experimental design of the production process and the statistical analysis of the process parameters effects. The RSM results were then used as objective functions of optimization method to find the process parameters for the global optimization of the response parameters. The results show that the sintering temperature significantly affects the density, and B_2O_3 have a considerable effect on the desired properties of Li-Zn ferrite, while other parameters have no such effect.

Key words: Li-Zn ferrite, response surface method, ferrite electrical properties, ferrite density.

INTRODUCTION

Lithium ferrites substituted with zinc have good magneto-dielectric properties which have been found to be useful in microwave application such as switches and phase shifters. Ferrites usually are synthesized by solid state reactions method because of its simplicity and applicability. In this method, the appropriate amount of oxides or carbonates are mixed together by milling, calcined, pressed and sintered at high temperatures. The structural characteristics of synthesized ferrites are strongly dependent on the processing parameters. Also, it is known that the intrinsic properties of ferrites, such as magnetization, initial permeability, Curie temperature and treatment and type of additives or substituted ions (Sattar et al., 2007).

Lithium ferrites require high sintering temperature (1200°C) to reach full density. Unfortunately, high sintering temperatures deteriorate electrical properties due to the volatilization of alkaline elements and resistivity depend on the chemical composition, heat formation of Fe^{2+} ions. Recently, several method have been attempted to resolve this problem. The sintering aids such as B_2O_3, Bi_2O_3, SiO_2 and ZnO were introduced to ferrite to decrease the sintering temperature due to formation of liquid phase during the sintering process. Decreasing sintering temperature would be helpful to prevent loss of volatile elements such as lithium. B_2O_3 is known as a sintering aid because of its low melting point. Fan et al. (2000) studied the effect of B_2O_3 on the microstructure and magnetic properties of Mn-Zn ferrite.

It had been reported that the properties of Mn-Zn ferrites are improved by the addition of B_2O_3. Rhim proposed that there is an optimum value for B_2O_3 (0.5 wt. %) to obtain suitable grain growth and sintering temperature. Higher amount of B_2O_3 lowered sintering behavior. Yuksel et al. (2008) carried out a detailed work on the effect of B_2O_3 as a sintering aid on the magnetic properties of $Ni_{0.4}Zn_{0.6}Fe_2O_4$ ferrites. They reported that the low value of B_2O_3 (below 0.01% mol), promoted significantly grain growth and densification behavior of ferrites.

However, investigation of effects of total parameters, including dopant type and its amount, milling parameters and sintering temperature on the properties of ferrite and optimization of them are very difficult and requires a

*Corresponding author. E-mail: majidkavan@gmail.com.

suitable experiment design. In order to organize the time and reduce the costs of the experiments, the experiment design methods have been used to minimize the number of experiments. There are several statistical approaches such as factorial, Response Surface Method (RSM) and Taguchi, which can be used for the design of sequential experiments in scientific researches. The experimental design methods reduce the costs of the experiments, but it is necessary to mention that these methods are only an approximation (Montgomery, 2001). Sharma et al. (2005) used a Taguchi L9 orthogonal design to obtain optimal magnetic properties in strontium ferrite sintered magnets. Vladikova et al. (1989) used experimental statistical modeling to obtain a series of substituted nickel ferrites with low magnetic and dielectric losses. The RSM has been widely applied in a variety of chemical processing and material manufacturing (Sahu et al., 2009, Chatterjee et al., 2007; Hou et al., 2007).

In this paper, the RSM method was used to recognize and investigate the effects of vital parameters of the production process of Li-Zn ferrite. RSM analyzes the relation between response variables and process variables and establish an approximation function between them. A quadratic polynomial function was generated for each response variable (density, dielectric constant and dielectric loss) and these functions were used in a multi variable optimization problem. The goal of optimization was maximizing of density and dielectric constant and minimizing of dielectric loss. B_2O_3 and nano silica were used as sintering aids and the time and speed of milling and the sintering temperature were considered as synthesis parameters to improve the properties of $Li_{.30}Zn_{.4}Fe_{2.30}O_4$ ferrite. In order to verify the validity of the design, the optimum results were used for synthesizing the samples in the lab. The properties of the samples were then measured and compared with the optimization results.

EXPERIMENTAL PROCEDURES

The samples with a composition of $Li_{.30}Zn_{.4}Fe_{2.30}O_4$ + x wt % B_2O_3 + y wt % nano SiO_2 (x=0-1-2 and y=0-2-4) were prepared by conventional solid state reaction method. High purity oxide and carbonate powders of Li_2CO_3 (99%, Merck, Germany), Fe_2O_3 (99.5%, Khorasan Powder Metallurgy Co., Iran), ZnO (99%, Merck, Germany), B_2O_3 (>99%, Merck, Germany) and nano SiO_2 (>99% purity, nanoamor, USA) were used as starting materials. Polyvinyl butyral (PVB, Sigma-Aldrich) was used as binder. First, the required materials were weighted in stoichiometric composition and then were mixed by an attrition mill, which equipped with hardened steel vial and balls in acetone alcohol medium. The mixture was calcined at 750°C for 3 h in air. Nano SiO_2 powder was dispersed with the aid of PVP - K15 solution. Sedimentation test was carried out to show the effect of dispersing process. Required amounts of nano SiO_2, B_2O_3 and 1.5 wt% of PVB solution (20 wt%) were added to mixture after calcination. The resulted mixture was milled again for 0.5 to 1.5 h at different rotation speeds (300 to 600 rpm). After evaporation of acetone at 50°C, dried powders were sieved and pressed into discs and toroid with diameter of 10 mm and thickness of 1.5 mm. The applied pressure was 15 KN/m^2. The specimens were sintered at 900 to 1200°C for 3 h in air. For calcination and sintering, both heating rates were 5°C/min and cooling was done in furnace. The density of the sintered samples was measured by the Archimedes method.

The microstructure of sintered samples was investigated using a Leica Cambridge, Stereoscan S 360 scanning electron microscopy (SEM). Grain size measurements were carried out from 10 different areas of the SEM images. X-ray diffraction (Brucker D8-ADVANCE diffractometer with Cu Kα radiation) was used for phase determination of samples. To measure the electrical properties, all samples were carefully polished to reach parallel surfaces and then electroded by silver paste and gold. The capacitance and tanδ were measured in 1 KHz by LCR meter (ALC-850 High Precision LCR Meter).

Response surface method

Basic concepts

Response surface method (RSM) is a statistical method, which was introduced by Box and Wilson (1951), and explores the relationships between several design variables and one or more response variables. The RSM has some advantages, as it is easy to estimate and apply, even when little is known about the process (Mead and Pike, 1975; Myers and Carter, 1989). A wide variety of works has been used in the design of experiments to perform statistical analysis regarding the effect of process parameters on the nature of the process. RSM method is an attracted optimum design tool in many practical engineering applications such as product process optimization. In general, the procedure of RSM consists in the following steps (Myers and Montgomery, 2002):

Step 1. Design and conduct a series of experiments to get adequate and reliable measurements from the interesting response.
Step 2. Develop mathematical models of the first and second order response surface with the best fittings.
Step 3. Represent the direct and interactive effects of the process parameters through two and three-dimensional plots.

Figure 1 presents the general framework of this study by using RSM methodology and optimization process.
If all independent variables to be measurable, continuous and controlled, the response surface can be expressed as follows:

Design variables: $X_i (i = 1,2,...,n)$

State variables: $Y = f(X_1, X_2,..., X_n) + \varepsilon$

where ε represents an error component and *Y* is the response. *f* is a function that approximates a problem with design variables and state quantities, using several analysis or experiment results.

In general, a simple polynomial model is adequate which can be used as function *f*. However, in complex systems and neural networks, nonlinear functions such as exponential function or Lagrange's interpolations can be used as response surfaces.

Design of experiments

Design of experiments is a technique for setting an efficient experiment point parameters. Full factorial design, orthogonal design and Central Composite Design (CCD) are most applicable methods in design of experiments.

A well-known RSM design called Central Composite Design

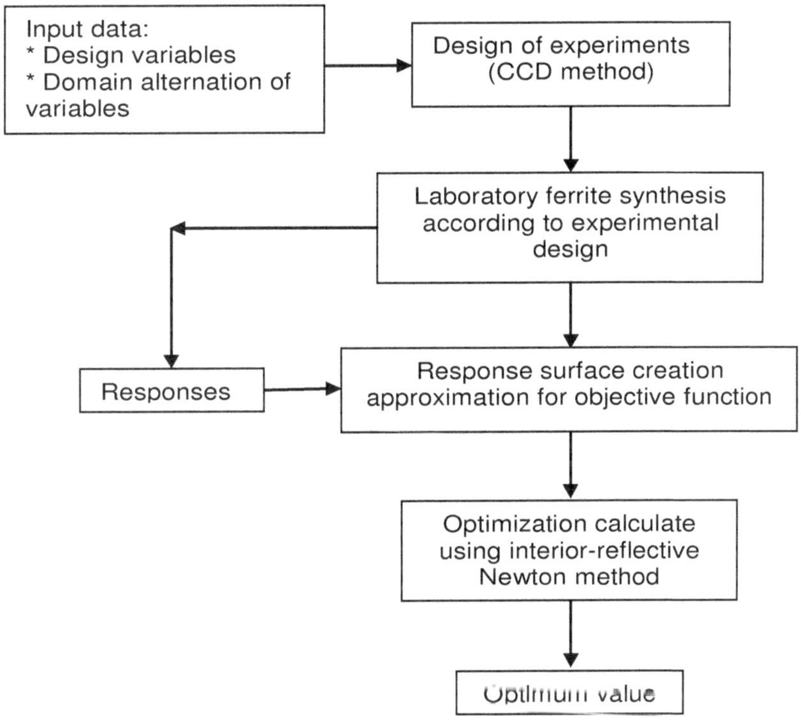

Figure 1. The general framework of RSM methodology and optimization process.

(CCD) was applied to analyze the variables for synthesizing Li-Zn ferrite in a conventional process. The central composite design was widely used for fitting a quadratic (2nd order) model which can be represented by the following equations according to Montgomery (1991).

$$Y = b_0 + \sum_{i=1}^{K} b_i X_i + \sum_{i=1}^{K} b_{ii} X_i^2 + \sum_{i=1}^{K}\sum_{j>1}^{K} b_{ij} X_i X_j + \varepsilon \qquad (2)$$

where Y is response, b_0, b_i, b_{ii} and b_{ij} are the regression coefficients for intercept, linear, quadratic and interaction terms, respectively, and X_i and X_j are the independent variables.

Least-squares method

The least-squares method was first described by Carl Friedrich Gauss around 1794 (Bjorck, 1996). Least square means that the overall solution minimizes the sum of the squares of the errors made in solving every single equation. The most important application is in data fitting. The best fit in the least-squares sense minimizes the sum of squared residuals, a residual being the difference between an observed value and the fitted value provided by a model. If equation was written with two variables for simplicity:

$$Y = b_0 + b_1 X_1 + b_2 X_2 + b_3 X_1^2 + b_4 X_2^2 + b_5 X_1 X_2 + \varepsilon \qquad (3)$$

With substitution, the second degree terms in Equation 3 by a single variable is as follows:

$$X_1^2 = X_3,\ X_2^2 = X_4,\ X_1 X_2 = X_5$$

Equation 3 is converted into linear expression and the linear least-squares method can be used easily for estimation response surface coefficients. Assuming that the number of experiments points is n and the number of design variables is k, the regression model can be represented by following equation:

$$Y = bX + \varepsilon$$

$$Y = \begin{bmatrix} y_1 \\ y_2 \\ \vdots \\ y_n \end{bmatrix} \quad X = \begin{bmatrix} 1 & x_{11} & \cdots & x_{1k} \\ 1 & x_{21} & \cdots & x_{2k} \\ \vdots & \vdots & \ddots & \vdots \\ 1 & x_{n1} & \cdots & x_{nk} \end{bmatrix} \quad b = \begin{bmatrix} b_1 \\ b_2 \\ \vdots \\ b_n \end{bmatrix} \quad \varepsilon = \begin{bmatrix} \varepsilon_1 \\ \varepsilon_2 \\ \vdots \\ \varepsilon_n \end{bmatrix} \qquad (4)$$

By minimizing the sum of squares of error, the unbiased estimator b' of the coefficient b is given by following equation:

$$b' = (X^T X)^{-1} X^T y \qquad (5)$$

Application of RSM in Li-Zn ferrite synthesis

As stated in the Introduction, several parameters may affect the output quality responses in the Li-Zn ferrite milling process. Therefore, the optimization of the mechanical milling process is complex. In this research for reducing the complexity the mill size, mill material, mill geometry, ball size and ball to powder ratio were fixed and was not considered as the controllable process parameters. In short, the five controllable process parameters were considered including the milling time, rotational speed of the mill, sintering temperature and amounts of B_2O_3 and nano silica additives. The choices of levels on these five parameters were performed based on the property of the milling process and the limits of the attrition mill and conventional method. For the parameter of milling time, the milling time should be limited at a reasonable range in order to achieve the milling efficiency and decrease the degree of contamination.

According to the engineers' experience, a reasonable range of

Table 1. The independent process variable and levels.

r		Level 1	Level 2	Level 3
X_1	Sintering temperature (°C)	900	1000	1100
X_2	Boric oxide additive (wt. %)	0	1	2
X_3	Nano Silica additive (wt %)	0	2	4
X_4	Milling time (h)	0.5	1	1.5
X_5	Rotation speed of mill (rpm)	300	450	600

Table 2. The central composite matrix and results.

S/No.	Actual level of parameter					Output response		
	X_1 (°C)	X_2 (wt. %)	X_3(wt. %)	X_4 (h)	X_5 (rpm)	Density	tanδ	k
1	1100	1	2	1.0	450	4.35895	0.527	12932.34
2	900	0	0	0.5	600	2.85703	2.319	9.17
3	1000	1	2	1.0	450	4.74064	0.381	2956.54
4	1000	1	2	1.0	600	4.52195	0.221	40460.72
5	1000	1	2	1.0	300	4.34624	0.611	34684.55
6	1000	1	2	1.0	450	4.47900	0.403	29560.64
7	1000	1	2	1.5	450	4.50161	0.331	27870.11
8	900	1	2	1.0	450	3.56100	0.301	2455.93
9	1000	0	2	1.0	450	4.29210	1.060	4814.65
10	1100	2	4	1.5	600	4.23986	0.281	30661.27
11	900	2	4	1.5	300	4.41818	0.583	16384.21
12	900	2	0	1.5	600	4.47405	1.050	9661.33
13	900	0	4	1.5	600	3.11033	1.970	179.68
14	1000	1	2	1.0	450	4.73000	0.283	26341.35
15	1100	0	4	1.5	300	4.14900	0.509	43240.76
16	1100	0	4	0.5	600	4.76300	0.418	52160.01
17	900	0	0	1.5	300	2.76776	2.311	9.89
18	900	0	4	0.5	300	2.80995	1.930	144.60
19	1000	1	2	0.5	450	4.55039	0.463	13306.79
20	1100	0	0	1.5	600	4.45536	0.342	25681.00
21	900	2	0	0.5	300	4.24907	0.840	40450.04
22	1000	1	4	1.0	450	4.58710	0.228	16235.01
23	1100	0	0	0.5	300	4.38625	1.810	174705.65
24	1100	2	4	0.5	300	4.28304	0.365	29754.56
25	1000	2	2	1.0	450	4.78100	0.850	298.98
26	1100	2	0	1.5	300	4.51047	0.854	67018.31
27	1000	1	2	1.0	450	4.74100	0.341	19433.37
28	1000	1	2	1.0	450	4.64000	0.332	24340.69
29	1000	1	2	1.0	450	4.58000	0.296	21455.02
30	900	2	4	0.5	600	4.10728	0.583	13284.62
31	1000	1	0	1.0	450	4.46440	0.788	32663.55
32	1100	2	0	0.5	600	4.43483	0.463	7713.28

milling time is between 0.5 to 2 h. Therefore, the levels of the milling time were set at 0.5, 1 and 1.5 h. The rotational speed levels of the agitator shaft were set based on the power limit of motor and the abrasion of the vane agitator. Consequently, the levels of the rotation velocity of the agitator shaft were set at 300, 450 and 600 rpm. The factor levels of additives weight ratio and sintering temperature, were set based on the engineers design experience.

The levels of the additives weight ratio were set at 0, 1 and 2 (wt.%) and 0, 2 and 4 (wt.%) for B_2O_3 and nano silica, respectively. The sintering temperature levels were set at 900, 1000 and 1100°C. The five process parameters, each one with three levels, and some interactions between the process parameters were investigated. The five process parameters and their factor levels are summarized in Table 1. The important output responses of the Li-Zn ferrite samples are the density and dielectric properties.

The layout of the Central Composite Design (CCD) and the responses are shown in Table 2. It also contains the three output responses for each experimental trail.

Table 3. The coefficients of empirical model (Y) are given by Equation 2.

	Actual level of parameter								
	Density			Tan δ			k		
	Coefficient	p	Remark	Coefficient	p	Remark	Coefficient	p	Remark
Const.	-61.079	0.000	Important	4.207	0.795		-1.190×10^{6}	0.184	
X_1	0.119	0.000	Important	1.86×10^{-3}	0.955		2.433×10^{3}	0.184	
X_2	4.210	0.000	Important	-4.703	0.000	Important	1.346×10^{5}	0.006	
X_3	0.150	0.421		-0.290	0.455		-1.443×10^{4}	0.484	Important
X_4	1.412	0.143		1.042	0.585		-4.293×10^{4}	0.672	
X_5	2.85×10^{-3}	0.463		6.69×10^{-3}	0.407		-4.833×10^{2}	0.267	
X_1^2	-5.52×10^{-5}	0.000	Important	9.20×10^{-7}	0.955		-0.812	0.363	
X_2^2	0.0243	0.759		0.550	0.006	Important	-1.325×10^{4}	0.150	
X_3^2	3.39×10^{-3}	0.864		0.025	0.534		2.158×10^{3}	0.335	
X_4^2	0.055	0.862		-0.031	0.962		1.908×10^{4}	0.589	
X_5^2	3.47×10-6	0.336		4.97×10-7	0.946		0.966	0.027	
$X_1 X_2$	-3.74×10^{-3}	0.000	Important	2.72×10^{-3}	0.001	Important	-150.049	0.001	
$X_1 X_3$	-1.40×10^{-4}	0.376		1.38×10^{-4}	0.670		-30.988	0.092	Important
$X_1 X_4$	-1.57×10^{-3}	0.025		-1.64×10^{-3}	0.220		-87.597	0.219	
$X_1 X_5$	1.08×10^{-6}	0.603		-9.55×10^{-8}	0.044		-0.686	0.011	
$X_2 X_3$	-0.030	0.068		0.017	0.590		2.185×10^{3}	0.220	
$X_2 X_4$	0.112	0.091		0.232	0.092		2.380×10^{4}	0.005	
$X_2 X_5$	-5.32×10^{-4}	0.024		5.19×10^{-4}	0.243		19.908	0.393	Important
$X_3 X_4$	-0.020	0.517		0.057	0.381		7.226×10^{3}	0.055	
$X_3 X_5$	5.26×10^{-5}	0.614		3.13×10^{-4}	0.164		51.225	0.001	
$X_4 X_5$	3.00×10^{-7}	0.999		4.56×10^{-4}	0.598		92.847	0.062	Important
R^2	98.4%						94.2%		

The RSM response functions were used to generate the objective function method to estimate the fitness values. In this study, the fitness function was generated in the manner density, and dielectric constant was maximized while dielectric loss was minimized after optimization. In fact, the output functions of RSM merge together and convert to a single multi variable objective function. The optimal set of parameters was found by solving the nonlinear optimization problem of:

$$\text{Min } \Phi\ (X_1, X_2, X_3, X_4, X_5)$$

$$\text{Constraints } \ 900 < X_1 < 1100 \quad 0 < X_2 < 1 \quad 0 < X_3 < 4$$
$$0.5 < X_4 < 1.5 \quad 300 < X_5 < 400$$

The objective function, Φ, was defined as follows:

$$\Phi = \frac{Y_3}{Y_1 \times Y_2} \qquad (6)$$

where Y_1 is the density, Y_2 is dielectric constant, Y_3 is the dielectric loss. Matlab software was used for programming of this study, and due to nonlinear multivariable optimization problem, the interior reflective Newton method was employed (Coleman and Li, 1994).

RESULTS AND DISCUSSIONS

Central composite design CCD was used in the experimental design of Li-Zn ferrite synthesizing. The MINITAB 13.2 software was used to analyze the experimental data. The quadratic model (Equation 2) was selected to correlate the relationship between the process parameters and the output responses. The estimated regression coefficients of quadratic model for density and dielectric properties and the most significant process parameters that affected the response process have been shown in Table 3.

Based on the results, boric oxide additive imposed the greatest effect on the density and electrical properties. In addition, temperature was found to have a significant effect on the density. Nano silica additive, milling time and rotation speed of mill, on the other hand, imposed the least effect on the response. Figure 2 shows the three-dimensional response surfaces, the combined effect of B_2O_3 additive and temperature on density (a), the combined effect of silica and temperature on δ (b), the combined effect of B_2O_3 and temperature on tan δ (c), the combined effect of rotation speed and milling time on density (d). Figure 2a shows how density varies with the

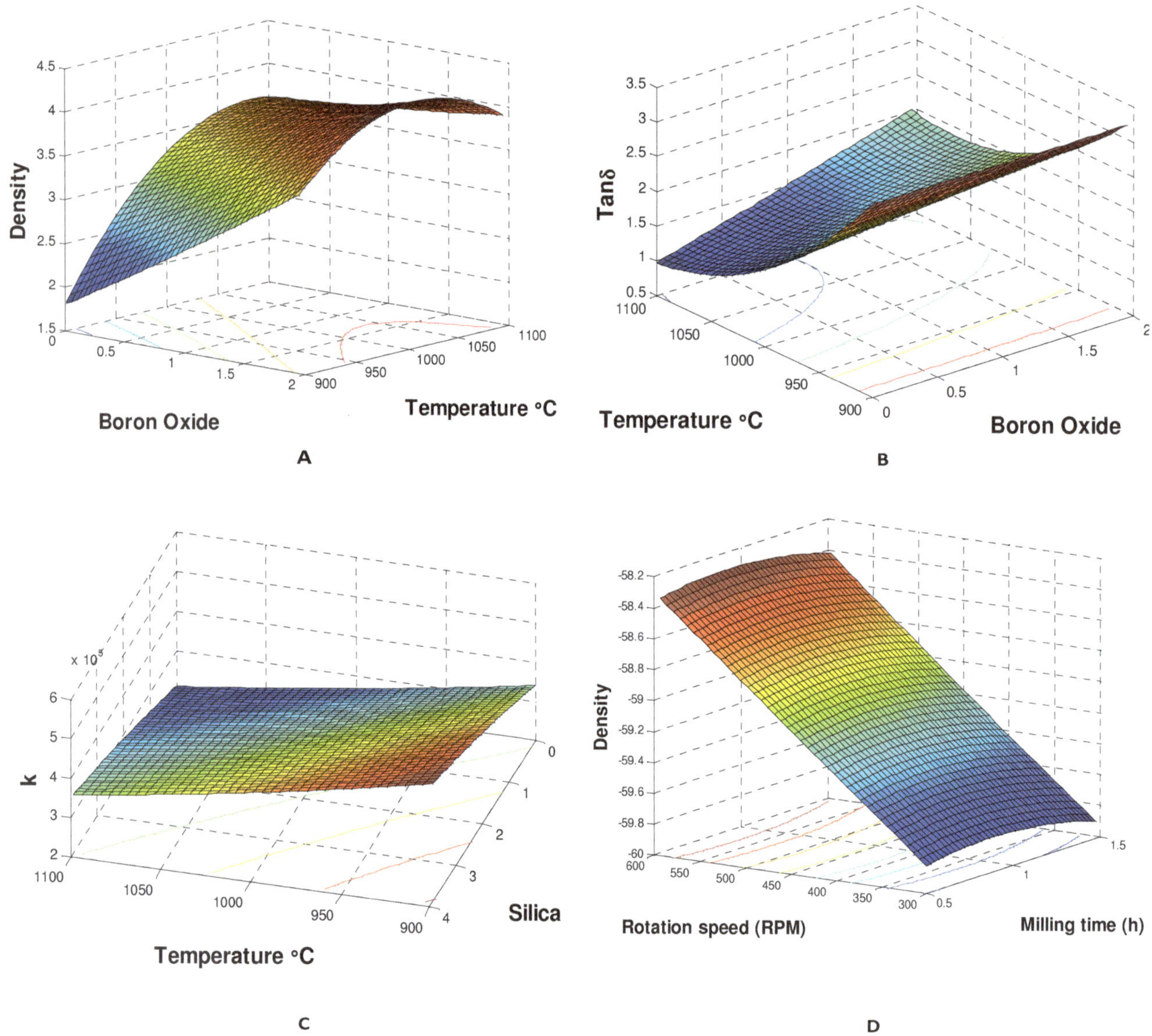

Figure 2. The combined effects of (A): Temperature and boric oxide additive on density, (B): Temperature and boric oxide additive on tan δ, (C): Temperature and silica additive on k, (D): Rotation speed and milling time on density.

variation of B_2O_3 additive and temperature. For example, the density decreases if the temperature approaches to maximum (1100℃) or minimum temperature (900℃), and so there is an optimum value for temperature in which the density maximizes. Similar result was observed by Liu et al. (2008) and Teo et al. (2008). They have reported decrease in density at high sintering temperature due to lithium volatility. The increase in the value of density is attributed to grain growth and liquid phase sintering mechanism. The densification temperature tends to shift to low temperature when there is an increase in the B_2O_3. Figure 2b shows that the tan δ decreases with increasing temperatures and at higher temperatures it increases with B_2O_3 additives. It can be clear from Figure 2c that there is a combined effect of temperatures and nano silica additives on K. It can be also seen that the K have a higher values at lower temperatures and it increased with SiO_2. These higher values of K at lower temperatures are accompanied by increase in dielectric loss tangent. The decrease in K with increasing temperatures may be due to either change the microstructure of ferrites or a decrease in number of porosity in the sample. It is clear from Figure 2d that the milling time and rotation speed have slight effect on the density.

In addition, dielectric properties of polycrystalline ferrite ceramics are affected by microstructure; therefore, B_2O_3 additives alter these properties while promoting grain

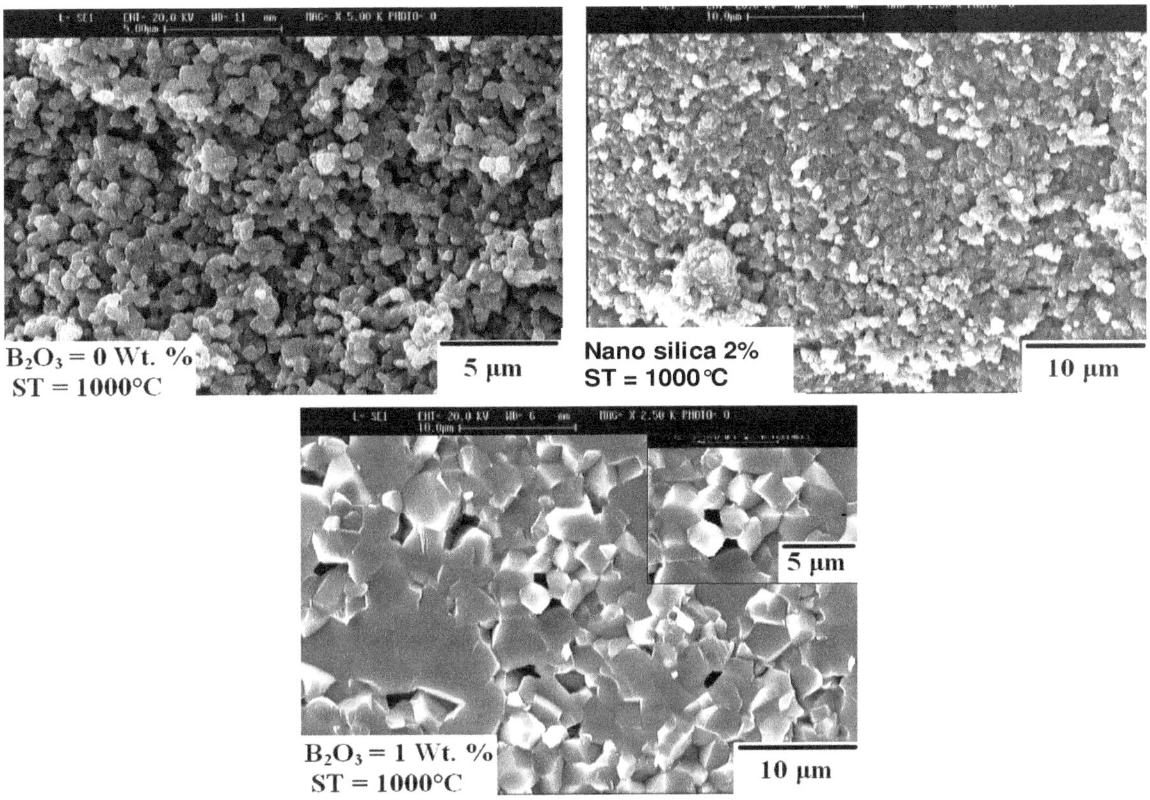

Figure 3. SEM images of the samples containing 0 wt.%, 1 wt.% B_2O_3 and 2 wt.% nano silica sintered at 1000°C for 1.5 h

Table 4. The results of optimum process parameters and experiments.

Variable	Process parameter					Output response		
	X_1 (°C)	X_2(wt. %)	X_3(wt. %)	X_4 (h)	X_5 (rpm)	Density	tan δ	K
Optimization	1065.27	1.167	2.412	0.853	457.16	4.681	0.124	2067.65
Experimental results	1065	1.16	2.4	0.85	450	4.673	0.317	1854.03

growth. The cross-sectional SEM images of the samples with 1 wt. % B_2O_3 and 2 wt. % nano SiO_2 sintered at 1000°C for 1.5 h are shown in Figure 3. It can be seen that the microstructure of Li-Zn ferrites was modified by adding B_2O_3, along with grain growth, dense and uniform microstructure were observed.

The output function of RSM method was used as an objective function (Equation 6) for the optimization. Table 4 represents the results of optimum process parameters and those obtained from the experiment.

The average deviation error (AD) between the optimum response of the design and the experimental data calculated by Equation 7 was 25.37%, suggesting a good consistency between the results.

$$AD = \frac{1}{3}\sum_{j=1}^{3}\left|\frac{Y_{exp.}(j) - Y_{opt.}(j)}{Y_{exp.}(j)}\right| \times 100 \quad (7)$$

where $Y_{exp.}$ is the experimental result, and $Y_{opt.}$ is optimum response.

Conclusions

The microstructure of the ceramic has an important role on improvement of properties; therefore, reaching the appropriate microstructure by using additives and decreasing the sintering temperature is a sensible method to increase the electrical and magnetic properties of Li-Zn ferrites. In this research, the effects of sintering temperature, B_2O_3 and nano-SiO_2 additive as aid sinter, milling time and RPM on the density and electrical properties of Li-Zn ferrites were modeled and parameters were optimized.

The response surface methodology based on five variables central composite design was used to determine the effect of parameters and fitting a quadratic model for each response. The optimum parameters were obtained via optimization process using quadratic models as objective function. Through analysis of the response surfaces derived from the models, temperature was found to have a significant effect on density. The addition of B_2O_3 greatly improved the electrical properties of $Li_{.30}Zn_{.4}Fe_{2.30}O_4$ samples. Comparison between the experimental results and RSM results for optimum parameters verify the validity of the design.

Notation

A, area of specimen, m^2; b_0, constant coefficient; b_i, linear coefficients; b_{ij}, interaction coefficients; b_{jj}, quadratic coefficients; Cu Kα, x-ray wavelength 1.541, A°; D, thickness of specimen, m; dexp, experimental density, gcm^{-3}; f, approximation function of response; K, dielectric constant; R, variance coefficient; T, sintering temperature, °C; X_1, sintering temperature, °C; X_2, boric oxide additive; X_3, nano silica additive; X_4, milling time, h; X_5, rotation speed of mill, rpm; X, independent variables; Y_1, density response, gcm^{-3}; Y_2, dielectric constant; Y_3, dielectric loss; Ω, weigh factor; P, density, gcm^{-3}; Tan δdielectric loss; ε, error component; ε', real part of dielectric constant; ε_0, permittivity of free space equal to 8.854×10^{-12} ($C^2.J.m^{-1}$); Φ, objective function;

REFERENCES

Bjorck A (1996). Numerical Methods for Least Squares Problems, SIAM.

Box GEP, Wilson, KB (1951). On the experimental attainment of optimum conditions (with discussion). J. R. Stat. Soc. Series B, 13(1): 1–45.

Chatterjee D, Oraon B, Sutradhar G, Bose PK (2007). Prediction of hardness for sintered HSS components using response surface method, J. Mater. Process Technol., 190: 123–129.

Coleman TF, Li Y (1994). On the Convergence of Reflective Newton Methods for Large-Scale Nonlinear Minimization Subject to Bounds, Math. Prog., 67: 189-224.

Fan J, Sale FR, (2000). The microstructures, magnetic properties and impedance analysis of Mn–Zn ferrites doped with B_2O_3. J. Eur. Ceram. Soc., 20(16): 2743-2751.

Hou TH, Su CH, Liu WL (2007). Parameters optimization of a nano-particle wet milling process using the Taguchi method, response surface method and genetic algorithm. Powder Technol., 173: 153–162.

Liu C, Lan Z, Jiang X, Yu Z, Sun K, Li L, Liu P (2008). Effects of sintering temperature and Bi_2O_3 content on microstructure and magnetic properties of Li-Zn ferrites. J. Magn. Magn. Mater., 320: 1335–1339.

Mead R, Pike DJ (1975). A review of response surface methodology from a biometric point of view. Biometrics, p. 803.

Montgomery DC (1991). Design and analysis of experiments. John Wiley & Sons.

Montgomery DC (2001). Design and Analysis of Experiments, 5th ed., John Wiley and Sons, New York, USA,

Myers K, Carter (1989). Response surface methodology. Technometrics, pp. 66–88.

Myers RM, Montgomery DC (2002). Response surface methodology, (2nd ed.). New York: Wiley.

Rhim SM, Hong S, Bak H, Kim OK (2000). Effects of B_2O_3 Addition on the Dielectric and Ferroelectric Properties of $Ba_{0.7}Sr_{0.3}TiO_3$ Ceramics. J. Am. Ceram. Soc., 83: 1145–48.

Sahu JN, Acharya J, Meikap BC (2009). Response surface modeling and optimization of chromium(VI) removal from aqueous solution using Tamarind wood activated carbon in batch process. J. Hazard. Mater., 172: 818–825.

Sattar AA, El-Sayed HM, Agami WR (2007). Physical and Magnetic Properties of Calcium-Substituted Li-Zn Ferrite. J. Mater. Eng. Perform., 16(5): 573-577.

Sharma P, Verma A, Sidhu RK, Pandey OP (2005). Process parameter selection for strontium ferrite sintered magnets using Taguchi L9 orthogonal design. J. Mater. Process. Technol., 168: 147–151.

Yuksel B, Kirtay S, Ozkan TO, Acikalin E, Erkalfa H (2008). The effect of B_2O_3 addition to the microstructure and magnetic properties of $Ni_{0.4}Zn_{0.6}Fe_2O_4$ ferrite. J. Magn. Magn. Mater., 320: 714–718.

The estimation of the cooling tower height by modeling the water and air contact situation in cooling tower falling film

Mortaza Gholizadeh* and Mosen Momayyeza

Department of Chemical Engineering, Islamic Azad University-Ahar Branch, Ahar, I.R Iran.

Deterioration of the filling material in traditional cooling towers is of serious concern. In this study, cooling towers of Arvand and Boualisia Petrochemical plants in the South of Iran are used as the filling material. The size of the cooling towers and outlet conditions are measured in a real situation and compared with heat and mass transfer correlations by modeling the towers and then calculating the tower heights. An experimental study to model the heat and mass balance equations and their relationship by tower height is conducted. The previous correlations found in the literature did not predict the relationship between these equations and tower height for the tested towers. A new formula by supplying the numerical methods was developed, and new variables defined. This correlation can predict the height of tower within an error of ±10%. The developed correlation is used along with theoretical modeling to predict the cooling tower outlet conditions within an error of ±5%.

Key words: Cooling tower, modeling, mass transfer coefficient, tower height.

INTRODUCTION

Cooling towers are one of the most widely equipment units used in cooling systems, which also consist of a network of heat exchangers in closed circuit that consume water only to make up for the inherent losses in the process. The thermal performance of cooling towers has vital importance in the operation of a process. Because of their relevance in the processing industry, there are many works in the literature that address cooling water systems, Moreover, specific aspects are studied, namely design of cooling towers, control and operation of towers; modeling and simulation of the thermal performance and mass transfer on the height of the tower. The highly integrated features of cooling water systems (a single tower usually supplies multiple users) produce strong interactions among the hydraulic and thermal and mass process variables. For instance, the overall point of the pump, which results from its characteristics and on the entire cooling system. In addition, the recycle water flow rate depends on the operating distribution of rates in the parallel branches is also a function of their resistance to flow that is determined from pipe diameters, equivalent lengths and adjustment of the valves in each of the pipeline segments (Soylemez, 2004; Kloppers and Kroger, 2005).

The operation of the system is even more complex at the thermal and mass levels. In each heat exchanger, a given heat load must be removed from process requirements. Furthermore, the inlet and outlet temperatures of the cooling water in the process heat exchangers must be within ranges that are compatible with the capacity of the cooling tower, as the total thermal load that is removed from the process units (through the heat exchangers) must be removed from the system in the cooling tower (Thomas and Houston, 1959a; Lowe and Christie, 1962; Thomas and Houston, 1959b).

On the other hand, the transferred mass of the water in the cooling tower to air changes the outlet temperatures and flow rates of air. The outlet water temperature at the cooling tower is determined by its performance including air velocity, rate, and temperature and transferred water mass from hot water to the air affecting the height of the cooling tower (Lebrun and Silva, 2002; Pannkoke, 1996; Badran, 2003).

*Corresponding author. E-mail: morteza_gh2000@yahoo.com.

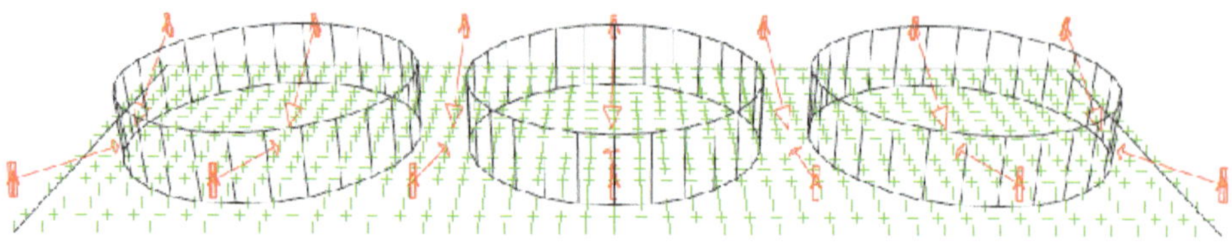

Figure 1. Arvand petrochemical plant cooling tower schematic.

Table 1. Inlet water condition for the supplied cooling towers.

	Data for Arvand tower	Data for Boualisina tower
Cooling water flow rate(m^3/h)	12,000	16,000
Density water@45°C(kg/m^3)	991	991
Hot water temperature (T1)(°C)	45	45
Site atmospheric pressure (bar)	1	1
Site elevation(m)	3.2	3.2
Dry bulb temperature of inlet air(°C)	50	50
Wet bulb temperature of inlet air (°C)	31	31
Relative humidity of inlet air(at 48°C)	30%	30%
Evaporation losses	1.9% of circulation water	1.9% of circulation water
Drift (wind age) losses	1.9% of circulation water	1.9% of circulation water
Drift (Wind age) losses	0.005% of circulation water	0.005% of circulation water
No. of cells	3	4

In a cooling tower, the water surface is extended by filling, which presents a film surface or creates droplets. The air flow may be cross flow or counter flow and caused by mechanical means, convection currents or by natural wind. In mechanical draft towers, air is moved by one or more mechanically driven fans to provide a constant air flow. The function of the fill is to increase the available surface in the tower, either by spreading the liquid over a greater surface or by retarding the rate of fall of the droplet surface through the apparatus. The fill should be strong, light and deterioration resistant. In this study, real cooling towers from Arvand and Boualisina petrochemical plants in the South of Iran were used as the filling material (Sirok et al., 2003; Kairouani et al., 2004; Kloppers and Kroger, 2005).

EXPERIMENTAL

The tested cooling tower is an induced draft counter flow type from Arvand and Boualisina petrochemical plants. The schematic diagram of the Arvandan tower is shown in Figure 1. Some of the applied cooling tower data are written as follows:

Inlet water conditions

All information is shown in Table 1.

Outlet water conditions

For both cooling towers, the outlet temperature was 35°C.

Main dimension of one cell of cooling tower

All dimensions are given in Table 2.

Condition of outlet air

With assuming that the air outlet temperature is the same as the temperature of hot water (113 ºF) and the condition of the air outlet is saturated, all other parameters were calculated and presented in Table 3. Furthermore, the calculation procedure for Arvand petrochemical plant cooling tower is written as:

Pv1 = Vapor pressure = 9.6 Kpa (at T1 = 45ºC) =1.4 psi (at T1 = 113ºF)

$$W1 = \text{Absolute humidity} = 18/29\ 29 \frac{Pv^1}{P-Pv1}$$

$$W1 = 0.622 \times 1.4/(14.5-1.4)$$

$$W1 = \text{Absolute humidity}=0.07 \frac{lb\ H_2O}{lb\ dry\ air}$$

hg1 = Saturated vapor enthalpy (at T1 = 45°C= 113°F) = 2582.43 kj/kg = 1110. 25 (BtU/lb of water)

$$h1 = 0.24\ (T1 - T0) + W1 \quad h1=0.24\ (113 - 32) + 0.07\ (1110.25) = 97.16 \frac{Btu}{lb\ dry\ air}$$

Condition of inlet air

The data are written in Table 4 and also the calculation for this part

Table 2. Main dimension of one cell of cooling tower.

	Data for Arvand tower	Data for Boualisina tower
Wetted area per cell(m^2)	242	242
Inner cell size(1×w) [m×m]	14.4×16.8	14.4×16.8
Cooling tower size (1×w) [m×m]	44.2×18	60×18
Top of basin curb	0.0	0.0
Depth of basin slab	3 m	3 m
Height of air inlet	5 m	5 m
Water distribution height	8.35 m	8.35 m

Table 3. Condition of outlet air

	Data for Arvand tower	Data for Boualisina tower
Pv_1 (at T_1 = 45°C)(kpa)	9.6	9.6
$W_1\left(\frac{\text{lb } H_2O}{\text{lb dry air}}\right)$	0.07	0.07
hg_1(at T_1 = 45°C)(BTU for lbm of water)	1110.25	1110.25
h_1(BTU for lbm of dry air)	97.16	97.16

of Arvand tower is written in below:

T2 = 31°C = 87.8°F

Pv_2 = 4.5 Kpa(at T=31°C) = 0.652 psi (at Tw=87.8°F)

$$W_2 = \frac{0.622\ 0.652}{14.5-0.652} = 0.03$$

$$hg_2 = 2557.35 \text{ kj/kg} = 1099.5 \frac{\text{Btu}}{\text{lb of water}}$$

$$h_2 = 0.24\,(T1 - T0) + Whg_2 = 0.24(87.8-32) + 0.03(1099.5) = 46.4\ 4 \frac{\text{Btu}}{\text{lb dry air}}$$

The equation for heat balance is given by:

$$Q = G\,(h_{hair\ outlet} - h_{hair\ inlet})$$

G = Mass flow rate of dry air $\frac{\text{(lb dry air)}}{\text{H}}$

Q = 138.1 MW = 130,893.5 Btu/Sec

$$G = \frac{130{,}893.5}{(97.16 - 46.4)} = 2578.7 \text{ lb dry air/sec}$$

Each cell air flow rate = 2578.7 ÷ 3 = 859.5(lb dry air/s) = 389.86 kg/s

Air specific volume @ 31°C = 0.872 m^3/kg

⇒389.86 kg/s × 0.872 m^3/kg = 340 m^3/s

$$L = 11{,}892{,}000 \text{ kg/h} = 26{,}217{,}340 \frac{\text{lb dry air}}{\text{h}}$$

⇒ G = 2578.7(lb dry air/s) = 9,283,320 (lb dry air/h)

$$\frac{L}{G} = \frac{26{,}217{,}340}{9{,}283{,}320} = 2.82$$ (It will be finalized by SPX company calculation)

340=V*14.4*16.8 ⇒ V=1.4m/s(air velocity)

The power of selected fan is = 165 Kw

Cooling tower design data

All information is shown in Table 5

Thermal design data of cooling tower

This data is presented in Table 6.

Mechanical design data of cooling tower

All relevant information is illustrated in Table 7.

Cooling tower dimension

Cooling tower dimentions are written in Table 8.

RESULTS AND DISCUSSION

When air flow passes a wetted surface, there is a transfer of sensible and latent heat. If there is a difference in temperature between the air and the wetted surface, heat will be transferred. If there is a difference in the partial

Table 4. Condition of inlet air.

	Data for Arvand tower	Data for Boualisina tower
T_2(°C)	31	31
Pv_2(kpa)	4.5	4.5
W_2 $\frac{\text{(lb } H_2O)}{\text{lb dry air}}$	0.03	0.03
hg_2(at T_1 = 45°C)(BTU for lbm of water)	1099.5	1099.5
h_2(BTU for lbm of dry air)	46.4	46.4
Q(BTU/S)	130,893.5	174,493.1
G(lb dry air / s)	2578.7	3,437.61
Each cell air flow rate($\frac{\text{(lb dry air)}}{\text{S}}$	859.5	859.5
Air specific volume (at 31°C)(m^3/kg)	0.872	0.872
L$\frac{\text{(lb dry air)}}{\text{h}}$	26,217,340	34,956,450
G $\frac{\text{(lb dry air)}}{\text{h}}$	9,283,320	12,375,396
The power of selected fan(kw)	165	165

Table 5. Cooling tower design data.

	Data for Arvand tower	Data for Boualisina tower
Number of cooling towers	1	1
Number of cells per cooling tower	3	4
Type of cooling tower	WIC240	WIC240
Working area per cell(m^2)	242	242
Arrangement of cooling tower	In line	In line
Number of air inlet sides/cell	2	2

pressure of water vapor in the air and that of the water, there will be a mass transfer. This transfer of mass causes a thermal energy transfer because if some water evaporates from the water layer, the latent heat of this vaporized water will be supplied to the air. The concept of enthalpy potential is a very useful one in quantifying the transfer of heat (sensible and latent) in those processes and components where there is a direct contact between the air and water (Kloppers and Kroger, 2005; Smrekar et al., 2006; Zhai and Fu, 2006; Lemouari, 2007). The expression for transfer of the total heat, dqt through a differential area, dA is expressed by Stoecker and Jones (1985):

$$dq_t = \frac{h_c dA}{C_{pm}}(h_i - h_a) \quad (1)$$

The name of enthalpy potential originates from the above equation because the potential for the transfer of the sum of the sensible and latent heats is the difference between the enthalpy of the saturated air at the wetted surface temperature hi and the enthalpy of the air stream h_a (Fredman and Saxén, 1995; Al-Nimr, 1998; Tan and Deng, 2003).

The rate of heat removed from the water is equal to the rate gained by the air, so the following expression can be written:

$$dq_t = m_a.dh_a = 4.19 m_w.dt \quad (2)$$

The heat transfer coefficient can be calculated by Equating Equations (1) and (2) and rearranging:

$$\frac{h_c A}{C_{pm} m_w} = 4.19 \int_{in}^{out} \frac{dt}{h_i - h_a} \quad (3)$$

However, A = aV and V = SZ, so Equation (3) can be written as:

Table 6. Thermal design data of cooling tower.

	Data for Arvand tower	Data for Boualisina tower
Duty(MW)	138.1	184.1
Water flow rate of cooling tower(m^3/h)	12,000	16,000
Water flow rate per cell(m^3/h)	4,000	4,000
Hot water temperature(°C)	45	45
Cold water temperature(°C)	35	35
Cooling range(K)	10	10
Inlet wet bulb temperature(°C)	31	31
Dry bulb temperature(°C)	50	50
Average ambient pressure(mbar)	1000	1000
Type of fill	Film fill, type FB20 or equal	Film fill, type FB20 or equal
Evaporation rate in % of water flow (%)	1.9	1.9
Evaporation rate(m^3/h)	228	304
Drift rate in % of water flow(%)	>0.005	>0.005

Table 7. Mechanical design data of cooling tower.

	Data for Arvand tower	Data for Boualisina tower
Fan type	Axial	Axial
No. of fans per cell	1	1
Fan diameter/no. of blades(m/pcs)	9.9/7	9.9/7
Output power at motor shaft(kw)	133	133
Motor size (750/1500 rpm) (kw)	38.165	38.165
Pumping head wet section(MWG)	9.85	9.85

Table 8. Cooling tower dimension.

	Data for Arvand tower	Data for Boualisina tower
Inner cell size (l ×W)(m×m)	14.4 ×16.8	14.4 ×16.8
Cooling tower size (l ×W)(m×m)	44.2 × 18	60 ×18
Top of basin curb(m)	0.0	0.0
Depth of basin slab(m)	3.0	3.0
Height of air inlet(m)	5.0	5.0
Water distribution height(m)	8.35	8.5
Fan deck height(m)	11.47	11.47
Height of fan stack(m)	2.67	2.67
Total higher of cooling tower(m)	14.34	14.34

Table 9. The calculated mass coefficient of the cooling tower and its height.

	a(m^2)	m'_w(kg/s)	m'_a(kg air/s)	K_x	V(m/s)	Z(m)
Data for Arvand tower	242	1111.11	389.86	1200.7	1.4	7.6
Data for Boualisina tower	242	1101	389.86	1197.8	1.5	8.25

$$\frac{h_c a Z}{C_{pm} m'_w} = 4.19 \int_{in}^{out} \frac{dt}{h_i - h_a} \qquad (4)$$

The relation between the heat and mass transfer coefficients is expressed by Reynold's analogy (Treybal, 1981):

$$\frac{h_c}{K_x C_{pm}} = Le^{\frac{2}{3}} \quad (5)$$

It is found that in most cases of air–water contact, the Lewis number Le can be considered to be unity as a good approximation (Kern, 1997):

$$\frac{h_c}{C_{pm}} = K_x \quad (6)$$

By substituting Equation (6) in Equation (4), the mass transfer coefficient can be expressed as:

$$\frac{K_x aZ}{m'_w} = 4.19 \int_{in}^{out} \frac{dt}{h_i - h_a} \quad (7)$$

The integration of Equation (7) is solved numerically by dividing the packed height into small segments starting from the bottom to the top of the tower with considering the relationship between air and water flow rates written in Equation 10 (Tan and Deng, 2003; Jin et al., 2007; Giorgia et al., 2009):

$$V = 0.3523.Z^{0.6781} \quad (8)$$

In addition, the design of a ceramic tile cooling tower requires heat and mass transfer correlations to estimate the tower height and to predict the water and air outlet conditions. Several workers have measured the heat and mass transfer coefficient in cooling towers. Thomas and Houston developed heat and mass transfer correlations using a tower of 2 m height and 0.3 m^2 cross section. They gave the following relations for the heat and mass transfer coefficients (Thomas and Houston, 1959):

$$h_c a = 3.0 m_w'^{0.26} m_a'^{0.72} \quad (9)$$

$$K_x a = 2.95 m_w'^{0.26} m_a'^{0.72} \quad (10)$$

In our calculations, firstly we calculated K_X from Equation 10 and then through Equation 7 and 8 we calculated Z and compare it with the real height. The result is written in Table 9. This correlation can predict the height of tower within an error of ±10%.

As we can see from the real data, the height of water-air contact for Arvand petrochemical plant cooling tower is 8.35 m that is comparable with our calculation result (7.6 m) by 10% error.

Conclusion

A mathematical model is used to simulate the effect of any change in operating conditions of cooling tower, especially the mass transferring coefficient, on the thermal performance of a cross flow tower. Available data was obtained using a numerical–experimental method which provides an insight on the current performance of the Arvand and Boualisina petrochemical plant towers. It is found that increasing the mass transferring coefficient between water and air, at constant dry bulb, the height of the tower will decrease considerably. The evaporation rate is increased as the dry bulb temperature increases and the rate of increase is almost constant at different wet bulb temperatures. The suggested model can predict the tower height with relating the contact situation between water and air.

Nomenclature: a, area of heat and mass transfer (m^2/m^3); **Cp,** specific heat (kJ/kg K); **h,** enthalpy (kJ/kg); **hc,** heat transfer coefficient (kW/m^2 K); **Kx,** mass transfer coefficient (kW/m^2 K); **Le,** Lewis number; M, flow rate (kg/s); **Q,** heat flux (kJ/kg); **Z,** tower height (m)x humidity ratio (kg water/kg dry air); **Pv,** vapor pressure; **W,** absolute humidity; **hg,** saturated vapor enthalpy; **G,** mass flow rate of dry air; **m,** superficial flow rate (mass velocity) (kg/m^2 s); **m,** flow rate (kg/s); **t,** contact time(s); **V,** air velocity.

Subscripts: a, air; **I,** inlet or interface; **o,** outlet; **v,** vapor; **w,** water.

REFERENCES

Al-Nimr MA (1998). Dynamic thermal behaviour of cooling towers. Energy Convers. Manage., 39: 631 – 6.

Badran AA (2003). Performance of cool towers under various climates in Jordan. Energy Build., 35: 1031 – 5.

Fredman T, Saxén H (1995). Modeling and simulation of a cooling tower. In European simulation multiconference, Prague, p. 66 – 70.

Giorgia F, Cortinovis, José L, Paiva, Tah W, Song, José MP (2009). A systemic approach for optimal cooling tower operation. Energy Conv. Manage., 29(14-15): 3124-3131.

Jin GY, Cai WJ, Lu L, Lee EL, Chiang AA. (2007). Simplified modeling of mechanical cooling tower for control and optimization of HVAC systems. Energy Convers. Manage., 48: 355 – 65.

Kairouani L, Hassairi M, Tarek Z (2004). Performance of cooling tower in south of Tunisia. Build. Environ., 39: 351–5.

Kern DQ (1997). Process heat transfer. (New York, McGraw-Hill), (36):2: 149-159.

Kloppers JC, Kroger DG (2005). The Lewis factor and its influence on the performance prediction of wet-cooling towers. Int. J. Therm. Sci., 44: 879 – 84.

Lebrun J, Silva CA (2002). Cooling tower—model and experimental validation. ASHRAE Trans., pp. 751–9.

Lemouari M, Boumaza M, Mujtaba IM (2007). Thermal performances investigation of a wet cooling tower. Appl. Therm. Eng., 27: 902 – 9.

Lowe HJ, Christie DG, (1962). Heat transfer and pressure drop data on cooling tower packings, and model studies of the resistance of natural draft towers to air flow. Inst. Mech. Eng. (Steam Group) Symposium Heat Trans., 113: 933.

Pannkoke T (1996). Cooling tower basics. Heat, Piping, Air Cond., 68: 137–55.

Sirok B, Blagojevic B, Novak M, Hocevar M, Jere F (2003). Energy and mass transfer phenomena in natural draft cooling towers. Heat. Transf. Eng., 24: 66 – 75.

Smrekar J, Oman J, Sirok B (2006). Improving the efficiency of natural draft cooling towers. Energy Convers. Manage., 47: 1086 – 100.
Soylemez MS (2004). On the optimum performance of forced draft counter flow cooling towers. Energ. Convers. Manage., 45: 2335 – 41.
Stoecker WF, Jones JW (1985). Refrigeration and air conditioning. (Singapore) .1: 74-81.
Tan K, Deng SA (2003). Numerical analysis of heat and mass transfer inside a reversibly used water cooling tower. Build Environ., 38: 91 – 7.
Thomas WJ, Houston P (1959a). Simultaneous heat and mass transfer in cooling towers. Brit. Chem. Eng., 160: 217.
Thomas WJ, Houston P (1959b). Simultaneous heat and mass transfer in cooling towers. Brit. Chem. Eng., 160: 217.
Treybal RE (1981). Mass transfer operations. (New York, McGraw-Hill).
Zhai Z, Fu S (2006). Improving cooling efficiency of dry-cooling towers under cross-wind conditions by using wind-break methods. Appl. Therm. Eng., 26: 1008 – 17.

Design and simulation reconfigurable liquid crystal patch antennas on foam substrate

S. Missaoui[1]*, A. Gharbi[1] and M. Kaddour[2]

[1]Electronics Laboratory, Department of Physics, Faculty of Science, Tunis El Manar, Tunisia.
[2]Sfax Faculty of Science, Tunisia.

In this article, the use of liquid crystal-based patch antenna is proposed. A first test device was, therefore, conceived in planar technology with a low-cost foam substrate. In order to have a reduced antenna structure, we have studied an antenna miniaturized by inserting slots. Moreover, the use of liquid crystal in the design of patch antennas allows adjustment of the resonance frequency. To benefit from liquid crystal anisotropy and thus obtain agility, the antenna response is modified by means of an external static electric field that tells on dielectric permittivity of the liquid crystals. The simulated results of the devices are compared with measured data, and good agreement is obtained.

Key words: Liquid crystals, patch antenna, agile structure, microwave.

INTRODUCTION

Wireless communications systems have growth drastically in the last years. For this reason, multifunction antennas have been developed to achieve multiple operation over several wireless services. The patch antennas used, for example, in portable equipment sometimes need to satisfy severe constraints on physical dimensions, and the need for miniaturization is called for. The antenna microstrip enjoy many advantages over their counterparts, such as low manufacturing cost via modern printed circuit technology, low profile, ease of integration with monolithic microwave integrated circuits (MMICs) (Thompson et al., 2006) and integrated passives, and the ability to be mounted on planar (Schussler et al., 2004), nonplanar, and rigid exteriors (Pozar, 1992). When designing planar antennas for wireless communications, it becomes necessary to have a microstrip antenna that is compact in size and able to be integrated with other devices. Up to now, several approaches have been proposed to study the patch antenna (Martin et al., 2005, 2004, 2003; Reed et al., 2001). One of the sought properties for this device is the possibility of external control. For decades, to achieve this objective, enormous efforts have been deployed for using new materials which have a better functionality. Among these materials, liquid crystals (LCs) (Dolfi, 1993; Zakharov and Mirantsev, 2003) are potentially useful. This material consists on a state of matter which has properties between those of a conventional liquid and those of solid crystals. In the nematic state, the rod like molecules float around as in a liquid phase however these are ordered in their orientation. Nevertheless, recent studies (Missaoui et al., 2011, 2010; Mueller et al., 2004; Splingart et al., 2001) have shown their dielectric anisotropy property. This property can be deduced from a permittivity tensor, depending on the direction of the applied electric field. The relative dielectric constant of a nematic LC can be changed by orienting the director of LC molecules relative to the excited radio frequency field polarization with an electro-static field.

In this paper, we propose patch antenna foam substrate using planar technology based on LCs materials. This choice allows the LC anisotropy property controlled by an electric field. We will compare the changes in frequency resonance caused by application of a DC voltage to two antennas of similar size, but containing LCs of different anisotropy. Finally, we will deal with the patch antenna size reduction realized by engraving inductive slots on the patch with the same requirement as above for the resonance frequency. Numerical results are compared with the existing data (Martin et al., 2003) to confirm the accuracy of the proposed analysis.

*Corresponding author. E-mail: sayed.elmissaoui@gmail.com

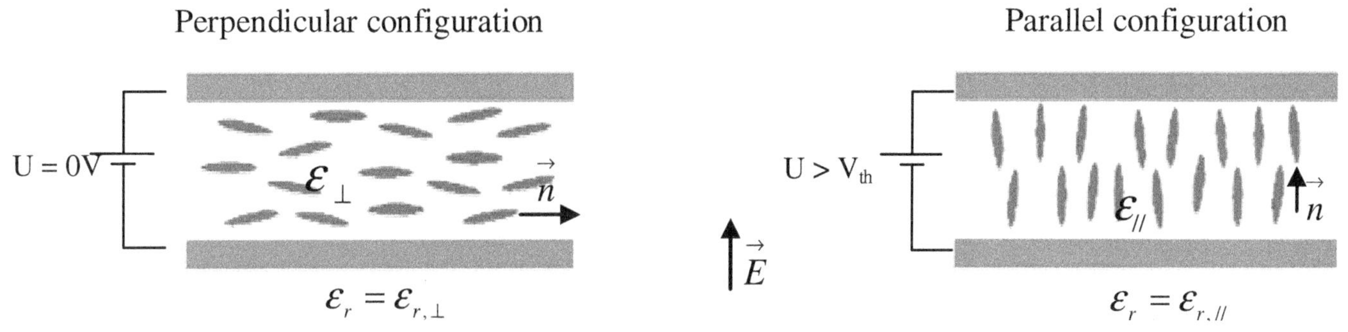

Figure 1. Configuration permittivities $\varepsilon_{r//}$ and $\varepsilon_{r\perp}$.

PROPERTIES OF LIQUID CRYSTALS

LCs is an anisotropic material showing both properties of a crystal and a liquid. All further explanations are related to nematic LCs which shows up to now the best dielectric properties at microwave and millimeter-wave frequencies. LC can be processed around 300°C, has excellent electrical properties, very low moisture absorption, light weight, mechanical stiffness, thermal stability (CTE = 0 to 30 ppm/°C), chemical resistance. Depending on the temperature, LC phase exists in a mesophase between a crystalline solid and an isotropic liquid. In this state, the material can flow like a liquid but at the same time, molecules have orientational order. The size of the molecule is typically a few nanometers. For this configuration, the electrical parameters of the LCs are defined as $\varepsilon_{\perp}$ and $\tan\delta_{\perp}$. The molecules can be rotated parallel to the RF field by applying a voltage between the conductors in order to create an electrostatic field in the LCs, thus changing the value of the permittivity and loss tangent to $\varepsilon_{//}$ and $\tan\delta_{//}$ respectively. The orientation with electric field is schematically presented in Figuro 1.

The dielectric anisotropy is defined as:

$$\Delta\varepsilon = \varepsilon_{//} - \varepsilon_{\perp} \quad (1)$$

and, analogously for the relative permittivity:

$$\Delta\varepsilon = \varepsilon_{r,//} - \varepsilon_{r,\perp} \quad (2)$$

The anisotropy can be positive or negative. All nematic LCs used in this work exhibit positive anisotropy. Therefore one can consider that the agility in frequency is a new application of these materials. The dielectric anisotropy of LC is very suitable for phase shifting purposes, since the change in permittivity is associated with a corresponding change in the phase of a wave propagating through the LC. Therefore, the most investigated applications for LCs at microwave frequencies were patch antennas devices.

SURVEY AND SIMULATION RECTANGULAR PATCH ANTENNA TO BASIS OF THE LIQUID CRYSTALS

Presentation of the rectangular patch antenna

A prototype nematic LC structure of high dielectric anisotropy from 2.7 to 2.9 is given by K15 (5CB) of Merck (Splingart et al., 2011). Then, focus will be on the dimensional requirements for the rectangular patch antenna in order to get a resonance frequency at 5 GHz. The geometry to be considered (Figure 2) consists to three elements: The ground plane, the 500-μm-high cavity realized with 500-μm-thick foam to house the capillarity-inserted LC, and the patch (Martin et al., 2003). Due to the close similarity of the dielectric constants of foam ($\varepsilon_r = 1.07, \tan\delta = 10^{-3}$) and air, the wave propagation is not affected, and thus the produced R.F. field is more concentrated in the LC. Application of a DC voltage to the feed line produces antenna agility. We arbitrarily set the resonance frequency to 5 GHz, which required a 20.7 × 21 mm rectangular patch antenna; these dimensions are issued from simulations.

Figuro 3 shows the design of the rectangular patch antenna dimensions based on LC. To get a 5-GHz resonance frequency, the patch antenna must be a 20.7 × 21 mm rectangle, with a cavity height and foam thickness of 500 μm. Resonance frequencies was varied through a superposition DC voltage to the microwave field.

SIMULATION RESULTS

In a first step, the HFSS (High frequency structure simulator), designed antenna was simulated to determine the patch size at a 5 GHz resonance frequency. Figure 4 shows that agility was obtained by varying the LC dielectric permittivity, established by dielectric characterisation, from 2.7 to 2.9. The simulated return loss is centred around 5 GHz. It can be seen that the return loss achieved -30 Db from 4.5 to 5.5 GHz. The resonance frequency variation (ΔFr) is 115 MHz, corresponding to a

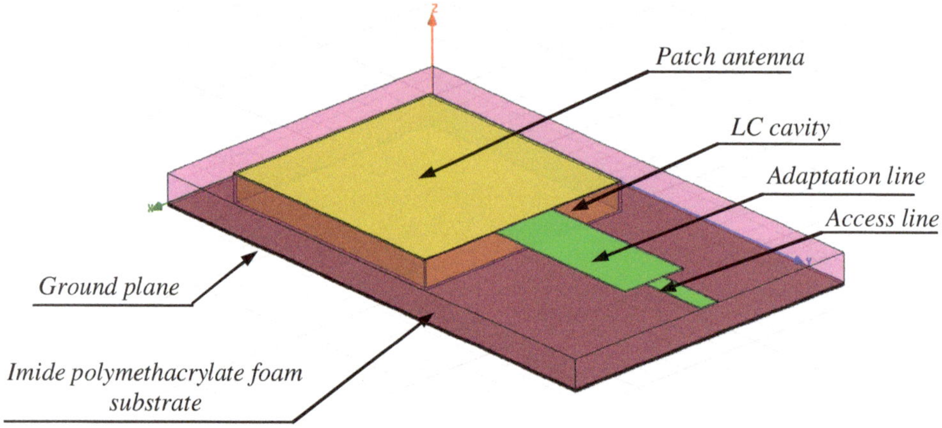

Figure 2. Design of the rectangular patch antenna (*HFSS*).

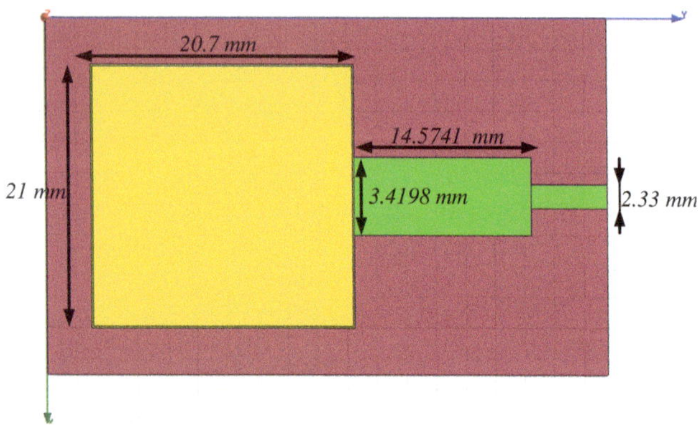

Figure 3. Rectangular patch antenna dimensions.

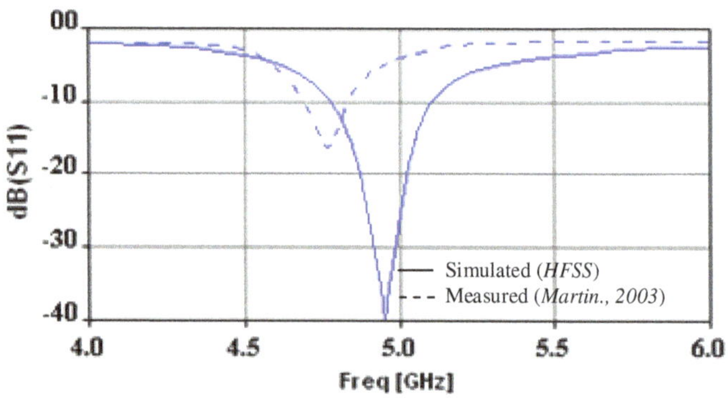

Figure 5. Simulated and measured return losses without applied DC voltage.

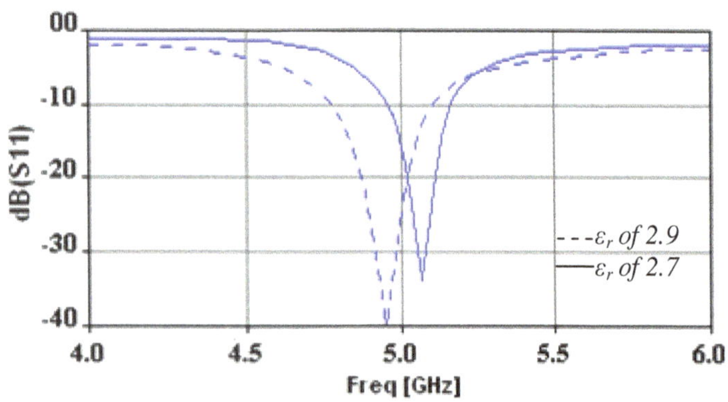

Figure 4. Simulated return losses for two different permittivities by (HFSS).

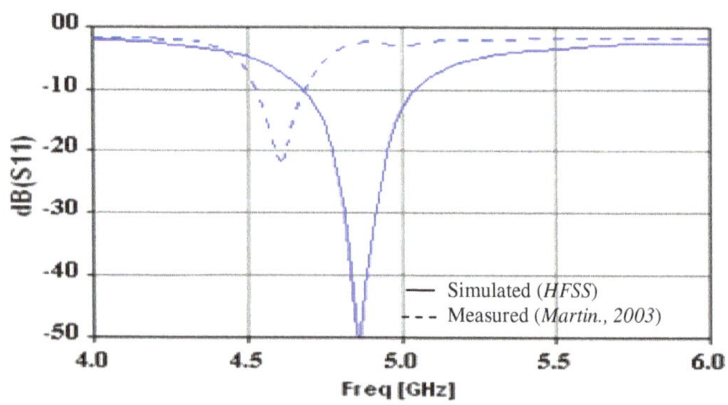

Figure 6. Simulated and measured return losses with applied 20 V DC voltage.

frequency agility of 2.3%.

Figures 5 and 6 depict the results of simulated and measured return losses with and without applied DC voltage and the dielectric permittivity liquid crystal is 2.9. It can be seen from Figure 5 that the simulated return loss achieved -30 dB from 4.5 to 5 GHz and the measured return loss achieved -10 dB from 4.5 to 5 GHz. The resonance frequency variation (ΔFr) between simulated and measured is 180 MHz corresponding to a frequency agility of 3.6%. The bandwidths simulated and measured at -10 dB are 314 and 141 MHz, respectively.

It can be seen from Figure 6 that the return loss simulated achieved -40 dB from 4.5 to 5 GHz and the measured return loss achieved -20 dB from 4.5 to 5 GHz. The

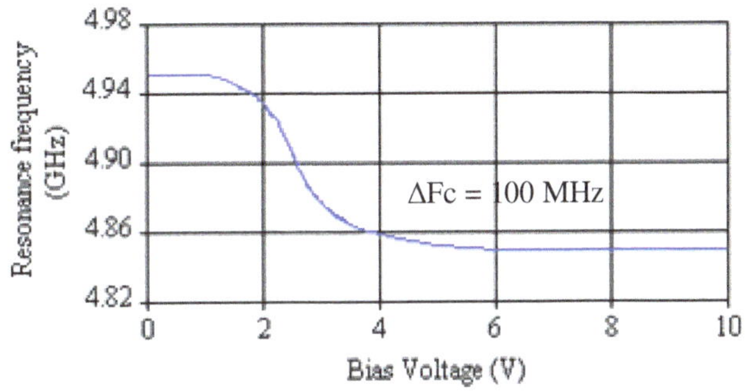

Figure 7. Variation of simulated resonance frequency versus applied DC voltage.

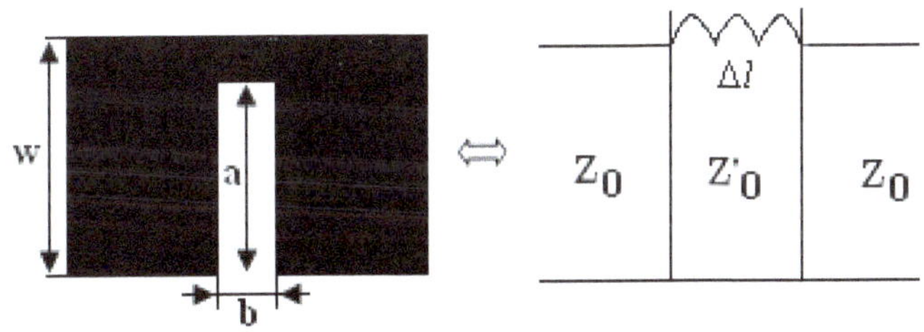

Figure 8. Representation of slot in a printed element and its equivalent circuit.

resonance frequency variation (ΔFr) between the simulated and measured is 250 MHz (5%). The bandwidths simulated and measured at -10 dB are, respectively, 352.56 and 160.25 MHz. This little variation between measurement and simulation data may result from a gap in the precision of the values found for the LC dielectric permittivities.

Figure 7 shows the results of resonance frequency versus bias voltage simulation. It is noticeable that the resonance frequency remains fairly constant for an applied electric field of 0 V, and 80% of the agility is obtained for a 5-V field amplitude.

NEW REDUCED SLOTS PATCH ANTENNA

The second structure has been conceived and affected in order to reduce the size antenna. This has been currently done through slots insertion (Martin et al., 2005). Here, we inserted slots in the printed element. Because of technological difficulties related to the foam substrate, we limited our antenna design to 7 slots. Figure 8 shows the equivalent circuit inductance, Δl, given by the following formula:

$$\Delta l = \frac{\mu_0 . \pi . h}{2} \times \left(1 - \frac{Z_0}{Z'_0} \times \sqrt{\frac{\varepsilon_{reff}}{\varepsilon'_{reff}}} \right) \quad (3)$$

where Z_0 is the characteristic impedance of the line of width w, Z_0' is the characteristic impedance of the line of width w-a, h is the substrate thickness, and ε_{reff}, ε'_{reff} are the effective dielectric permittivities of lines of w and w-a widths.

Figures 9 and 10 shows the new design reduced patch antennas that we conceived by HFSS software and the antenna dimensions based on LC. The LC height is kept to 500 µm to allow comparison with the first antenna studied.

Figure 11 shows the simulated return losses for two different permittivities. It can be seen that the return loss achieved -30 dB from 5 to 5.5 GHz. The resonance frequency variation (ΔFr) is 100 MHz, corresponding to a frequency agility of 2%.

Figures 12 and 13 depict the results of simulated and measured return losses with and without applied DC voltage. It can be seen from Figure 12 that the simulated return loss achieved -35 dB from 5 to 5.5 GHz and the measured return loss achieved -5 dB from 5 to 5.5 GHz. The resonance frequency variation (ΔFr) between simulated and measured is 235 MHz corresponding to a frequency agility of 4.7%.

It can be seen from Figure 13 that the return loss simulated achieved -35 dB from 4.5 to 5 GHz and the measured return loss achieved -15 dB from 5 to 5.5 GHz. The resonance frequency variation (ΔFr) between the simulated and measured is 133 MHz (2.66 %).

Figures 14 depict a resonance frequency centered at 5 GHz as expected, and a threshold voltage of approximately 2 V (LC command voltage). It also evidences that a voltage of about 5 V produces 80% agility.

We noticed that for a DC voltage applied, the simulation resonance frequency is decreased of a value of 89 MHz in relation to the result without DC voltage.

The change in second structure (from 20.7 × 21 to 13 × 14 mm rectangular patch) while adding the insertion slots, leads to modified results, in Figures 5 and 6, the stimulated curves are in right hand of measured curves but in Figures 12 and 13, the stimulated curves are in left hand of measured curves.

CONCLUSION

Two structures based on foam substrate and the liquid crystals were designed and simulated. Resonance frequency variation confirmed the frequency adjustment capability of LC-based devices; the extent of adjustment depends on LC anisotropy. In the first and second design, the simulation reflection return losses has been greatly improved by about 10 dB respectively 2.14 dB, along with the variation of the simulation resonance frequency of 100 MHz respectively 89 MHz, both before and after applying a continuous voltage. The accuracy of simulation was verified by comparison with experimental

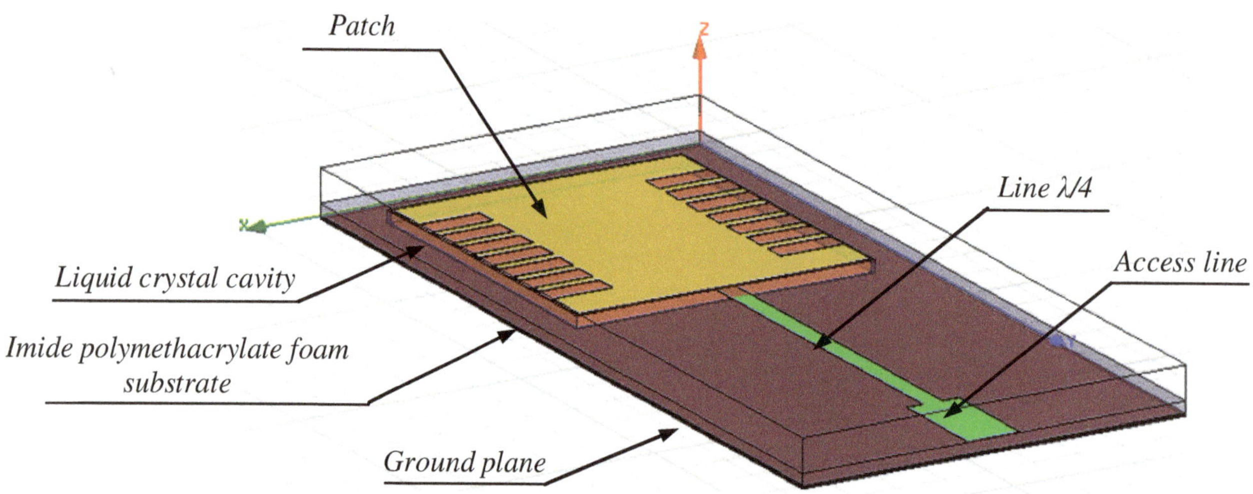

Figure 9. New design slots patch antenna by (HFSS).

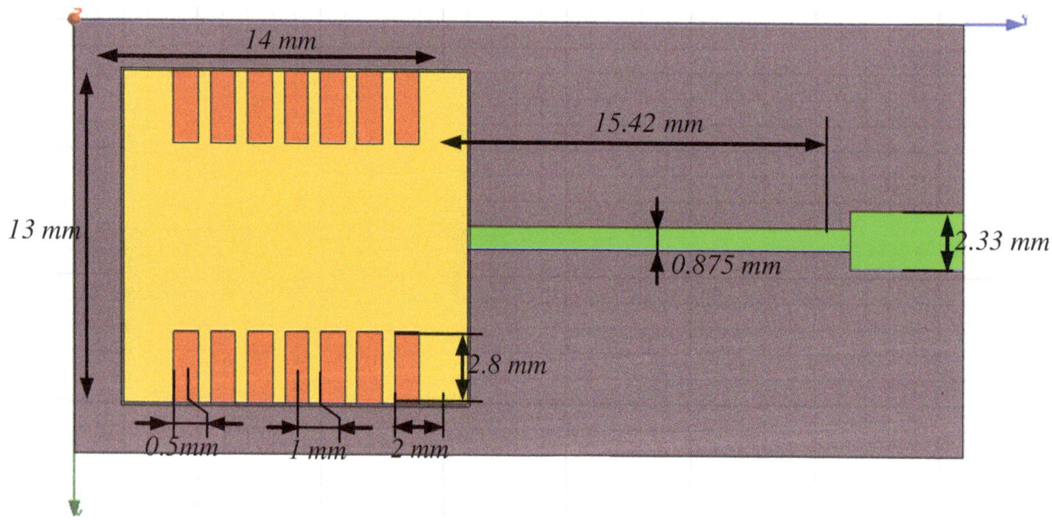

Figure 10. New structure reduced patch antenna dimensions.

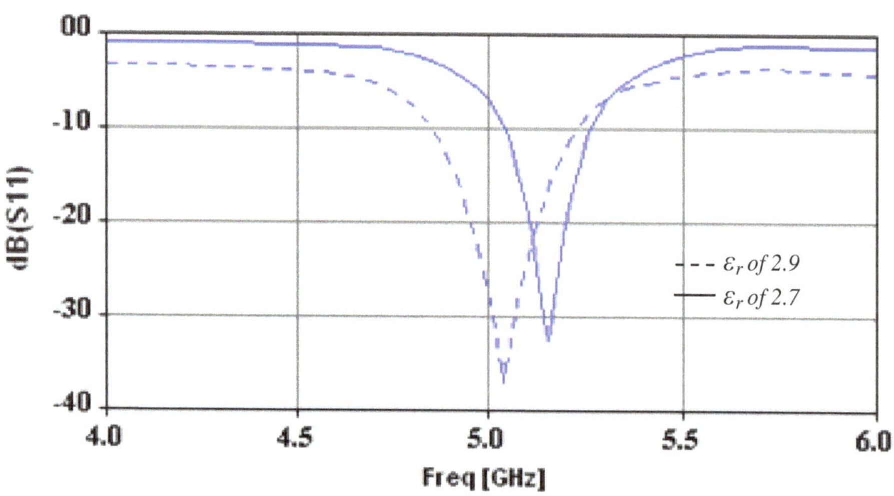

Figure 11. Simulated return losses for two different permittivities.

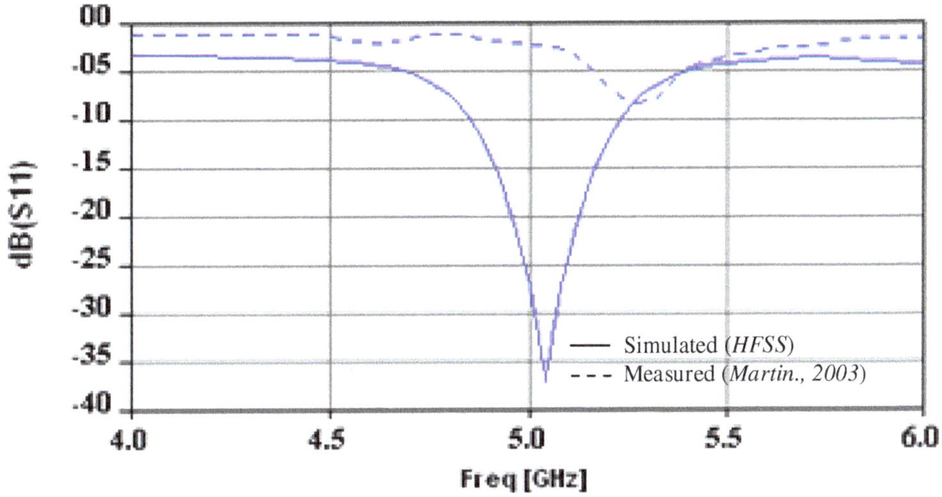

Figure 12. Simulated and measured return losses without applied DC voltage.

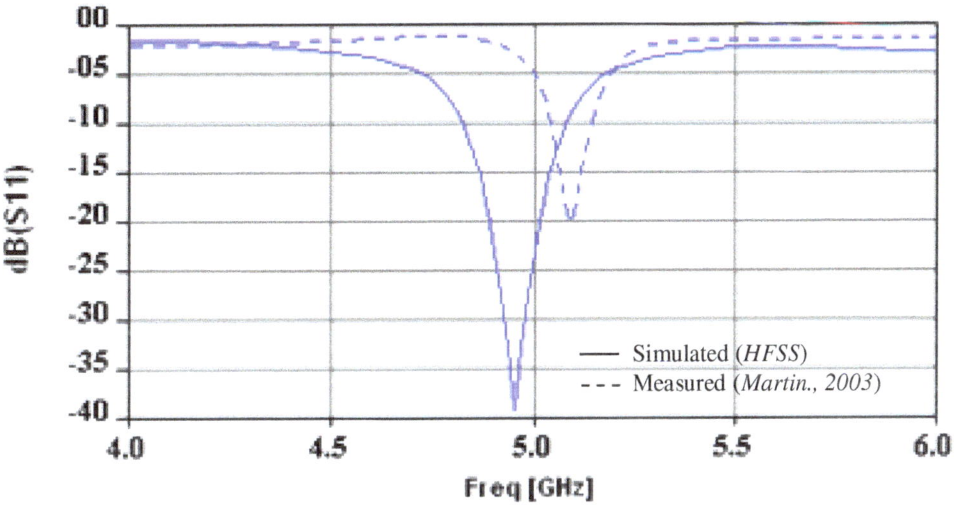

Figure 13. Simulated and measured return losses with applied 20 V DC voltage.

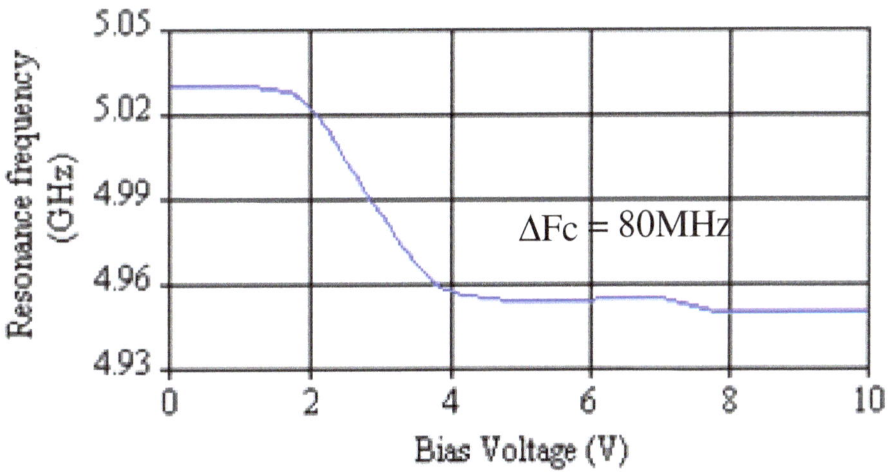

Figure 14. Variation of simulated resonance frequency versus applied DC voltage.

data. The observation of the variation results confirms the potential frequency agility of the devices that use LCs.

REFERENCES

Dolfi D, All AD (1993). "Liquid Crystals Microwave Phase Shifter", Elect. Lett., 29(10): 926-928.

Martin N, Laurent P, Person C, Gelin P, Huret F (2004). "Size reduction of a liquid crystal-based, frequency-adjustable patch antenna". Microwave conference, 34th European, 2: 825-828.

Martin N, Laurent P, Person C, Le Roy M, Perennec A, Gelin P, Huret F (2005). "Influence of design liquid crystal-based devices on the agility capability". IEEE MTTS Int. Microw. Symp. Dig., 4: 1835-1838.

Martin N, Laurent P, Person CH, Gelin PH, Huret F (2003). "Patch Antenna Adjustable in Frequency Using Liquid Crystal", Proc. 33th Eur. Microw. Conf., 2: 699-702.

Missaoui S, Kaddour M, Gharbi A (2010). "Development and design of multilayer antennas based on liquid crystals", IOP Conference Series: Mater. Sci. Eng., (13): 1.

Missaoui S, Kaddour M, Gharbi A (2011). "Design and Simulation of Tunable Phase Shifters Based on Liquid Crystals", Electromagnetics, 31(4): 285-293.

Mueller S, Scheele P, Weil C, Wittek M, Hock C, Jakoby R (2004). "Tunable passive phase shifter for microwave applications using highly anisotropic liquid crystals", Proc. IEEE MTT-S Intl. Microw. Symp. Dig., 2: 1153-1156.

Pozar D (1992). "Microstrip antennas," Proc. IEEE., 80(1): 79-91.

Reed S, Desclos L, Tenet C, Toutain S (2001). "Patch antenna size reduction by means of inductive slots", Microw. Opt. Technol. Lett., 29(2): 79–81.

Schussler M, Freese J, Jakoby R (2004). "Design of compact planar antennas using LH Transmission Lines", IEEE Int. Microw. Symp., Fort Worth, Texas, USA., pp. 209-212.

Splingart B, Tentillier N, Huret F, Legrand C (2001). "Liquid crystals applications to RF and microwave tunable components", Mole. Crystals Liquid Crystals, 368: 183-190.

Thompson DC, Tentzeris MM, Papapolymerou J (2006). "Packaging of MMICs in multilayer LCP substrates", IEEE Wirel. Compon. Lett. Microw., 16: 410-412.

Zakharov AV, Mirantsev LV (2003). "Dynamic and dielectric properties of liquid crystals", Phys. Solid State, 45: 183-188.

Algorithm and computational complexity of biochemistry

Lutvo Kurić

Bosnia and Herzegovina, Novi Travnik, Kalinska 7. E-mail: lutvokuric@yahoo.com.

The modern science mainly treats the biochemical basis of sequencing in bio-macromolecules and processes in biochemistry. One can ask weather the language of biochemistry is the adequate scientific language to explain the phenomenon in that science. Is there maybe some other language, out of biochemistry, that determines how the biochemical processes will function and what the structure and organization of life systems will be? The research results provide some answers to these questions. They reveal to us that the process of sequencing in bio-macromolecules is conditioned and determined not only through biochemical, but also through cybernetic and information principles. This paper discusses new methods for developing biochemistry technologies, in particular more advanced digital technology based on programming, cybernetics, and informational laws and systems, and how this new technology could be useful in biochemistry, medicine, bioinformatics, genetics, and other natural sciences.

Key words: Biochemistry, algorithm, discrete codes, periodic table, codes 19 and 7, code 931

INTRODUCTION

Many studies have indicated that analysis of protein sequence codes and various sequence-based prediction approaches, such as predicting drug-target interaction networks (He et al., 2010), predicting functions of proteins (Hu et al., 2011; Kannan et al., 2008), analysis and prediction of the metabolic stability of proteins (Huang et al., 2010), predicting the network of substrate-enzyme-product triads (Chen et al., 2010), membrane protein type prediction (Cai and Chou, 2006; Cai et al., 2003, 2004), protein structural class prediction (Cai et al., 2006; Ding et al., 2007), protein secondary structure prediction (Chen et al., 2009; Ding et al., 2009b), enzyme family class prediction (Cai et al., 2005; Ding et al., 2009a; Wang et al., 2010), identifying cyclin proteins (Mohabatkar, 2010), protein subcellular location prediction (Chou and Shen, 2010a; 2010b; Kandaswamy et al., 2010; Liu et al., 2010), among many others as summarized in a recent review (Chou, 2011), can timely provide very useful information and insights for both basic research and drug design and hence are widely welcome by science community. The present study attempted to develop a novel sequence-based method for studying insulin in hopes that it may become a useful tool in the relevant areas.

METHODS

What we did is the following: We translated the physical and chemical parameters from the language of biochemistry into the digital language of programmatic, cybernetic and information principles. This we did by using the adequate mathematical algorithms. By using chemical-information procedures, we calculated the numerical value for the information content of molecules. What we got this way is the digital picture of the phenomenon of biochemistry. These digital pictures reveal to us a whole new dimension of this science. They reveal to us that the biochemical process is strictly conditioned and determined by programmatic, cybernetic and information principles.

RESULTS

The mathematical balance in groups of chemical elements from X to Y exist with the help of the codes 19 and 7.

Algorithm

[(ΣB(X1,2,3,n) × A) - (ΣA(X1,2,3,n) × B)] = [(ΣB (Y1,2,3,n) × A) - (ΣA (Y1,2,3,n) × B)]

Sr> atomic number = 38;

Te> atomic number = 52;

[(S7(**38**) x 19) - (S19(**38**) x 7)]} = [(S7(**52**) x 19) - (S19(**52**) x 7)]

S7(38) = 245; S19(38) = 551;
S7(52) = 343; S19(52) = 817;

[(245x19)-(551x7)]} = [(343x19)-(817x7)];

(4655-3857) = (6517-5719);
798 = 798;

Figure 1. In these examples, using codes 19 and 7, we have established a mathematical balance of the atomic numbers of the following elements: Cs and Rn and Sr and Te.

A = 19; b = 7;
[(S7(X1,2,3,n) × 19) - (S19(X1,2,3,n) × 7)]} = [(S7(Y1,2,3,n) × 19) - (S19(Y1,2,3,n) × 7)]}

X1,2,3,n = Atomic numbers 1,2,3,n
Y1,2,3,n = Atomic numbers 1,2,3,n

Example 1

Cs> atomic number = 55;
Rn> atomic number = 86;
[(S7(55) × 19) - (S19(55) × 7)]} = [(S7(86) × 19) - (S19(86) × 7)]
S7(55) = (49 + 50 + 51 + 52 + 53 + 54 + 55) = 364;
S19(55) = (37 + 38 + 39 + 40 + 41 + 42 + 43 + 44 + 45 + 46 + 47 + 48 + 49 + 50 + 51 + 52 + 53 + 54 + 55) = 874;
S7(86) = 581; S19(86) = 1463;
[(364 × 19) - (874 × 7)] = [(581 × 19) - (1463 × 7)];
(6916 - 6118) = (11039 - 10241);
798 = 798;

Example 2

This example is given in Figure 1. A similar balance is established among all the other chemical elements. Although those elements have different atomic numbers, these numbers, when put into correlation with codes 19 and 7, given the same mathematical result, which is number 798. This goes for all the sequences in chemistry. All the sequences, those with identical as well as those with different numerical values, when put into correlation with codes 19 and 7, give one result only. In this way, a global mathematical balance is established among sequences in nature. This means that the mathematical balance can be established even when sequences are not in balance.

Number 798, as the result of the said mathematical

{[(SB(X1,2,3,n) x A)] - [(SA(X1,2,3,n) x B)]} **+ (AB)} = ABA;**
A=19; B = 7;

{[(S7(X1,2,3,n) x A)] - [(S19(X1,2,3,n) x B)]} **+ (AB)} = (19x7x19);**

(19x7x19) = 931;

931 > 9 3 1 >3^2 3^1 3^0;
S = Groups of atomic numbers 1,2,3,n
X1,2,3,n = Atomic numbers 1,2,3,n

Figure 2. Discrete code $\mathbf{3^2 3^1 3^0}$ inter connecting some chemical elements.

balance, is in correlation with numbers 19 and 7. This can be seen in the following example:

798 = [(19 × 7) + (19 × 7) + (19 × 7) + (19 × 7) + (19 × 7) + (19 × 7)];

Codes 19 and 7 demonstrate that the process of sequencing in biomacromolecules is really conditioned and determined, not only by bio-chemical, but also cyber-information rules.

Discret code 3^2 3^1 3^0

In digital pictures of Biochemistry, physical and chemical parameters are in a strict submission to programmed, cyber and information rules. In some examples, chemical elements are connected through the discrete code 931, which is transformed into 3^2 3^1 3^0 code. The code we can found using the algorithm given in Figure 2.

The connection is established through various parameters. These are: Odd and even values, primary and secondary values, positioning of chemical elements in a given group of elements, etc.

In the table shown in Figure 3, we have a group of 25 chemical elements. These elements are interconnected through the discrete code $3^2 3^1 3^0$. This code connects them as shown in Figure 4

Each atomic number has its own analogue form. That form is one of constants in the programmed, cyber and information systems in biochemistry. The analogue code together with the code is yet another proof that the process of sequencing in the biomacromolecules is really conditioned and determined, not only by biochemical, but also cyber-information rules.

In the example shown in Figure 5, the atomic numbers of the group of chemical elements in question are interconnected through the discrete code $\mathbf{3^2 3^1 3^0}$. This connection is not a visible one, so we do not know if it exists. This connection can be established using inner and outer atomic numbers and their analogue codes, which can clearly be observed in the previous example.

In the example shown Figure 5, the chemical elements

8 O	9 F	15 P	24 Cr	25 Mn
26 Fe	30 Zn	36 Kr	43 Tc	44 Ru
45 Rh	46 Pd	47 Ag	49 In	59 Pr
60 Nd	61 Pm	62 Sm	63 Eu	64 Gd
65 Tb	79 Au	87 Fr	88 Ra	93 Np

Figure 3. Group of chemical elements in the chemical elements table.

8 O	9 F	15 P	24 Cr	25 Mn
26 Fe	30 Zn	36 Kr	43 Tc	44 Ru
45 Rh	46 Pd	47 Ag	49 In	59 Pr
60 Nd	61 Pm	62 Sm	63 Eu	64 Gd
65 Tb	79 Au	87 Fr	88 Ra	93 Np

Figure 4. Inner and outer atomic numbers.The blue shading is for the outer, and green for inner atomic numbers of the group of chemical elements in question.
Inner atomic numbers = (30 + 36 + 43 + 46 + 47 + 49 + 61 + 62 + 63) = 437;
Outer atomic numbers = (8 + 9 + 15 + 24 + 25 + 26 + 44 + 45 + 59 + 60+ 64 +65+79+ 87+88+93) = 791;
Inner code = 437; Analog inner code = 734; 734||437 Outer code = 791; Analog outer code = 197; 197||791 (734 + 197) = Discret code 931 > $3^2 3^1 3^0$

8 O	9 F	15 P	24 Cr	25 Mn
26 Fe	30 Zn	36 Kr	43 Tc	44 Ru
45 Rh	46 Pd	47 Ag	49 In	59 Pr
60 Nd	61 Pm	62 Sm	63 Eu	64 Gd
65 Tb	79 Au	87 Fr	88 Ra	93 Np

Figure 5. Even and odd atomic numbers. The blue shading is for the even, and green for odd atomic numbers of the group of chemical elements in question. Even atomic numbers = 488; Odd atomic numbers = 740; Even code = 488; Analog even code = 884; Odd code = 740; Analog odd code = 047; (884 + 47) = Discret code 931 > $3^2 3^1 3^0$.

are interconnected through even and odd atomic numbers and their analogue values. The connection between them is the discrete code 931.

In the example shown in Figure 6, chemical elements in even and odd columns are connected through their atomic numbers and analogue codes.

In the example in Figure 7 the chemical elements within the diagonals and those out of diagonals are connected through the discrete code 931.

There is over 10 million of similar combinations of connecting of elements. There are also other groups of chemical elements that are also interconnected through the discrete code 931. Some examples are given in Figure 8.

It is important to note that we are here talking only about the macro theoretical concept of sequencing in Biochemistry and that we are not talking about the biochemical empirical image of these elements.

Discrete code in the constitution of atoms

Discrete code 931 we se in the constitution of atoms. Figure 9 show some examples

As we can see, the electrons in constitution of atoms are really interconnected through the discrete code 931.

8 O	9 F	15 P	24 Cr	25 Mn
26 Fe	30 Zn	36 Kr	43 Tc	44 Ru
45 Rh	46 Pd	47 Ag	49 In	59 Pr
60 Nd	61 Pm	62 Sm	63 Eu	64 Gd
65 Tb	79 Au	87 Fr	88 Ra	93 Np

Figure 6. Even and odd rank. The blue shading is for the chemical elements with the odd category in the group of elements (1,3,5...25), and green for chemical elements with even category in the same group of chemical elements (2,4,6...24). Even rank = (9+24+26+36+44...+ 88) = 587; Odd rank = (8+15+25+30+43...+ 93) = 641; Even rank code = 587; Analog even rank code = 785; Odd rank code = 641; Analog odd rank code = 146; (785 +146) = Discret code 931 > $3^2 3^1 3^0$

8 O	9 F	15 P	24 Cr	25 Mn
26 Fe	30 Zn	36 Kr	43 Tc	44 Ru
45 Rh	46 Pd	47 Ag	49 In	59 Pr
60 Nd	61 Pm	62 Sm	63 Eu	64 Gd
65 Tb	79 Au	87 Fr	88 Ra	93 Np

Figure 7. Even and odd column. The blue shading is chemical elements positioned into odd columns (1,3,5), and green for the elements in the even columns (2,4). Even columns = (9 + 30 + 46 + 61...+ 88) = 492; Odd columns = (8 + 26 + 45 + 60...+ 93) = 736; Even columns code = 492; Analog even rank code = 294; Odd columns code = 736; Analog odd rank code = 637; (294 +637) = Discret code 931 > $3^2 3^1 3^0$.

8 O	9 F	15 P	24 Cr	25 Mn
26 Fe	30 Zn	36 Kr	43 Tc	44 Ru
45 Rh	46 Pd	47 Ag	49 In	59 Pr
60 Nd	61 Pm	62 Sm	63 Eu	64 Gd
65 Tb	79 Au	87 Fr	88 Ra	93 Np

Figure 8. Diagonals. The blue shading is for the elements out of diagonals, and green for the elements within the diagonals.
Atomic numbers within diagonals = 435; Analog code = 534; Atomic numbers out of diagonals = 793; Analog code = 397; (534 +397) = Discret code 931 > $3^2 3^1 3^0$.

Digital periodic table

Mendeleyev assumed that the atomic mass is the most important feature of an atom, so he arranged the table of elements in accordance with this assumption. Today's science is familiar with the fact that the Biochemistry its own language that can be described through the theory of systems and cybernetics, and which functions within specific rules. That new language of Biochemistry enables us to design new Tables of Elements where we will introduce the criterion of the theory of systems and cybernetics for establishing of the order of elements. That could be done as given in Figure 10. Mathematical position of the chemical elements in Periodic table is given in Figure 10.

In the first part of the digital table of elements are the elements from the period 1,2,3,4 and 5, and in the second part are the elements from the sixth and seventh period. Chemical elements from this table are interconnected through the codes: 42, 219 and 692.

Digital code 346

Chemical elements from the sixth and seventh period interconnected through the code 346. Some examples are given in Figure 11.

Making a sequence of all phenomena in Periodic system Table is conducted according to the exact mathematical laws (for such descriptions we can use theory

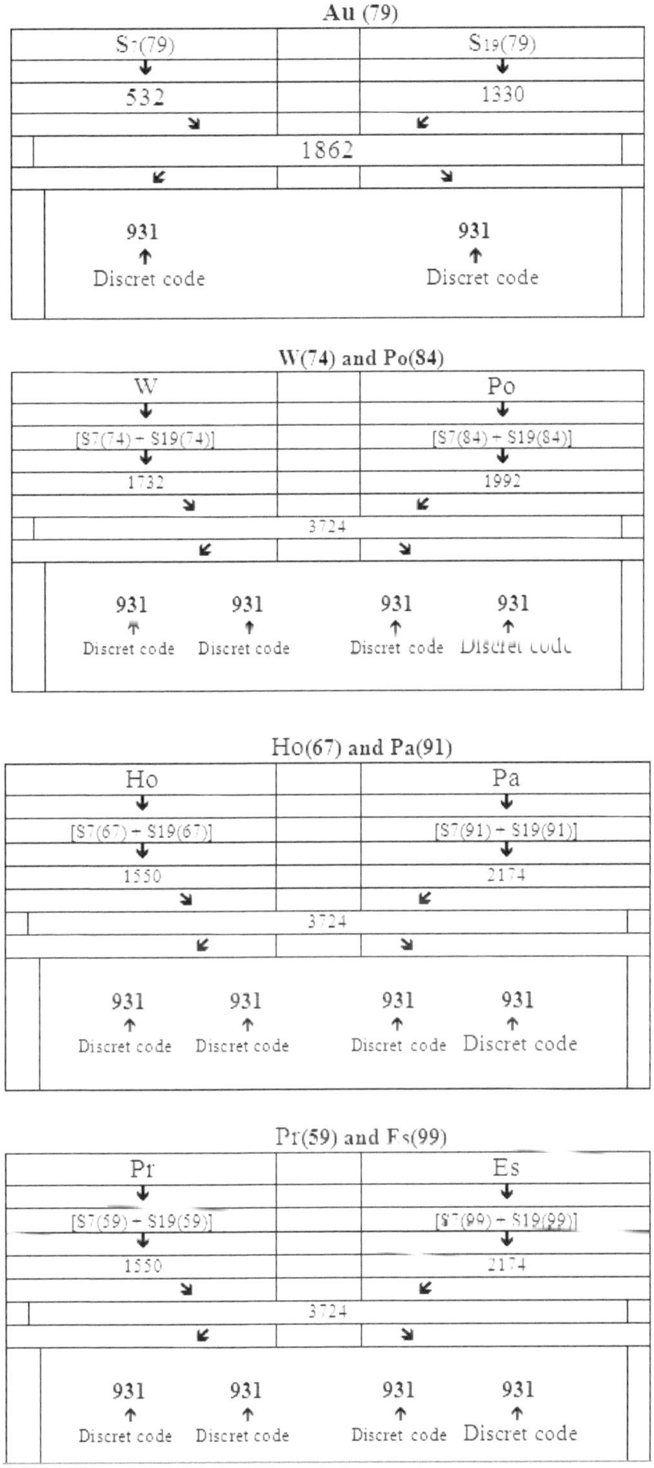

Figure 8. Chemical elements from the previous example interconnected through the discrete code 931.

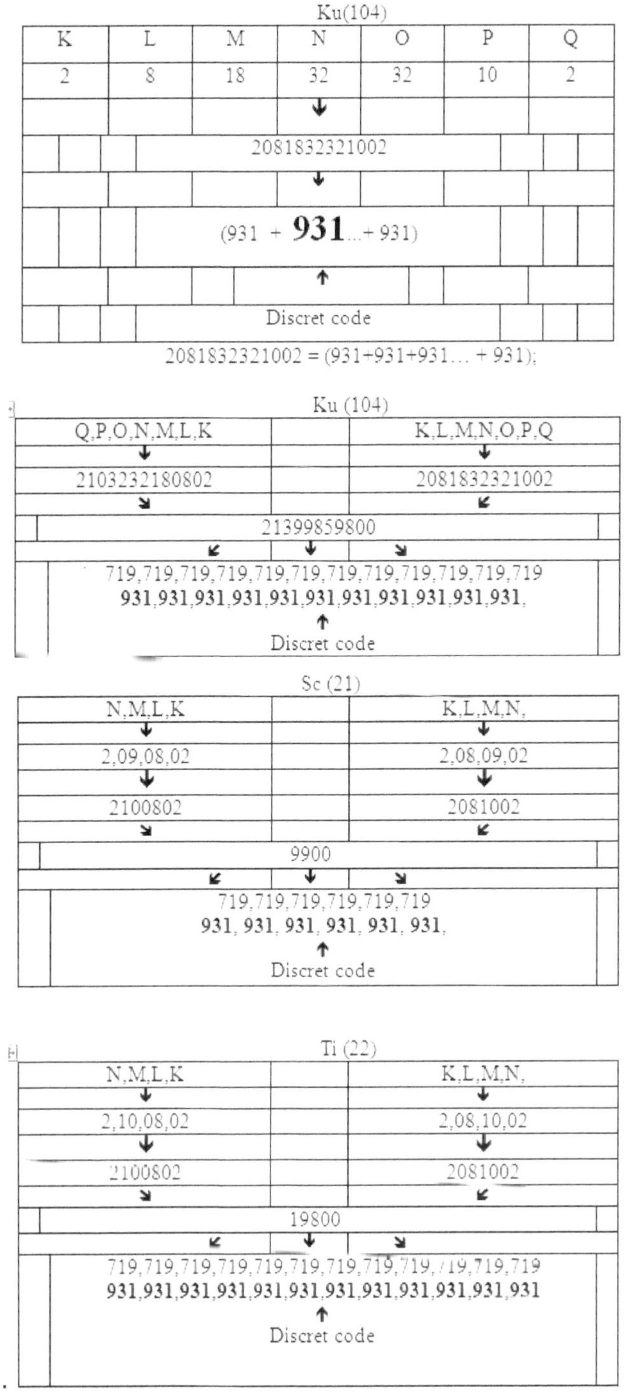

Figure 9. Discrete code in the constitution of atoms.

of systems and cybernetics). Within the digital pictures in biochemistry, the physical and chemical parameters are in a strict compliance with programmatic, cybernetic and information principles.

Conclusion

The process of sequencing in bio-macromolecules is conditioned and determined not only through Biochemical, but also through cybernetic and information principles. The digital pictures of Biochemistry provide us with cybernetic and information interpretation of the

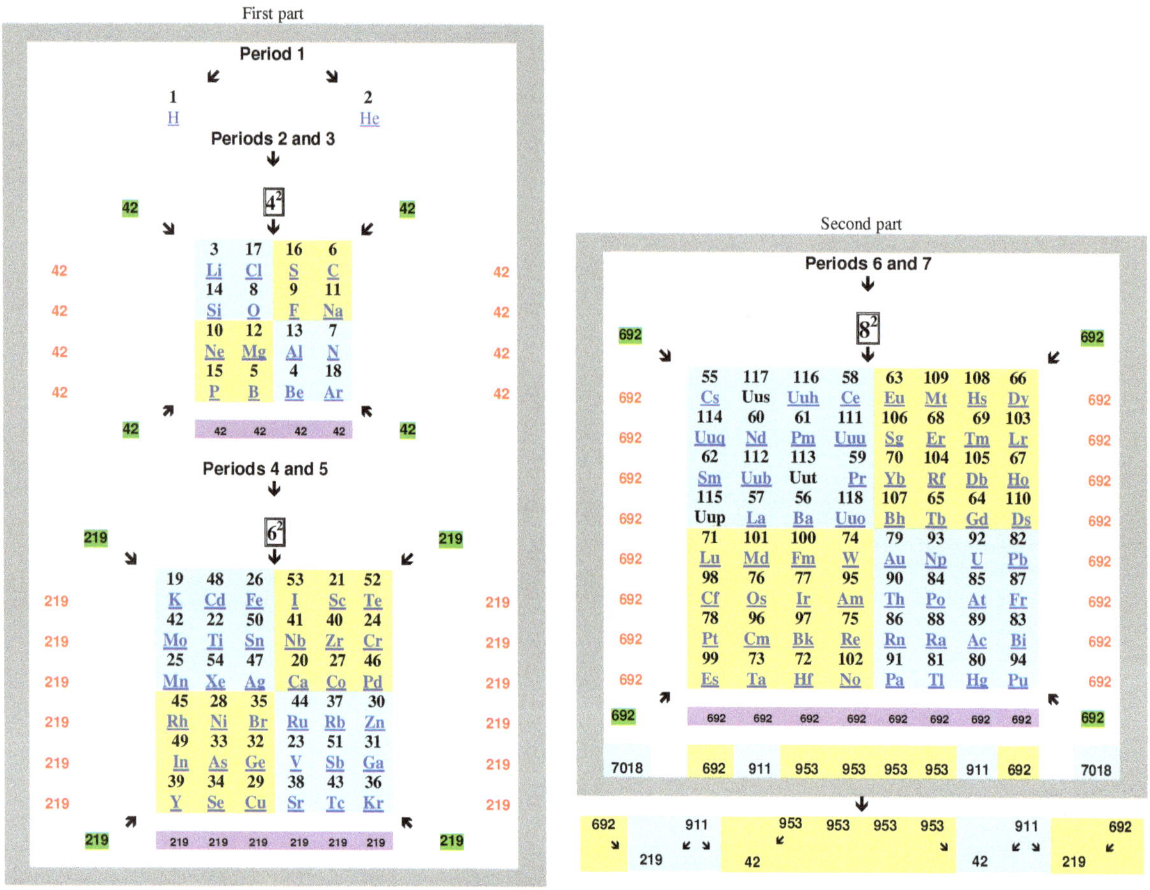

Figure 10. Mathematical position of chemical elements in digital tables of elements.

Fragment 1

55	**117**	**116**	**58**	**346**
Cs	Uus	Uuh	Ce	
114	**60**	**61**	**111**	**346**
Uuq	Nd	Pm	Uuu	
62	**112**	**113**	**59**	**346**
Sm	Uub	Uut	Pr	
115	**57**	**56**	**118**	**346**
Uup	La	Ba	Uuo	
346	**346**	**346**	**346**	

Diagonal A = 346; Diagonal B = 346;
Determinants 4 x 4 = 0;

Fragment 2

63	**109**	**108**	**66**	**346**
Eu	Mt	Hs	Dy	
106	**68**	**69**	**103**	**346**
Sg	Er	Tm	Lr	
70	**104**	**105**	**67**	**346**
Yb	Rf	Db	Ho	
107	**65**	**64**	**110**	**346**
Bh	Tb	Gd	Ds	
346	**346**	**346**	**346**	

Diagonal A = 346; Diagonal B = 346;
Determinants 4 x 4 = 0.

Fragment 3

71	**101**	**100**	**74**	**346**
Lu	Md	Fm	W	
98	**76**	**77**	**95**	**346**
Cf	Os	Ir	Am	
78	**96**	**97**	**75**	**346**
Pt	Cm	Bk	Re	
99	**73**	**72**	**102**	**346**
Es	Ta	Hf	No	
346	**346**	**346**	**346**	

Diagonal A = 346; Diagonal B = 346;
Determinants 4x4 = 0.

Fragment 4

79	**93**	**92**	**82**	**346**
Au	Np	U	Pb	
90	**84**	**85**	**87**	**346**
Th	Po	At	Fr	
86	**88**	**89**	**83**	**346**
Rn	Ra	Ac	Bi	
91	**81**	**80**	**94**	**346**
Pa	Tl	Hg	Pu	
346	**346**	**346**	**346**	

Diagonal A = 346; Diagonal B = 346;
Determinants 4x4 = 0.

Figure 11. chemical elements from the sixth and seventh period interconnected through the code 346 scientific facts. Now we have the exact scientific proofs that there is a genetic language that can be described by the theory of systems and cybernetics, and which functions in accordance with certain principles.

REFERENCES

Cai YD, Chou KC (2006). Predicting membrane protein type by functional Domain composition and pseudo amino acid composition. J. Theor. Biol., 238: 395-400.

Cai YD, Zhou GP, Chou KC (2003). Support vector machines for predicting membrane protein types by using functional domain composition. Biophys. J., 84: 3257-3263.

Cai YD, Zhou GP, Chou KC (2005). Predicting enzyme family classes by hybridizing gene product composition and pseudo-amino acid composition. J. Theor. Biol., 234: 145-149.

Cai YD, Feng KY, Lu WC, Chou KC (2006). Using LogitBoost classifier to predict protein structural classes. J. Theor. Biol., 238: 172-176.

Cai YD, Pong-Wong R, Feng K, Jen JCH, Chou KC (2004). Application of SVM to predict membrane protein types. J. Theor. Biol., 226: 373-376.

Chen C, Chen L, Zou X, Cai P (2009). Prediction of protein secondary structure content by using the concept of Chou's pseudo amino acid composition and support vector machine. Protein Peptide Lett., 16: 27-31.

Chen L, Feng KY, Cai YD, Chou KC, Li HP (2010). Predicting the network of substrate-enzyme-product triads by combining compound similarity and functional domain composition. BMC Bioinformatics, 11: 293.

Chou KC (2011). Some remarks on protein attribute prediction and pseudo amino acid composition (50th Anniversary Year Review). J. Theor. Biol., 273: 236-247.

Chou KC, Shen HB (2010a). Cell-PLoc 2.0: An improved package of web-servers for predicting subcellular localization of proteins in various organisms. Nat. Sci., 2: 1090-1103 (openly accessible at http://www.scirp.org/journal/NS/).

Chou KC, Shen HB (2010b). Plant-mPLoc: A Top-Down Strategy to Augment the Power for Predicting Plant Protein Subcellular Localization. PLoS ONE, 5: e11335.

Ding H, Luo L, Lin H (2009a). Prediction of cell wall lytic enzymes using Chou's amphiphilic pseudo amino acid composition. Protein Peptide Lett., 16: 351-355.

Ding YS, Zhang TL, Chou KC (2007). Prediction of protein structure classes with pseudo amino acid composition and fuzzy support vector machine network. Protein Peptide Lett., 14: 811-815.

Ding YS, Zhang TL, Gu Q, Zhao PY, Chou KC (2009b). Using maximum entropy model to predict protein secondary structure with single sequence. Protein Peptide Lett., 16: 552-560.

He ZS, Zhang J, Shi XH, Hu LL, Kong XG, Cai YD, Chou KC (2010). Predicting drug-target interaction networks based on functional groups and biologicalfeatures. PLoS ONE, 5: e9603.

Hu L, Huang T, Shi X, Lu WC, Cai YD, Chou KC (2011). Predicting functions of proteins in mouse based on weighted protein-protein interaction network and protein hybrid properties. PLoS ONE 6, e14556.

Huang T, Shi XH, Wang P, He Z, Feng KY, Hu L, Kong X, Li YX, Cai YD, Chou KC (2010). Analysis and prediction of the metabolic stability of proteins based on their sequential features, subcellular locations and interaction networks. PLoS ONE, 5: e10972.

Kandaswamy KK, Pugalenthi G, Molle S, Hartmann E, Kalies KU, Suganthan PN, Martinetz T (2010). Prediction of Apoptosis Protein Locations with Genetic Algorithms and Support Vector Machines Through a New Mode of Pseudo Amino Acid Composition. Protein Peptide Lett., 17: 1473-1479.

Kannan S, Hauth AM, Burger G (2008). Function prediction of hypothetical proteins without sequence similarity to proteins of known function. Protein Peptide Lett., 15: 1107-1110.

Liu T, Zheng X, Wang C, Wang J (2010). Prediction of Subcellular Location of Apoptosis Proteins using Pseudo Amino Acid Composition: An Approach from Auto Covariance Transformation. Protein Peptide Lett., 17: 1263-9.

Mohabatkar H (2010). Prediction of cyclin proteins using Chou's pseudo amino acid composition. Protein Peptide Lett., 17: 1207-1214.

Wang YC, Wang XB, Yang ZX, Deng NY (2010). Prediction of enzyme subfamily class via pseudo amino acid composition by incorporating the conjoint triad feature. Protein Peptide Lett., 17: 1441-1449.

7

CuO-CeO_2 catalytic systems destined for CO removal synthesized by means of the Pechini method: An evaluation of the structures obtained

L. S. Neiva*, H. M. C. Andrade and L. Gama

Academic Unit of Engineering of Materials, Federal University of Campina Grande – 58109-970, Avenue Aprígio Veloso, 882 – Bodocongó – Campina Grande, PB – Brazil.

In this work, catalytic systems based on CuO-CeO_2 were synthesized by the Pechini method. These catalytic systems were prepared without the need to perform a step of impregnation of active species as is often seen in works of catalyst preparation. In the synthesis procedure, the CuO-CeO_2 catalytic systems were synthesized with 0.1 and 0.5 mol of CuO. The CuO-CeO_2 catalytic systems were characterized by X-ray diffraction (XRD) with the Rietveld refinement method, Energy Dispersive X-ray spectrometry (EDX), Scanning Electron Microscopy (SEM), textural analysis, thermogravimetry and agglomerates size distribution. According to the results, the CuO concentration value (0.1 or 0.5 mol) inserted into the CuO-CeO_2 catalytic system showed great influence on the structural characteristics observed in the X-ray diffraction results. However, the morphologic and textural features as well as the results of agglomerates size distribution did not show significant changes to the variation of CuO concentration inserted in the catalytic systems developed in this work by the Pechini method. On the other hand, it was observed that these catalytic systems presented a good porous structure, which is favorable for the catalytic application in PROX process. The next stage of this work is the catalytic test in the PROX reaction.

Key words: CuO-CeO_2 catalytic system, Pechini method, CO removal, PROX process.

INTRODUCTION

Cerium oxide (CeO_2) has earned intensive interest in the past decade because it plays a vital role in emerging technologies for environmental and energy-related application (Sun et al., 2004; Trovarelli, 1996; Carrettin et al., 2004; Park et al., 2000). Among a variety of oxides, ceium oxide is particularly of interest in a variety of catalytic processes. For instance, ceria has been used as catalyst or support in the selective oxidation, fine chemicals synthesis, methane steam reforming, phenol hydrogenation at atmospheric pressure and solid oxide fuel cell applications (Chojnscki et al., 1991; Yee et al., 1999; Putna et al., 1995, Craciun et al., 1998; Oh and Sinkevitch, 1993).

One of the important applications of ceria is its use in the area of environmental catalysis, particularly in the design of new generation of three-way catalysts for simultaneously controlling NO, CO and hydrocarbon emissions in oxygen-rich conditions. This is because ceria has large oxygen storage capacity and the ability of fastly transferring bulk oxygen to its surface (Crucq, 1991; Fornasiero et al., 1999). The former may uptake oxygen under fuel-lean conditions and the latter may release oxygen under reducing conditions, thus allowing a ceria-containing catalyst to efficiently work in a relatively wide window of oxygen partial pressure. In addition, ceria may also improve metal dispersion on it and promote surface and bulk oxygen reducibility of the support when it is used as a catalyst additive (Leitenburg et al., 1997; Kondarides et al., 1998).

In the last years, much research efforts have been focused on finding and developing a catalyst which is suitable to selectively oxidize CO (CO removal), also known as preferential oxidation of CO. Such a catalyst

*Corresponding author. E-mail: lsoutoneiva@yahoo.com.br.

should be active for the oxidation of CO, and inactive for the undesired oxidation of H_2, in the temperature range of 80 to 250 °C. Supported noble metal catalysts especially Pt-group and Au catalysts, have been extensively studied (Avgouropoulos and Ioannides, 2007). However, the high cost of precious metals has encouraged researchers around the world to look for alternative catalysts. In particular, it has been found that the $CuO–CeO_2$ catalytic system is highly active and exceptionally selective for the preferential oxidation of CO (Haruta, 1997; Mariño et al., 2004; Marques et al., 2006; Kandoi et al., 2004; Martínez-Arias et al., 2006). A strong synergetic effect between copper and ceria has been reported to be responsible for the high activity of these catalysts compared to conventional Cu-based catalysts (Mariño et al., 2005).

In the literature, noble metal catalysts (Au, Pt and Ru) and oxide catalyst of $CuO–CeO_2$ have been developed for PROX process. Gold-based catalysts are active at low reaction temperature (~ 80 °C) but their activities are seriously affected by CO_2 and excessive H_2O in hydrogen-rich gas (Avgouropoulos et al., 2002). Pt and Ru catalysts are active at medium reaction temperature (~ 140 °C) but relatively low in selectivity of CO oxidation (<50%) (Manaslip and Gulari, 2002; Han et al., 2004). Comparably, $CuO–CeO_2$ catalysts have advantages of low cost and high selectivity of CO oxidation.

The chemical synthesis method based on polyesters obtained from citrates was developed by Pechini (Pechini, 1967) and is intensively used for the synthesis of several polycation metallic oxides (Lessing, 1989; Error and Anderson, 1986). This Pechini method is based on chelation or complexation of cations by hydroxycarboxylic acid such as citric acid. The source of cations can be diverse. The chelation process occurs during the mix of cations and citric acid in aqueous solution. This solution, with no particle in the suspension, is mixed with a glycol such as ethylene glycol and heated up to 90 to 100 °C until complete water elimination. Polymerization occurs at 120 to140 °C through the polyesterification reaction among citrate salt and ethylene glycol. Thus the formation of transparent solid polyester resin with no phase segregation is observed in this condition. The general idea of this method is to obtain a random distribution of cations at atomic level in a solid resin. After this stage the polyester is decomposed to eliminate the excess organic material and the cation is oxidized to form the desirable stoichiometric phase.

In this work, the $CuO-CeO_2$ catalytic system was synthesized by the Pechini method. The obtained catalytic systems were subject to various characterizations techniques to analyse structural characteristics for the knowledge of the same. After this research, these same synthesized catalytic systems in the work will be subject to tests, specifically in the preferential oxidation of CO reaction (PROX) in the bench scale.

MATERIALS AND METHODS

The $CuO-CeO_2$ catalytic systems were synthesized by the Pechini method using the following precursors: citric acid, cerium nitrate, copper nitrate, and ethylene glycol, all of them of high purity. Initially a solution using distilled water, nitrates (metal ions sources), citric acid and ethylene glycol, was prepared under constant agitation at a temperature of 70 °C. After adding all the mentioned precursors, the obtained solution was heated at a temperature of polyesterification reaction (~ 120 °C) to form a resin. The resin formed had undergone a pre-calcination stage (~ 400 °C) for removing organic matter excess from the ethylene glycol, generating as product a carbonized material still containing traces of organic matter. This product was broken, sieved in a 325 mm mesh, and then subjected to a calcination stage at 700 °C to eliminate the remaining organic matter, and also the formation of the desired phase ($CuO-CeO_2$) in the developed material structure. The final product obtained after the calcination stage was the desired catalytic system, which proceeded to the characterizations stage. Thus, the catalytic system based on $CuO-CeO_2$ was obtained without the necessity of performing an additional step of impregnation for incorporation of active species that in this case is copper oxide. The $CuO-CeO_2$ catalytic system was fully prepared in only step that is, the synthesized Pechini method was able to promote the preparation of this catalyst system without requiring additional steps.

In this work, two catalytic systems based on $CuO-CeO_2$ were synthesized; one containing 0.1 mol of CuO and another 0.5 mol of CuO. The samples of the $CuO-CeO_2$ catalytic system containing 0.1 mol of CuO was called P1, and the samples of the $CuO-CeO_2$ catalytic system containing 0.5 mol of CuO was called P5. The samples of the catalytic systems were submitted to the following characterization techniques: structural analysis by X-ray diffraction (XRD) using a Shimadzu XRD 6000 diffractometer (with CuKα radiation, λ = 0.15418 nm) combined with Rietveld refinement realized through the Maud Software; chemical analysis by Energy Dispersive X-ray spectrometry (EDX) using a Shimadzu EDX-GP spectrometer; morphological analysis by means of Scanning Electron Microscopy (SEM) in a ZEISS DSM 940 microscope; textural analysis by BET method using N_2 gas in Quantachrome Nova 3200 analyser; thermogravimetric analysis was performed on an analyzer TA instrument, model SDT 2960. This thermogravimetric analysis was performed at a temperature range of between 25 and 1000 °C using a heating rate of 10 °C min^{-1} and an inert atmosphere of N_2 and the agglomerates size distribution was determined using a CILAS 1064 L D particle size analyzer.

RESULTS AND DISCUSSION

Figure 1 shows the x-ray patterns of the data collected before (referencial or theoretical) and after (refined) Rietveld refinement referring to $CuO-CeO_2$ catalytic systems synthesized by the Pechini method defined as P1 and P5. According to the x-ray patterns referring to P1 sample, it was observed that it is possible to synthesize $CuO-CeO_2$ catalytic system with 0.1 mol of CuO and to obtain the formation of structure crystalline composed by CeO_2 as a single phase, that is, it is possible to obtain the desired structure ($CuO-CeO_2$) without the presence of segregated phases composed by CuO by means of Pechini method synthesis. The x-ray patterns referring to P5 sample shows that the concentration of 0.5 mol of CuO was sufficient to favor the formation of the

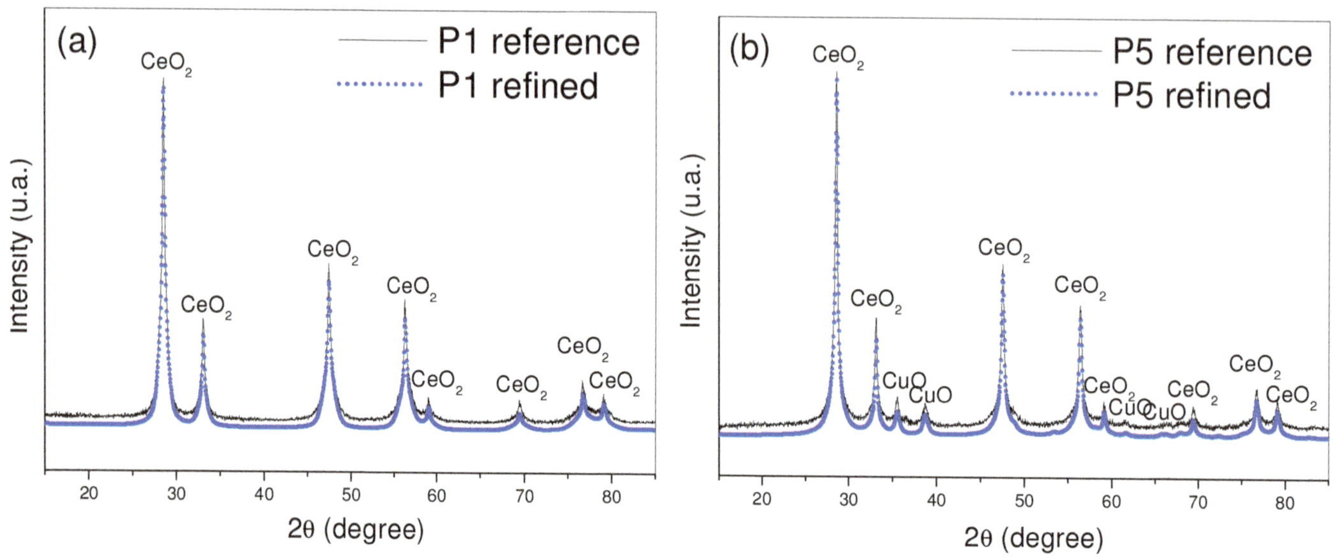

Figure 1. X-ray diffraction patterns before and after Rietveld refinement for the P1 and P5 catalytic systems synthesized by the Pechini method.

Table 1. Results of the quantification of the phases and others structural characteristics determined by the Rietveld method for the P1 and P5 catalytic systems synthesized by the Pechini method.

Sample	Phase	Crystalline system	Percentage mass (%)	Crystallite size (nm)
P1	CeO_2	Cubic	70.3	13.3
	CeO_2	Cubic	29.7	65.5
P5	CeO_2	Cubic	48.1	7.3
	CeO_2	Cubic	37.9	55.5
	CuO	Monoclinic	14.0	20.6

segregated phase composed by CuO over the structure of the CuO-CeO_2 catalytic system.

The structural characteristics determined by means of the Rietveld method for the CuO-CeO_2 catalytic systems synthesized by the Pechini method are shown in Table 1. According to the quantification of the phases formed in the analyzed structures, it can affirm that the predominant phase in both catalytic systems (P1 and P5) is the ceria phase (CeO_2). In the structure of the P1 sample containing 0.1 mol of CuO, it was not possible to detect the presence of the segregated phase of CuO. On the other hand, the P5 sample containing 0.5 mol of CuO presented the segregated phases composed by CuO over its structure with a massic percentage of 14%. This is an indication that the 0.1 mol of CuO concentration is within the solubility limit of CeO_2 atomic structure which in this case acts as a host matrix structure, that is, the structure of cerium oxide can incorporate 0.1 mol of CuO so that no remnant of copper oxide will be in an amount sufficient to form segregated phases over the structure of CeO_2. The same cannot be affirmed with respect to the P5 sample containing 0.5 mol of CuO. Therefore, the concentration of CuO contained in the P5 sample certainly exceeds the solubility limit of the structure of cerium oxide. The proof is the presence of segregated phases composed by CuO in the synthesized structure.

It was also observed in Table 1 that both catalytic systems synthesized in this work presented in their structures the formation of two types of majority phases; both composed of CeO_2, but with different crystalline sizes. This difference in the crystalline sizes is a consequence of the calcinations step at 700°C used in the synthesis procedure of the Pechini method. The calcination step promotes the coalescence of some crystals to form larger crystals, and thus, the final product can present phases composed by the same oxide but with different crystallite sizes. The phases formed in the structure of the P5 catalytic system belongs to different crystal systems (cubic and monoclinic) as can be seen in Table 1.

The chemical analysis results determined by EDX technique are presented in Table 2 and it shows that the

Table 2. Chemical composition determined by EDX technique for the P1 and P5 catalytic systems synthesized by the Pechini method.

Sample	CeO_2 (%)	CuO (%)	ZnO (%)	NiO (%)	LOI (%)
P1	90.97	7.89	0.18	0.12	0.84
P5	65.67	33.41	0.23	0.09	0.60

Figure 2. Micrographs obtained by SEM showing the morphology of the P1 and P5 catalytic systems synthesized by the Pechini method.

$CuO-CeO_2$ catalytic systems synthesized by the Pechini method are predominantly composed of CeO_2 and CuO. However, it was detected that the presence of the oxides, ZnO and NiO were probably incorporated into these catalytic systems by means of contaminants present in reagents, containers or ovens used in the synthesis procedure, or during the handling of these samples in the EDX analysis. The percentage of Loss on Ignition (LOI) was determined at the 1000°C.

Figure 2 illustrates the results of morphological evaluation performed by Scanning Electron Microscopy (SEM) in the $CuO-CeO_2$ catalytic systems synthesized by the Pechini method. The micrographs in Figure 2 show their respective images with a magnification of 30.000X.

According to the micrographs shown in Figure 2, the $CuO-CeO_2$ catalytic systems synthesized by the Pechini method presented a strong tendency to the state of particles agglomeration, forming noticeably porous structures. This morphology aspect is very appropriate from the point of view that potential catalytic properties are expected of these materials. It is worth pointing out that these catalytic systems present a relatively homogeneous size distribution of agglomerates. There appeared to be no significant differences between the morphological characteristics presented by the P1 and P5 catalytic systems therefore, it can be rationalized that the incorporation process of CuO in different concentrations (0.1 and 0.5 mol) did not alter the morphological characteristics of these materials. Similar morphological characteristics were found in several other studies related to the preparation of cerium dioxide for application in catalytic reactions by different synthesis methods (Triki et al., 2009; Polato et al., 2005; Sun et al., 2007; Neiva et al., 2010).

Figure 3 shows N_2 adsorption-desorption isotherms referring to the textural analysis of $CuO-CeO_2$ catalytic systems synthesized by the Pechini method. According to the IUPAC classification (Sing et al., 1985), these isotherms can be considered as type III, characterizing porous materials with a predominantly macroporous nature. The hysteresis loop presented on these isotherms is of type H3, especially between $P/P0 = 0.45$ and 0.95 which reveals the presence of pores generally formed by the association of non-rigid agglomerates of particles (Figueiredo and Ribeiro, 1987).

Table 3 shows the values of the textural characteristics such as surface area (S_{BET}) determined by BET method, and values of pore volume (V_P) and medium pore diameter (D_P) determined by BJH method, all referring to $CuO-CeO_2$ catalytic systems called P1 and P5 synthesized by the Pechini method.

According to the values presented in Table 3, the medium pore diameter (D_P) of the $CuO-CeO_2$ catalytic systems synthesized in this work are of macroporous dimensions ($D_P > 50$ nm), which confirm the porosity type revealed in the isotherms presented in Figure 3.

In Table 3, it was also observed that the values of surface area, pore volume, and medium pore diameter

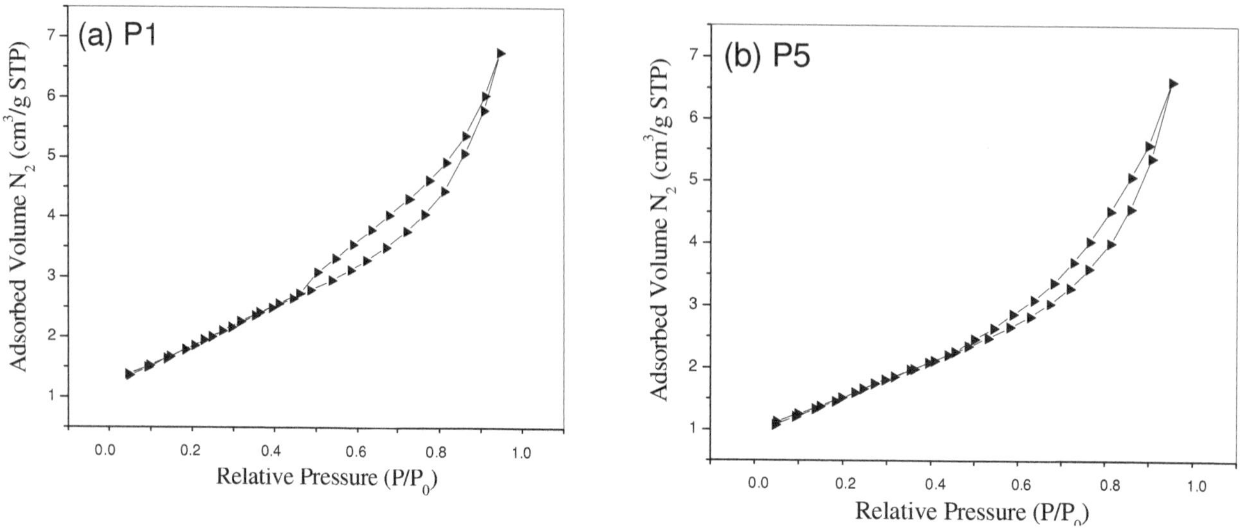

Figure 3. N_2 adsorption-desorption isotherms related to the P1 and P5 catalytic systems synthesized by the Pechini method.

Table 3. Surface area, pore volume and medium pore diameter determined for the P1 and P5 catalytic systems synthesized by the Pechini method.

Sample	S_{BET} (m^2/g)	V_P (nm)	D_P (nm)
P1	16	0.02204	387
P5	14	0.02251	383

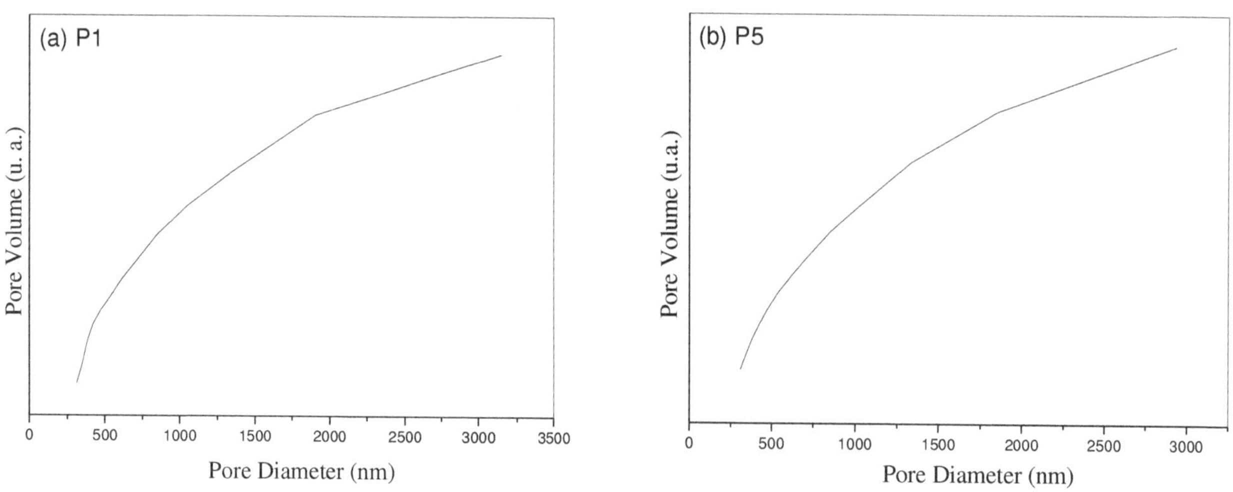

Figure 4. Graphs of pore diameter distribution of the P1 and P5 catalytic systems synthesized by the Pechini method.

did not present significant variations for the CuO concentrations incorporated in the synthesis of the CuO-CeO_2 catalytic systems (P1 and P5) by the Pechini method. This finding is viewed positively, since the CuO substance was inserted into the material structure, probably causing the formation of catalytic active sites without generating great changes in the textural characteristics of the structures.

Figure 4 presents graphs of pore size distribution referring to the CuO-CeO_2 catalytic systems called P1 and P5 synthesized by the Pechini method. The graphs of pore size distribution of the P1 and P5 catalytic systems synthesized by the Pechini method show that these materials present fully macroporous structures, as had been revealed by the isotherms shown in Figure 3 and the data shown in Table 3. Other researches

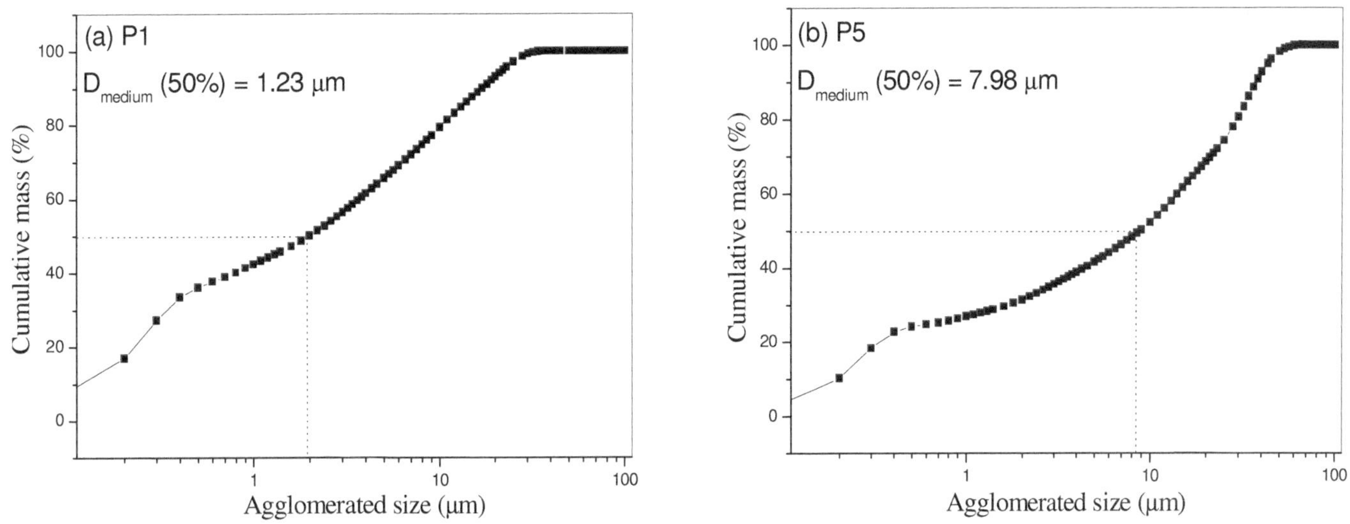

Figure 5. Agglomerates size distribution of the P1 and P5 catalytic systems synthesized by the Pechini method.

referring to CuO-CeO_2 catalytic system destined for preferential oxidation of CO has also been shown textural characteristics very similar to the observed for the CuO-CeO_2 catalytic systems obtained in this work (Qiang et al., 2007; Moretti et al., 2008).

According to the thermogravimetric analysis performed on samples of the catalytic system synthesized in this work using a temperature range of between 25 and 1000°C, the samples showed no weight loss throughout the temperature range studied. The calcination step at 700°C experienced by these samples during their synthesis was certainly sufficient to eliminate any concentration of organic matter and humidity, therefore, these samples remained inert resulting in mass loss during the thermogravimetric analysis.

Figure 5 shows the curves of agglomerates size distribution for the P1 and P5 catalytic systems. According to Figure 5, it is possible to observe that the curves of agglomerate size distribution present similar profiles between the systems under study. However, the P1 catalytic system presented a lower medium diameter ($D_{50\%}$) of agglomerates. Thus, it is possible to affirm that the particles' agglomeration capacity is directly related to the CuO concentration value inserted in the structure of these materials, that is, the particles agglomeration capacity is directly related to the CuO concentration value inserted in the structure of cerium oxide.

The study of the granulometric behavior of compound materials by single or mixed oxides can lead to a determination of suitable size of the grain for use as a control parameter for a specific application of the material, according to the elements or substances that constitute the material as well as with the conditions for obtaining them. A detailed analysis of this behavior can guarantee obtaining a catalytic system based on CuO-CeO_2 with more uniform and standardized structure.

Conclusions

According to the results we can affirm that it is possible to synthesize catalytic systems based on CuO-CeO_2 by means of the Pechini method, without the need for a step of impregnation of the active species catalysis as is often seen in researches on the preparation of catalysts. Also it can be affirmed that the catalytic system based on CuO-CeO_2 can incorporate 0.1 mol of CuO without forming segregated phase composed by CuO over the structure of the material when synthesized by the Pechini method. On the otherhand, the phases of quantification realized by the Rietveld method showed that the CuO-CeO_2 catalytic system containing 0.5 mol of CuO allowed the formation of a structure with mass percentage of 14% of the segregated phase composed of CuO. The CuO concentration value incorporated in the CuO-CeO_2 catalytic system synthesized by the Pechini method did not lead to significant alterations in the morphological and textural characteristics of these materials. On the other hand, it is possible to affirm that the particle agglomeration capacity of these catalytic systems is directly related to the CuO concentration value inserted in their structures.

REFERENCES

Avgouropoulos G, Ioannides T (2007). Adsorption and reaction of CO on CuO-CeO_2 catalysts prepared by the combustion method, Catal. Lett., 116: 15-22.

Avgouropoulos G, Ioannides T, Papadopoulou C, Batista J, Hocevar S, Matralis HK (2002). A comparative study of Pt/γ-Al_2O_3, Au/α-Fe_2O_3 and CuO–CeO_2 catalysts for the selective oxidation of carbon monoxide in excess hydrogen. Catal. Today, 75: 157-167.

Carrettin S, Concepcion P, Corma A, Nieto JML, Puntes VF (2004). Nanocrystalline CeO_2 Increases the Activity of Au for CO Oxidation by Two Orders of Magnitude. Angew. Chem. Int. Ed., 43: 2538-2540.

Chojnscki T, Krause K, Schimidt LD (1991). Microstructure and reactivity of Pt-Ce and Rh-Ce particles on silica. J. Catal., 128: 161-185.
Craciun R, Shereck B, Gorte RJ (1998). Kinetic studies of methane steam reforming on ceria-supported Pd, Catal. Lett., 51: 149-153.
Crucq A (1991). Catalysis and Automotive Pollution Control II, Elseiver Science Publisher, Amsterdam, Oxford, New York, Tokyo. Stud. Surface Sci. Catal., p. 71.
Error NG, Anderson HU (1986). Polymeric Precursor Synthesis of Ceramic Materials. Muter. Rex Sot. Symp. Proc., 73: 571-578.
Figueiredo JL, Ribeiro FR (1987). Catálise Heterogênea, Ed. Fundação Calouste Gulbenkian, Lisboa. In Portuguese, pp. 94-99.
Fornasiero P, Ka˘spar J, Graziani M (1999). On the rate determining step in the reduction of CeO_2-ZrO_2 mixed oxides, Appl. Catal. B: Environ., 22: 11-14.
Han YF, Kinne M, Behm RJ (2004). Selective oxidation of CO on Ru/γ-Al_2O_3 in methanol reformate at low temperatures. Appl. Catal. B: Environ. 52: 123-134.
Haruta M (1997). Size-Dependency and Support-Dependency in the Catalysis of Gold. Catal. Today, 36: 153-166.
Kandoi S, Gokhale A, Grabow L, Dumesic J, Mavrikakis M (2004). Why Au and Cu are more selectivity than Pt for preferential oxidation of CO at low temperature? Catal. Lett., 93: 93-100.
Kondarides DI, Zhang ZL, Verykios XE (1998). Chlorine-Induced Alterations in Oxidation State and CO Chemisorptive Properties of CeO_2-Supported Rh, J. Catal., 176: 536-544.
Leitenburg CD, Trovarelli A, Ka˘spar J (1997). A Temperature-Programmed and Transient Kinetic Study of CO_2Activation and Methanation over CeO_2Supported Noble Metals. J. Catal., 166: 98-107.
Lessing PA (1989). Mixed-cation oxide powders via polymeric precursors, Ceramic Bulletin, 68 (5): 1002-1008.
Manaslip A, Gulari E (2002). Selective CO oxidation over Pt/alumina catalysts for fuel cell applications, Appl. Catal. B: Environ. 37: 17-25.
Mariño F, Descorme C, Duprez D (2004). Noble metal catalysts for the preferential oxidation of carbon monoxide in the presence of hydrogen (PROX), Appl. Catal. B: Environ., 54: 59-66.
Mariño F, Descorme C, Duprez D (2005). Supported base metal catalysts for the preferential oxidation of carbon monoxide in the presence of excess hydrogen (PROX). Appl. Catal. B: Environ., 58: 175-183.
Marques P, Ribeiro N, Schmal M, Aranda D, Souza M (2006). Selective CO oxidation in the presence of H_2 over Pt and Pt-Sn catalysts supported on niobia, J. Power Sources,158: 504-508.
Martínez-Arias A, Hungría B, Munuera G, Gamarra D (2006). Preferential oxidation of CO in rich H_2 over CuO/CeO_2: Details of selectivity and deactivation under the reactant stream, Appl. Catal. B: Environ., 65: 207-216.
Moretti E, Lenarda M, Storaro L, Talon A, Montanari T, Busca G, Rodrigues-Castellón E, Jiménez-López A, Turco M, Bagnasco G, Frattini R, (2008). One-step synthesis of a structurally organized mesoporous CuO-CeO_2-Al_2O_3 system for the preferential co oxidation, Appl. Catal. A: Gen., 335: 46-55.
Neiva LS, Leal E, Costa ACFM, Mascarenhas AJS, Andrade HMC, Gama L (2010). Evaluation of the Cu dopping effects in CeO_2 catalytic supports obtained by combustion reaction, Mater. Sci. Forum, 660-661, 889-903.
Oh S, Sinkevitch R (1993). Carbon Monoxide Removal from Hydrogen-Rich Fuel Cell Feed Streams by Selective Catalytic Oxidation. J. Catal., 142: 254-262.
Park S, Vohs JM, Gorte RJ (2000). Direct oxidation of hydrocarbons in a solid-oxide fuel cell. Nature, 404: 265-267.
Pechini MP (1967). Method of preparing lead and alkaline earth titanates and niobates and coating methods using the same to form a capacitor. U.S. Patent 3,330,697, July 11.
Polato CMS, Henriques CA, Neto AA, Monteiro JLF (2005). Synthesis, characterization and evaluation of CeO_2/Mg,Al-mixed oxides as catalysts for SOx removal, J. Mol. Catal. A: Chem., 241: 184-193.
Putna ES, Stubenrauch J, Vohs JM, Gorte RJ (1995). Ceria-Based Anodes for the Direct Oxidation of Methane in Solid Oxide Fuel Cell. Langmuir, 11: 4832-4837.
Qiang G, Meiling W, Yuan L, Xue B (2007). Mesoporous CeO_2-Supported Co_3O_4 Catalysts for CO Preferential Oxidation in H_2-Rich Gases, Chin. J. Catal., 28: 953-957.
Sing KSW, Everett DH, Haul RAW, Moscou L, Pierotti RA, Rouquerol J, Siemieniewska T (1985). Pure and Applied Chemistry, IUPAC Recommendations, 57: 4: 603.
Sun C, Li H, Chen L (2007). Study of flowerlike CeO_2 microspheres used as catalyst supports for CO oxidation reaction, J. Phys. Chem. Solids, 68: 1785-1790.
Sun CW, Li H, Wang ZX, Chen LQ, Huang XJ (2004). Synthesis and characterization of polycrystalline CeO2 nanowires. Chem. Lett., 133: 662-663.
Triki M, Ksibi Z, Ghorbel A, Medina F (2009). Preparations and characterization of CeO_2-TiO_2 support for Ru catalysts: Application in CWAO of p-hydroxilbenzoic acid. Micr. Mesop. Mat., 117: 431-435.
Trovarelli A (1996). Catalytic Properties of Ceria and CeO_2-Containing Materials. Catal. Rev. Sci. Eng., 38: 439-520.
Yee A, Morrison SJ, Idriss H (1999). A Study of the Reactions of Ethanol on CeO_2 and Pd/CeO_2 by Steady State Reactions, Temperature Programmed Desorption, and *In Situ* FT-IR. J. Catal., 186: 279- 295.

Equilibrium modeling for a downdraft biomass gasifier for cotton stalks biomass in comparison with experimental data

Koroneos C.[1*] and Lykidou S.[2]

[1]Laboratory of Heat Transfer and Environment, Aristotle, University of Thessaloniki, Box 483, 54124 Thessaloniki, Greece
[2]Department of Chemistry, Laboratory of Organic Chemistry Technology, Aristotle University of Thessaloniki, 54124 Thessaloniki, Greece

Equilibrium modeling has been used to predict the composition of the fuel gas produced in a downdraft gasifier for cotton stalks biomass at 800°C with the moisture content from 0 to 30%. The calorific value of the fuel gas has been determined. The predicted values are in good agreement with the experimental data. The moisture content in the cotton stalks biomass and the variability in calorific value, have been investigated. It has been found that the calorific value of the producer gas decreases with the increase in moisture content of the cotton stalks.

Key words: Gasification, biomass, downdraft gasifier, modeling.

INTRODUCTION

Biomass is one of the most promising renewable energy sources. Biomass is in various forms and it is abundant in many areas of the world. Due to its abundance, its energy content, and the low emissions to the atmosphere, it could play a major role in meeting world energy demand (Gao et al., 2008).

Biomass gasification is one of the effective technologies for its thermo chemical conversion. The conversion is achieved by reactions between a feed gas and a feedstock. The primary goal of biomass gasification is optimal energy conversion of the solid biomass into a combustible gaseous product known as producer gas. In a gasifier, there are three main thermal conversion layers: the combustion zone, the pyrolysis zone and the reduction zone. In a downdraft fixed bed, the biomass undergoes combustion, pyrolysis and gasification, one after the other (Gao et al., 2008).

Wang and Kinoshita (1993) quoted that the char reduction reactions are slow due to the fact that they are surface phenomena and the overall biomass gasification rate can be controlled by the char reduction zone. The reaction temperature, residence time and thus volume of the reaction zone decide the reaction rates and their direction and thus the calorific value of gas as well as the efficiency of conversion. In order to obtain the optimal energy conversion, it is important to understand the chemistry of heterogeneous and homogeneous reactions taking place in the downdraft biomass gasifier and specifically in the reduction zone.

Giltrap et al. (2003) developed a model of the reduction zone of a downdraft biomass gasifier to predict the composition of the producer gas under steady state operation.

The accuracy of the model is limited by the availability of data on the initial conditions at the top of the reduction zone. Moreover, it is assumed that the char reactivity factor (CRF), a factor representative of the char reactivity which is a key variable, is constant throughout the reduction zone.

Babu and Sheth (2006) suggested an exponentially varying char reactivity factor in order to predict better simulation of the temperature profile in the reduction reaction zone. They have also compared the role of

*Corresponding author. e-mail: koroneos@aix.meng.auth.gr

constant, linearly varying, and exponentially varying char reactivity factors on the model predictions.

Melgar et al. (2007) combine the chemical equilibrium and thermodynamic equilibrium of the global reaction to predict the final composition of the producer gas.

Srivastava et al. (1996) predict the concentration profiles in the cases of pyrolysis of different biomass materials in isothermal and non-isothermal conditions.

Koufopanos et al. (1989) presented a two step mechanism scheme for describing the kinetics of biomass pyrolysis, and Babu and Chaurasia (2003) attempted to find the optimum kinetic modeling parameters for the pyrolysis process.

Zainal et al. (2001) have studied an equilibrium gasifycation model based on equilibrium constants, to simulate the gasification process in a downdraft gasifier. They have confirmed that the residence time of the reactants can be considered to be high enough to reach chemical equilibrium.

In this work, the equilibrium model of Zainal et al. (2001) of the gasification process is used as a basis in order to predict the composition of the fuel gas produced and its associated calorific values. The equilibrium model is used for the gasification of cotton stalks at 800°C with the moisture from 0 to 30%. The model predictions are compared with the experimental data reported by Kantarelis et al. (2009)

ANALYSIS

The equilibrium model assumes that all the reactions are in thermodynamic equilibrium. It is expected that the pyrolysis product burns and achieves equilibrium in the reduction zone before leaving the gasifier.

The reactions that take place are the following:

$$C + CO_2 \leftrightarrow 2CO \quad (1)$$

$$C + H_2O \leftrightarrow CO + H_2 \quad (2)$$

$$C + 2H_2 \leftrightarrow CH_4 \quad (3)$$

Equations 1 and 2 can be combined to give the shift reaction

$$CO + H_2O \leftrightarrow CO_2 + H_2 \quad (4)$$

For Equation 3 (methane formation) the equilibrium constant is:

$$K_1 = \frac{P_{CH4}}{(P_{H2})^2} \quad (5)$$

The equilibrium constant for the shift reaction is:

$$K_2 = \frac{P_{CO2}P_{H2}}{P_{CO}P_{H2O}} \quad (6)$$

The chemical formula of cotton stalks biomass, based on a single atom of carbon is (Kantarelis et al., 2009):

$CH_{1,51}O_{0,83}$

The general equation can be written as follows:

$$CH_{1,51}O_{0,83} + wH_2O + mO_2 + 3.76mN_2 \leftrightarrow x_1H_2 + x_2CO + x_3CO_2 + x_4H_2O + x_5CH_4 + 3.76mN_2 \quad (7)$$

where w is the amount of water per mol of cotton stalks; m is the amount of oxygen per mol of cotton stalks, and x_1, x_2, x_3, x_4 and x_5 are the coefficients of constituents of the products.

If MC is taken to be the content of moisture per mole of biomass, then

$$MC = \frac{mass\_of\_water}{mass\_of\_dry\_biomass}(100\%) \Rightarrow MC = \frac{18w}{24+18w}(100\%)$$

Then

$$w = \frac{26.8MC}{18(1-MC)}$$

Since the moisture content is known, the value of w becomes a constant.

From Reaction (7), there are six unknowns, x_1, x_2, x_3, x_4, x_5, the species of the product, and m, the oxygen content for the reaction. Therefore, six equations are required based on the following:

Carbon balance:

$$1 = x_2 + x_3 + x_5 \quad (8)$$

Hydrogen balance:

$$2w + 1.51 = 2x_1 + 2x_4 + 4x_5 \Rightarrow w + 0.755 = x_1 + x_4 + 2x_5 \quad (9)$$

Oxygen balance:

$$w + 0.83 + 2m = x_2 + 2x_3 + x_4 \quad (10)$$

Equilibrium constant from methane formation (Equation (5)):

$$K_1 = \frac{x_5}{x_1^2} \quad (11)$$

Equilibrium constant from shift reaction (Equation (6)):

$$K_2 = \frac{x_1x_3}{x_2x_4} \quad (12)$$

The gasification process is assumed to be adiabatic. The equation for the heat balance is:

$$H^o_{fwood} + w\,(H^o_{fH2O(l)} + H_{(vap)}) + mH^o_{fCO2} + 3{,}76mH^o_{fN2} = x_1H^o_{fH2} + x_2H^o_{f}CO + x_3H^o_{fCO2} + x_4H^o_{fH2O(vap)} + x_5H^o_{fCH4} + \Delta T(x_1C_{pH2} + x_2C_{pCO} + x_3C_{pCO2} + x_4C_{pH2O} + x_5C_{pCH4} + 3.76mC_{pN2}) \quad (13)$$

Since the values of H^o_{fH2}, H^o_{fN2} and H^o_{fO2}, are zero at ambient temperature, equation (13) reduces to:

$$H^o_{fwood} + w(H^o_{fH2O(l)} + H_{(vap)}) = x_2H^o_{fCO} + x_3H^o_{fCO2} + x_4H^o_{fH2O(vap)} + x_5H^o_{fCH4} + \Delta T(x_1C_{pH2} + x_2C_{pCO} + x_3C_{pCO2} + x_4C_{pH2O} + x_5C_{pCH4} + 3.76mC_{pN2}) \quad (14)$$

where, H^o_{fwood}, is the heat of formation of wood; $H^o_{fH2O(l)}$, is the heat of formation of liquid water; $H_{(vap)}$, is the heat of vaporization of water; $H^o_{fH20(vap)}$, is the heat of formation of water vapor; H^o_{fCO}, H^o_{fCH4}, are heats of formation of the gaseous products;

C_{pH2}, C_{pCO}, C_{pCO2}, H^o_{fCH4}, are heats of formation of the gaseous products; C_{pH2}, C_{pCO}, C_{pCO2}, C_{pH2O}, C_{pCH4}, C_{pN2}, are specific heats of the gaseous products; $\Delta T=T_2-T_1$, T_2, are the gasification temperature at the reduction zone; T_1, is the ambient temperature at the reduction zone. Equation (13) can be simplified to be:

$$H^o_{fwood} + w(H^o_{fH2O(l)} + H_{(vap)}) = x_2H^o_{fCO} + x_3H^o_{fCO2} + x_4H^o_{fH2O(vap)} + x_5H^o_{fCH4} + \Delta T(x_1C_{pH2} + x_2C_{pCO} + x_3C_{pCO2} + x_4C_{pH2O} + x_5C_{pCH4} + 3.76mC_{pN2}) \quad (15)$$

Where, $dH_{(for\ any\ gas)}$, is the heat formation + enthalpy change

$$dH_{(for\ any\ gas)}=H^o_f + \Delta H, \quad \Delta H=\Delta T(C_{p(g)}), \quad (16)$$

$$dH_{H2O(l)} = H^o_{fH2O(l)} + H_{(vap)},$$

$$dH_{biomass} = H^o_{biomass}$$

Equations (8) to (12) and (15) represent six equations with six unknowns. Two of the Equations, (11) and (12), are nonlinear equations while the rest are linear equations. The aforementioned system of equations can be reduced to three sets of equations, one linear and two nonlinear equations.

Based on Equation (8), the following holds

$$x_5 = 1- x_2 - x_3 \quad (17)$$

From Equation (9),

$$x_4 = w + 0.755 - x_1 - 2x_5 \quad (18)$$

Substitution of Equation (17) into (18) gives

$$x_4 = w + 0.755 - x_1 - 2(1 - x_2 + x_3) \Rightarrow x_4 = -x_1 + 2x_2 + 2x_3 + w - 1.245 \quad (19)$$

From Equation (10), (20)

$$M = 1/2(x_2+2x_3 + x_4 - w - 0.83)$$

Substitution of Equation (18) into (20) gives:

$$M = 1/2(x_2 + 2x_3 - x_1 + 2x_2 + 2x_3 + w - 1{,}245 - w - 0.83)$$

$$\Rightarrow m = 1/2(-x_1 + 3x_2 + 4x_3 - 2.075) \quad (21)$$

From Equation (11),

$$x_5 = x^2{}_1K_1 \quad (22)$$

Substitution of Equation (17) into (22) gives:

$$1-x_2- x_3 = x_1^2K_1 \Rightarrow x_1^2K_1 + x_2 + x_3 - 1 = 0 \quad (23)$$

From Equation (12),

$$x_1x_3 = x_2x_4K_2 \quad (24)$$

Substitution of Equation (18) into (24) gives:

$$x_1x_3 - x_2(-x_1 + 2x_2 + 2x_3 + w - 1{,}245)K_2 = 0$$

$$\Rightarrow - K_2(x_1x_2)+(w-1{,}245)K_2(x_2)+2K_2(x_2)^2-(x_1x_3)+2K_2(x_2x_3) = 0 \quad (25)$$

Substitution of Equations (17), (19) and (21) into (15) gives:

$$x_1dH_{H2} + x_2dH_{CO} + x_3dH_{CO2} + (-x_1 + 2x_2 + 2x_3 + w-1.245)dH_{H2O}(g) + (1 - x - x_3)dH_{CH4} + 3.76*1/2*(-x_1 + 3x_2 + 4x_3 - 2{,}08)dH_{N2} - dH_{wood} - wdH_{H2O(l)} - 0$$

$$\Rightarrow (dH_{H2} - dH_{H2O(g)} - 1.88dH_{N2})x_1 + (dH_{CO}+2dH_{H2O(g)}dH_{CH4} + 5.64dH_{N2})x_2 + (dH_{CO2} + 2dH_{H2O(g)} - dH_{CH4} + 7.52dH_{N2})x_3 (dH_{H2O(g)} - dH_{H2O(l)})w + dH_{CH4} - 1.245dH_{H2O(g)} - 3.9104dH_{N2} - dH_{wood} = 0 \quad (26)$$

To simplify Equation (26), the known constants are simplified as follows:

$$A = dH_{H2} - dH_{H2O(g)} - 1{,}88dH_{N2},$$

$$B = dH_{CO} + 2dH_{H2O}(g) - dH_{CH4} + 5.64dH_{N2},$$

$$C = dH_{CO2} + 2dH_{H2O}(g) - dH_{CH4} + 7.52\, dH_{N2},$$

$$D = dH_{H2O(g)} - dH_{H2O(l)}$$

$$E = dH_{CH4} - 1.245dH_{H2O(g)} - 3.91dH_{N2} - dHwood$$

Therefore, Equation (26) is simplified as follows:

$$Ax_1 + Bx_2 + Cx_3 + Dw + E = 0 \quad (27)$$

The systems of remaining equations are 3, two nonlinear Equations (23) and (25), and one linear Equation (27).

$$x_1^2K_1 + x_2 + x_3 - 1 = 0 \quad (23)$$

$$- K_2(x_1x_2) + (w-1.245)K_2(x_2) + 2K_2(x_2)^2 - (x_1x_3) + 2K_2(x_2x_3) = 0 \quad (25)$$

$$Ax_1 + Bx_2 + Cx_3 + Dw + E = 0 \quad (27)$$

The set of equations is solved using trial and error method.

The equation for calculating the heat of formation for 1 mol of solid biomass of cotton stalks ($CH_{1.51}O_{0.83}$) , which contains solid carbon, hydrogen and oxygen, is:

$$C_{(sol)} + 0.76H_2(g) + 0.415O_{2(g)} \rightarrow CH_{1,51}O_{0.83} \quad (28)$$

The reaction cannot occur in reality. The formation of ($CH_{1,51}O_{0,83}$) is based on the following reactions:

$C + O_2 \rightarrow CO_2$ $\Delta H_c = -393509$

$0.7\ 55H_2 + 0.38O_2 \rightarrow 0.755H_2O$ $\Delta Hc = 0.755(-241818)$

$CO_2 + 0.755H_2O \rightarrow CH_{1.51}O_{0.83} + 1{,}945O_2$ $\Delta H_C = 449568$

$C + 0.755H_2O + 0{,}415O_2 \rightarrow CH_{1,51}O_{0,83}$ $\Delta H_f = -126513.59$ kJ/Mol

Therefore, the heat of formation is – 126513.59 kJ/Mol. At constant pressure, the dependence of specific heat on temperature is given by the empirical equation (Perry, 1984)

$$Cpmh = R\left(A + BTam + \frac{C}{3}(4T_{am} - T_1T_2) + \frac{D}{T_1T_2}\right) \quad (29)$$

Where, $T_{am} = (T_1 + T_2)/2$ is the arithmetic mean temperature; A, B, C and D are the constants for the properties of the gases concerned (Perry, 1984), and R is the universal gas constant (8.314 J/molK).

Substituting the values of A, B, C and D for the gases concerned, the specific heat of the gas can be obtained as follows:

$C_{pH2O} = 37.36$ kJ/kmolK

$C_{PH2} = 29.63$ kJ/kmolK

$C_{PCO} = 31.16$ kJ/kmolK

$C_{PCO2} = 48.33$ kJ/kmolK

$C_{PN2} = 30.75$ kJ/kmolK

$C_{PCH4} = 56.55$ kJ/kmolK

Therefore, following the gasification process, the temperature of the gas increases from $T_1 = 298K$ to the gasification temperature $T_2 = 1073$ K. The enthalpy change ΔH can be obtained using Equation (30)

$$\Delta H = C_{pmh}(T_2 - T_1) \quad (30)$$

Therefore

$\Delta H_{H2O} = 28954$ kJ/mol

$\Delta H_{H2} = 22963.25$ kJ/mol

$\Delta H_{CO} = 24304$ kJ/mol

$\Delta H_{CO2} = 37455.75$kJ/kmol

$\Delta H_{N2} = 23831.25$ kJ/mol

$\Delta H_{CH4} = 43826.25$ kJ/mol

The term $dH_{(of\ any\ gas)}$ in Equation (16) can be determined from the value of the enthalpy change ΔH, of the gases. The equilibrium constant K is a function of temperature only, and is written as follows (Perry, 1984)

$$-RT\ln K = \Delta G^{\circ} \quad (31)$$

where: ΔG° is the standard Gibbs function of formation (Perry, 1984) and ΔH° is the heat of formation. The dependence of ΔG° to the temperature T, can be expressed as follows:

$$\frac{d(\Delta G^0 / RT)}{dT} = \frac{-\Delta H^0}{RT^2} \quad (32)$$

The manipulation of Equation (31) leads to:

$$\frac{\Delta G^0}{RT} = -\ln K$$

$$\frac{\Delta G^0}{RT} = -\ln K$$

Therefore,

$$\frac{d\ln K}{dT} = \frac{\Delta H^0}{RT^2} \quad (33)$$

Equation (33) gives the effect of temperature on the equilibrium constant. If ΔH° is negative, that is, if the reaction is exothermic, the equilibrium constant will be reduced if the temperature increases. The equilibrium constant K, increases with temperature, for an endothermic reaction.

Solution of Equation 33 leads to:

$$\ln K = \int \frac{\Delta H^0}{RT^2} dT + I \quad (34)$$

where, I is the constant of integration.

ΔH^o is given from the following equation (Perry,1984)

$$\frac{\Delta H^0}{R} = \frac{J}{R} + (\Delta A)T + \frac{\Delta B}{2}T^2 + \frac{\Delta C}{3}T^3 - \frac{\Delta D}{T} \quad (35)$$

where, J is a constant and ΔA, ΔB, ΔC, ΔD are the coefficients for determining specific heat.

Substitution of Equation (35) into (34) and integrating gives:

$$\ln K = \frac{-J}{R} + (\Delta A)\ln T + \frac{\Delta B}{2}T + \frac{\Delta C}{6}T^2 + \frac{\Delta D}{2T^2} + I \quad (36)$$

From Equation (31), $\Delta G^o = -RT\ln K$. The multiplication of Equation (36) with the term (-RT) gives:

$$\Delta G^0 = J - RT\left(\Delta A \ln T + \frac{\Delta B}{2}T + \frac{\Delta C}{6}T^2 + \frac{\Delta D}{2T^2} + I\right) \quad (37)$$

Equations (35) to (37) will be used to find the equilibrium constant for any reaction temperature T since the specific heat is known and the constants J and K can be determined. Using Equation (35), we can determine the constant J at the temperature of 298K, where the value ΔH^o is known. The constant I is determined from Equation (36) or (37) at the temperature at which the values of lnK or ΔG^o are known, normally at 298K.

Two equilibrium equations are required to determine the equilibrium constants K_1 and K_2. K_1 is the equilibrium constant for the reaction of Equation (3) and is solved as follows:

ΔA, ΔB, ΔC and ΔD can be obtained from the data of heat capacity. For the reaction from Equation (3),

$$C + 2H_2 \leftrightarrow CH_4$$

$$\Delta = CH_4 - C - 2H_2$$

The equations to determine the values of ΔA, ΔB, ΔC and ΔD can be written as:

$\Delta A = A_{CH4} - A_c - 2A_{H2}$,

$\Delta B = B_{CH4} - B_c - 2B_{H2}$,

$\Delta C = C_{CH4} - C_c - 2C_{H2}$

$\Delta D = D_{CH4} - D_c - 2D_{H2}$

Therefore, from the data of heat capacity (Perry, 1984)

ΔA = 1,702 - 1,771- (2)(3.249) =1,702 – 1.771 – 6.498 = -6.567

$\Delta B = (9{,}081 - 0.771 - (2)0.422)*10^{-3} = 7{,}466*10^{-3}$

$\Delta C = (-2{,}164 - (0 - 2)(0)(10^{-6}) = 2{,}164(10^{-6})$

$\Delta D = (0.867 - 0.1666)(10^5) = 0.701(10^5)$

Calculation of the constant J and I for Equations (35) and (37) at 298K, requires the values for ΔH^o_{298} and ΔG^o_{298}. This data is available from the heat of formation data (Perry, 1984) and the Gibbs energy of formation (Perry, 1984).

$\Delta H^o_{298} = (\Delta H^o_{298})_{CH4} - (\Delta H^o_{298})_c - 2(\Delta H^o_{298})_{H2}$

$\Delta H^o_{298} = -74520 - 0 - 2(0) = -74520$ J/mol

$\Delta G^o_{298} = (\Delta G^o_{298}) - (\Delta G^o_{298})_c - 2(\Delta G^o_{298})_{H2}$

$\Delta G^o_{298} = -50460 - 0 - 0 = -50460$

Substituting the known values into Equation (36) for T = 298K, gives;

$$\frac{-74520}{8.314} = \frac{J}{R} + (-6.657)(298.15) + \frac{7.466\times10^{-3}}{2(298.15)^2} + \frac{(-2.164\times10^{-6})}{3}\times(298.15)^3 - \frac{0.701(10^5)}{298.15}$$

ð J/R = -7083.18

ð J= -58889.6

Substituting the known values into Equation (37) for T = 298 K gives

-50460 = -58889.6 -

$$8.314\times 29815\left[(-6.567)\ln 29815 + \frac{7.466\times10^{-3}}{2}(29815) + \frac{(-2.164\times10^{-6})}{6}(29815)^2 + \frac{0.701\mathrm{X}10^5}{2(29815)^2} + I\right]$$

I = 32.544

Therefore, the general equation for $\ln K_1$ (using Equation (36) is:

$$\ln K_1 = \frac{7082848}{T} + (-6{,}567)\ln T + \frac{7.466x10^{-3}}{2}T + \frac{-2.164x10^{-6}}{6}T^2 + \frac{0.701x10^{-5}}{2(T)^2} + 32541 \quad (38)$$

The equilibrium constant K_1 for any temperature T can be obtained by substituting the temperature T into Equation (38). A similar procedure is used to determine the equilibrium constant K_2 for the reaction of Equation (6), that is;

$$CO + H_2O \leftrightarrow CO_2 + H_2$$

After going through the calculation steps, the general

Table 1. The composition of the producer gas (%v/v) against the moisture content for cotton stalks biomass gasification.

Moisture content (wet basis) (%)	H_2(%v/v)	CO (%v/v)	CO_2 (%v/v)	CH_4 (%v/v)	N_2 (%v/v)
0	17.17	35.093	2.016	0.224	35.093
10	18.3	29.93	4.93	0.35	33.8
20	20.34	24.34	8.34	0.67	31.34
30	22.77	16.62	13.23	0.92	30.15

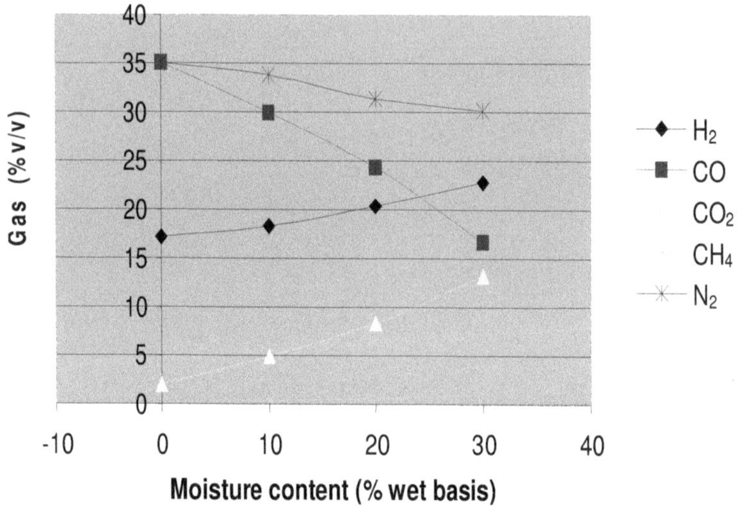

Figure 1. Effect of moisture content in the cotton stalks on gas composition at 800°C.

equation $\ln K_2$ is obtained as follows:

$$\ln K_2 = \frac{5872.45}{T} + 1.86\ln T - 2.69\times10^{-4}T - \frac{58200}{T^2} - 18 \qquad (39)$$

Similarly, the equilibrium constant K_2 for any temperature T can be obtained by substituting the value of temperature T into Equation (39).

The lower heating value of the produced gas was calculated by the following equation (Lv et al., 2004).

LHV = $(30[CO] + 25.7[H_2] + 85.4\ [CH_4] + 151.3[C_mH_n])4.2/1000$ MJ/Nm^3 (40)

RESULTS AND DISCUSSION

The equilibrium model (Zainal et al., 2001) previously described is used for the gasification of cotton stalks biomass. The results which have been obtained are presented in Table 1.

The effect of moisture content in the cotton stalks on the composition of the producer gas when the bed is working at 800°C is shown in Figure 1. It can be observed from the figure that the composition of hydrogen in the fuel gas increases continuously with the moisture content. A similar trend is also observed for the carbon dioxide. On the other hand, the percentage of carbon monoxide reduces for the same variation of moisture content. The composition of the inert nitrogen is almost constant with moisture content. The same is also observed for the composition of the methane.

The comparison between the predicted values and the experimental data can be seen in Figure 2.

The variation of calorific values for cotton stalks against moisture content is shown in Figure 3. It is observed that the calorific values decrease with increase in the moisture content, as expected.

Conclusion

In this work, it was attempted to apply an equilibrium model of gasification to the gasification process of cotton stalks. The equilibrium model was used for the gasification of cotton stalks at 800°C with the moisture from 0 to 30% in order to predict the composition of the fuel gas produced and its associated calorific values. The results obtained from the model are in close agreement with the actual measured values. The conclusions reached are the following:

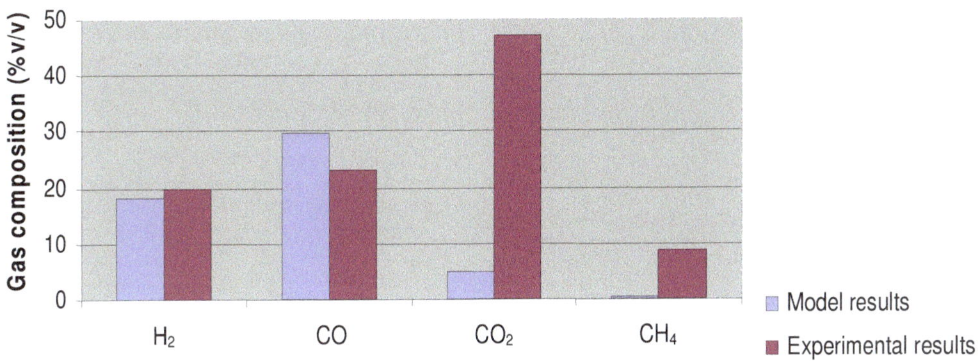

Figure 2. Comparison between model and experimental results for cotton stalks biomass and gasification temperature of 800°C.

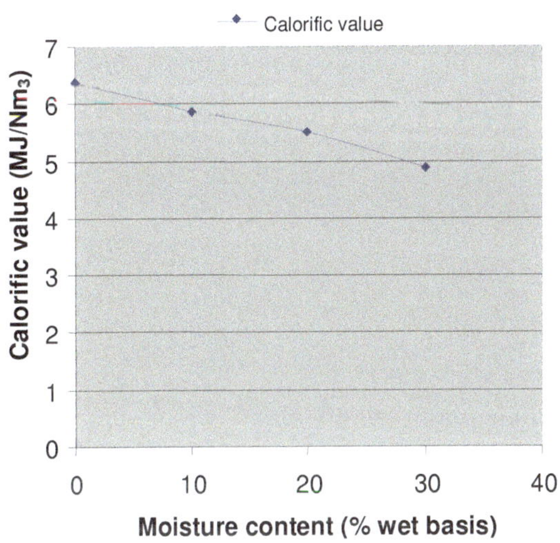

Figure 3. Calorific value against moisture content at 800°C for cotton stalks biomass.

1. The content of hydrogen in the produced gas increase with moisture content
2. The carbon monoxide content in the produced gas decreases with moisture content.
3. The methane content in the produced gas increases linearly with the moisture content
4. The calorific value of the produced gas decreases with the increase in moisture content due to the decrease of CO content and the increase of CO_2 content
5. There exists a good agreement between the model results and the experimental results for the hydrogen and carbon monoxide and not satisfying for carbon dioxide and methane.

The last point of the conclusions can be explained by taking into consideration the following: (a) The model assumes that gasification reaction rates are fast enough and residence time is sufficiently long to reach the equilibrium state. The kinetics of gasification reactions are not simple. In the gasification process and the thermal pyrolysis process, homogeneous gas phase and heterogeneous gas-solid reactions take place, so that thousands of chemical reactions may occur. The steam and carbon dioxide reforming reactions of char are kinetically limited at temperatures lower than 1000°C. Furthermore, the product gas of fluidized bed gasifiers generally contains tar, which is not considered in equilibrium models, and furthermore, it contains many hydrocarbons (especially methane), many more than the ones predicted. (b) The gasifier is regarded as a perfectly insulated apparatus, heat losses are neglected. In practice, gasifiers have heat losses to the environment. (c) Perfect mixing and a uniform temperature are assumed for the gasifier. In practice are observed different hydrodynamics, depending on the design of the gasifier.

Nomenclature: w, amount of water per kmol of biomass; **M,** amount of oxygen per kmol of biomass; x_1, coefficient of constituent of the hydrogen; x_2, coefficient of constituent of the carbon monoxide; x_3, coefficient of constituent of the carbon dioxide; x_4, coefficient of constituent of the water; x_5, coefficient of constituent of the methane; K_1, equilibrium constant from methane formation (1/s); K_2, equilibrium constant from shift reaction; H^o_{fwood}, the heat of formation of wood (kJ/kmol); $H^o_{fH2O(l)}$, the heat of formation of liquid water (kJ/kmol); $H_{(vap)}$, the heat of vaporization of water; $H^o_{fH2O(vap)}$, the heat of formation of water vapor; H^o_{fCO}, the heat of formation of the carbon monoxide; H^o_{fCO2}, the heat of formation of the carbon dioxide; H^o_{fCH4}, the heat of formation of the methane; C_{pH2}, the specific heat of the hydrogen (kJ/kmol); C_{pCO}, the specific heat of the carbon monoxide (kJ/kmol); C_{pCO2}, the specific heat of the carbon dioxide (kJ/kmol); C_{pH2O}, the specific heat of the water (kJ/mol); C_{pCH4}, the specific heat of the methane (kJ/mol); C_{pN2}, the specific heat of the nitrogen; T_1, the ambient temperature at the reduction zone (K); C_{pmh}, the average specific heat over the temperature change; $\mathbf{T_{am}}$, the arithmetic mean temperature; A,B,C,D, the constants for the properties of the gases concerned; R, universal gas constant (8.314 J/molK); $\mathbf{\Delta G^o}$, the standard Gibbs function of formation; I, J, constants of integration; **ΔA,ΔB,ΔC,ΔD,** the coefficients for determining specific heat.

REFERENCES

Babu BV, Chaurasia AS (2003). Modeling, simulation and estimation of optimum parameters in pyrolysis of biomass. Energy Convers. Manag., 44(13): 2135-58.

Babu BV, Sheth PN (2006). Modelling and simulation of reduction zone of downdraft biomass gasifier: effect of char reactivity factor. Energy Convers. Manag., 47(15-16): 2602-11.

Gao N, Li A (2008). Modeling and simulation of combined paralysis and reduction zone for a downdraft biomass gasifier. Energy. Convers. Manag. 49: 3483-3490.

Giltrap DL, McKibbin R, Barners GRG (2003). A steady state model of gas-char reactions in a downdraft gasifier. Solar Energy, (74): 85-91.

Kantarelis E, Zabaniotou A (2009). Valorization of cotton stalks by fast pyrolysis and fixed bed air gasification for syngas production as precursor of second generation biofuels and sustainable agriculture. Bioresour. Technol. 100: 942-947

Koufopanos AC, Maschio G, Lucchesi (1989). A. Kinetic modelling of the pyrolysis of biomass and biomass components. Can. J. Chem. Eng. 67(1): 75-84

Lv PM, Xiong ZH, Chang J, Wu CZ, Chen Y, Zhu JX (2004). An experimental study on biomass air-steam gasification in a fluidized bed. Bioresour. Technol., 95: 95-101.

Melgar A, Perez JF, Laget H, Horillo A (2007). Thermochemical equilibrium modelling of a gasifying process. Energy Convers. Manage., 48(1): 59-67

Perry RH, Green DW (1998). Perry's Chemical Engineers Handbook, 7th MGH International Edition, NY. Section 9

Srivastava VK, Sushil, Jalan RK (1996): Prediction of concentration in the pyrolysis of biomass material-II. Energy. Convers. Manag., 37(4): 85-91.

Wang Y, Kinoshita CM (1993). Kinetic model of biomass gasification. Solar Energy, 51(1): 19-25

Zainal ZA, Ali R, Lean CH, Seetharamu KN (2001). Prediction of performance of a downdraft gasifier using equilibrium modelling for different biomass materials, Energy Conver. Manag., 42: 1499-1515

Fracture behavior of cryogenically solidified aluminum-alloy reinforced with Nano-ZrO_2 metal matrix composites (CNMMCs)

Joel Hemanth

Rajiv Gandhi Institute of Technology, R. T. Nagar Post, Hebbal, Bangalore-560032, Karnataka, India.
E-mail: joelhemanth@hotmail.com.

Fracture behavior of LM-13 aluminum alloy reinforced with different wt.% of Nano-ZrO_2 particles, solidified under various shallow cryogenically cooled chills (at -80°C), cast using DMD (disintegrated melt deposition) technique followed by hot extrusion were investigated. The size of the particles dispersed varies from 50 to 80 nm and amount of addition varies from 3 to 15 wt.% in steps of 3 wt.%. Microstructural studies of the Nano-composite developed indicate that there is uniform distribution of the reinforcement in the matrix and mechanical properties reveal that presence of Nano-ZrO_2 particles as dispersoid (up to 12 wt.%) and chill thickness (25 mm) has improved significantly the strength, hardness and fracture toughness with slight reduction in ductility. Fractography of the specimens showed that the fracture behavior of matrix alloy has changed from ductile intergranular mode to cleavage mode of fracture. Finally heat capacity of the shallow cryo-chill is identified as an important parameter which affects mechanical properties.

Key words: Composites, mechanical, fracture, Nano, chill, shallow cryogenic, microstructure.

INTRODUCTION

It is indeed to measure the strength of material to estimate the toughness property, because strength is usually measured by the deformation area which is influential on the property of the material (Bowden and Tabor, 1964). Reports of measurement of mechanical properties at cryogenic temperature are few and the properties at these conditions are unknown fully (Petty, 1960; Oku and Usui, 1971). But recently several researches have been carried out on ferrous and non ferrous materials to fill the gap (Bensely et al., 2006; Huang and Liao, 2003). Hence it is important that the mechanical properties of the composite developed are evaluated at cryogenic temperatures. In particularly, the fracture properties of this composite need to be investigated in detail (Ogata et al., 2002; Ono et al., 2002; Yuri et al., 2002), since fatigue fracture has been found in automobiles and it is considered to cause the failure of many automobile and structural components. And it has been also reported that sub-surface crack initiation is associated with the microstructure rather than specific defects such as inclusions (Umezawa et al., 1990). Since sub-surface crack initiation tends to occur at low stress amplitudes and high numbers of cycles, whereas fatigue crack initiation at the specimen surface occurs at high stress amplitudes to reduce the fatigue properties in the longer life region (Murakami, 2002; Shiozawa et al., 2002; Nishijma and Kanazawa, 1999). A high level combination of strength and toughness is required for, components for space application components which need to sustain large fatigue forces (Suemune et al., 1998). On the other hand Nano-structured materials inherit relatively large interfacial energy (Gleiter, 2000). Recently mechano-synthesis of Cu-AL_2O_3 Nano-composite has been reported by Seung and Wang (2005).

As aerospace technology continues to advance, there is a rapidly increasing demand for advanced materials with high mechanical and thermal capabilities for such ultra high applications (Opeka et al., 2004). Its application also stretched to automobile, electronic and computer industries to replace the existing materials including plastics (Luo, 1995). The early 1990s are considered to be the renaissance for Al as structural material due to environmental concerns, increasing safety and comfort levels. A significant improvement in the properties of Al alloys, reduced fuel consumption because of light weight

Table 1. Chemical composition of matrix material (Al-alloy LM 13).

Elements	Zn	Mg	Si	Ni	Fe	Mn	Al
% by wt	0.5	1.4	12	1.5	1.0	0.5	Bal

and hence it has made huge demand from automobile industry (Saravanan and Surappa, 2000; Hassan and Gupta, 2002). This growing requirements of materials with high specific mechanical properties with weight savings has fueled significant research activities in recent times targeted primarily for further development of Al based composites (Singh and Tsai, 2003; Guy, 1967; Gupta et al., 2000). A recent industrial review revealed that there are hundreds of components from structural to engine in which aluminum alloy is being developed for variety of applications (Yamamoto and Sasamoto, 2000). It is also predicted that for Al alloys the demand increased globally at an average rate of 20% every year (Awasthi and Wood, 1988). It is noticed that the limited mechanical properties (strength and hardness) of Al and its alloys adversely affect its applications in automobile and aerospace industries (Lai, 2000; Wang and Yang, 2002). This remains one of the major concerns in its fabrication to suit its application in recent days.

Search of open literature indicates that, so for number of Al based MMCs including chilled MMCs (Hemanth, 2003, 2005, 2007, 2008) are being developed but no work has been done in this field. Hence the present research is undertaken to fill the void and to investigate the integrated properties of Al-alloy/ ZrO_2 CNMMCs. Among all the reinforcements used in Al based composites only Nano size particulates has shown their potential superiority in improving mechanical properties, microstructure with noticeable weight savings (Lloyd, 1994).

Disintegrated melt deposition (DMD) technique (Tham and Cheng, 1999; Zhang and Guo, 2002) has been chosen in this research which remains an attractive choice among all the solidification processing routes due to its capability of bringing together the advantages of both spray processing and conventional casting. Accordingly, the primary aim of the present investigation is to synthesize Al alloy based composite reinforced with Nano-ZrO_2 particulates using DMD technique. The composites thus obtained were hot extruded and characterized for their micro structural, mechanical, thermal and electrical properties. Particular emphasis is placed to study the effect of presence of Nano particulates and its increasing amount on the microstructure and mechanical response of standard LM 13 Al alloy.

EXPERIMENTAL PROCEDURE

In this research Nano-ZrO_2 particles dispersed in Al alloy (chemical composition of LM 13 the matrix material are shown in Table 1), fabricated by DMD technique, solidified under cryogenically cooled chills followed by hot extrusion were investigated. Copper end chills of thickness 5, 10, 15, 20 and 25 mm in which arrangements were made to circulate liquid nitrogen (at -80°C) to study the effect of heat capacity on mechanical and microstructural behavior (arrangement shown in Figure 1). The size of the Nano-ZrO_2 particles dispersed varies from 50 to 80 nm and the amount addition varies from 3 to 15 wt.% in steps of 3 wt.%. The primary processing consists of synthesis of monolithic and Nano-sized ZrO_2 reinforced Al composites (Al/Nano-ZrO_2) containing five different weight % of ZrO_2 was carried out using DMD technique. Synthesis of the composite involved heating of Al alloy in a graphite crucible up to 720°C using resistance furnace to which the preheated reinforcement (to 250°C) was added and stirred well by an impeller which rotates at 450 rpm to create vortex to get uniform distribution of the reinforcement. This treated Al alloy containing Nano-ZrO_2 particles were made to solidify under the influence of cryo-cooled chills of different thickness set in AFS standard dry sand mold. In secondary processing, the deposited monolithic and Nano sized ZrO_2 in Al matrix are extruded in a hydraulic press at 260°C.

Properties of reinforcement material

Properties of the reinforcement (Nano-ZrO_2) are as follows: Density: 2.6 gm/cc, melting point: 1860°C, UTS: 425 MPa, VHN: 150, Young's Modulus: 98 GPa., Size: 50 to 80 nm, Density: 2.6 gm/cc (Nano-ZrO_2 was supplied by Nano structured and Amorphous Materials, Inc, USA)

Testing procedure

Microstructural characterization were conducted on polished CNMMC specimens (etched with Keller's reagent) using OLYMPUS metallographic microscope to investigate morphological characteristics of grains, reinforcement distribution and interfacial integrity between matrix and the reinforcement. Hitachi S4100 field emission scanning electron microscope was used to analyze the fractured surfaces. Vickers microhardness measurements were made on polished specimens using Matsuzawa MXT50 digital hardness tester using 25 gf indention load in accordance with ASTM E18-94 standard. Tension tests were performed on American Foundry men Society (AFS) standard tensometer specimens in accordance with ASTM E8M-01 standard using Instron 8516 machine. Fracture toughness tests were performed using a closed loop INSTRON servo-hydraulic material testing system. This method involves 3-point bend testing (in accordance with ASTM E 399 1990 standard) of machined specimen which have been pre-cracked by fatigue. All the above tests were conducted at an ambient temperature of 28°C.

RESULTS AND DISCUSSION

Results of the investigation on mechanical properties reveal that adding reinforcement content up to 12 wt.% enhances mechanical properties and addition above this limit detoriates mechanical properties. Also, 25 mm thick chill has pronounced effective on mechanical properties

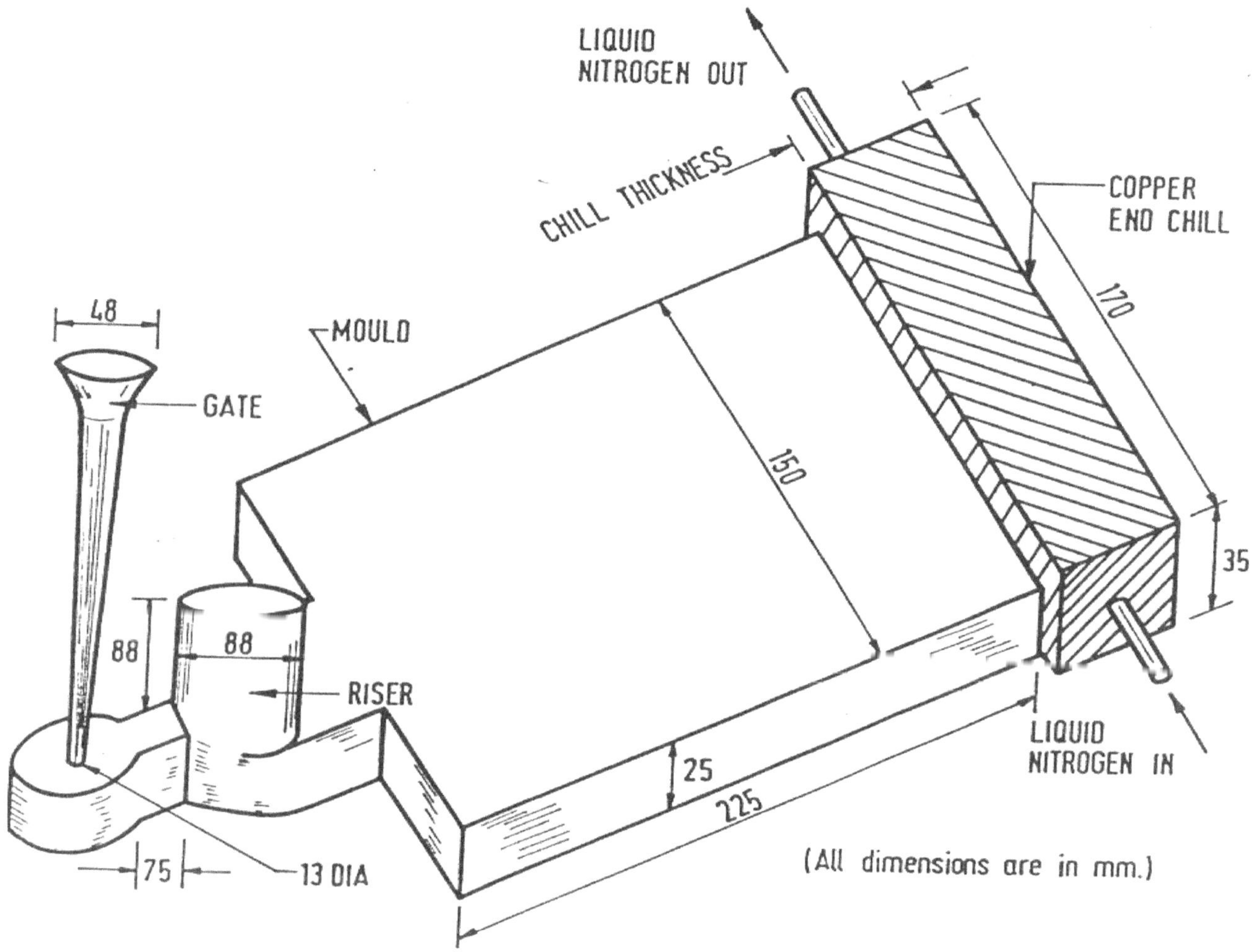

Figure 1. Experimental set up (AFS standard mold containing cryogenic end chill block).

because of its high volumetric heat capacity. Hence present discussion Is focused mainly on CNMMCs containing 12 wt. % dispersoid cast using 25 mm thick cryo-chill block.

For all the tests specimens were taken from the chill end, because it is observed that, farther away from the chill the specimen is taken, the lower are the mechanical properties. This could be because the rate of chilling is lower, when the specimen is farther away from the chill.

Synthesis of CNMMCs

Synthesis of Al –alloy based CNMMCs was successfully accomplished by DMD process followed by cryo-chilling and hot extrusion. The condition prevailed during melting, processing, dispersion, deposition and solidification under cryo conditions (as mentioned in experimental procedure) was instrumental in the prevention of micro porosity, segregation or agglomeration of reinforcement due to effect of its gravity has indicated its suitability of stirring and the realization of good solidification conditions during deposition to produce sound chilled castings. The results, in essence, indicate the feasibility of the DMD process along with cryo chilling followed by hot extrusion act as a potential fabrication technique for Nano-ZrO_2 composites.

Cryogenic effect on solidification

The optical photomicrographs in Figures 2, 3 and 4 show that the matrix microstructure of cryogenically chilled Nano composites is finer than that of the unchilled matrix alloy. Figures 2 and 3 show microstructure of un chilled as cast composite and extruded composite containing 12 wt.% reinforcement and Figure 4 shows microstructure of CNMMC containing 12 wt.% dispersoid cast using 25 mm thick chill. When the melt of the CNMMC solidifies under cryogenic conditions, the sub-zero temperature of the chill and the hot melt come in contact experiences severe super-cooling. This results from a high rate of heat transfer and rapid cooling of the melt in chilled CNMMC samples. Hence, the critical nucleus size of the

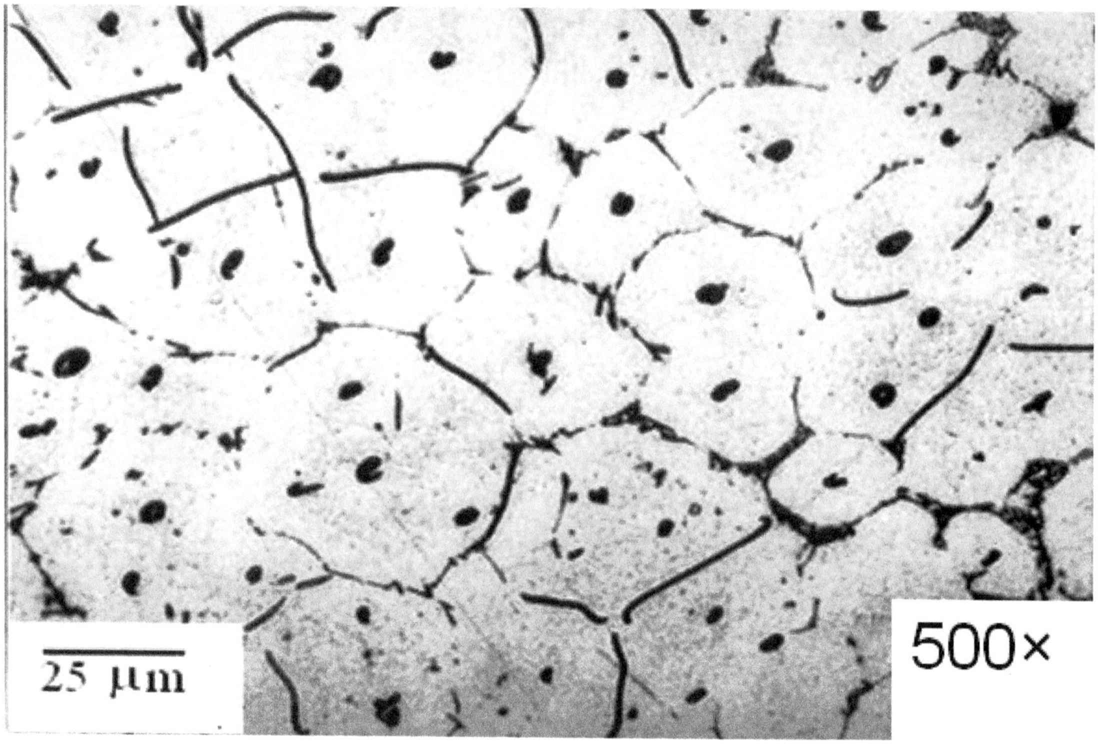

Figure 2. Optical microstructure of as cast un chilled Nano-composite containing 12 wt.% reinforcement (500 × magnification).

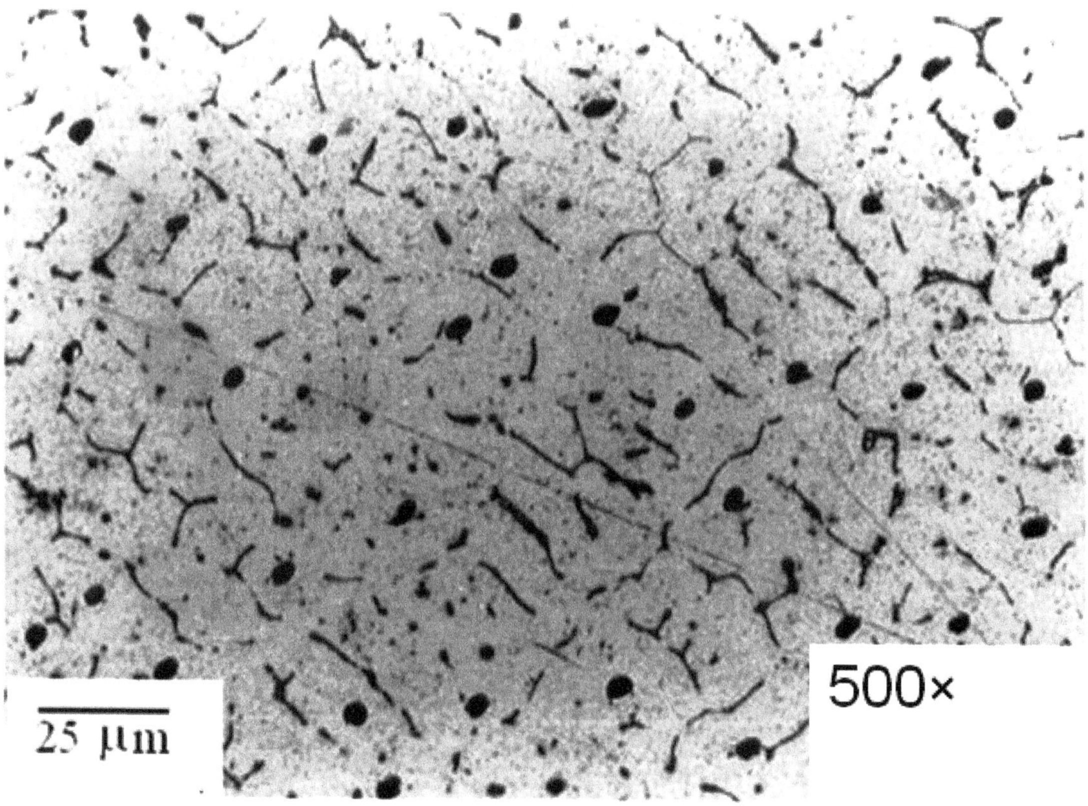

Figure 3. Optical microstructure of extruded un chilled Nano-composite containing 12 wt.% reinforcement (500 × magnification).

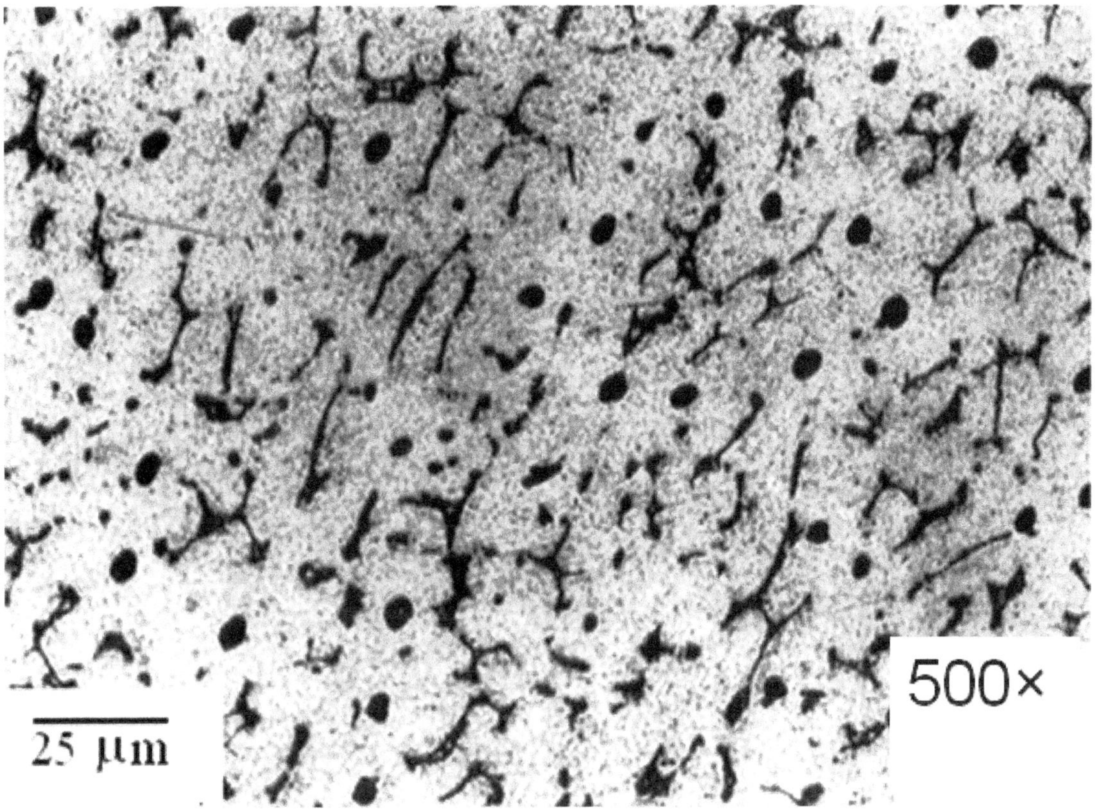

Figure 4. Optical microstructure of extruded CNMMC containing 12 wt.% dispersoid cast using 25 mm thick cryo-chill (500 × magnification).

solidified melt is reduced and a greater number of nuclei are generated, causing a finer microstructure. Additionally, because of the rapid cooling of the melt and stirring the dispersoid particles do not have time to settle down due to density differences between the matrix melt and the dispersoid, and these results in a more uniform distribution of Nano-ZrO_2 particles in the matrix. The cryogenic effect during solidification causes stronger bonding between the matrix and the dispersoid. This may be attributed to the fact that the wettability between the particles and the matrix was improved with cryogenic effects. This uniform distribution of particles and the finer matrix structure and chilling (Figure 4) lead to improved mechanical properties of the CNMMCs as compared with the un-chilled composite (Figures 2 and 3). Thus the strong bonding between the dispersoid and the matrix causes more effective load transfer. This in turn reduces the possibility of pullout of particulates from the matrix. Hence, any failure of the CNMMC would more likely be due to particle failure rather than particle pullout from the matrix.

Micro structural characterization

Micro structural characteristics of extruded cryo-chilled Nano-composites are discussed in terms of distribution of reinforcement and reinforcement matrix interfacial bonding. Microstructural studies conducted on the extruded Nano-composites containing 12 wt.% dispersoid cast using 25 mm thick chill revealed uniform distribution of the reinforcement with limited extent of clusters, good reinforcement-matrix interfacial integrity, significant grain refinement with minimal porosity (Figure 4). This is due to the gravity of ZrO_2 associated with judicious selection of stirring parameters (vortex route), good wetting of pre heated reinforcement by the matrix melt. Metallography studies of the extruded samples revealed that the matrix is recrystallized completely. Grain reinforcement in case of Nano-composites can primarily be attributed to capability of Nano-ZrO_2 particulates to nucleate Al grains during directional solidification and restricted growth of recrystallized Al grains because of presence of finer reinforcement and chilling. Interfacial integrity between matrix and the reinforcement was assessed using scanning electron microscope of the fractured surfaces to analyze the interfacial de-bonding at the particulate-matrix interface. Here also the result revealed that a strong bond exists between the interfaces as expected from metal/oxide systems (Gilchrist, 1989).

Micro and macro tests reveal that, high rate heat transfer During solidification (cryo-chilling) of the composite in this

Table 2. Mechanical properties (strength and hardness) of CNMMCs and matrix alloy cast using cryo chill of thickness 25 mm.

Property (wt% ZrO_2)	Hardness (HV)	0.2% Yield strength (MPa)	UTS, (MPa)	Ductility (%)
3	111	127	245	17.7
6	118	135	253	16.4
9	129	161	262	15.9
12	138	186	277	14.1
15	131	171	268	13.2
Matrix alloy (LM 13)	58	120	130	18

investigation leads to strong bonding of the dispersoid and the matrix. This may be one of the main reasons for increase of strength, hardness and fracture toughness of the composite developed. The result of micro structural studies of CNMMCs however did not reveal presence of any micro-pores or shrinkage cavity. After the secondary process (extrusion) also, there was no evidence of any microstructural defects

Mechanical properties (hardness and strength)

The results of micro hardness test conducted on extruded CNMMCs samples revealed an increasing trend in matrix hardness with increase reinforcement content up to 12% and chill thickness up to 25 mm (Table 2 for mechanical properties). Results of hardness measurements revealed that an increase in the reinforcement content and chill thickness leads to a significant increase in the hardness and can be attributed primarily to: presence of harder ZrO_2 ceramic particulates in the matrix, a higher constraint to the localized deformation during indentation due to their presence and reduced grain size due to chilling. In ceramic-reinforced composite, there is generally a big difference between the mechanical properties of the dispersoid and those of the matrix. This result in incoherence and a high density of dislocations near the interface between the dispersoid and the matrix. Precipitation reactions are accelerated because incoherence and the high density of dislocations act as heterogeneous nucleation sites for precipitation.

UTS is higher for all the CNMMCs as compared against the matrix alloy. It is further seen that for CNMMCs the values of these parameters increase as the dispersoid content increases up to 12 wt. %, beyond which the trend reverses. Conversely, percentage elongation (measure of ductility) decreases monotonically as dispersoid content increases. Table 2 shows the UTS near the chill end for different composites cast using cryogenic chills of different thicknesses. As the chill thickness increases, UTS also increases, confirming that the heat capacity of the chill significantly enhances the UTS. Results of the ambient temperature tensile tests also revealed significant improvement in 0.2% yield strength (YS) (Table 2) for CNMMCs developed with little adverse affect on ductility.

Increase in tensile strength (reinforcement up to 12%) is also attributed to increase in grain boundary area due to grain refinement, built of thermal stresses (due to cryo-chilling) at the interface due to different coefficient of thermal expansion (CTE) and effective transfer of applied tensile load to the uniformly distributed well bonded reinforcement (strength of ceramic materials lies much higher than metallic materials (Srinivasan and Gupta, 1999) as the UTS of the composite developed is 46% higher than that of the matrix alloy. Under the applied stress, increasing the amount of grain boundaries acts as obstacle to the dislocation movement and end up with dislocation pile up at the grain boundary region (Reed, 1964). Again, multi-directional thermal stress induced during processing easily starts multi-gliding system under applied stress so that dislocations were found developing and moving in several directions. These multi-glide planes agglomerate under the applied stress forms grain boundary ledges. As applied load increases, these ledges act as obstacle to dislocation movement resulting in pile-ups. The coupled effect of these two obstacles lead to increase in the strength of the composite (Hemanth, 2002).

Fracture studies conducted on fractured tensile specimens indicate that the change in fracture mode from ductile to a typical cleavage mode could not improve ductility of the Nano-composite due to presence of ceramic reinforcement. As compared with the matrix (Al alloy) ductility, the ductility of the CNMMCs are reduced (by 78.33% for 12 wt.% addition of reinforcement) due to presence of Nano-ZrO_2 particles, which is a contrary effect.

Fracture toughness

Fracture toughness tests performed on 3 point bend (containing chevron notch) specimen which was pre cracked by fatigue loading. The manner in which stress response varies with cycles and plastic-strain amplitude is an important feature of the low-cycle fatigue process. The cyclic stress required for pre-cracking the specimen provide useful information pertaining to the mechanical stability of the intrinsic micro structural features during reverse plastic-straining, coupled with an ability of the material to distribute the plastic strain over the entire volume, are the two key factors governing the cyclic response of a material.

Table 3. Fracture toughness of CNMMCs and matrix alloy.

Copper chill block thickness (mm)	Fracture toughness (K_{1C}) (MPa) √m (near the chill block end)				
	3 wt% dispersoid	6 wt% dispersoid	9 wt% dispersoid	12 wt% dispersoid	15 wt% dispersoid
25	11	13	14	16	12
20	9	11	12	13	10
15	9	10	11	11	08
10	6	7	7	08	06
05	04	05	05	06	06

Fracture toughness of the matrix alloy LM 13 is 3.8 MPa √m.

Effect of dispersoid content on fracture toughness

The fracture toughness of the composite chilled using different type of chills and dispersoid content are shown in Table 3. Comparing these results, it can be seen that increasing the reinforcement content and chilling seems to have an effect on the fracture toughness of the material. It is also observed that CNMMC containing 12 wt. %, cast using 25 mm thick chill block invariably has the highest fracture toughness. Further it is observed that, at reinforcement content beyond 12 wt. %, the values register a decrease in fracture toughness

Fracture toughness results of the cryo-chilled Nano-composite indicate that presence of hard ZrO_2 has a pronounced effect and it increases as dispersoid content increases up to 12 wt.%. However, it is not customary to assume that reinforced materials are weaker than un-reinforced materials since in most cases it is quite reverse. The results of this study have also shown that, as reinforcement content increases along with the chilling and extrusion process, the matrix becomes dense, stronger and accommodate the reinforcement rigidly and this could lead to a greater strength and fracture toughness of the composite compared with the monolithic alloys. The mechanisms which control the variation of fracture toughness of chilled Al-Nano-ZrO_2 composites are dependent upon both microstructure and strain range. The possible micro mechanisms controlling the fracture behavior during cyclic loading are ascribed to the following synergistic influences (Hemanth, 2002, 2009):

(a) Load transfer between the soft aluminum matrix and the hard Nano-ZrO_2 particle reinforcement.
(b) A pre-existing high dislocation density in the aluminum matrix caused by the presence of the hard, brittle ZrO_2 particles.
(c) Hardening effect arising from constrained plastic flow and tri-axibility in the aluminum matrix is due to the presence of Nano-ZrO_2 reinforcement and cryo-chilling. As a result, the particles resisting the plastic flow of the matrix develops an average internal stress or back stress
(d) Dislocations arising from competing influences of back stresses in the plastically deforming composite matrix is due to plastic relaxation by the formation of dislocation loops around ZrO_2 particles.
(e) Residual stresses generated and dislocations arising are due to mismatch in thermal expansion coefficients (due to chilling) between the components of the composite, that is, the soft aluminum matrix and the hard reinforcing ZrO_2.

Fracture surface analysis

Fracture surface analysis revealed different topographies for the composites containing different weight percentage of ZrO_2 particles. Results of the fracture surface analysis conducted on fracture toughness specimens of FCC structured LM 13 Al alloy samples revealed large dimples along with large amount of plastic deformation indicating ductile fracture. The fracture surfaces also exhibit fine and shallow dimples, indicating that the fracture is ductile.

Scanning electron micrographs of the fracture surface of the CNMMCs tested are shown in Figures 5 and 6. Examination of these fracture surface features in SEM at low magnification is to identify the fatigue and final fracture regions, and at higher magnification in the fatigue is region to identify areas of micro-crack initiation and early crack growth, and the over-loaded region to identify the fine scale fracture features. Fracture surfaces revealed different topographies for the composites containing different weight percentage of ZrO_2 particles cast using different type of cryo-chill blocks.

Fracture surface in the case of CNMMCs containing 6 wt.% (Figure 5) reinforcement, cast using 25 mm thick cryo chill block reveal mixed mode fracture and CNMMCs containing 12 wt.% reinforcement revealed cleavage type of fracture (Figure 6) due to the presence of excessive Nano-ZrO_2 particulates. It can also be seen that regions of clustered particles for reinforcement above 12 wt.%, (Figure 7) are sensitive to premature damage in the composite and the large particles seems to be prone to fracture and hence registered a reduction in the fracture toughness value (Table 3). Fracture mode of the matrix alloy which changed form ductile to cleavage type (in case of CNMMC) was dominated with microcrack nucleation and propagation as shown in Figure 6.Scanning electron microscope observations of CNMMCs suggest that, at higher reinforcement content (15 wt.%, Figure 7), void nucleation may also takes place at large precipitates

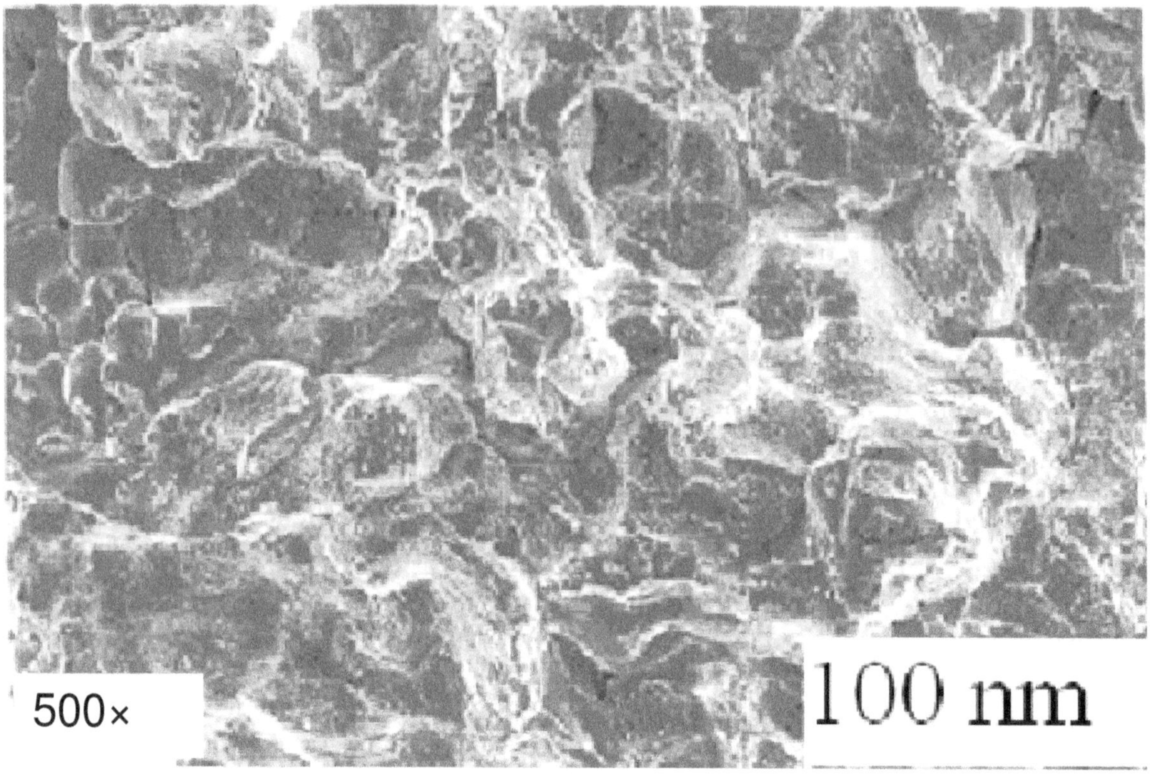

Figure 5. SEM fractograph of CNMMC (6 wt.% ZrO_2) cast using 25 mm thick chill block.

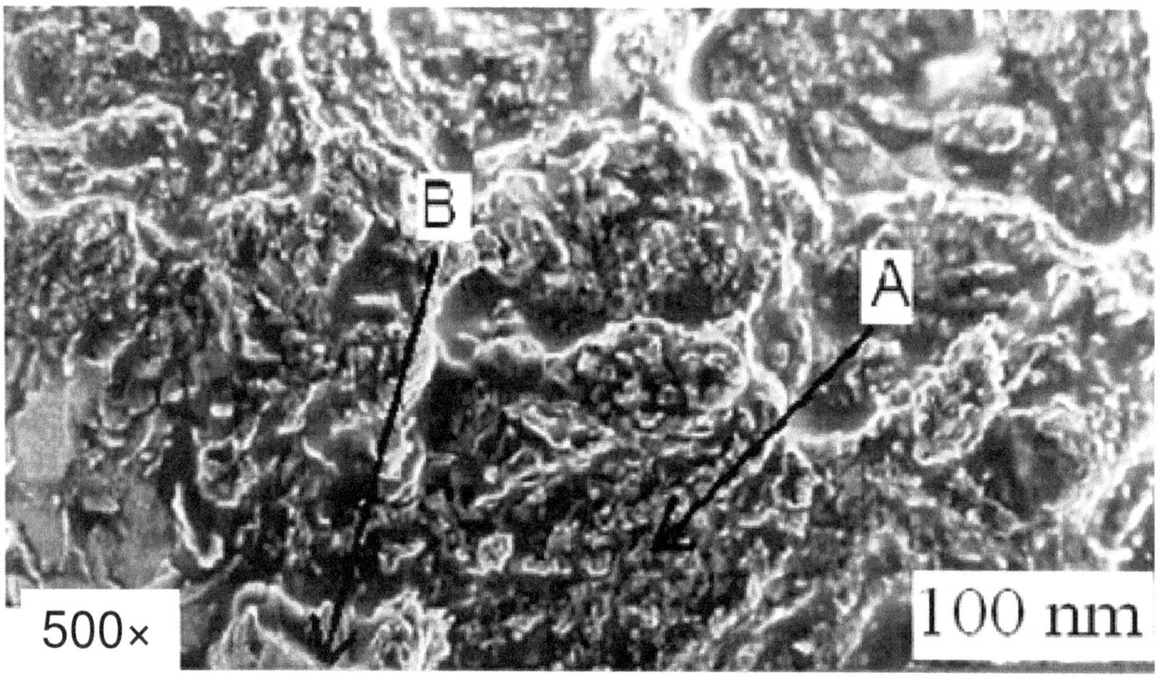

Figure 6. SEM fractograph of CNMMC (12 wt.% ZrO_2,) cast using 25 mm chill block.

in addition to the matrix/particle interface, while at lower reinforcement(3 wt.%, Figure 8) rates the fracture seems to occur by breakage of the particles.

Fracture surface analysis of CNMMC containing 9 wt.% dispersoid (Figure 9) exhibit predominantly that fractured particles (dispersoid) and the matrix material with coarse

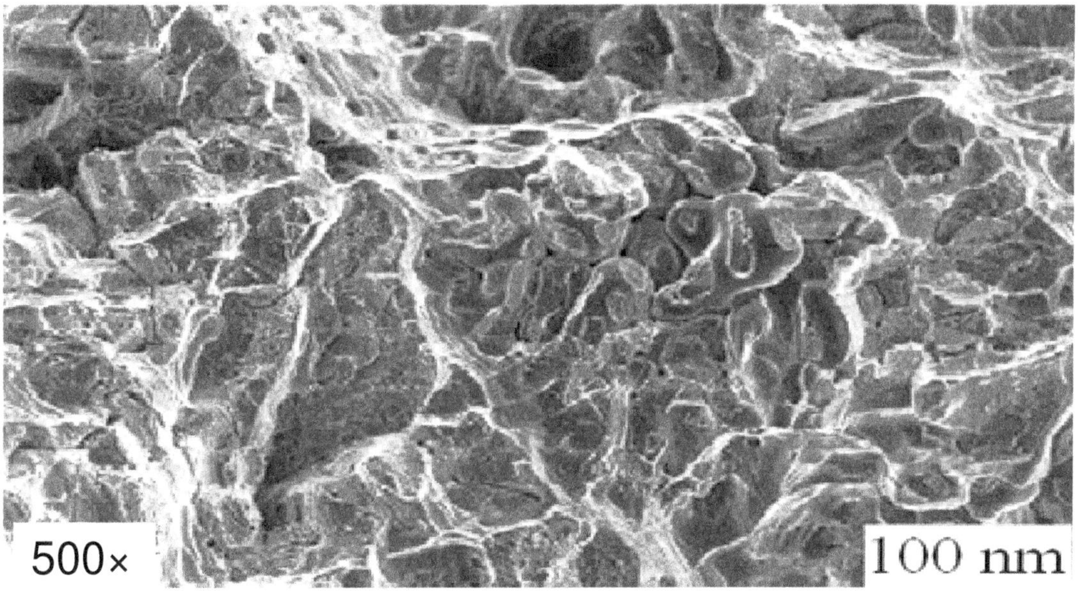

Figure 7. SEM fractograph of CNMMC (15 wt.% ZrO_2,) cast using 25 mm chill block.

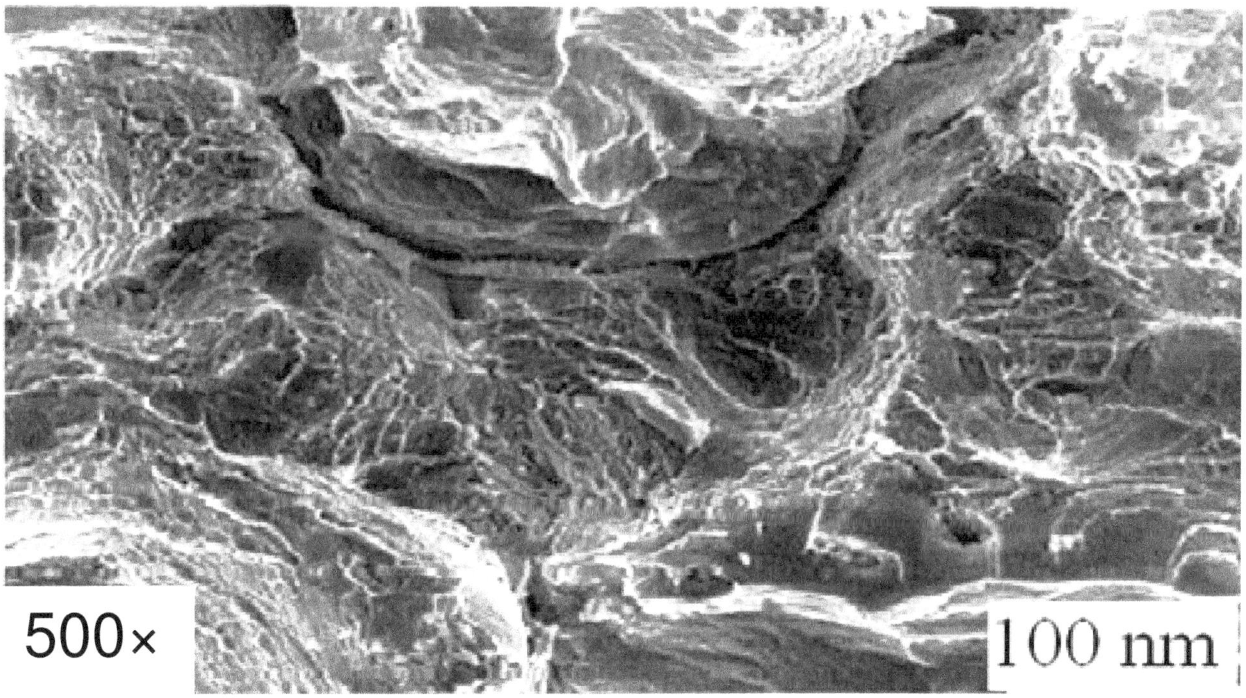

Figure 8. SEM fractograph of CNMMC (3 wt.% ZrO_2,) cast using 25 mm chill block.

dimples, suggests that fracture is of mixed mode type. Close examination of the fractured surface indicate that, most dimples were associated with the matrix material (Figure 10). Massive separations of the particle indicate that failure of the material is initiated by fracture of the particles rather debonding between the matrix and reinforcements. SEM fractographs of the cryo-chilled Nano-composite tested for fracture toughness show a dimpled appearance with a large amount of plastic deformation of the matrix. It was found that the extent of fracture toughness of the composite with increasing the chilling rate as structure becomes fine. Particle fracture was found in some locations (Arrow A in Figure 6) in the particles clusters as well in large particles (Arrow B in

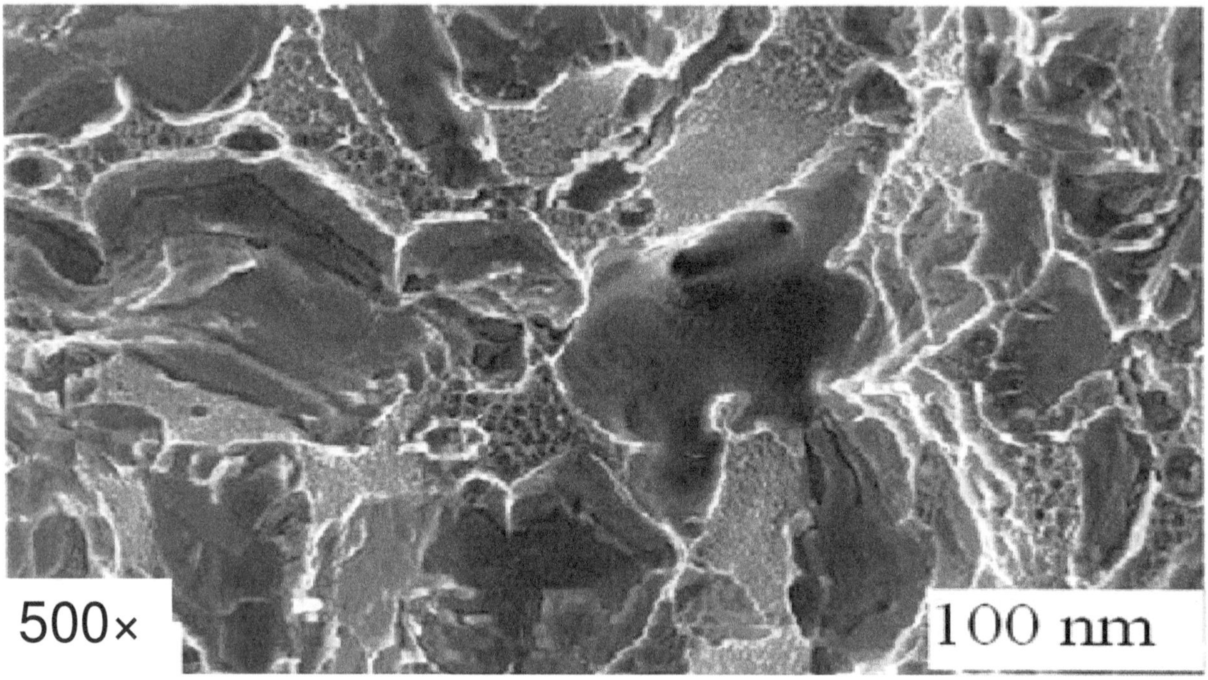

Figure 9. SEM Fractograph of CNMMC (9 wt.% ZrO_2,) cast using 25 mm chill block.

Figure 6) present in the matrix somewhat away from the fracture surface, at higher chilling rates. The microstructural features shown in Figure 6 which is a sample containing 12 wt.% dispersoid cast using cryo-chill of 25 mm thick indicates that ZrO_2 particles have fractured during straining and fracture could have occurred in a brittle manner as a result of stress concentrations and propagation of the cracks.

Therefore, cyclic fracture of the un-chilled as cast matrix alloy on a macroscopic and microscopic scale, exhibited ductile fracture with isolated micro-cracks (Figure 5, optical microstructure of Matrix alloy). Observations of CNMMC containing 6 wt.% disperoid revealed large areas of the fracture surface to be covered with a bimodal distribution of dimples, indicative of mixed mode ductile rupture (Figure 5). A dimple is a half void which form at the coarse second-phase particles, namely the iron and silicon inter-metallic contained in the aluminum matrix. However, growth of the void is limited by competing and synergistic influence of reinforcing particles, chilling effect and the cyclic ductility of the composite microstructure. Finally composite containing 12 wt.% dispersoid showed cleavage indicating that the fracture is towards brittle.

Effect of chilling rate on fracture toughness

The experimental results of the fracture toughness tests done on castings cast using different cryo chill blocks are shown in Table 3. It can be seen that changing the chill thickness seems to have an effect on the fracture toughness of the material. This implies that increasing the rate of chilling by increasing the heat capacity of the chill block tends to result in an increase in the fracture toughness of the material. Of more significant effect is the temperature of liquid nitrogen (-80°C) and the chill thickness. It can be seen that if all other factors are kept constant, castings cryo-chilled using 25 mm thick copper block invariably has the highest fracture toughness followed by 20, 15, 10 and 5 mm thick cryo-chilled composites in that order. It is also observed that, farther away from the chill the specimen is taken, the lower is the fracture toughness. This could be because, the farther away from the chill the specimen is, the lower is the rate of chilling. This agrees with the deductions made earlier that increasing the rate of chilling tends to result in an increase in the fracture toughness of the material (Hemanth, 2002).

Conclusions

DMD technique coupled with cryo-chilling followed by hot extrusion used for the synthesis Nano-ZrO_2 composites reveal the following: Microstructural analysis showed fine grain refinement, fairly uniform distribution of reinforcement (stirring with vortex route at high speed) with minimal porosity (cryo-chilling effect). Mechanical characterization of composite cast using 25 mm thick cryo chill block containing 12 wt.% dispersoid revealed that the presence of Nano-ZrO_2 particulates in cryo-chilled Al matrix significantly improved hardness (by 42%), strength (by 46%) and fracture toughness (by

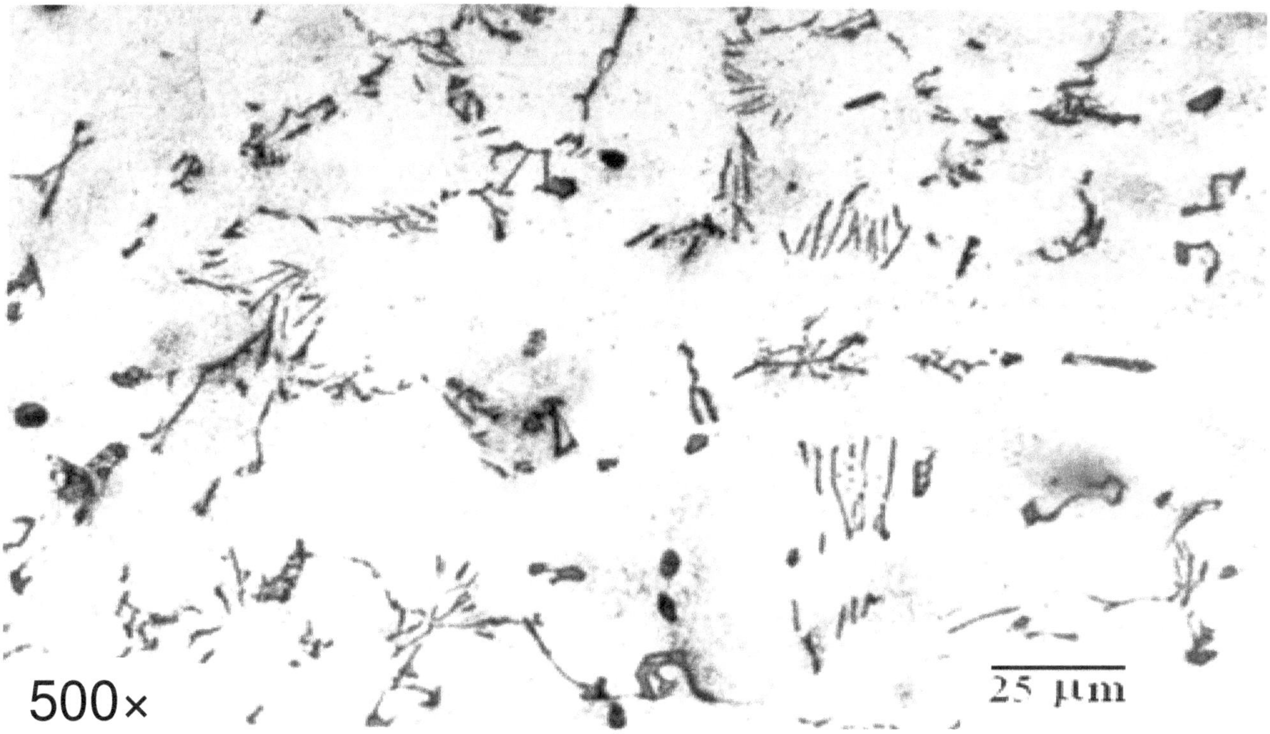

Figure 10. Optical microstructure of matrix alloy (LM 13) (500 × magnification).

23%) but with slight reduction in ductility (by 28%). Fractography analysis revealed that fracture behavior of FCC structured Al matrix alloy was changed from ductile mode of fracture to cleavage mode because of presence of ZrO_2 particulates.

ACKNOWLEDGEMENT

The author wishes to thank Miss. C.J. Sowmya, B.C.A., M.C.A., his colleague for her kind cooperation, help, encouragement and support especially in preparing and embedding the graphics.

REFERENCES

Awasthi S, Wood JL (1988). 'Mechanical properties of extruded ceramic reinforced Al based composites', Adv. Ceramic Mater., 35: 3449-3458.

Bowden FP, Tabor D (1964). 'The friction and lubrication of solids. part II, Oxford, UK, Oxford press', 34: 320-331.

Bensely A, Prabhakaran, Mohan LDA, Nagarajan (2006). G 'Enhancing the wear resistance of case carburized steel (En353) by cryogenic treatment, Cryogenics, 45: 747-754.

Gleiter H (2000). 'Grain refinement of Al-Si alloys', Acta Mater., 48: 1-29

Guy AG (1967) 'Elements of Physical Metallurgy', Addison-Wesley; USA, pp. 78- 85

Gilchrist JD (1989). 'Extraction Metallurgy, 3rd Edition, Pergamon press; Great Britain, 8: 148-155.

Hassan SF, Gupta J (2002). 'Development of high strength Al based composites using elemental Ni particles', J. Mater. Sci., 37: 2467-2477.

Huang JY, Liao XZ (2003) 'Microstructure of cryogenic treated M2 tool steel, Mater. Sci. Eng., A(339): 241-244.

Hemanth J (2003). 'Effect of high rate heat transfer during casting on the strength, hardness and wear behavior of Al-Quartz particulate MMCs', J. Eng. Manuf., Part-B(127): 651-662.

Hemanth J (2005). 'Tribological behavior of cryogenically treated Al-12%Si alloy/B4C Composites', WEAR, Elsevier Sc., 258: 1732-1745.

Hemanth J (2007). 'Cryogenic effects during solidification on the wear behavior of Al alloy/glass MMCs', J. Compos. Mater., Part-A(38): 395-1402.

Hemanth J (2008). 'Production, Mechanical and Thermal properties of chilled Al-Quartz (SiO2p) castable MMC for automotive applications', SAE International World congress paper 2008 no. 2008-01-1093, Detroit, Michigan, USA, pp. 1-8.

Hemanth J (2002). 'Fracture toughness and wear behavior of Al-boron carbide Particulate composite cast using metallic and non metallic chills' Mater. Des. Elsevier Sci., 23: 41-50.

Hemanth J (2009). 'Quartz (SiO2p) reinforced chilled metal matrix composites (CMMCs) for automotive applications, Mater. Des. Elsevier Sci., UK, 30: 323-329.

Gupta M, Lai MO, Saravanaranganathan D (2000). 'Synthesis, microstructure and property characterization of disintegrated melt deposited Al/SiC composites', J. Mater. Sci., 35: 2155-2164.

Luo A (1995). 'Processing, microstructure and mechanical properties of cast Mg MMC', Metallogr. Mater. Trans., 26: 2445-2453.

Lai MO (2000). 'Development of ductile Mg composite material using Ti reinforcement', J. Mater. Sci., 31: 2155-2167.

Lloyd DJ (1994). 'Particle reinforced Al-Mg metal matrix composites', Int. Mater. Rev., 39: 3-10.

Murakami Y (2002). 'Metal fatigue: Effects of small defects and non-metallic inclusions', Oxford: Elsevier Sci. Ltd., 35: 123-134.

Nishijma S, Kanazawa K (1999). 'Fatigue fracture of Engineering Materials. Struct. Mater. Des., 22: 581- 593.

Oku T, Usui T (1971). 'Neutron irradiation effects on the correlation of flow stress with hardness of some BCC metals', J. Nucl. Mater., 40: 93-103.

Ogata T, Isawa M, Wu K, Leo F (2002). 'In proceedings of 23rd International symposium on space technology and science, Japan,

45: 658-669.
Ono Y, Furgi H, Sawada J (2002). 'T Collected abstracts of 2002 Autumn meeting of the Japan Inst. Metals, 12: 89-99.
Opeka MM, Talmy IG, Zaykoki JA (2004). 'Mechanical properties of ZrB_2 composites', J. Mater. Sci., 32: 5887-5894.
Petty ER (1960). 'A low temperature inflection in the temperature dependence of hardness of pure metals', J. Inst. Met., 89: 123-129.
Reed HRE (1964). 'Physical metallurgy principles, 2nd Edition, D. Van Nostrand Company, UK, pp. 56-69.
Shiozawa K, Lu L, Ishihara S (2002). 'Fatigue fracture of Engg. Material Structures' Mater. Des., 24: 781-789.
Suemune K, Sakamoto TOT, Ozaki T (1988). 'Manufacture and properties of nitrogen containing Cr-Mn and Cr-Ni austenitic steels for cryogenic use', Adv. Cryo Eng., 34A: 123-132.
Seung J, Wang H (2005). 'Mechanical and Microstructural properties of Al-Zr02 ceramic Composites', Mater. Sci. Eng., 405A: 140-150.
Saravanan RA, Surappa MK (2000). 'Fabrication and characterization of Al-SiCp particle Composite', Mater. Sci. Eng., 108: 276-285.
Singh A, Tsai AP (2003). 'Quasicrystal strengthened Mg-Zn-Y alloys by extrusion', Scr. Mater., 49: 417-426.
Srinivasan TS, Gupta M (1999). 'Microstructure and grain growth behavior of al alloy MMC processed by disintegrated melt deposition', J. Mater. Eng. Perform., 8: 473-489.
Tham LM, Cheng L (1999). 'Influence of processing parameters during disintegrated melt deposition processing on near shape synthesis of Al based MMC', Mater. Sci. Technol., 15: 1139-1148.
Umezawa O, Nagai K, Ishikawa K (1990). 'Microstructure and mechanical behavior of Al based composites, Mater. Sci. Eng., 129: 217-228.
Wang CR, Yang JM (2002). 'Materials characterization by scanning electron microscopy, Mater. Chem. Phys., 14: 272-286.
Yamamoto T, Sasamoto IM (2000). 'Extrusion of al based composites', J. Mater. Sci. Lett., 19: 1053-1064.
Yuri T, Ono Y, Ogata T (2002). 'Composite structures and properties, J. Jpn. Inst. Metals., 66: 1287-1295.
Zhang GJ, Guo QG (2002). 'Mechanical and microstructural properties of high strength ceramic based composites', J. Nucl. Mater., 301: 187-195.

Investigation of turbulence mixed convection in air filled enclosures

Hamid Reza Goshayeshi[1]* and Mohammad Reza Safaiy[2]

[1]Islamic Azad University-Mashhad Branch (IAUM), Mashhad, Iran.
[2]Young Researchers Club, Islamic Azad University-Mashhad Branch, Iran.

Mixed convection heat transfer is a process where natural and forced convection happens simultaneously. The effect of temperature difference in natural convection and momentum difference results in mechanical ventilation or pressure difference that result in indoor and outdoor flow. When used in an optimum way, it is one of the newly subjects in HVAC industry which is a prominent example of mixed convection heat transfer. In the present study, first, the laminar flow of natural convection of air in a room with forced convection results in mechanical ventilation and then the calculated results are compared with the results of other researchers. After showing validation of calculations, aforementioned flow is solved as a turbulent flow, using the valid turbulence models RNG $K-\varepsilon$, standard $K-\varepsilon$ and RSM. To solve governing differential equations for this kind of flow, method of finite volume was used. This method is a special kind of residual weighted method. The results show that turbulence intensity is limited to vertical walls and boundary condition, such that flow in center region enclosure is turbulence and laminar is close to walls. Moreover, it is discovered that when Richardson number increases the maximum local Nusselt decreases; and therefore heat transfer is decreased with increasing Ri number.

Key words: Mixed convection, turbulence, Richardson number, turbulent intensity, wall shear stress, enclosures.

INTRODUCTION

The heat transfer phenomenon, in which both natural and forced convections simultaneously exist, is called mixed convection. Mixed convection heat transfer occurs when its buoyant flow matters in a forced flow or when the effect of forced flow matters in a buoyant flow. Prevailing dimensionless numbers which are used to determine this type of flow are as follows: Grashof number (Gr), Reynolds number (Re), Rayleigh number (Ra), Prandtl number (Pr) and Richardson number (Ri). Richardson number is calculated by dividing natural convection effect by forced convection effect, and expressed by: Ri=Gr/Re2 and from a physical standpoint, the Richardson number is the ratio of the convection to the forced convection.

In limit case where Ri→0 or Ri→∞, dominant heat transfers are forced and natural convections, respectively. Mixed convection plays an important role in heat transfer from fan coils or package radiators into the room air, and in heat transfer from cooling systems into the air. Atmospheric and oceanic movements and the relevant heat transfer processes important in environmental sciences also involve mixed convection. Mixed convection is also employed for cooling modern electric and electronic systems (transistors, computers, transformers). In manufacturing and double-layer ceilings, mixed convection phenomenon can be used to calculate thermal loss.

There are different methods for creating mixed convection. One way includes the entry of hot (or cold) fluid from one side, passing isothermal walls, and exiting from the other side. In this case we could evaluate and compare the forced convection effect caused by the entry and exit of the fluid. Some scientists have applied thermal flux on the way fluid passes through the channel and also

*Corresponding author. E-mail: hamidreza97@yahoo.co.uk.

studied the effects. Among the studies, we can mention the ones done by Rahman et al. (2007), Saha et al. (2006), and Saha et al. (2008). Another method for creating mixed convections is to move enclosure walls in the presence of hot (cold) fluid inside the enclosure. This creates shear stresses and provides thermal and hydrodynamic boundary layers in the fluid inside the enclosure, and eventually creates forced convection flows in it. Numerous studies have been conducted in this field so far. For example, Oztop and Dagtekin (2004) studied a two-dimensional and square-shaped enclosure with vertical isothermal moving walls and insulated horizontal walls. In this work, different situations have been considered concerning the movement of vertical walls, and 0.01≤Ri≤100 has been presupposed. Heat transfer rate has also been expressed in the form of Nusselt numbers. The results of this work suggest that in low Richardson values, if the moving walls move in the opposite directions, heat transfer from enclosure is much higher than when the walls slide converge. Basak et al. (2009) studied the mixed convection flow inside a square enclosure with left and right cold walls, insulated moving upper wall, and fixed lower hot wall by using finite element method. They suggested that by increasing Gr, with Pr and Re fixed, recirculation power will improve. In 2007, Sharif studied the laminar mixed convection in inclined rectangular enclosures with aspect ratio of 10, by using Fluent6. He let the Rayleigh number variable to be between 105 to 107, and Reynolds number fixed, to be 408.21. The fluid he used was water with Prandtl as 6, and the enclosure inclination angle to the horizon varied between 0 to 30°. The mentioned enclosure had hot moving upper wall, cold fixed lower wall, and adiabatic left and right walls. His study showed that local Nusselt number heightened by increasing the enclosure inclination angle.

Although progression of different sciences in the last decade has provided much subtle laboratory measuring tools and application of modern methods like parallel processing has enabled us efficiently use numerical analysis methods, analysis of turbulent flows inside the enclosure is still a challenging topic in fluid mechanics. This is because it is too difficult to reach ideal adiabatic wall condition in experimental situation. It is also so difficult to measure low speeds in mixed convection process using present sensors and probes. Although numerical methods like DES, LES, and DNS have seen much progression, it is still hardly possible to predict the stratification in the core of the enclosure. Non-linearity and coupling of the governing equations have also made the calculations complicated and time consuming. That is while in designing large enclosures, Rayleigh number is usually large, and so the flow nature is turbulent. The complexity of calculations in mixed convection situation has made scientists study the flow only in natural convection, among which we can mention the studies of Salat et al. (2004), Aounallah et al. (2007), Bessaih and Kadja (2000), Ampofo (2004), and Xaman (2005). Tian and Karayiannis (2000) started an experimental study in South Bank University which was followed by Ampofo and Karayiannis in 2003. The data in this work were experimental benchmark data of natural convection flow inside a square enclosure, and were used for other studies. Whereas Peng and Davinson (2001) studied the mentioned flow by using LES, and Omri and Galanis (2007) used the SST k-ω to study the flow. Ampofo (2005) expanded his previous study in 2005 and tested the same enclosure with partitions and horizontal conducting walls and compared the results with his previous findings. Hsieh and Lien (2004) used turbulence models of steady RANS like Low-Re k-ε, and numerically analyzed the works done by Tian and Karayiannis (2000) and Betts and Bokhari (2000). In the present work, turbulent natural convection flow inside the enclosure is modeled first, and the results have been compared with the studies of Tian and Karayiannis (2000), Ampofo and Karayiannis (2003), Peng and Davinson (2001), Hsieh and Lien (2004), Omri and Galanis (2007). After the calculations are made valid, the Laminar mixed convection flow is solved in the square enclosure first, in comparison to Sharif (2007) and eventually turbulence mixed convection flow is modeled for the first time, using turbulence models like standard k-ε, RNG k-ε and RSM.

PROBLEM FORMULATION

For modeling the flow, continuity, momentum, energy, and turbulence equations have been studied. The properties have been considered as fixed. Density is calculated vertically by using variable density parameter for ΔT>30 and Boussinesq approximation for ΔT<30. The governing equations are as follow:

Continuity equation:

$$\frac{\partial u}{\partial x}+\frac{\partial v}{\partial y}=0 \tag{1}$$

Momentum equations in X and Y direction:

$$\frac{\partial u}{\partial t}+u\frac{\partial u}{\partial x}+v\frac{\partial u}{\partial y}=-\frac{1}{\rho}\frac{\partial p}{\partial x}+\frac{\partial}{\partial x}(\upsilon+\upsilon_t)(2\frac{\partial u}{\partial x})+\frac{\partial}{\partial y}(\upsilon+\upsilon_t)(\frac{\partial u}{\partial y}+\frac{\partial v}{\partial x}) \tag{2}$$

$$\frac{\partial v}{\partial t}+u\frac{\partial v}{\partial x}+v\frac{\partial v}{\partial y}=-\frac{1}{\rho}\frac{\partial p}{\partial x}+g\beta(T-T_m)+\frac{\partial}{\partial y}(\upsilon+\upsilon_t)(2\frac{\partial v}{\partial y})+\frac{\partial}{\partial x}(\upsilon+\upsilon_t)(\frac{\partial v}{\partial x}+\frac{\partial u}{\partial y}) \tag{3}$$

Energy equation:

$$\frac{\partial T}{\partial t}+u\frac{\partial T}{\partial x}+v\frac{\partial T}{\partial y}=$$
$$\frac{\partial}{\partial x}(\frac{v}{\Pr}+\frac{v_t}{\sigma_T})\frac{\partial T}{\partial x}+\frac{\partial}{\partial y}(\frac{v}{\Pr}+\frac{v_t}{\sigma_T})\frac{\partial T}{\partial y} \quad (4)$$

Now for k-ε model, we will have:
Turbulent kinetic energy transport equation for k-ε model:

$$\frac{\partial k}{\partial t}+u\frac{\partial k}{\partial x}+v\frac{\partial k}{\partial y}=\frac{\partial}{\partial x}(\upsilon+\frac{\upsilon_t}{\sigma_k})\frac{\partial k}{\partial x}$$
$$+\frac{\partial}{\partial y}(\upsilon+\frac{\upsilon_t}{\sigma_k})\frac{\partial k}{\partial y}+P_k+G_k-\varepsilon \quad (5)$$

Dissipation of turbulent kinetic energy transport equation for k-ε model:

$$\frac{\partial \varepsilon}{\partial t}+u\frac{\partial \varepsilon}{\partial x}+v\frac{\partial \varepsilon}{\partial y}=\frac{\partial}{\partial x}(\upsilon+\frac{\upsilon_t}{\sigma_\varepsilon})\frac{\partial \varepsilon}{\partial x}$$
$$+\frac{\partial}{\partial y}(\upsilon+\frac{\upsilon_t}{\sigma_\varepsilon})\frac{\partial \varepsilon}{\partial y}+C_1\frac{\varepsilon}{k}P_k+$$
$$C_2\frac{\varepsilon^2}{k}+C_3\frac{\varepsilon}{k}G_k-R_\varepsilon \quad (6)$$

The eddy viscosity obtained from Prandtl- Kolomogorov relation:

$$\upsilon_t=C_\mu f_\mu\frac{k^2}{\varepsilon} \quad (7)$$

The stress production term, P_k, is modeled by:

$$P_k=\upsilon_t\left[2\left(\frac{\partial u}{\partial x}\right)^2+2\left(\frac{\partial v}{\partial x}\right)^2+\left(\frac{\partial u}{\partial y}+\frac{\partial v}{\partial y}\right)^2\right] \quad (8)$$

The buoyancy term, G_k, is defined by:

$$G_k=-g\beta\frac{\upsilon_t}{\sigma_t}\frac{\partial T}{\partial y} \quad (9)$$

We will also have the following for RNG k-ε:

$$R_\varepsilon=\frac{C_\mu\rho\eta^3\left(1-\frac{\eta}{\eta_0}\right)}{1+\beta\eta^3}\frac{\varepsilon^2}{k} \quad (10)$$

That:

$$\eta=\frac{Sk}{\varepsilon} \quad (11)$$

The main difference between standard k-ε and RNG k-ε methods is in the ε equation, such that we can say the model RNG k-ε is the very same standard k-ε model, whose analytical formulas for turbulent Prandtl numbers have been improved. This is while these values in standard k-ε model are gained experimentally.
For RSM model, the turbulence equations are as follows:
Reynolds stress transport equations:

$$\frac{D}{Dt}\left(\overline{u_i'u_j'}\right)=\frac{\partial\left(d_{ijk}\right)}{\partial x_k}+P_{ij}+G_{ij}+\phi_{ij}-\varepsilon_{ij} \quad (12)$$

That:

$\frac{D}{Dt}\left(\overline{u_i'u_j'}\right)=\frac{\partial\left(\overline{u_i'u_j'}\right)}{\partial t}+\bar{u}_k\frac{\partial\left(\overline{u_i'u_j'}\right)}{\partial x_k}$: Advection (By Mean Flow)

$P_{ij}=-\left[\left(\overline{u_i'u_k'}\right)\frac{\partial \bar{u}_j}{\partial x_k}+\left(\overline{u_j'u_k'}\right)\frac{\partial \bar{u}_i}{\partial x_k}\right]$: Production (By Mean Strain)

$G_{ij}=\left(\overline{u_i'f_j'+u_j'f_i'}\right)$: Production (By Body Force)

$\phi_{ij}=\overline{\frac{P'}{\rho}\left(\frac{\partial u_i'}{\partial x_j}+\frac{\partial u_j'}{\partial x_i}\right)}=\overline{\frac{2P'}{\rho}S_{ij}}$: Pressure-Strain Correlation

$\varepsilon_{ij}=2v\overline{\frac{\partial u_i'}{\partial x_k}\frac{\partial u_j'}{\partial x_k}}$: Dissipation

$d_{ijk}=v\frac{\partial\left(\overline{u_u'u_j'}\right)}{\partial x_k}-\overline{\frac{P'}{\rho}\left(u_i'\delta_{jk}+u_j'\delta_{ik}\right)}-\overline{u_i'u_j'u_k'}$: Diffusion

Turbulent kinetic energy transport equation for RSM model:

$$\frac{Dk}{Dt}=\frac{\partial d_i^{(k)}}{\partial x_i}+P^{(k)}+G^{(k)}-\varepsilon \quad (13)$$

That:

$P^{(k)}=-\overline{u_i'u_j'}\frac{\partial \bar{u}_i}{\partial x_j}$: Production (By Mean Strain)

$G^{(k)}=\overline{u_i'f_i'}$: Production (By Body Force)

$\varepsilon=v\overline{\left(\frac{\partial u_i'}{\partial x_j}\right)^2}$: Dissipation

$d_i^{(k)}=v\frac{\partial k}{\partial x_i}-\frac{\overline{P'u_i'}}{\rho}-\frac{1}{2}\overline{u_j'u_j'u_i'}$: Diffusion

Table 1. Coefficients for RNG k-ε turbulent model (Rahman et al., 2007).

Coefficient	Value
C_μ	0.0845
σ_k	1
σ_ε	1.3
C_1	1.42
C_2	1.68
η_0	4.38
β	0.012
K	0.41

Table 2. Coefficients for Standard k-ε turbulent model (Rahman et al., 2007).

Coefficient	Value
C_μ	0.0845
σ_k	1
σ_ε	1.3
C_1	1.42
C_2	1.68

Table 3. Coefficients for RSM turbulent model (Rahman et al., 2007).

Coefficient	Value
$c_2^{(w)}$	0.3
$c_1^{(w)}$	0.5
C_S	0.22
C_1	1.8
C_2	0.6
C_3	2.5

Except the terms Convection and Production in Reynolds stress transport equation, all the other terms have contributed in introducing a series of correlations, which have to be identified according to some known and unknown quantities, so that the equation system can be configured.

Diffusion term:

$$-\overline{u_i'u_j'u_k'} = C_S \frac{k}{\varepsilon}\overline{u_k'u_l'}\frac{\partial \overline{u_i'u_j'}}{\partial x_l} \tag{14}$$

Redistribution term:

$$\phi_{ij} = \phi_{ij}^{(1)} + \phi_{ij}^{(2)} + \phi_{ij}^{(w)} \tag{15}$$

That:

$$\phi_{ij}^{(1)} = -C_1 \frac{\varepsilon}{k}\left(\overline{u_i'u_j'} - \frac{2}{3}k\delta_{ij}\right)$$

$$\phi_{ij}^{(2)} = -C_2\left(P_{ij} - \frac{1}{3}P_{kk}\delta_{ij}\right)$$

$$\phi_{ij}^{(w)} = \left(\tilde{\phi}_{kl}n_k n_l \delta_{ij} - \frac{3}{2}\tilde{\phi}_{ik}n_j n_k - \frac{3}{2}\tilde{\phi}_{jk}n_i n_k\right)\psi$$

$$\tilde{\phi}_{ij} = -C_1^{(w)}\frac{\varepsilon}{k}\overline{u_i'u_j'} + C_2^{(w)}\phi_{ij}^{(2)}$$

$$\psi = \frac{k^{\frac{3}{2}}/\varepsilon}{C_3 y_n}$$

y_n is the distance from the wall. The role of the terms $\Phi_{ij}^{(2)}$, $\Phi_{ij}^{(1)}$ is to return isotropy (or terminating anisotropic flow with distributing kinetic energy of Reynolds' huge stresses among the stresses of smaller size). The terms $\Phi_{ij}^{(1)}$ and $\Phi_{ij}^{(2)}$ are called "return to isotropy" and "isotropization of production", respectively. The term $\Phi_{ij}^{(w)}$ is named as "wall reflection term".
For Dissipation term, we have:

$$\varepsilon_{ij} = \frac{2}{3}\delta_{ij}\varepsilon \tag{16}$$

The constants in the above relations have been presented in Table 1 for RNG k-ε, Table 2 for standard k-ε, and in Table 3 for RSM models.

In order to solve the differential equation that governs the flow, the finite volume method, which is explained in detail by Patankar (1980) was used. This method is a specific case among the residual of weighting methods. In this approach, the computational field is divided into some control volumes in a way that each node is surrounded by a control volume, and that control volumes have no volumes in common. The differential equation is then integrated on each control volume. Profiles in pieces which show changes (of a certain quantity like temperature, velocity, etc.) among the nodes, were used to calculate the integrals. The result is discretization equation, which includes quantities for a group of nodes (Patankar, 1980).

RESULTS

Grid independency

Grids designed to cover control volumes are square meshes provided on physical domain with different distances in order to reach independence. The mentioned mesh-independence for each turbulence

Table 4. Some meshes used for solving the problem.

	RNG k-ε			Standard k-ε			RSM		
Ri=0.1	57×57	101×101	161×161	61×61	119×119	181×181	75×75	119×119	201×201
Ri=1	68×68	135×135	201×201	81×81	160×160	241×241	90×90	160×160	270×270
Ri=10	75×75	150×150	225×225	90×90	175×175	270×270	105×105	180×180	300×300

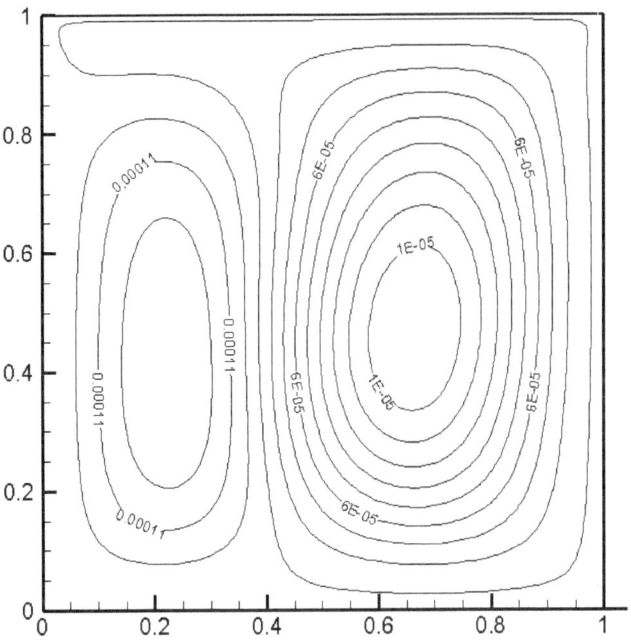

Figure 1. Stream function with Re=10, in comparison with Basak et al. (2009) experiment.

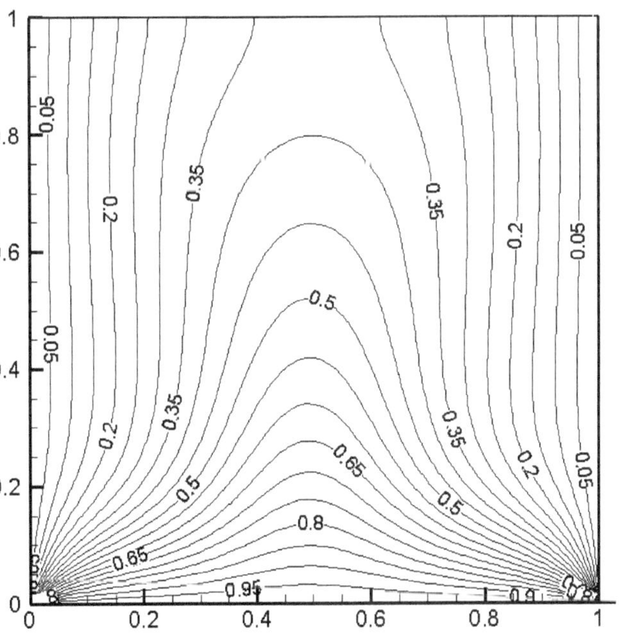

Figure 2. Temperature Contour with Re=1, in comparison with Basak et al. (2009) experiment.

model and any different Ri has been separately calculated. Table 4 shows some meshes used in this study.

Evaluating laminar mixed convection flow inside a square enclosure

In order to compare this work with the results of Basak et al. (2009], Pr and Gr have been considered as 0.7 and 10^4, respectively. 1<Re<100 has also been changed. Figures 1 and 2 present stream function and temperature contour in comparison with Basak et al. (2009). The acceptable harmony among the figures validates the accuracy of this study (Sharif, 2007)

Evaluation of turbulent mixed convection flow inside the enclosure

In order to prove the accuracy of the calculations, firstly, natural convection flow is solved into rectangular and square enclosures which are studied by other scientists like Tian and Karayiannis (2000), Ampofo and Karayiannis (2003), Peng and Davinson, (2001) and Omri and Galanis (2007), and then after the results are validated (Figures 3 and 4), the flow is solved inside the enclosure of mixed convection. In this case Ri varies from 0.1 to 10.

Figure 5 shows the schematics of the problem and also details of the Rectangular and Square Enclosures have been shown in Table 5.

Figure 6 illustrates the turbulent intensity for y/L=0.5 in a different Ri and turbulent models. It can be seen that turbulence intensity is limited to right and left walls and it reached a peak at the centre of cavity.

Figure 7 shows the Nusselt number counter on hot wall. As a result of this figure, it is clear that the Nusselt number is maximum for the case of Ri=0.1 and it moderately decreases due to increase in the value of Ri. It means that Ri=0.1, the forced convection governs fluid treatment. So the rate of heat transfer from cavity increased as a result of less Richardson number. Moreover, the rate of heat transfer by natural convection will grow if the Richardson number will increase. In large Richardson numbers, the natural convection is a major parameter of heat transfer in a cavity. The rate of heat transfer by mix and forced convection is much more than natural convection.

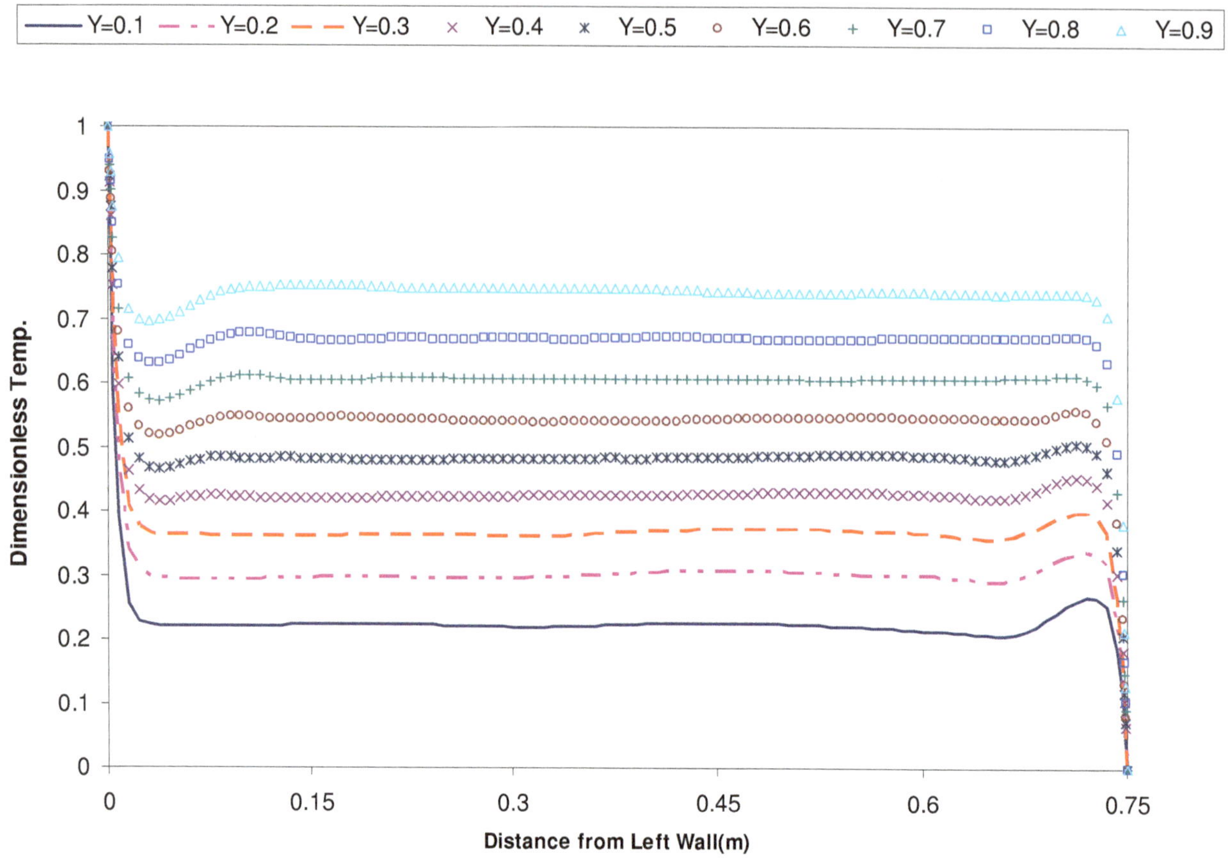

Figure 3. Temperature distribution at different height, in comparison with Tian and Karayiannis (2000) experiment.

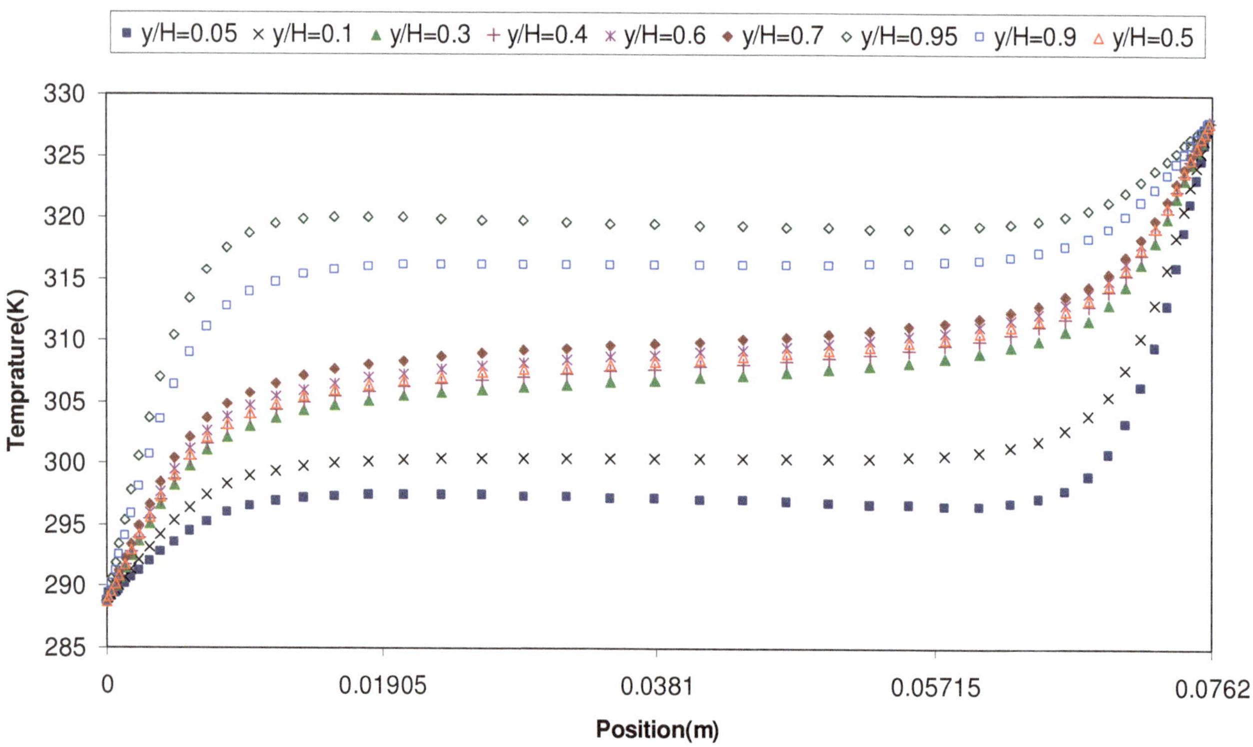

Figure 4. Variation of temperature in $Ra=1.43\times10^6$, in comparison with Betts and Bokhari (2000) experiment.

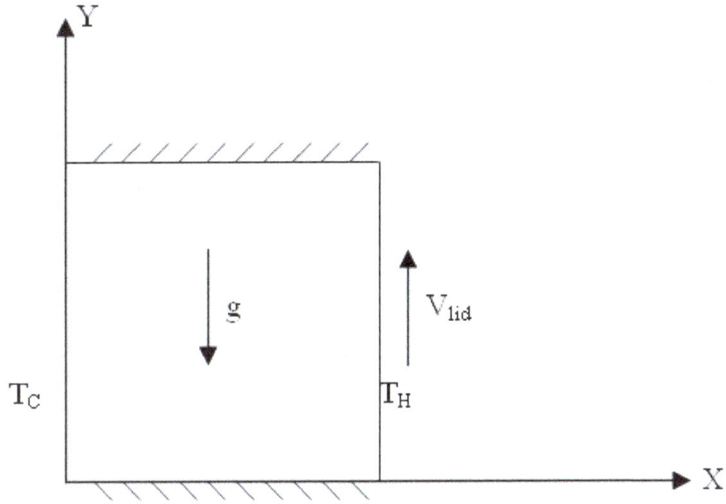

Figure 5. Schematic of the problem.

Table 5. Details of the rectangular and square enclosures.

Variable	Betts and Bokhari	Tian and Karayiannis
Rayleigh number	1.43×10^{6}	1.58×10^{9}
Length of enclosure (m)	2.18	0.75
Wide of enclosure (m)	0.076	0.75
Left wall temp.	15.6	50
Right wall temp.	54.7	10
Prandtl number	0.697	0.707

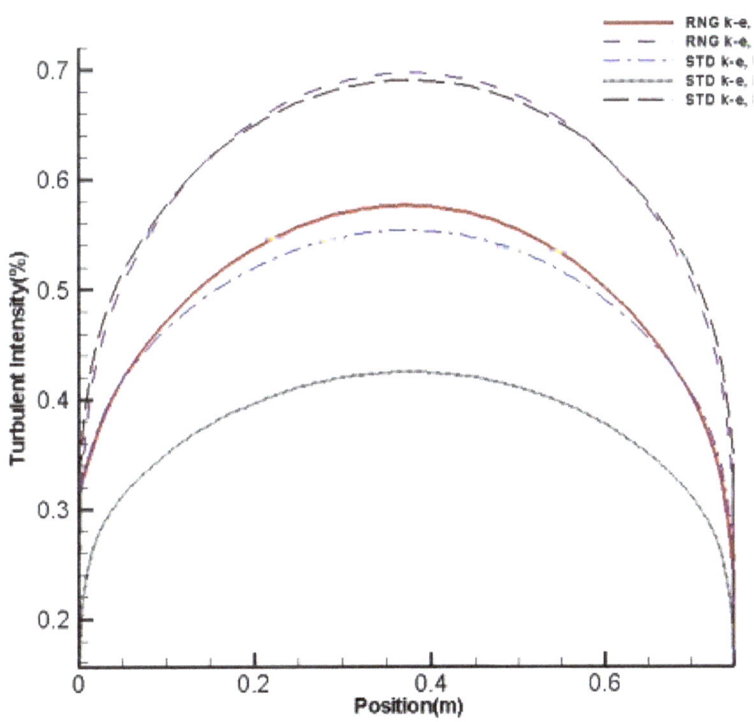

Figure 6. Turbulent Intensity at y/H=0.5.

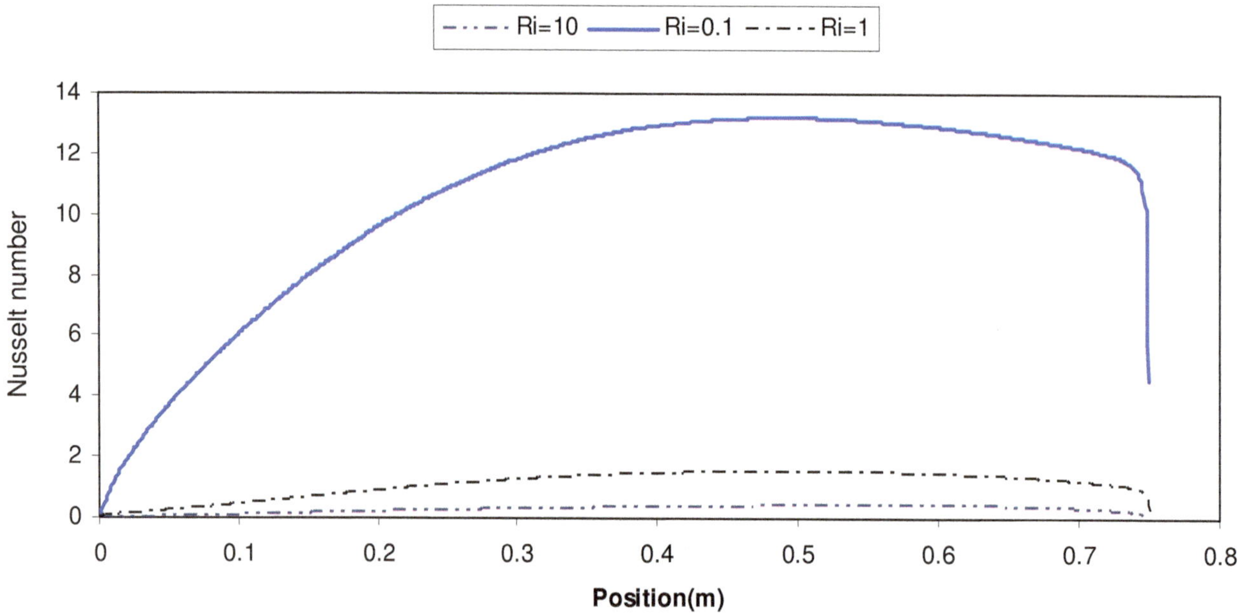

Figure 7. Nusselt number along the hot wall

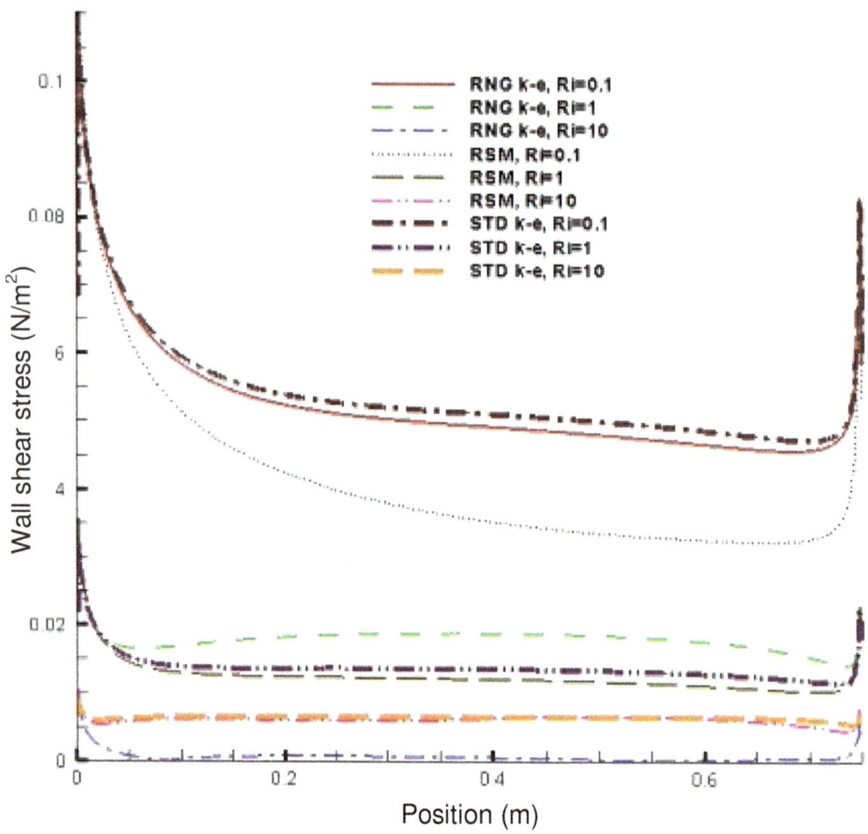

Figure 8. Wall shear stress at y/H=0.5.

The wall shear stress for y/H=0.5 and also in the different Ri and turbulence models is illustrated in Figure 8. It can be seen from this figure that, in a boundary layer of the left wall, due to the non-existence of slip condition, the value of shear stress hugely increased linearly; therefore, the value of shear stress decreased moderately at the boundary layer of the right wall, and also at this point, its value linearly increased hugely due to lack of slip condition. Another important point is that, if the value of Ri increases, the effect of fluid viscosity is low. As a result, it is reasonable to reduce the value of shear stress.

Conclusions

Mixed convection usually induced in enclosures or cavities heating elements on one of its walls or both is important from both theoretical and practical point of view. In fact, this configuration can be encountered in various engineering application. Numerous studies related to mixed convection in enclosures have been reported in order to investigate the heat transfer and fluid flow in such geometries. The following conclusions can be drawn from the results of the present work:

1) In large Richardson numbers, the natural convection is a major parameter of heat transfer in a cavity.
2) Heat transfer is decreased with increasing Ri number.
3) When the Reynolds number increases, the circulation of flow vortices increases and becomes stronger, making the forced convection effect more dominant for different values of Richardson numbers.
4) If the value of Ri increases, the effect of fluid viscosity is low.

Nomenclature

u, v: velocities in x and y directions (m/s); x, y: Cartesian coordinates (m); W :height of the cavity (m); P pressure (N/m^2); T temprature (K); t Time (Sec); g gravitational acceleration (m^2/s); K Turbulent kinetic energy transport (m^2/s^2); k, $(W/m.k)$ thermal conductivity; Re, Reynolds number; Ri, Richardson number; Gr, Grashof number; Nu Nusselt number; Pr ,Prandtl number, Ra, Rayleigh number; ε ,Dissipation of turbulent kinetic energy transport (m^2/s^3); υ_t turbulent kinematic viscosity (m^2/s); σ_T turbulent thermal diffusivity (m^2/s); β thermal expansion coefficient (1/K); ρ density (kg/m^3); υ (m^2/s) kinematic viscosity; h =hot wall; c cold wall; m mean; lid Lid.

REFERENCES

Rahman MM, Alim MA, Mamun MAH (2007). Numerical Study of Opposing Mixed Convection in a Vented Enclosure, ARPN J. Eng. Appl. Sci., 2: 25-35.

Saha S, Saha G, Ali M (2006). Combined Natural and Forced Convection inside a Two-Dimensional Multiple Ventilated Rectangular Enclosure, ARPN J. Eng. Appl. Sci., 1: 23-35.

Saha S, Mamun AH, Hossain MZ (2008). Mixed Convection in an Enclosure with Different Inlet and Exit Configurations, J. Appl. Fluid Mech., 1: 78-93.

Oztop HF, Dagtekin I (2004). Mixed Convection In Two-Sided Lid-Driven Differentially Heated Square Cavity, Int. J. Heat Mass Transf,, 47: 1761-1769.

Basak T, Roy S, Sharma PK (2009). Analysis of Mixed Convection Flows within a Square Cavity with Uniform and Non-Uniform Heating of Bottom Wall, Int. J. Therm. Sci., pp. 85-93.

Sharif MAR (2007). Laminar mixed convection in shallow inclined driven cavities with hot moving lid on top and cooled from bottom, Appl. Therm. Eng., 27: 1036-1042.

Salat J, Xin S, Joubert P (2004). Experimental and numerical investigation of turbulent natural convection in a large air-filled cavity, Int. J. Heat Fluid Flow, 25: 824-832.

Aounallah M, Addad Y, Benhamadouche (2007). Numerical investigation of turbulent natural convection in an inclined square cavity with a hot wavy wall, Int. J. Heat Mass Transf., 50: 1683-1693.

Bessaih R, Kadja M (2000). Turbulent natural convection cooling of electronic components mounted on a vertical channel, Appl. Therm. Eng., 20: 141-154.

Ampofo F (2004). Turbulent natural convection in an air filled partitioned square cavity, Int. J. Heat Fluid Flow, 25: 103-114.

Xaman J, Alvarez G, Lira L (2005). Numerical study of heat transfer by laminar and turbulent natural convection in tall cavities of facade elements, Energy Build., 37: 787-794.

Tian YS, Karayiannis TG (2000). Low Turbulence Natural Convection in an Air Filled Square Cavity. Int. J. Heat Mass Transf., 43: 849-866.

Ampofo F, Karayiannis TG (2003). Experimental benchmark data for turbulent natural convection in an air filled square cavity, Int. J. Heat Mass Transf., 46: 3551-3572.

Peng SH, Davinson L (2001). Large Eddy Simulation for Turbulent Buoyant Flow in a confined cavity, Int. J. Heat Fluid Flow, 22: 323-331.

Omri M, Galanis Ni (2007). Numerical analysis of turbulent buoyant flows in enclosures: Influence of grid and boundary conditions, Int. J. Therm. Sci., 46: 727-738.

Ampofo F (2005). Turbulent natural convection of air in a non-partitioned or partitioned cavity with differentially heated vertical and conducting horizontal walls. Exp. Therm. Fluid Sci., 29: 137-157.

Hsieh KJ, Lien FS (2004). Numerical Modeling of Buoyancy-Driven Turbulent Flows Enclosures, Int. J. Heat Fluid Flow, 25: 659-670.

Betts PL, Bokhari IH (2000). Experiments On Turbulent Natural Convection In An Enclosed Tall Cavity, Int. J. Heat Fluid Flow, 21: 675-683.

Patankar SV (1980). Numerical Heat Transfer and Fluid Flow, Hemisphere Washington DC, pp. 83-84.

11

Short review on the crystallization behavior of PET/clay nanocomposites

Ranjana Sharma*, Himani Joshi and Purnima Jain

School of Applied Sciences, Netaji Subhas Institute of Technology, Sector- 3, Dwarka, New Delhi – 110078.

In recent years, the area of polymer nanocomposites based on organically modified clay has attracted much interest in academia and industry because of their peculiar and fascinating properties. Nanomaterials are presently considered to be high potential filler material, with high aspect ratio and extremely large surface area. They improve mechanical and physical properties of polymers. Lighter, thinner, stronger and cheaper structures are the goals of materials science and engineering applications nowadays. Appropriate addition of nanoparticles to a polymer matrix can enhance its performance by simply focussing on the nature and properties of the nanoscale filler. So, the properties of polymer nanocomposites depend greatly on the chemistry of polymer matrix, nature of nanofillers and the way in which they are prepared. Various analysis techniques were used to characterize the dispersion and the properties of the nanocomposites, using XRD, TGA, SEM, DSC, TEM, DMTA, etc. Several polymers such as polypropylene, polyamide 6, polyethylene, polyethylene terephthalate (PET), polybutylene terephthalate (PBT), polylactic acid (PLA), etc., have been tested as matrices. This paper emphasises on the overview of the crystallization properties of PET/clay nanocomposites by different authors. It has been reported that nanoparticles of clay dispersed in PET polymer matrix act as nucleating agent. In this way, they enhance the crystallization rates of PET/clay nanocomposites.

Key words: Nanocomposites, polyethylene terephthalate (PET), clay, crystallization.

INTRODUCTION

The reinforcement of polymers using fillers is common in the production and processing of polymeric materials. Fillers play important roles in modifying the desirable properties of polymers and reducing the cost of their composites (Tjong, 2006; Pettrain et al., 2008). The interest in new nanoscale fillers has grown rapidly in the last two decades. This interest arises from the fact that when these nanoscopic fillers are used instead of conventional reinforcing material, the composite materials exhibit better performances, such as, mechanical, thermal stability, dimensional stability, flame retardant and gas barrier properties etc., at relatively small loadings (Gonzalez et al., 2006; Kusmono et al., 2008; Roohollah and Ahmad, 2008; Nirukhe and Shertukde, 2009).

Polymer nanocomposite is a new class of composite materials derived from nanoparticles and it is a two phase material where one of the phases is in the nanometer range in at least one dimension being less than 100 nm (Giannelis, 1996; Wang et al., 2004; Lei et al., 2006; Manias, 2007).

The most common filler used in nanocomposites is nanoclay, other fillers like carbon nanofibers, carbon nanotubes, graphene, and core shell materials are also found very commonly in polymer nanocomposites. In recent years, polymer/clay nanocomposites have emerged as one of the most promising nanocomposites. They have attracted great academic and industrial interest because of enhancement in properties. These composites offer better mechanical, thermal, barrier and chemical resistance properties as compared to conventional composites. Nanocomposites have high potential for applications in many fields such as automative and packaging industries. With relatively low

*Corresponding author. E-mail: prnm.jain@gmail.com.

loading of nanometer-sized particles, they show dramatically increased stiffness, heat distortion temperature (HDT), dimensional stability, gas barrier properties, electrical conductivity and flame retardancy (Suprakas and Musami, 2003; Ahmadi et al., 2004; Lei et al., 2006; Hong et al., 2008; Miray and Ulku, 2008; Baniasadi et al., 2010).

The concept of polymer/clay nanocomposites systems were first introduced by researchers from Toyota who discovered the possibility of building exfoliated nanocomposites from nylon 6 and organophilic clay, known as nylon 6-clay hybrid (NCH) via in situ polymerization. The NCH exhibits various superior properties such as high strength, high modulus, and high heat resistance compared to nylon 6 (Naoki et al., 1996, 2000). In addition to this, another observation was proposed by Vaia et al. (2003) to represent the possibility of the making melt-mix polymers with layered silicates, without the use of organic solvents. Today, efforts are being conducted globally, using almost all types of polymer matrices. Until now, many kinds of polymer/clay nanocomposites have been researched upon using polycarbonate, polystyrene, polyimide, epoxy, polymethyl methacrylate (PMMA), polyethylene, polypropylene, polyamide, polyurethane, polyethylene oxide, polyethylene terephalate, polybutylene terephalate, and so on, as matrices (Naoki et al., 2000; Li et al., 2001; Suprakas and Musami, 2003; Ahmadi et al., 2004; Defeng et al., 2006; Quang and Donald, 2006).

Polymer-clay nanocomposites use smectite type clays as fillers such as hectorite, montmorillonite (MMT) or synthetic mica and all minerals with layered structures. Among various kinds of natural and synthetic clays, the most common nanoscopic filler is derived from MMT clay. It was found naturally in layered silicate structure with high surface area, about 750 m^2/gm and high aspect ratio and have a cross sectional area of 100 m^2 which is very small compared to conventional fillers (Akane and Arimitsu, 2006; Clois and Gary, 2006). The properties of clay based nanocomposites polymers are influenced by the type of clay, choice of clay pre-treatment, the selection of polymer component and the way in which the polymer is incorporated into the nanocomposites (Yeh et al., 2004; Tjong, 2006). In general, the improvements in the properties may be attributed to the following factors: Large surface area; High aspect ratio; Submicroscobic dispersion of the clay in the polymeric matrix (Young-Wook et al., 2005); Ionic bond between organic polymer and inorganic clay (Mahmood and Musa, 2002; Nicolas et al., 2007).

The performance of polymer-clay nanocomposites strongly depends on the breaking up of clay particles in the polymer matrix. The dispersion of nanofillers, that is, clay in the polymers, is rather poor due to their incompatibility with polymers. Therefore, organic surfactant and compatibilizer additions are required to improve the dispersion of these nanofillers in polymeric matrices. In order to obtain good interfacial adhesions and mechanical properties, the hydrophilic clay needs to be modified prior to introduce in most polymer matrices which are organophilic (Garcia-Lopez et al., 2003; Suprakas and Musami, 2003; Benetti et al., 2005).

Recently, a new class of materials, PET/clay nanocomposites, has been developed successfully. PET is an engineering semi crystalline thermoplastic polymer with low cost and high performance, which has wide applications in the form of films, fibres and non- fibres. Film applications mainly include food and beverage bottles (In Yee et al., 2004; Mazeyar et al., 2010). PET is intrinsically a polymer with slow crystallization rate and high melting temperature. So, it has not been considered for applications involving high speed processing, such as injection moulding. To accommodate typical industrial injection moulding process, additives (most often nucleating agents) are introduced to enhance the PET crystallization (Cheng et al., 2003, 2004; Yaming et al., 2004; Tong et al., 2004; Yee et al., 2004).

PET/clay nanocomposites have attracted much attention because of its superior thermal resistance, mechanical properties, low permeability, chemical resistance and mechanical performance. The studies on PET/clay nanocomposites are mainly focused on its synthesis, effect of organoclay content on crystallization behavior, nanoscale morphology, thermal behavior and melting behavior. The presence of nanometer sized clay layers in a PET polymer matrix not only creates an enormous interfacial area but also increases crystallization rate, O_2 permeability, wear resistance, flexural, tribological, thermal and mechanical behaviour of PET (Mazeyar et al., 2010). Because of the complexity of the PET/clay system, the effect of clay on the crystallization performance of PET has not been properly addressed (Yaming et al., 2004; Tong et al., 2004; Guohu et al., 2004; Ali et al., 2010). In this article, PET/clay nanocomposites have been studied with regard to their crystallization properties by different authors.

CRYSTALLIZATION BEHAVIOR OF PET/ORGANOCLAY NANOCOMPOSITES

Studies on semi crystalline polymer-clay nanocomposites describe that the nanosized clay particles affect the crystallization behavior and the crystal structure of the matrix polymer (Young-Wook et al., 2005). Qi et al. (1999) reported that PET/clay nanocomposites had a three times faster crystallization rate than that of pure PET. This indicates that nanoparticles of clay dispersed in matrix acted as nucleating agent. Crystallization behavior of PET/clay nanocomposites is of great importance from both a theoretical and industrial standpoints. Many authors studied the effect of the presence of clay on the isothermal and non-isothermal crystallization behavior of PET/clay nanocomposites. Non-isothermal crystallization can broaden and supplement the knowledge of the crystallization behavior of polymers.

Cheng et al. (2003) studied the nucleating effect of montmorillonite on crystallization of PET nanocomposite by solution intercalation method. They prepared four compositions with varying amount of organoclay (wt %) and neat PET and carried out a comparative study. Layered silicate used in this study was sodium MMT with cationic exchange capacity (CEC) of 114 mEq/100 g (abbreviated as 11 K). Clay was organo-modified with intercalation agent cetyltrimethylammonium chloride (CMC). Nucleating effect of organoclay is investigated using DSC analysis. The crystallization rate of PET in the PET/11 K-M (organoclay) nanocomposites increases with the organoclay loadings from 1 to 10 wt %. Maximum increase in crystallization rate was observed in the 10 wt % organoclay. Further increases in the content of organoclay lead to decrease in crystallization rate. It is due to the agglomeration of organoclay in the PET matrix.

The degree of super cooling ($\Delta T = T_m - T_c$) may be a measurement of a polymer's crystallinity. ΔT is proportional to the free energy of melting, which is thermodynamic driving force of crystallization (Seung et al., 2000). The smaller the value of ΔT, the higher the overall crystallization rate. The ΔT values for the PET/11 K-M nanocomposites are smaller than that of neat PET and 90/10 PET/11 K-M exhibits the smallest ΔT. These findings indicate that 11 K-M based PET nanocomposites have heterogeneous nucleation which offers enormous surface area and hence accelerate the crystallization rate of PET during nucleation.

Cheng et al. (2004) studied the effects of organoclay loading on the crystallization behavior and thermal stability of PET/clay nanocomposites. In their study, layers of sodium MMT with cationic exchange capacity (CEC) value of 87 mEq/100 g (abbreviated as 8 K) were intercalated by cetylpyridinium chloride (CPC) instead of CMC. Again, DSC analysis showed that clay behaves as nucleating agent and enhances the crystallization rate of PET. They calculated half crystallization time ($t_{½}$) using equation:

$$t_{½} = (T_{on} - T_c)/\chi$$

where, T_{on} is the crystallization onset temperature, T_c is temperature where the exothermal shows a peak and χ is the cooling rate (°C/min). $t_{1/2}$ represents the time for the system to reach 50% of relative degree of crystallinity and it is used to evaluate the rate of crystallization of different systems (Calcagno et al., 2007; Jiann-Wen, 2008).

The $t_{½}$ values of nanocomposites are smaller than that of neat PET. The 90/10 PET/8 K-P (organoclay) exhibits the smallest $t_{½}$ (1.45 min) and its crystallization rate is 1.55 times than that of pure PET. On the other hand 90/10 PET/11K-M (organoclay) exhibits the smallest $t_{½}$ (1.35 min) and its crystallization rate is 1.67 times than that of pure PET.

Wang et al. (2004) studied the non-isothermal melt crystallization behavior of pristine PET and PET/clay nanocomposites of different viscosities by using DSC, according to different kinetic models namely, Avrami analysis modified by Jeziorny, the Ozawa model and method developed by Mo. Non- isothermal crysatllization dynamics show that the nanocomposites of PET have greater crystallization rates than that of pure PET.

Crystallization morphology and isothermal crystallization kinetics of PET/clay naocomposites were studied by Wan et al. (2004). In an isothermal crystallization process, initially, some crystallites in the nanocomposite are rod shaped and later, exhibited three dimensional growths. Neat PET has typical spherulitic superstructure with a spherulitic size of about 20 μm, whereas, nanocomposites are irregularly shaped with a size of 5 μm, and they interlock with others, without clear boundaries. Avrami equation was used in a kinetic analysis of the isothermal crystallization process, that is:

$$X_c(t) = 1\text{-}\exp(-kt)^n$$

where $X_c(t)$ is the weight fraction of the crystallites at time t, k is the crystallization rate constant; and n is the Avrami exponent. k and n are very useful for determining the crystallization mechanism.

Phang et al. (2004) studied the heterogeneous nucleation effects of clay layers on cold and melt crystallization and melting behavior of PET. They prepared PET/clay nanocomposites by simple melt compounding with different clay loadings. DSC heating-cooling cycles showed that the cold and melt crystallization enthalpies of the nanocomposites increase with increasing clay content. This indicates that, the nanoclay fillers induce the formation of crystals as heterogeneous nucleation sites, thus, promoting the crystallization rate and degree of crystallinity in the nanocomposites.

Calcagno et al. (2007) prepared PET nanocomposites using montmorillonite with different organic modifiers (cloisite 15A, 30B, and 10A) in order to evaluate the effect of clay organic modifies on the morphology and crystallization properties of the nanocomposites. The T_c of the PET nanocomposites was higher than that of pure PET. Higher T_c of the PET nanocomposites indicated that the clay nucleated the PET crystallization process, and the nucleating effect was higher when cloisite 10A was used. This is due to the presence of the aromatic ring in the modifier as in the PET chain, which favours the diffusion process, resulting in more sheets of the clay and more crystallization nuclei. The morphology of crystallization was observed by POM.

Calcagno et al. (2008) studied PP/PET blends system based on nanocomposites. They studied the crystallization behaviour and morphology of nanocomposites based on PP/PET blends. For this purpose, blend containing PP, PET and MMT were prepared in a twin screw extruder. Polymer blending is a convenient route for the development of new polymeric materials with excellent properties of more than one polymer. The PP/PET blend has importance due to some engineering applications. Several studies have been demonstrated that, clay, acting as a nucleating agent in the nanocomposites, increases the crystallization rate of PP/PET. However, the presence of clay in both blends (with and without compatibilizer, (PP-g-MA)) resulted in a decrease of crystallization rate of the PET phase. The crystallinity degree of PP was higher in the blends when the start point to PP crystallization.

Guan et al. (2004) studied the crystallization behavior of PET/clay nanocomposites, prepared through in situ polymerization method by using techniques DSC, POM and SEM. They concluded that different surface modification of MMT resulted in variation in crystallization behavior. Discrepant crystallization behavior was due to morphology of clay into the matrix, surface shielding effect of small organic surfactant molecule and the nucleating affects of metallic derivative released from MMT. So, the surface modification of clay affects the crystallization rate of PET nanocomposites.

After Phang et al. (2004), Ali et al. (2010) studied the cold and melt crystallization behaviours of PET/clay nanocomposites, using natural MMT (Na MMT) and organo-modified montmorillonite (org-MMT) with the same amount of inorganic content. Two nanocomposites samples were prepared by melt processing. DSC analysis revealed that dispersed clay layers enhanced nucleation by increasing the interfacial surface energy. From the kinetics study, it was reported that the melt crystallization rate of the sample containing Na-MMT was higher than org-MMT at cooling rate. On the basis of their results, they have concluded that the organic ammonium groups in the org-MMT decrease the crystallization rate of PET chains possibly by affecting the chain diffusion and folding.

Conclusion

In the PET/clay nanocomposites, clay particles play a nucleating role and have strong interactions with PET molecular chains. Nanoclays enhance the nucleation, possibly by decreasing the interfacial surface energy. Higher clay contents in PET phase retard the crystallization rate of PET due to the agglomeration of the organoclay in the PET matrix.

ACKNOWLEDGEMENT

Authors would like to thank Netaji Subhas Institute of Technology, New Delhi for financial support.

REFERENCES

Ahmadi SJ, Huang YD, Li W (2004). Review, Synthetic routes, properties and future applications of polymer-layered silicate nanocomposites. J. Mater. Sci., 39: 1919-1925.

Akane O, Arimitsu U (2006). Twenty Years of Polymer-Clay Nanocomposites. Macromol. Mater. Eng., 291: 1449–1476.

Ali D, Nevra E, Goksu S, Huseyin D, Ahmet K (2010). Nonisothermal crystallization kinetics of Poly(ethylene terephthalate)/clay nanocomposites prepared by melt processing. Polym. Compos., pp. 1056-1066.

Ashwini BN, Shertukde VV (2009). Preparation and Characterization of Poly(butylene terephalate) Nanocomposites with Various Organoclays. J. Appl. Polym. Sci., 113: 585-592.

Baniasadi H, Ramazani ASA, Javan SN (2010). Investigation of in situ polypropylene/clay nanocomposites properties and comparing to melt blending method. Mater. Des., 31: 76-84.

Benetti EM, Causin V, Marega C, Marigo A, Ferrara G, Ferraro A, Consalvi M, Fantinel F (2005). Morphological and Structural characterization of polypropylene based nanocomposite. Polymer, 46: 8275-8285.

Calcagno CIW, Mariani CM, Teixeira SR, Mauler RS (2007). The effect of organic modifier of the clay on the morphology and crystallization properties of PET nanocomposites. Polymer, 48: 966-974.

Calcagno CIW, Mariani CM, Teixeira SR, Mauler RS (2008). The role of the MMT on the morphology and mechanical properties of PP/PET blends. Compos. Sci. Technol., 68: 2193-2200.

Cheng FO, Mong TH, Jia RL (2003). The nucleating effect of montmorillonite on crystallization of PET/Montmorillonite nanoocomposite. J. Polym. Res., 10: 127-132.

Cheng FO, Mong TH, Jia RL (2004). Synthesis and crystallization of Poly(ethylene terephalate) nanocomposites with organoclay. J. Appl. Polym. Sci., 91: 140-145.

Clois EP, Gary WB (2006). Physical properties of polymer/clay nanocomposites. Curr. Opin. Solid State Mater. Sci., 10: 73–80.

Defeng W, Chixing Z, Wei Y, Fan X (2006). Effect of Blending Sequence on the Morphologies of Poly(butylene terephalate)/Epoxy/Clay Nanocomposites by a Rheological Approach. J. Appl. Polym. Sci., 99: 340-346.

Garcia-Lopez D, Picazo O, Merino JC, Pastor JM (2003). Propylene-Clay naocomposites: Effect of compatibilizing agents on clay dispersion. Eur. Polym. J., 39: 945-950.

Giannelis EP (1996). Polymer-layered Silicate Nanocomposites. Adv. Mater., 8: 29-35.

Gonzalez I, Eguiazabal JI, Nazabal J (2006). Nanocomposites based on polyamide 6/maleated styrene-butylene-co-ethylene-styrene blend: Effects of clay loading on the morphology and mechanical properties. Eur. Polym. J., 42: 2905-2913.

Guohu G, Chuncheng L, Xuepei Y, Yaonan X, Xiaoqing L, Dong Z (2004). New Insight into the Crystallization Behavior of Poly(ethylene terephthalate)/Clay Nanocomposites. J. Polym. Sci. Part B: Polym. Phys., 46: 2380-2394.

Hong CK, Kim MJ, Oh SH, Lee YS, Nah C (2008). Effects of propylene -g-(maleic anhydride/styrene) compatibilizer on mechanical and rheological properties of polypropylene/clay naocomposites. J. Ind. Eng. Chem., 14: 236-242.

Jiann-Wen H (2008). Poly(butylene terephalate)/Clay Nanocomposite compatibilized with Poly(ethylene-co-glycidyl methacrylate). II. Nonisothermal Crystallization. J. Polym. Sci: Part B: Polym. Phys., 46: 564-576.

Ke Y, Long C, Qi Z (1999). Crysatllization, properties and crystal and nanoscale morphology of PET-clay Nanocomposites. J. Appl. Sci., 71: 1139-1146.

Kusmono ZA, Mohd I, Chow WS, Takeichi T, Rochmadi (2008). Compatibilizing effect of SEBS-g-MA on the mechanical properties of different types of OMMT filled polyamide 6/polypropylene nanocomposites. Composites: Part A(39): 1802-1814.

Lei SG, Hoa SV, Ton-That M-T (2006). Effect of clay types on the processing and properties of polypropylene nanocomposites. Compos. Sci. Techonol., 66: 1274-1279.

Mahmood M, Musa RK (Dec. 2002). Polymer-Clay Nanocomposites Based on Blends of Polyamide-6 and Polyethylene. Can. J. Chem. Eng. 80: 1083-1092.

Manias E (2007). “Nanocomposites”. Nat. Mater., 6: 9-11.

Mazeyar P, Siamak M, Abosaeed R, Mohamad-Esmail Y (2010). Effect

of the Addition of Modified Nanoclays on the Surface Properties of the Resultant Polyethylene Terephthalate/Clay Nanocomposites. Polymer-Plastics Technol. Eng., 49: 874–884.
Miray M, Ulku Y (2008). Processing and properties of modified polyamide 66-organoclay naocomposites. J. Appl. Sci., 108: 3890-3900.
Naoki H, Hirotaka O, Makoto K, Arimistu U (1998). Preparation and Mechanical Properties of Polypropylene-Clay Hybrids Using a Maleic Anhydride-Modified Propylene Oligomer. J. Appl. Polym. Sci., 67: 87-92.
Naoki H, Hirotaka O, Makoto K, Arimistu U (2000). Preparation and Mechanical Properties of Polypropylene-Clay Hybrids Based on Modified Polypropylene Organophilic Clay. J. Appl. Polym. Sci., 78: 1918-1922.
Nicolas F, Elodie S, Michel S, Christian B, Jean-Marc L, Jacques D. (2007). Polymer-based nanocomposites: Overview, applications and perspectives. Progr. Organic Coatings, 58: 87-95.
Pettrain V, Frontini PM, Rodriguez VJR, Lopes MD, Valelzuela FD (2008). Polyethylene/(organo-montmorillonite) composites modified with ethylene/methacrylic acid copolymer: Morphology and Mechanical properties. Composites: Part A(39): 1822-1828.
Quang TN, Donald GB (2006). Preparation of polymer-clay nanocomposites and their properties. Adv. Polym. Technol., (25)4: 270-285.
Roohollah DF, Ahmad RSA (2008). Melt preparation of properties of toughened Polyamide 66 with SEBS-g-MA and their nanocomposites. Mater. Des., 29: 105-111.
Seung WH, Hee SM, Jong SB, Eui SY, Seung SIM (2000). Synthesis and Crystallization Behaviors of Modified PET Copolymers. Fibers Polym., 1(2): 76-82.
Suprakas SR, Musami O (2003). Polymer/layered silicate nanocomposites: A review from preparation to processing. Prog. Polym. Sci., 28: 1539-1641.
Tjong SC (2006). Structure and mechanical properties of polymer nanocomposites. Mater. Sci. Eng., 53: 73-197.
Tong W, Ling C, Yang CC, Xuehong L (2004). Crystalline morphology and isothermal crystallization kinetics of Poly(ethylene terephalate)/Clay nanocomposites. J. Appl. Polym. Sci., 94: 1381-1388.
Yaming W, Changyu S, Haimei L, Qian L, Jingbo C (2004). Nonisothermal melt crystallization kinetics of Poly(ethylene terephalate)/Clay nanocomposites. J. Appl. Polym. Sci., 91: 308-314.
Yee P, Pramoda KP, Tianxi L, Chaobin H (2004). Crystallization and melting behavior of polyester/clay nanocomposites. Polym. Int., 53: 1282-1289.
Yeh W, Feng BC, Yann CL, Kai CW (2004). Melt processing of propylene/clay nanocomposites modifies with maleated propylene compoatibilize. Composites: Part B(35): 111-124.
Young-Wook C, Suksoo K, Youngbin K (2005). Poly(butylene terephalate)-clay nanocomposites prepared by melt intercalation: Morphology and thermomechanical properties. Polym. Int., 54: 348-353.

A kinetic and mechanistic study on the oxidation of arginine and lysine by hexacyanoferrate (III) catalysed by iridium (III) in aqueous alkaline medium

Anjali Goel* and Ruchi Sharma

Department of Chemistry, KGM, Gurukul Kangri University, Post - Jwalapur, Hardwar-249407 (UK) India.

The kinetics of Ir (III) catalysed oxidation of some amino acids like arginine and lysine by hexacyanoferrate (abbreviated as HCF) (III) ions in aqueous alkaline medium at constant ionic strength 0.5 mol dm^{-3} and temperature 35ºC has been studied spectrophotometrically. The reactions exhibit 2:1 stoichiometry and follow first order kinetics in [HCF (III)] and [alkali]. The dependence of the rate on substrate concentration has been found to be of Michaelis-Menten type. The ionic strength of the reaction mixture shows positive salt effect on the reaction rate. To calculate thermodynamic parameters, the reactions have been studied at four different temperatures between 35 to 50ºC. A complex mechanism involving the complex formation between catalyst and the substrate has been proposed. Keto acids; ε-guanidino-α-oxo valeric acid and 6 – amino-α-oxo caproic acid have been identified chromatographically and spectroscopically as the final product of oxidation of arginine and lysine, respectively. Based on the kinetic data and product analysis a reaction mechanism is proposed.

Key words: HCF (III), iridium (III), oxidation, arginine, lysine.

INTRODUCTION

Oxidation reactions are of fundamental importance in nature, and are key transformations in organic synthesis. Oxidation of α- amino acids is of great importance both from chemical point of view and its bearing on the mechanism of amino acids metabolism (Devra Vijay, 2005). Amino acids have been oxidised by a variety of reagents under different experimental conditions (Goel and Sharma, 2010; Goel et al., 2002; Gupta et al., 1986; Harihar et al., 1999; Hemmige et al., 2003; Jeffery et al., 1996; Jose et al., 2006).

It was shown by the previous workers that the oxidation of α - amino acids by hexacyanoferrate (III) poceeds very slowly in the absence of any catalyst while it follows a complex kinetics in the presence of a catalyst (Kapoor et al., 1991). The oxidation rate was improved by the use of some metal ions: Os, Ru and Ag (Mahadevappa et al., 1982; Mahanti and Laloo, 1990). These reactions follow a complex kinetics in the presence of catalyst (Naik et al., 2008; Patil et al., 2009).

Thus to understand about the catalysis of Ir (III) in oxidation of amino acids by HCF (III) and to explore the mechanism of these oxidations in aqueous alkaline medium, two amino acids like arginine and lysine have been selected as substrate for oxidation.

EXPERIMENTAL

All chemicals and reagents used were of AR grade. All solutions and reaction mixture were prepared in double distilled water. Absorbance was recorded on Systronic UV-vis spectrophotometer. λ_{max} for the reaction mixture was 420 nm at which the absorbance was noted only in the period in which the λ_{max} did not change and no precipitate/turbidity appeared. $IrCl_3$ $10H_2O$ (SRL) was prepared by dissolving the sample in dil. HCl. The final strength of iridium trichloride was kept 3.35 × 10^{-3} mol dm^{-3}. The kinetic experiments

*Corresponding author. E-mail: anjaligoel10@gmail.com.

were carried out by mixing the required quantity of amino acid solution maintained at constant temperature with solution of HCF(III), NaOH, KCl and iridium (III) chloride kept at the same temperature. The mixture and stock solution of amino acid was then clamped in a thermostate at 35 ± 0.1°C. After 0.5 h, a required amount of amino acid solution was added to the mixture and stirred to start the reaction. Aliquots were withdrawn from the reaction mixture after repeated intervals of 5 min and the absorbance was recorded. Initial rates $(dA/dt)_i$ were evaluated after 5 min from the start of the reaction by plane mirror method and pseudo first order rate constant 'k_1' were calculated by Guggenheim's method.

The stoichiometry of the reaction was studied by estimating the amount of HCF (II) ions produced after definite interval of time with standard solution of ceric (IV) sulphate using ferroin as redox indicator. Estimation of the residual oxidant showed that one mole of substrate consumed two moles of hexacyanoferrate (III), corresponding to the stoichiometry.

$$RCH(NH_2)COOH + 2[Fe(CN)_6]^{-3} + 2OH^- \rightarrow RCOCOOH + NH_3 + 2[Fe(CN)_6]^{-4} + H_2O.$$

Using the same experimental conditions that were used for the kinetic determinations, solutions of substrate and oxidant, and NaOH (ionic strength adjusted by the addition of the requistic amount of KCl) were mixed and kept at atmospheric conditions for about 24 h. The main reaction products were identified as keto acid and ammonia. Ammonia was identified by Nessler's reagent (Patil et al., 2009) and keto acid by the following methods. The reaction mixture was extracted with diethyl ether and then concentrated. The concentrated extract was subject to TLC which shows the presence of single white product, keto acid (Puttaswamy and Vaz, 2001; Salem 2006). The concentrated extract was evaporated at room temperature. A solid residue was left. It was analysed by melting point determination, spot test analysis and I. R. spectroscopy. The melting point of the two ketoacids- ε - guanidino - α -oxo valeric acid and 6 – amino - α -oxo caproic acid separated are 214 and 211°C, respectively (Literature value 216 and 212°C, respectively. The I R bands at frequency 1644 cm^{-1}(carbonyl group) and 1632 cm^{-1} (acid group) for arginine and 1635 cm^{-1}(carbonyl group) and 1614 cm^{-1} (acid group) for lysine shows the presence of keto acid group in the final products extracted.

RESULTS AND DISCUSSION

Kinetic experiments were made at different concentration of one reactant keeping the concentration of others constant.

The concentration of substrate (lysine and arginine, abbreviated as S) was varied in the range of $1 \times 10^{-3} - 10 \times 10^{-3}$ mol dm^{-3} at 35°C keeping all other reactants concentration constant (Table 1). The data presented in Table 1 shows first order dependence on lower concentration of substrate which tends to be zero order at its higher concentration. The effect of $[OH^-]$ on the rate of reaction was studied at constant [S], [HCF (III)] and ionic strength at 0.5 mol dm^{-3} at 35°C (Table 1). A perusal of data reveals first order kinetics with respect to OH^- concentration. The [HCF (III)] was varied in the range $2 \times 10^{-4} - 7 \times 10^{-4}$ mol dm^{-3} at fixed [S], $[OH^-]$ and ionic strength. The k_1 values indicate that the order in [HCF(III)] is unity (Table 1). The effect of iridium (III) concentration on the oxidation of amino acids was also studied. The results presented in Table 1 clearly reveal first order kinetics with respect to Ir (III) concentration in the concentration range of $0.67 \times 10^{-5} - 3.01 \times 10^{-5}$ mol dm^{-3}.

The effect of ionic strength was studied by varying the [KCl] in the reaction mixture. The ionic strength of the reaction medium was varied from 0.5 to 0.7 mol dm^{-3} at constant [HCF (III)], [S], and $[OH^-]$. The positive salt effect was found (Table 2).

To calculate the thermodynamic parameters, the effect of temperature on the rate of the oxidation was studied by carrying out the reaction at four temperatures 35, 40, 45 and 50°C. Linear plots of log k_1 vs 1/T indicate that the reaction obeys Arrhenius equation. The values of energy of activation Ea, enthalpy of activation $\Delta H^{\#}$, entropy of activation $\Delta S^{\#}$ and free energy of activation $\Delta F^{\#}$ were evaluated as given in Table 3. The value of $\Delta F^{\#}$ are approximately the same for both the reactions studied, indicating that common mechanism is followed in these amino acids oxidation. Large negative values of entropy of activation suggest the formation of a solvated and charged transition state.

Mechanism

Keeping in view of the above experimental results, the following mechanistic path has been suggested for the oxidation of arginine and lysine by hexacyanoferrate (III) in aqueous alkaline medium in presence of Ir (III)

$$RCHNH_2COOH + OH^- \underset{}{\overset{K}{\rightleftharpoons}} RCHNH_2COO^- + H_2O \quad (1)$$

$$RCHNH_2COO^- + IrCl_6^{3-} \overset{K_1}{\rightleftharpoons} \underset{C_1}{[RCHNH_2COO\ldots IrCl_6]^{4-}} \quad (2)$$

$$[RCHNH_2COO\ldots IrCl_6]^{4-} + Fe(CN)_6^{3-} \overset{K_2}{\rightleftharpoons} \underset{C_2}{[RCHNH_2COO^-\ldots IrCl_6\ldots Fe^{3+}]} \quad (3)$$

$$[RCHNH_2COO^-\ldots IrCl_6\ldots Fe^{3+}] \xrightarrow[slow]{k} Cl^- + IrCl_5^{4-} + Fe^{2+} + product \quad (4)$$

$$IrCl_5^{4-} + 2Fe(CN)_6^{3-} + H_2O \xrightarrow{fast} [IrCl_5.H_2O]^{2-} + 2\,Fe(CN)_6^{4-} \quad (5)$$

$$[IrCl_5.H_2O]^{2-} + Cl^- \xrightarrow{fast} IrCl_6^{3-} + H_2O \quad (6)$$

Table 1. Effect of [S] [NaOH] [HCF (III)] on reaction rate at Temperature = 35 ± 0.1°C, μ=0.5 mol dm^{-3}.

[S] x10^3 (mol dm^{-3})	[NaOH] (mol dm^{-3})	[HCF(III)]x10^4 (mol dm^{-3})	[Ir(III)] x10^5 (mol dm^{-3})	k_1 x 10^4(sec^{-1}) Lysine	Arginine
1.0	0.40	3.00	3.35	2.68	2.68
2.0	0.40	3.00	3.35	3.83	3.07
3.0	0.40	3.00	3.35	4.66	4.20
4.0	0.40	3.00	3.35	5.75	4.60
5.0	0.40	3.00	3.35	6.90	5.37
6.0	0.40	3.00	3.35	7.67	6.14
7.0	0.40	3.00	3.35	8.06	6.52
8.0	0.40	3.00	3.35	8.44	6.90
9.0	0.40	3.00	3.35	8.82	7.29
10.0	0.40	3.00	3.35	9.21	7.67
3.0	0.10	3.00	3.35	1.92	2.67
3.0	0.20	3.00	3.35	3.07	3.83
3.0	0.30	3.00	3.35	3.83	4.60
3.0	0.40	3.00	3.35	4.98	6.14
3.0	0.50	3.00	3.35	5.75	6.52
3.0	0.40	2.00	3.35	3.07	3.83
3.0	0.40	3.00	3.35	4.20	4.66
3.0	0.40	4.00	3.35	4.70	5.75
3.0	0.40	5.00	3.35	5.37	5.75
3.0	0.40	6.00	3.35	6.14	6.14
3.0	0.40	7.00	3.35	6.52	6.52
3.0	0.40	3.00	0.67	6.14	10.74
3.0	0.40	3.00	1.01	6.90	3.83
3.0	0.40	3.00	1.34	7.29	4.60
3.0	0.40	3.00	1.67	7.67	5.75
3.0	0.40	3.00	2.09	8.44	6.52
3.0	0.40	3.00	2.34	9.21	8.06
3.0	0.40	3.00	2.67	9.59	8.82
3.0	0.40	3.00	3.01	9.97	9.59

Table 2. Effect of ionic strength on reaction rate, [HCF(III)] = 3.0 × 10^{-4} mol dm^{-3}, [S] =3.0 × 10^{-3} mol dm^{-3}, [Ir (III)] = 3.35 × 10^{-5} mol dm^{-3}, Temperature = 35 ± 0.1°C.

Parameter	k_1 x 10^4 (sec^{-1})	
μ x 10^1 mol dm^{-3}	Lysine	Arginine
0.5	3.07	2.30
0.55	4.22	3.07
0.60	5.76	4.22
0.65	7.29	5.37
0.70	9.60	5.75

as catalyst.

It is reported that Ir (I) and Ir (II) are the stable species of iridium, but in alkaline medium $[IrCl_6]^{3-}$ is the only reacting species of iridium (Singh et al., 1994; Srivastava et al., 1980). It is reported that the oxidation of amino acids involves the cleavage of N-H and C-H bond in the rate determining step (Shukla and Upadhyay, 2008). Based on the above facts it is assumed in the present study that substrate anion forms a loose bonded complex 'C_1' with iridium trichloride. The carbonyl oxygen of acid is most likely involved in the formation of complex C_1. In

Table 3. Activation parameters, [S] = 3.0 × 10^{-3} mol dm^{-3}, [HCF (III)] = 3.0 × 10^{-4} mol dm^{-3}, [Ir(III)] = 3.35 × 10^{-5} mol dm^{-3}, μ = 0.5 mol dm^{-3}.

T/K (± 0.1°)C	k_1 x 10^{-4} Lysine	Arginine
308	2.30	3.07
313	3.07	4.22
318	4.60	5.75
323	6.14	7.67
Ea kcal mol^{-1}	11.5	12.1
$\Delta H^{\#}$ kcal mol^{-1}	10.7	11.5
$\Delta S^{\#}$ e.u.	-28.3	- 25.9
$\Delta F^{\#}$ kcal mol^{-1}	18.3	19.6
A x10^7 l $mol^{-1}sec^{-1}$	1.16	3.21

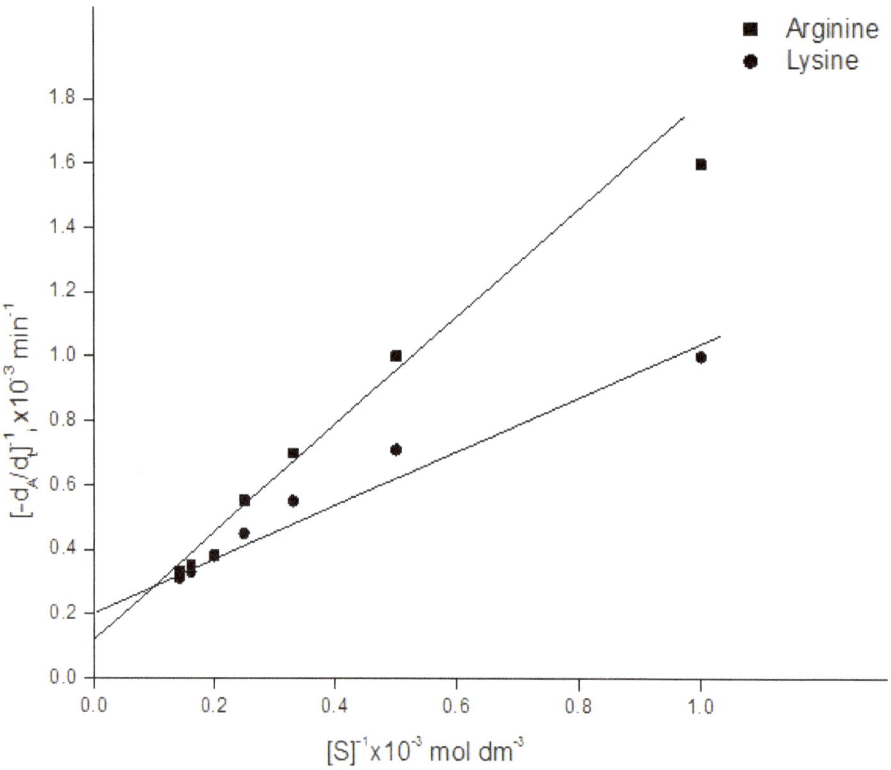

Figure 2. Plot between $rate^{-1}$ Vs $[S]^{-1}$

the next step the complex 'C_1' then combines with HCF (III) through electron abstraction to form complex (C_2). The complex (C_2) then slowly disproportionates into Ir^{1+} and HCF (II) along with final product. Ir^{1+} is reoxidized to Ir^{3+} by two moles of HCF (III) via one electron transfer process. The metal ion complexes with organic substrate to make the electron transfer easier (Shukla and Upadhyay, 2008).

The formation of the complex was proved kinetically by Michaelis – Menten plot that is, a non - zero intercept of the plot of 1/rate vs. 1/ [S] (Figure 1). The complex formation between oxidant and substrate was also reported in literature (Richard and Emmett, 1958).

The reaction rate(r) is measured in terms of rate of disappearance of HCF (III). According to step (III) the rate of disappearance of HCF (III) would be

$$rate = \frac{-d[HCF(III)]}{dt} = k\,[C_2] \qquad (i)$$

Now the total concentration of Ir^{3+} will be

$$[Ir^{3+}]_t = [Ir^{3+}] + [C_1] + [C_2] \qquad (ii)$$

Substituting the values of $[Ir^{3+}]$, (C_1) and (C_2) in Equation (i)

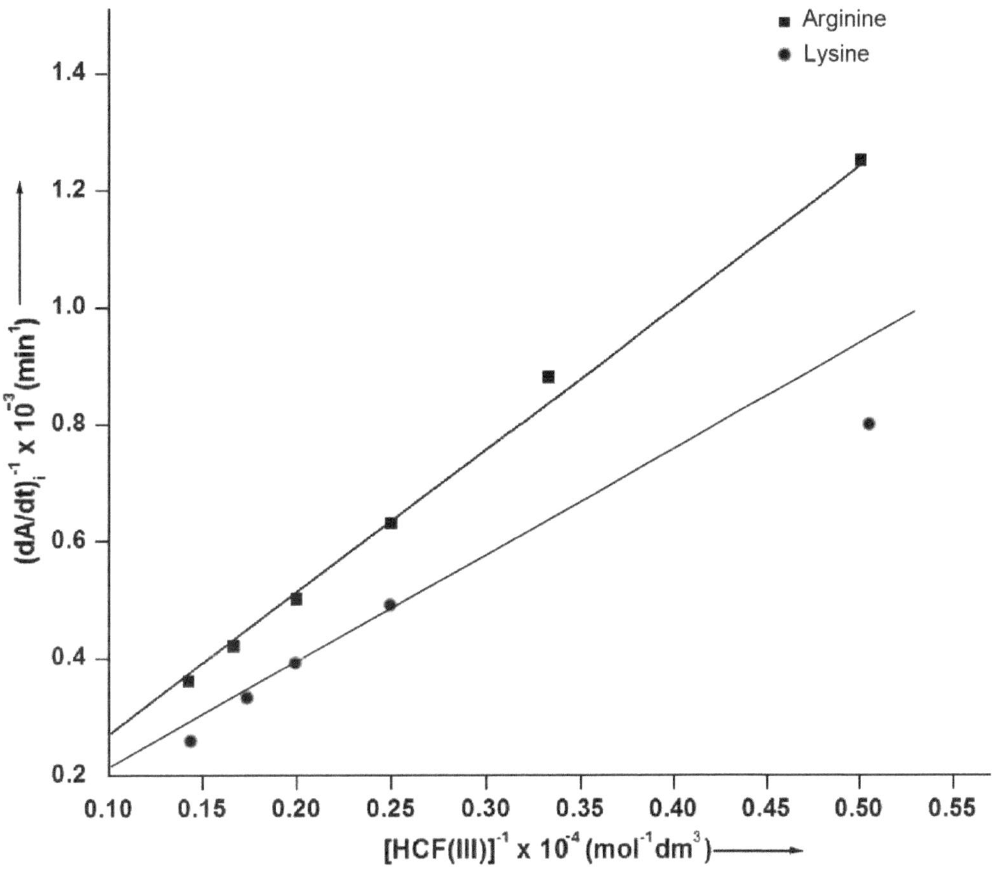

Figure 1. Plot between rate^{-1} Vs [HCF(111)]$^{-1}$

$$rate = \frac{kK_1'K_2\,[HCF(III)]\,[Ir^{3+}]_t\,[S]\,[OH^-]}{1 + K_1'[S][OH^-] + K_1'K_2[S][OH^-][HCF(III)]} \tag{iii}$$

Where $K_1' = KK_1$

At low concentration of HCF (III), S and OH$^-$, eq. (iii) reduces to

$$r = k\,K_1'\,K_2\,[HCF\,(III)]\,[Ir^{3+}]_t\,[S]\,[OH^-] \tag{iv}$$

The rate law (iv) clearly accounts for the first order kinetics with respect to HCF (III), organic substrate, hydroxide ion and catalyst at their lower concentrations. In order to verify this law (iii) at higher concentration of above said reactants it could be re-written as-

$$\frac{1}{r} = \frac{1}{kK_1'K_2\,[HCF(III)]\,[Ir^{3+}]_t\,[S]\,[OH^-]} + \frac{1}{kK_2\,[HCF(III)]\,[Ir^{3+}]_t} + \frac{1}{k\,[Ir^{3+}]_t} \tag{v}$$

This Equation (v) indicates that the plot of 1/rate vs. 1/[HCF(III)] & 1/rate vs. 1/[S] should give a straight line with positive intercept at 1/rate axis. Such plots are presented in Figures 1 and 2 for arginine and lysine, respectively. A close examination of these figures clearly indicate that these are evidently straight lines with positive intercept at 1/rate axis. The rate constant of slow step k, the ionization constant K_1' of the first step and the formation constant of the complex 'K_2' of the second step of all organic substrate at four different temperatures (35, 40, 45 and 50°C) were calculated from intercept and slope of the straight line plots between rate^{-1} vs. [HCF(III)]$^{-1}$ and rate^{-1} vs. [S]$^{-1}$ for both the organic substrate (arginine and lysine). Data are presented in Tables 4a and b for equilibrium constants and thermodynamic parameters, respectively. The constancy in K_1', K and K_2 values clearly validates the derived rate law equation on the basis of proposed mechanism. A comparison of the later values with those obtained for the slow step of the raection shows that these values mainly refer to the rate limiting step.value of $\Delta S\#$ suggests that the intermediate complex is more ordered than the reactants.

The observed modest activation energy and sizeable entropy of activation supports a complex transition state in the reaction. The observed modest enthalpy of activation and a relatively low value of entropy of activation indicate that the oxidation presumably occurs via an inner – sphere mechanism.

Table 4a. Effect of temperature on equilibrium constants.

Temperature ± 0.1	Arginine		Lysine	
(°K)	K_1'	K_2	K_1'	K_2
308	24.0	520.8	124.6	2.69
313	58.3	666.6	144.3	2.74
318	70.8	793.5	180.8	2.94
323	78.6	1235.0	196.7	2.99

Table 4b. Thermodynamic parameters

Parameter	Arginine		Lysine	
	value from K_1'	value fromK_2	value from K_1'	value from K_2
Ea (kcal mol^{-1})	8.01	9.81	33.55	10.2
$\Delta S^{\#}$ (e.u.)	-27.92	-16.36	56.22	-15.32
$\Delta H^{\#}$ (kcal mol^{-1})	7.36	9.17	32.91	9.573
$\Delta F^{\#}$ (kcal mol^{-1})	16.18	14.33	15.17	14.40

REFERENCES

Devra Vijay (2005). Kinetics and mechanisem of Silver (I) catalyzed oxidation of alanine by cerium (IV) in perchloric acid medium. J. Ind. Chem. Soc., 82: 290-294.

Goel A, Sharma S (2010). Mechanistic study of the oxidation of L-phenyl alanine by hexacyanofferrate (III) catalyzed by Iridium (III) in aqueous alkaline medium. Trans. Met. Chem., 35:549-554.

Goel A, Verma GR, Singh HS (2002). Kinetics and mechanisem of Iridium (III) Chloried catalyzed oxidation of ethylene glycol and methyl glycol by hexacyanofferrate (III) in aqueous alkaline medium J. Indian Chem. Soc., 79:665.

Gupta N, Naik RM, Nigam PC (1986). Kinetics and mechanisem of pentacyanohydroxoferrate (II) formation of FeL complex (L= triethyenetetraaminehexaacetic acid and 2-hydroxy-1,3 diamino propanetetraacetic acid. Ind. J. Chem., 25A:39.

Harihar AL, Kanbhari MR, Nandebewoor ST (1999). Kinetics and mechanisem of N-bromo succinimide oxidation of L-arginine in aqueous alkaline medium J. Indian Chem. Soc., 76:128.

Hemmige SY, Chapuradoddi RR, Kikkeri NM (2003), Kinetics and mechanisem of L- Isoleucine and L-Ornithine Hydrochloride by Sodium N-Bromobenzenesulphonamide in Perchloric Acid Medium. Turk J. Chem., 27:571-580.

Jeffery GH, Bassett JM, Mendham J, Denny RC (1996). Vogel"s Test book of quantitative chemical analysis (Fifth ed.), ELBS, Longman, U K, (a) pp. 455, (b) pp. 371.

Jose TP, Nandibewoor ST, Tuwar SM (2006). Kinetics and mechanisem of the oxidation of vanillin by Hexacyanoferrate (III) in aqueous alkaline medium. J. Sol. Chem., 35(1):51-62.

Kapoor RC, Mehrotra RN, Vajpai SK, Chaudhary P (1991). Kinetics and mechanisem of the oxidation by metal ions.PartXI (1) Kinetics of Osmium (VIII) –catalyzed oxidation of Glycol by alkaline hexacyanofferrate (III), Trans. Met. Chem., 16:65.

Mahadevappa DS, Rangappa KS, Gouda NNM, Thimmegouda B (1982). Kinetics and mechanistic studies of the oxidation of Arginine and Histidine by Sodium N-Chloro-4-methyl benzene sulphonamide (Chloramine-T) Iin Hydrochloric Acid Medium. Int. J. Chem. Kinet., 14:1183.

Mahanti MK, Laloo D (1990). Kinetics of oxidation of amino acids by alkaline hexacyanofferrate (III), J. Chem. Soc., Dalton. Trans., 311-313.

Naik PN, Chimatadar SA, Nandibewoor ST (2008). A kinetics and mechanistic study of oxidation of tyrosine by Chromium (VI) in aqueous perchloric acid medium. Trans. Met. Chem., 33: 405-410.

Patil RK, Chimatadar SA, Nandibewoor ST (2009), Oxidation of thiosulphate by hexacyanofferrate (III) in aqueous perchloric acid medium- A kinetic and mechanistic study. Ind. J. Chem., 48A:357-361.

Puttaswamy, Vaz N (2001). Proc. Kinetics of oxidation of acidic amino acids by Sodium N-bromobenzenesulphonamide in acid medium-A mechanistic approach. Indian Acad. Sci. Chem. Sci., 113:325-332.

Richard J, Emmett L (1958). A manual of paper chromatography and paper electrophoresis. 2nd edition academic press inc. Publisher – New York.

Salem JKJ (2006). Kinetics of the oxidation of Phenylhydrazine by [Fe(CN)-36] in water –in-oil Microemulsions. J. Dispersion. Sci. Technol., 27:795-798.

Shukla R, Upadhyay SK (2008). Non–ionic micellar inhibition on the rate of oxidation of 1-istidine by alkaline hexacyanofferrate (III), Ind. J. Chem., 47A:551-555.

Tween 60 – amido black 10B – ascorbic acid system: Studies of photogalvanic effect and solar energy conversion

K. R. Genwa* and C. P. Sagar

Department of Chemistry, Jai Narain Vyas University, Jodhpur – 342005, India.

Solar energy conversion and storage by use of photogalvanic cell was studied in Tween 60 – amido black – ascorbic acid system. Photogalvanic effect was studied in photogalvanic cell consisting of ascorbic acid as a reductant, Amido black 10B as photosensitizer and Tween 60 as surfactant in aqueous solution. The photopotential and photocurrent generated were 996.0 mV and 420 μA respectively. The observed power of the cell was 84.68 μW at its power point. The conversion efficiency and fill factor was determined as 1.62 and 0.38% respectively. The cell performance was 130 min. in dark. A mechanism for the photogeneration of electricity has also been proposed. The effect of different parameters (pH, diffusion length, electrode area, light intensity) on the electrical output of the cell were also been observed.

Key words: Photopotential, photocurrent, conversion efficiency, amido black 10B, Tween 60, ascorbic acid.

INTRODUCTION

The photogalvanic cell produces an electromotive force (voltage) upon absorbing photon. This cell works on photogalvanic effect. Rideal and Williams (1925) first of all reported the photogalvanic effect but it was systematically investigated by Rabinowitch (1940A and B). Later on this field get momentum by the work of some other workers (Hann et al., 1973; Clark and Eckert, 1975; Albery and Archer, 1977; Goldstein et al., 1980; Hamdi and Aliwi, 1996). Construction of a photogalvanic cell using photosynthetic reaction center complex and Cytochrome c2 from Rhodosprillium rubrum by Matsumoto et al. (2001). Progress in conversion of solar energy directly to electricity by means of the multilayer ion-thionine photo-galvanic system is). Spectral studies on the interaction of safarnine T, phloxin, fluorescien with inorganic ions their photogalvanic effect was studied by Ghosh and Bhattacharya (2002). Bisquert et al. (2004) have reviewed the physical chemical principal of dye-sensi-tized solar cells, whereas Meyer (2005) has presented the molecular approaches to solar energy conversion.

Singh and Mishra (2005) also reported studies of photochemical conversion. Photogalvanic studied for solar energy conversion in fuchsine basic-EDTA system reported by Ameta et al. (2006). Recently, some new photogalvanic systems with appreciable outputs reported using various dyes with surfactants (Meena and Singh, 2006; Chhagan Lal, 2007; Genwa and Mahaveer, 2008; Meena et al., 2009; Genwa et al., 2009; Genwa and Kumar, 2010; Yadav and Lal, 2011). Present system is the effort to observe the effect of neutral surfactant (Tween 60) on the electrical output of the cell in presence of Amido Black 10B as photosensitizer in photogalvanic cell for solar energy conversion.

MATERIALS AND METHODS

Amido Black 10B (LOBA Chemie pvt. Ltd, Mumbai), Tween 60 (LDH), Ascorbic acid (SISCO, Mumbai) and NaOH (RANKEM) were used in the present work. Structure and properties of Amido Black 10B, Tween 60 are shown Figure 1. All solutions were prepared by direct weighing in double distilled water and kept in amber cloured container to protect them from direct light. Photogalvanic effect of dye was studied using H- shaped glass tube which consist known amount of the solution of Amido Black 10B, Tween 60, Ascorbic acid, NaOH and distilled water so as to keep total volume of the

*Corresponding author. E-mail: krg2004@rediffmail.com.

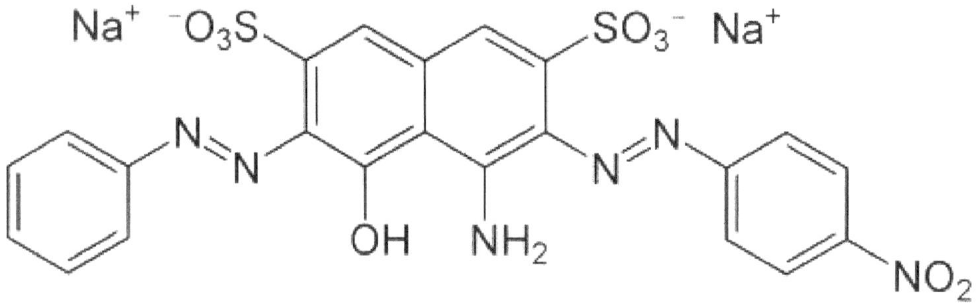

(a)

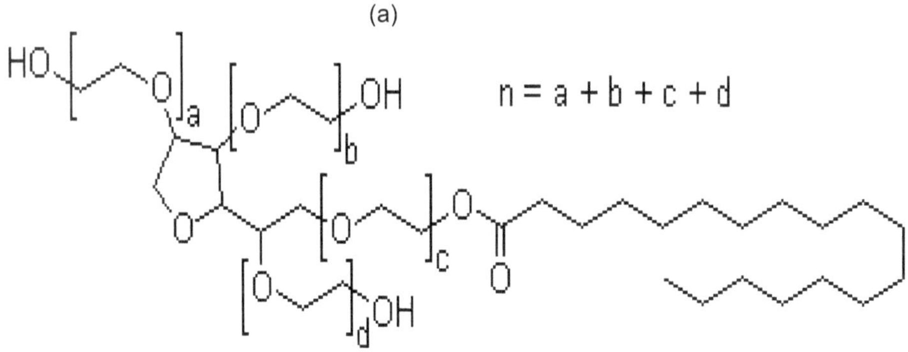

(b)

Figure 1. (a) Structure of Amido Black 10B; Solid crystalline black power, Molecular formula - $C_{22}H_{14}N_6Na_2O_9S_2$, Molecular weight - 616.49, Soluble in water, Maximum absorption (λ_{max}) – 610 nm (b) Tween 60 (Polyoxyethylene sorbitan monostearate); A pale yellow semisolid liquid, molecular formula - $C_{24}H_{46}O_6(C_2H_4O)_n$, Molecular weight – 1312, Soluble in water.

mixture always 25.0 ml. NaOH is used to maintained the pH of solution. A Platinum electrode (1.0×1.0 cm^2) was dipped in one limb and a saturated calomel electrode (SCE) immersed in another limb of the H- tube. In present system, we have used two electrode system; platinum electrode as a working electrode and SCE as counter electrode. The terminals of the electrodes were then connected to a digital pH meter. The entire system was first placed in dark until a stable potential was reached, then the limb containing platinum electrode was exposed to the light source (projector Tungsten lamp). The light intensity was varied by employing tungsten lamp of different wattage. A water filter was placed between the illuminated chamber and the light source to cut off thermal radiations. The photochemical bleaching of dye (Amido Black 10B) were studied potentiometrically. Absorption spectra of dye-surfactant combination have been taken by using Systronics Spectrophotometer 106 with the matched pair of silica cuvetts (path length 1 cm). All spectral measurements were duplicated in a constant temperature water bath maintained with in ±1°C and mean values were processed for data analysis. Over all experimental setup of cell assembly (Genwa and Kumar, 2010) is shown in Figure 2.

RESULTS AND DISCUSSION

Absorption properties of Amido Black 10B – Tween 60

Amido black10b shows absorption peak (λ_{max}) in visible region with maximum at 610 nm. Maximum absorption was recorded at Amido Black 10B - Tween 60 combination of concentration 1.1×10^{-5} M + 0.80×10^{-3} M. The changes observed in the spectra shown in Figure 3.

Effect of variation of pH

Photogalvanic cell containing Amido Black 10B – Tween 60 – Ascorbic acid system was found to be quite sensitive to pH of the solution. The photopotential and photocurrent is increased with increase pH value (in alkaline range) of the cell. At pH 11.17, a maxima was obtained. On further increase in pH, there was a decrease photopotential and photocurrent. The optimum electrical output was obtained at particular pH value (11.17). It may be due to better availability of reductant's donor form at that pH value. The results showing the effect of pH are summarized in Table 1.

Effect of variation of dye (Amido Black 10B), surfactant (Tween 60), and reductant (Ascorbic acid) concentration

Results showing the effect of dye, surfactant and reductant concentration are given in Table 2. It was observed that the

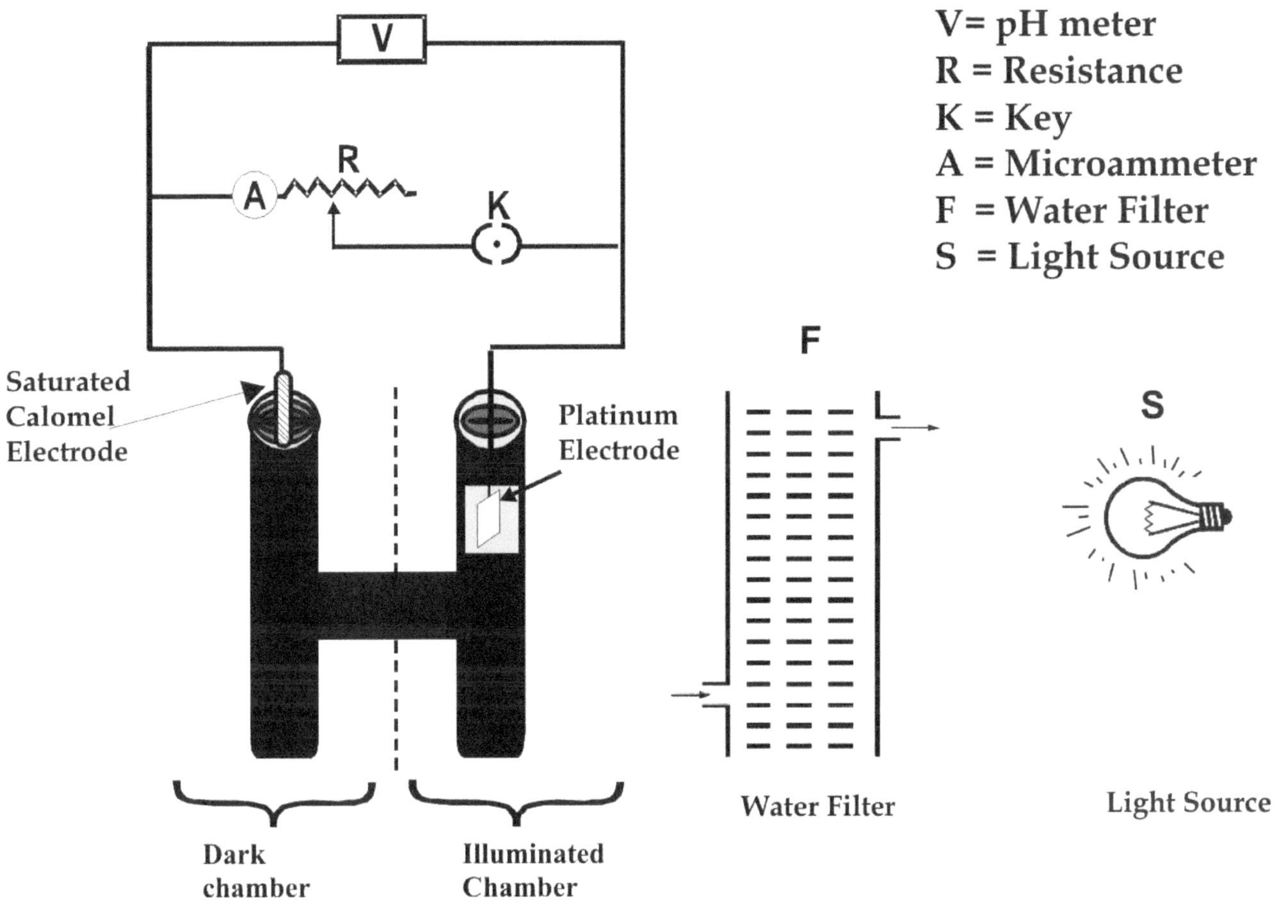

Figure 2. Set up of photogalvanic cell assembly.

photopotential and photocurrent of Amido Black 10B – Ascorbic acid system for the better performance of the photogalvanic cell proper concentration of dye was needed. The Photopotential and photocurrent increased with increase in Amido Black 10B concentration. A maxima was obtained at certain dye concentration (1.1×10^{-5} M). On further increase in the concentration of dye, a fall in the electrical output was observed. At the lower concentration range of dye, there are only a few dye molecules to absorb the major portion of the light and therefore, there was a low electrical output, while higher concentration of dye again resulted in a decrease in electrical output due to intensity of light reaching the molecule near the electrode decrease due to absorption of the major portion of the light by the dye molecules present in the path, therefore corresponding fall in the electric output of the cell.

Tween 60 was used as a Surfactant in the system. The photopotential and photocurrent of the cell was increased on increasing the concentration of Tween 60 reaching a maximum value at 0.8×10^{-3} M. On further increase in Tween 60 concentration, a fall in electrical output of cell was observed. Most important properties of surfactants (miceller systems) are the ability to solubilize a variety of molecules and substantial catalytic effect on chemical reaction in the cell.

The electrical output of the cell was affected by the variation of reductant (ascorbic acid) concentration. It was observed that photopotential and photocurrent was found to increase to maximum value and then decrease in electrical output because fever reductant molecules were available for electron donation to dye molecules. Higher concentration of reductant again resulted in a decrease in electrical output, because large numbers of reductant molecules hinders the dye molecule reaching close to electrode in the desired time limit.

Effect of diffusion length and electrode area

The diffusion length is the distance between the two electrodes in H-Cell. The effect of variation of diffusion length on the current parameters of the cell (i_{max}, i_{eq}) and initial rate of generation of photocurrent are studied using H-Shaped cell of different diameters. i_{max} found to increase with diffusion length. It is observed that in the first few minutes of illumination there is sharp increase in the photocurrent. The conductivity of electroactive species

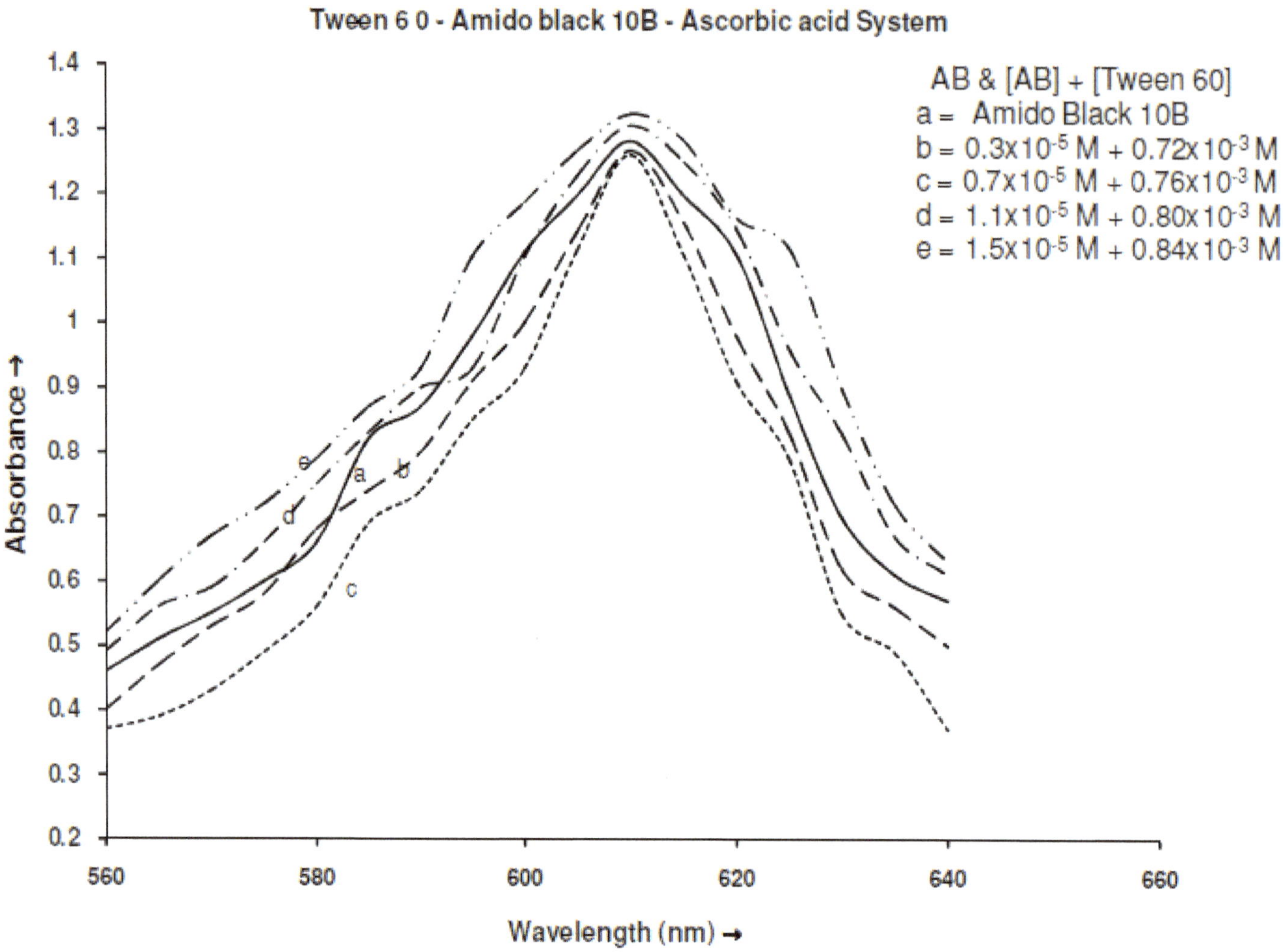

Figure 3. Absorption spectrum of dye and Dye + Surfactant.

Table 1. Effect of pH.

pH	Photopotential (mV)	Photocurrent (μA)
11.09	812	334
11.13	884	385
11.17	996	420
11.21	943	407
11.25	892	386

[a]Tween 60 = 0.80 × 10^{-3} M, [b]Amido Black 10B = 1.1 × 10^{-5} M, [c]Ascorbic Acid = 1.72 ×10^{-3} M, [d]Light Intensity = 10.4 mW cm^{-2}, [e]Temp. = 303 K

depends on its population between electrodes. As diffusion length increased, the volume of dye solution and intern population of dye molecule(Amido Black 10B) increased leading higher i_{max} the electroactive nature of dye/dye$^-$is provide by the fact that i_{max} increase with diffusion length. There for it may be concluded that the main electro active species are the leuco or semi form of dye- and the dye in the illumination and the dark chamber respectively. The reductant its oxidation products act only as electron carriers in the path the result are summarized in Table 3.

Effect of light intensity

The effect of light intensity was studied by using tungsten

Table 2. Effect of variation of Amido Black 10B, Tween 60, ascorbic acid concentration.

Concentration	Photopotential(mV)	Photocurrent(μA)
	[Amido Black 10B] × 10^{-5} M	
0.3	794	295
0.7	902	389
1.1	996	420
1.5	882	381
1.9	775	309
	[Tween 60] × 10^{-3} M	
0.72	758	248
0.76	889	366
0.80	996	420
0.84	865	329
0.88	780	279
	[Ascorbic acid] × 10^{-3} M	
1.64	715	238
1.68	891	356
1.72	996	420
1.76	860	284
1.80	810	231

[a]Light Intensity = 10.4 mW cm^{-2}, [b]Temp. = 303 K, [c]pH = 11.17.

Table 3. Effect of diffusion length.

Diffusion length D_L(mm)	Maximum Photocurrent i_{max}(μA)
50	477
55	489
60	500
65	511
70	522

[a]Tween 60 = 0.80 × 10^{-3} M, [b]Amido Black 10B = 1.1 × 10^{-5} M, [c]Ascorbic Acid = 1.72 × 10^{-3} M, [d]Light Intensity = 10.4 mW cm^{-2}, [e]Temp. = 303 K, [f]pH = 11.17

lamps of different wattage. It was observed that photocurrent showed a linear increasing behavior with the increase in light intensity whereas photopotential increase in logarithmic manner. This increasing behavior of electrical output with increase in light intensity was due to increase in number of photons with increase in light intensity, The observed results showing dependence of light intensity on photopotential and photocurrent of the cell are reported in Figure 4.

Current – voltage (i-V) characteristics

The open circuit voltage (V_{oc}) and short circuit current (i_{sc}) of the photogalvanic cell were measured under the continuous illumination of light, with the help of digital pH meter (keeping the circuit open) and a micro ammeter (keeping the circuit closed), respectively. The potential and current in between this two extreme values (V_{pp} and i_{pp}) were recorded with the help of a carbon pot (log 407K) connected in the circuit of micro ammeter, through which an external load applied on it. i-V characteristics of the cell containing Amido Black 10B – Tween 60 – Ascorbic acid system is shown as a curve in Figure 5. It was observed that i-V curve deviated from its regular rectangular shape. A point in the i-V curve, called power point (pp), was determined where the product of current and potential was maximum. With the help of i-V curve, the fill-factor was calculated as 0.38 using the formula:

$$\text{Fill-factor } (ff) = \frac{V_{pp} \times i_{pp}}{V_{oc} \times i_{sc}} \qquad (1)$$

Storage capacity (Performance) and conversion efficiency of the cell

The performance was determined in term of $t_{1/2}$ i.e. the time required in fall of the power output to its half at power point in dark. It was obtained with time- power curve

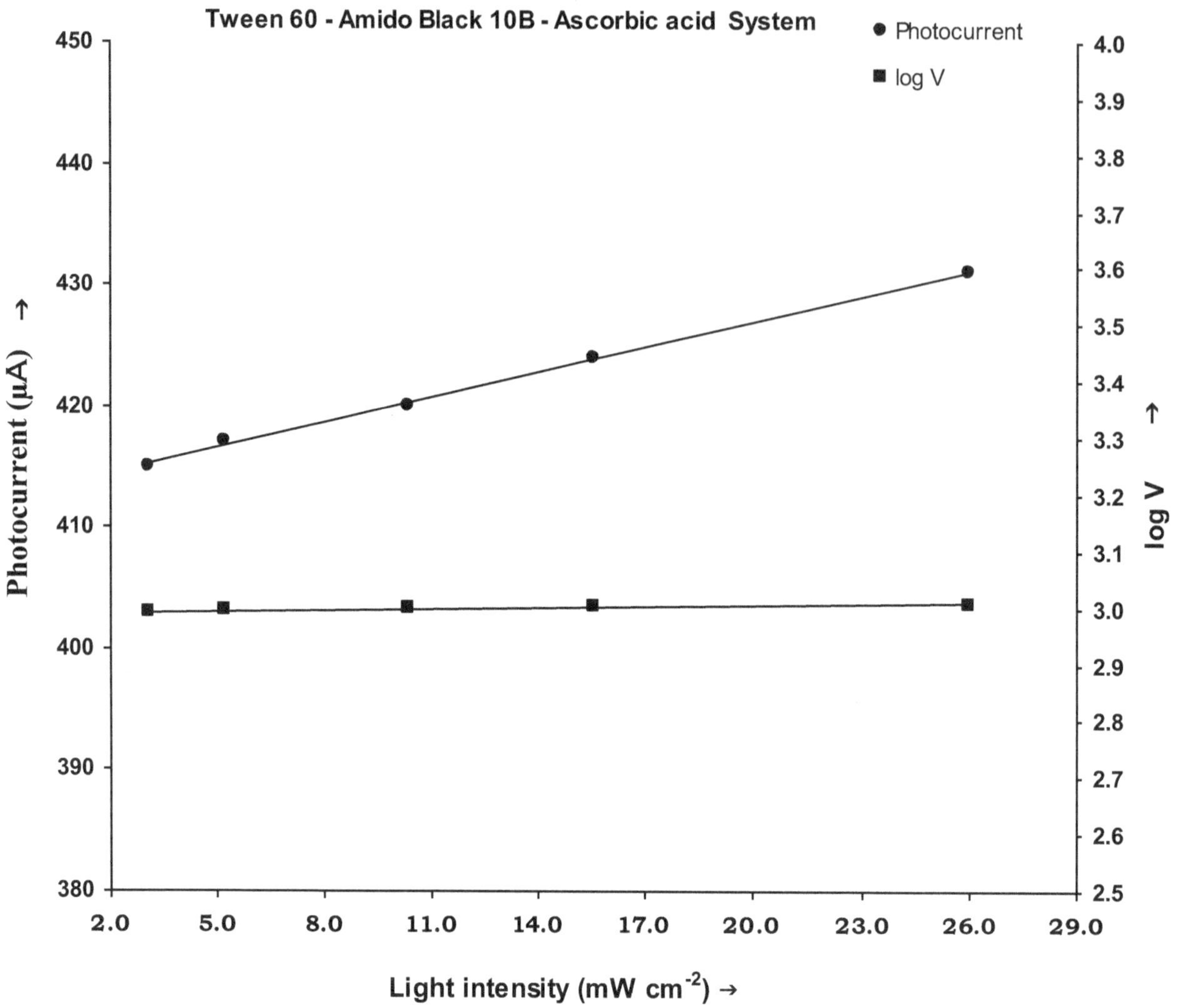

Figure 4. Variation of photocurrent and photopotential (log V) with light intensity.

(Figure 6). It was observed that the cell can be used in dark at its power point for 130 min using Amido Black 10B – Tween 60 – Ascorbic acid system in photogalvanic cell. The conversion efficiency of the cell was determined as 1.62% using the formula:

$$\text{Conversion efficiency} = \frac{V_{pp} \times i_{pp}}{A \times 10.4 \text{ mWcm}^{-2}} \times 100\% \qquad (2)$$

where, V_{pp}, i_{pp} and A are the potential at power point, current at power point and electrode area, respectively.

Role of Surfactant (Tween 60)

The surfactant head can be charged (anionic or cationic), dipolar (zwitterionic), or non-charged. The Tween 60 has non charged head (neutral). The micellization is the important property of surfactant, by this surfactant are able to solubilize a variety of molecules and substantial catalytic effect on many chemical reactions (Fendler and Fendler, 1975). Photoinduced electron transfer processes in miceller system are potentially important for efficient energy conversion and storage due to surfactant micelles help to achieve the separation of photoproducts by the hydrophilic- hydrophobic interaction of the products with the micellar interface (Genwa and Sonel, 2009). Dye has tendency of interacting with surfactant micelle of opposite charges. In the opposite charge dye and surfactant, a strong dye-surfactant complex is formed in which dye molecule is surrounded by surfactant micelles in some regular geometry which retards the intermolecular twisting resulting in an enhancement of fluorescence. Groenen et al. (1984) studied an Iron-thionin photogalvanic cell with addition of neutral micelles (TX-100), Jana (2000) studied the molecular interaction of Phenazine dyes in aqueous solution of Triton X-100 and observed that dye form charge transfer or electron-doner-

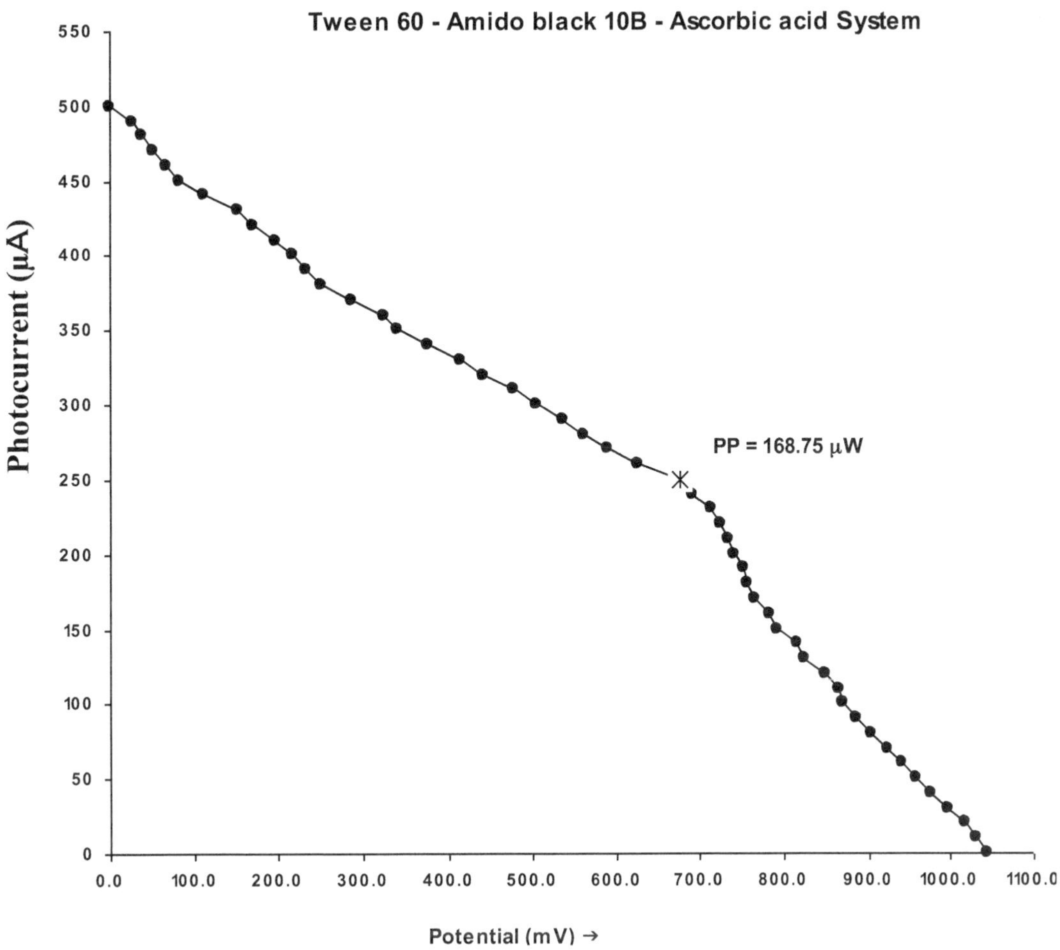

Figure 5. Current – potential (i-V) curve of the cell.

acceptor complex with surfactant (Triton X-100).

Mechanism

When certain dyes are excited by the light in the presence of electron donating substance (reductant), the dyes are rapidly changed into colorless form. The dye now acts as on powerful reducing agent and can donate electron to other substance and reconverted to its oxidised state (Oster, 1962). On the basis of earlier studies, a tentative mechanism in the photogalvanic cell may be proposed as follows:

Illuminated chamber

On irradiation, dye molecule get excited:

$$D \longrightarrow D^* \quad (1)$$

Ground state First excited state

The excited dye molecule accepts an electron from reductant and converted into semi or leuco form of dye, and the reductant into its excited state:

$$D^* + R \longrightarrow D^-(\text{semi or leuco}) + R^+ \quad (2)$$

Triplet state

At platinum electrode

The semi or leuco form of dye loses an electron and converted into original dye molecule:

$$D^- \longrightarrow D + e^- \quad (3)$$

Dark chamber

At counter electrode:

$$D + e^- \longrightarrow D^- \ (\text{semi or leuco}) \quad (4)$$

Finally leuco/semi form of dye and oxidized form of reductant combine to give original dye and reductant molecule and the cycle will go on:

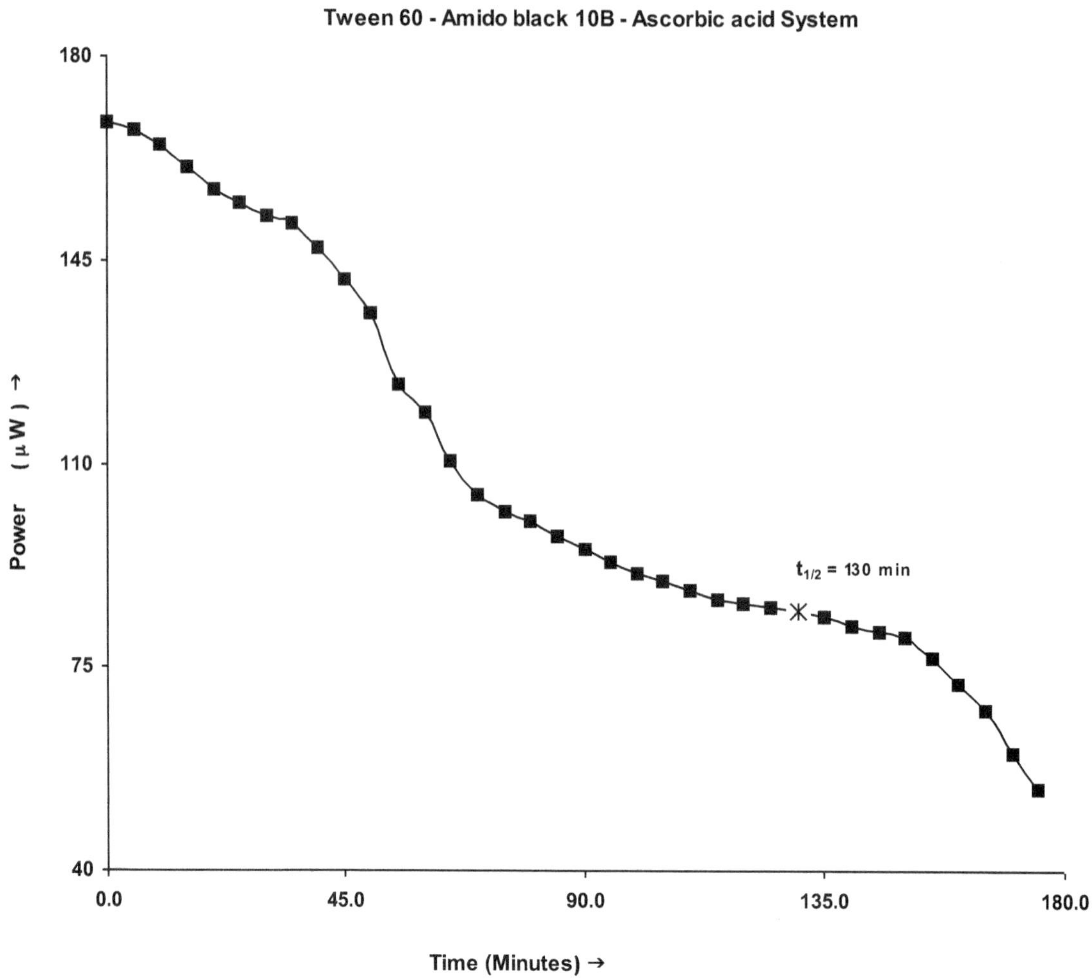

Figure 6. Time-power curve of the cell.

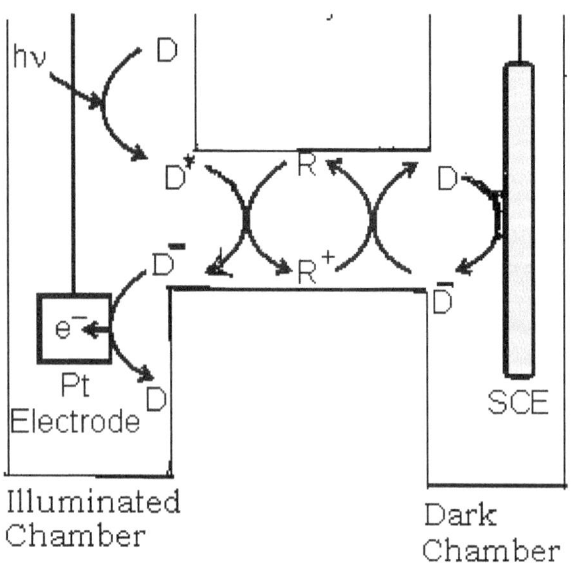

Figure 7. Schematic diagram of cell reaction mechanism. SCE=saturated Calomel Electrode. D=Dye (Photosensitizer). R=Reductant. ē=Electron.

$$D^- + R^+ \longrightarrow D + R \qquad (5)$$

Where D, D*, D^-, R and R^+ represents the dye, its excited form, reduced form of dye, reductant and its oxidized form respectively. Schematic representation of mechanism is shown in Figure 7.

Conclusions

The photogalvanic conversion of solar energy has attracted attention of scientists towards solar energy conversion and storage. This cell undergoes cyclical charging and discharging process. The charging of cell occurs only in presence of illuminating source. The discharging of cell takes place only when we apply the external circuit for electron transfer. As long as there is no external circuit, the cell will keep light energy stored. The photogalvanic cell have inbuilt storage capacity and stored energy can be used in absence of light where as photovoltaic cells needs extra hardware as batteries for energy storage, photogalvanic cells are economic than photovoltaic cells because low cost materials are used in these cells. The Conversion efficiency, storage capacity and fill factor are recorded as 1.62%, $t_{1/2}$130 min. and 0.38 respectively in Tween 60 – Amido Black 10B – Ascorbic acid system. These values of electrical outputs are more than values reported (0.048,17 min, and 0.24) in system with neutral surfactant Tween 80 (Gangotri et al., 1994) present work has given an impetus to explore the more suitable selection of surfactant, reductant and photosensitizer in order to its practical implementation.

ACKNOWLEDGEMENTS

The authors are thankful to Head, Department of Chemistry for providing necessary facilities and Dr. Pooran Koli for critical discussion and suggestions.

REFERENCES

Albery WJ, Archer MD (1977). Optimum efficiency of photogalvanic cell for solar energy conversion. Nature, 270: 399-402.

Ameta SC, Punjabi PB, Vardia J, Madhwani S, Choudhary S (2006). Use of bromophenol red-EDTA system for generation of electricity in a photogalvanic cell. J. Power Sources, 159: 747-751.

Bisquert J, Cahen D, Hodes G, Riihle S, Zaban A, (2004). Physical chemical principles of photovoltaic conversion with nanoparticales, mesoporous dye sensitized solar cell. J. Physical Chem. B., 108: 8106-8118.

Clark WDK, Eckert JA (1975). Photogalvanic cells. Solar energy, 17: 147- 150.

Fendler JH, Fendler EJ (1975). Catalysis in Micellar and Macromoleculer systems. Academic press, New York.

Gangotri KM, Kalla P, Lal C, Regar OP, Meena R (1994). Use of Tween 80 in photogalvanic cell for solar energy conversion and storage: Toluidine blue-Glucose System. J. Ind. Council Chem., 10: 19-22.

Genwa KR, Mahaveer (2008). Photogalvanic cell: A new approach for green and sustainable Chemistry. Solar Energy Meter. Sol. Cells, 92: 522-529.

Genwa KR, Kumar A, Sonel A (2009). Photogalvanic solar energy conversion: Study with photosensitizer toluidine blue and malachite green in presence of NaLS. Appl. Energy, 86: 1431-1436.

Genwa KR, Kumar A (2010). Role of Rhodamine B in photogalvanic generation using anionic surfactant in liquid phase photoelectrochemical cell for solar energy conversion and storage. J. Indian. Chem. Soc., 87(8): 933-939.

Ghosh JK, Bhattacharya SC (2002). Spectral studies on the interaction of safranine T, phloxin, fluorescien with inorganic ions and their photogalvanic effect. J. Indian Chem. Soc., 79: 225-230.

Goldstein S, Jaenicke S, Levanon H (1980). Photogalvanic cell based on the photolysis of rubidium anions in THF. Chem. Phys. Lett., 71(3): 490-493.

Groenen EJJ, De Groot MS, De Ruiter R, De Wit N (1984). Triton X-100 micelles in the ferrous/thionine photogalvanic cell. J. Phys. Chem., 88: 1449-1454.

Hamdi ST, Aliwi SM (1996). The photogalvanic effect of Fe(II)- beta-diketonate/thionine systems in aqueous acetonitrile', Chemie ISSN Monatshefte fuer 0026-9247, Springer, 127: 339-346.

Hann RA, Read G, Rosseinky DR, Wassey P (1973). Photogalvanic output from cells containing organic dyes. Nature Phys. Sci., 244: 126-127.

Jana AK (2000). Solar cells based on dyes. J. Photochem. Photobiol. A: Chem., 132: 1-17.

Chhagan L (2007). Use of mixed dyes in a photogalvanic cell for solar energy conversion and storage: EDTA–thionine–azur-B system. J. Power Sources, 164: 926-930.

Matsumoto K, Shuji F, Yoichi M, Masanuri W, Tatsuo E (2001). Concentration of photogalvanic cell using photosynthetic reaction center complex and Cytochrome c2 from Rhodospirillum rubrum. Denki Kagaku oyobi Kogyo Butsuri Kagaku, 69: 69: 340.

Meena RC, Kumari M, Pachwaria RB (2009). Role of rose Bangal as a photosensitizer for solar energy conversion and storage; Rose Bangal-Ascorbic acid system. Int. J. Energy Res., 31: 1- 9.

Meena RC, Singh G (2006). Use of some dyes for solar energy conversion and storage. Int. J. Chem. Sci. 4: 463-470

Oster G (1962). The Chemical effect of light, lasers and light. In; Arthur LS, WH Freeman and Co.:93.

Rabinowitch E (1940A). The photogalvanic effect Part I: The photogalvanic properties of the thionine-iron system. J. Chem. Phys., 8: 551.

Rabinowitch E (1940B). The photogalvanic effect Part II: The photogalvanic properties of the thionine-iron system. J. Chem. Phys., 8: 560.

Rideal EK, Williams DC (1925). Photogalvanic effect. J. Chem. Soc., 127: 258.

Singh K, Mishra SSD (2005). J. Indian Chem. Soc., 80: 401.

Yadav S, Lal C (2011). Efficient solar energy conversion and storage through photogalvanic cell based on EDTA: Brilliant green and fast green system. Int. J. Green Energy, 8(2): 265-274.

Meyer GJ (2005). Molecular approaches to solar energy conversion with coordination compound anchored to semiconductor surface. Inorganic Chem., 44: 6852-6864.

14

Factors controlling the structural properties of carbon nitride films deposited by electrochemical method

M. A. Moharram[1], F. H. Abd El-kader[2], Kh. I. Grais[1], M. G. Khafagi[1] and Fathia Mamdouh[1]*

[1]Spectroscopy Department, Physics Division, National Research Center, El Behooth Street, 12622 Doki, Giza, Egypt.
[2]Faculty of Science, Physics Department, Cairo University, Egypt.

Polycrystalline carbon nitride thin films were deposited on Si (100) substrates by electrolysis of methanol-urea solution at atmospheric pressure and low temperature. The effect of the process parameters, namely deposition time, concentration and applied voltage on growth and bonding state of carbon nitride (CN_x) thin films was studied using Fourier transform infrared (FTIR) spectroscopy, grazing incidence X-ray diffraction (GIXRD) and scanning electron microscope (SEM). The analysis of infrared (IR) spectra revealed that the nitrogen content in the film increased with increasing both urea concentration and the applied voltage. It found from GIXRD patterns that the crystalline structure of carbon nitride films at moderate nitrogen content changed from amorphous phase to polycrystalline α-C_3N_4, and β- C_3N_4 phases. At relatively higher nitrogen content, a new phase of CN_x appeared which increased gradually and became the more predominant phase. SEM images of the films indicated that the surface of the deposited films consists of homogeneous grains of ~0.2 μm and these grains became more closely packed with increasing nitrogen content.

Key words: Carbon nitride, thin films, structure properties.

INTRODUCTION

Ever since the first-principle pseudopotential calculation of novel superhard crystalline α-C_3N_4 and β-C_3N_4 by Liu and Cohen (1989) and Cohen (1985), the synthesis of these phases has been a challenging research in modern science. So far, five different structures of C_3N_4 have been proposed (Teter and Hemley, 1996): the hexagonal α-C_3N_4 and β-C_3N_4, the cubic and pseudocubic C_3N_4 phases and the graphitic C_3N_4. The properties of their high hardness, high thermal conductivity, high atomic density and wide band gap make them have wide potential applications in various fields, such as good cutting tools, excellent resistance to corrosion and wear, heat-dissipative substrates for devices to be used in short wavelength photodiodes. Over the past few years, different kinds of amorphous carbon nitride films have been synthesized. However, the methods known for depositing these films, such as particle-beam-assisted laser processing (Zocco et al., 1999), magnetron sputtering (Lopez et al., 1998) and ion-beam-assisted deposition (Hammer et al., 2000), were all vapor deposition. Compared to vapor deposition techniques, electrodeposition from a liquid phase has advantages such as the simplicity of the apparatus and the fact that this process can be performed at atmospheric pressure and near room temperature. Depositing the films at low temperatures also decreases the thermal stresses that can develop in the films and in the substrates as they are cooled. The electrodeposition process is also attractive because of its low production cost and easy scalability. Only in recent years have a few attempts been made to synthesis carbon nitride films by electrodeposition from various organic liquids under high voltages. Among these, Fu et al. (1999) reported the synthesis of an amorphous carbon matrix containing some mixed polycrystalline phases of α-C_3N_4 and β-C_3N_4. More recently, Kundoo et al. (2003) succeeded in depositing crystalline carbon nitride films by electrolysis of methanol-urea solution.

In carbon nitrides, carbon and nitrogen are bonded in different configurations, depending on the hybridization of

*Corresponding author. E-mail: F_nrc@yahoo.com.

Abbreviations: GIXRD, Grazing incidence X-ray diffraction; **CN_x,** carbon nitride; **FTIR,** fourier transform infrared; **SEM,** scanning electron microscope.

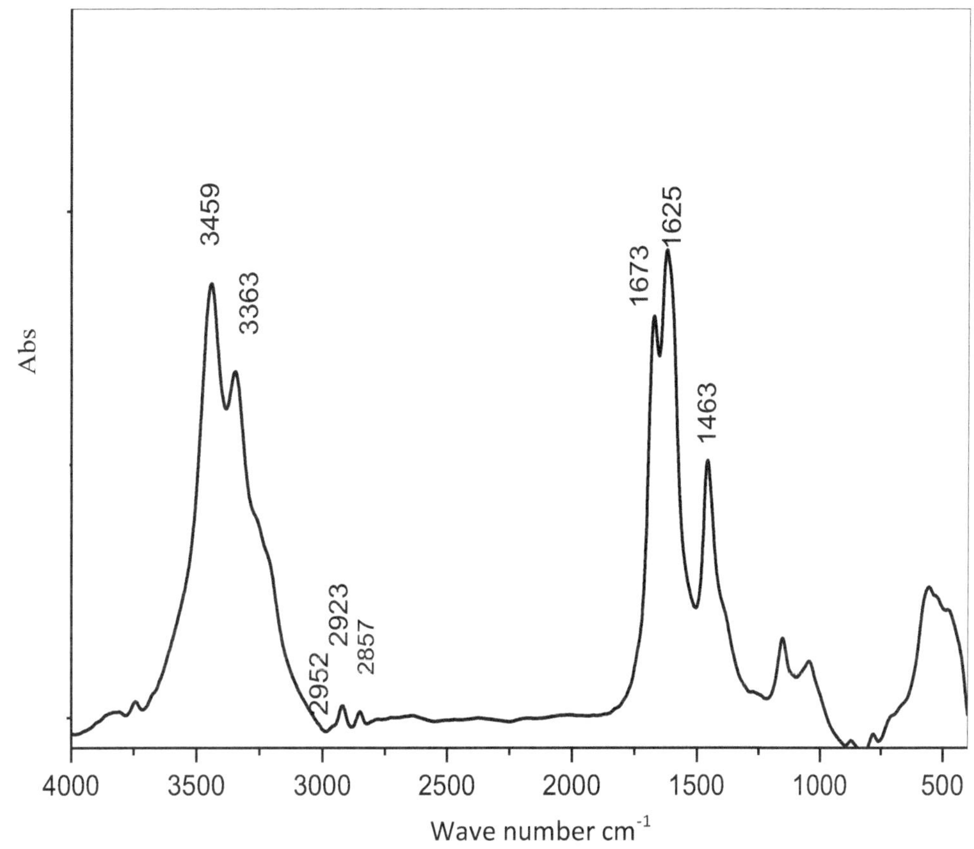

Figure 1. FTIR spectrum of sample deposited under concentration 2:1000 of urea to methoanol at 800 V for 10 h.

carbon (sp, sp^2 or sp^3). These bondings are known to have a major influence on the properties of the carbon nitride films. Therefore, study of the bonding in carbon nitrides prepared by electrochemical means is important.

In this paper, carbon nitride films were deposited on a silicon substrate by the electrolysis of methanol-urea solution under different conditions. Fourier transformation infrared (FTIR) spectrometry, grazing incidence X-ray diffraction (GIXRD), scanning electron microscope (SEM) with energy dispersive X-ray (EDX) were used in assessing the effects of the process parameters on the nature of the bonds in the film, and morphology of the deposited films.

Experimental

Preparation of carbon nitride (CN_x) films

A simple electrolytic cell system was used to prepare the CN_x films as used elsewhere (Kundoo et al., 2003; Yan et al., 2004). A substrate of Si (100) with a sheet resistance of about 7-13 Ωcm^{-2} was mounted on the negative electrode. A graphite plate was used as the counter-electrode. The span between the electrodes in all cases was kept as 6 mm. Urea ($CO(NH_2)_2$, purity above 99.9%) dissolved in analytically pure methanol (CH_3OH, purity above 99.5%) was used as the electrolyte. The electrolytic solution was prepared by mixing 0.075 (first), 0.15 (second) and 0.3 g (third) urea per 100 ml methanol. Prior to deposition, Si (100) was immersed in an aqueous solution of 5%HF for a few minutes to remove the native oxide layer, sequentially cleaned in an ultrasonic bath with deionized water, ethanol, and acetone, and then placed in the electrolytic solution, with an area of 1.0×1.5 cm^2 submerged to allow the deposition of the CN_x film at a temperature of about 60 + 2°C, a direct current (dc) power-supply voltage of 0-1200V by applying three different voltages 800, 1000 and 1200 V.

Measurements

FTIR was used in the mid range 4000-400 cm^{-1} on Jasco FT/IR-430. X'Pert panalatical GIXRD was performed in the range 10-90° using CuKα as X-ray source, operated at 40 kV/30 mA and an incidence angle of 0.5°. SEM of the model JEOL JXA-840A Electron probe micro analyzer was used to observe the morphology of the films.

RESULTS AND DISCUSSION

FTIR spectroscopy

The FTIR spectrum of the sample deposited with the concentration of mole ratio 2:1000 of urea to methanol under 800 V for 10 h is shown in Figure 1. The spectrum

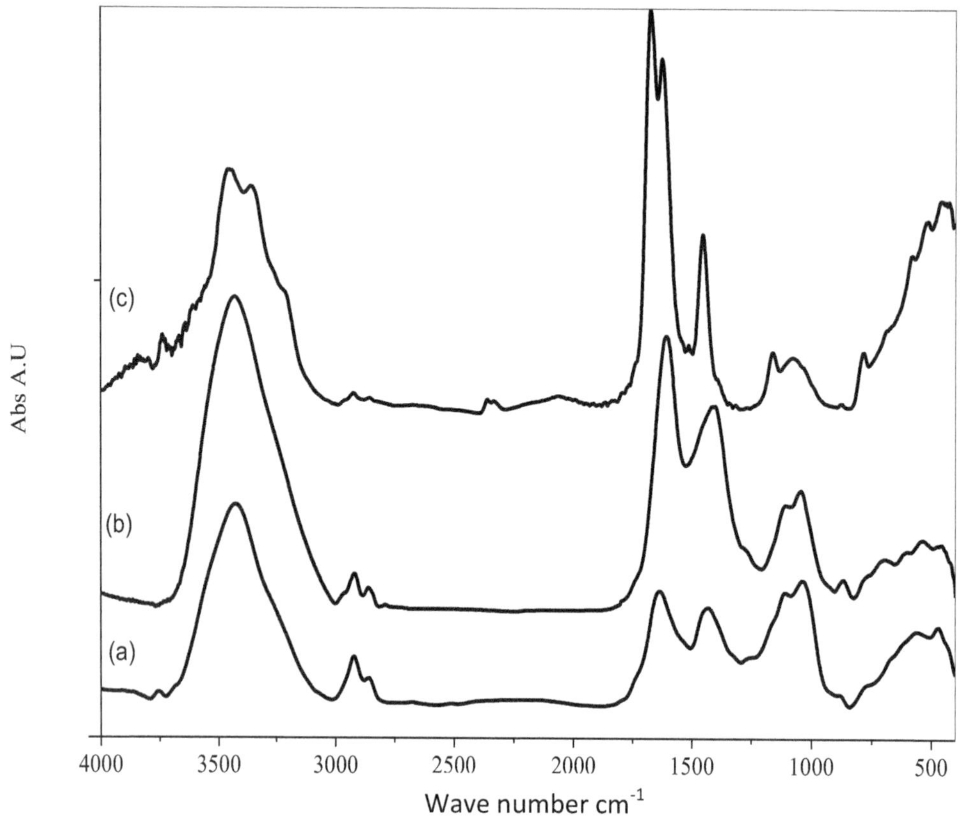

Figure 2. FTIR spectra of sample deposited with first (a) and third (c) concentrations under 800 V for 6 h.

exhibits absorption bands at 3459, 3363, 2952, 2923, 2857, 1673, 1625 and 1463 cm^{-1}. The assignments of these frequencies to their functional groups were carried out by comparison with the assignments reported in text books and literature. The peaks at 3459 and 3363 cm^{-1} are attributed to N-H stretching vibrations (Kundoo et al., 2003; Yan et al., 2004). The peaks at 2923 and 2857cm^{-1} are assigned as CH_2 asymmetric stretching vibration and CH_3 symmetric stretching vibration respectively (Kundoo et al., 2003; Yan et al., 2004). It can also be observed that, there is weak shoulder at 2952 cm^{-1} which may be due to CH_3 asymmetric stretching vibration. The spectrum exhibits a peak at 1625 cm^{-1}; this peak is due to the C=N stretching vibration (Majumdar et al., 2008). The presence of the peak at 1673 cm^{-1} indicates the existence of C=C bond. There is no significant peak attributed to C≡N. The peak at 1463 cm^{-1} is due to overlapping of C-H bending vibration and C-N single bond stretching vibration (Kundoo et al., 2006). The presence of these peaks proves that CN_x films were deposited on Si substrate.

FTIR spectra of carbon nitride films prepared at various concentrations of urea to methanol first, second and third concentrations at a constant voltage (800 V) for constant time (6 h) are shown in Figure 2. Visual examination of the spectra revealed reveals the following:

(1) In the region of the N-H stretching vibration, the spectrum of the sample deposited under the first concentration shows only a single broad band centered at 3436 cm^{-1}. The spectrum of the sample deposited under the second concentration displays the same features as before. On the other hand, the spectrum of the sample deposited under the third concentration exhibits two peaks in this region at 3436 and 3355 cm^{-1}.
(2) In the region of the C-H stretching vibrations the spectra of the samples deposited under both the first and the second concentrations show two definite absorption bands at 2923 and 2858 cm^{-1}, while in the case of the third concentration the absorption in this region appears very weak.
(3) In the region from 1700 to 1300 cm^{-1} the spectrum of the sample deposited under the first concentration shows two broad bands at 1634 and 1434 cm^{-1}, while in the case of the second concentration the spectrum displays these two peaks at 1604 and 1411 cm^{-1}. In case of the third concentration the recorded spectrum displays three bands at 1673, 1623 and 1454 cm^{-1}. The absence of both C-H stretching and C-H bending vibrations from the spectrum of the sample deposited at the third concentration resulted in the presence of a single stretching peak at 1454 cm^{-1} corresponding to the absorption of C-N stretching only.

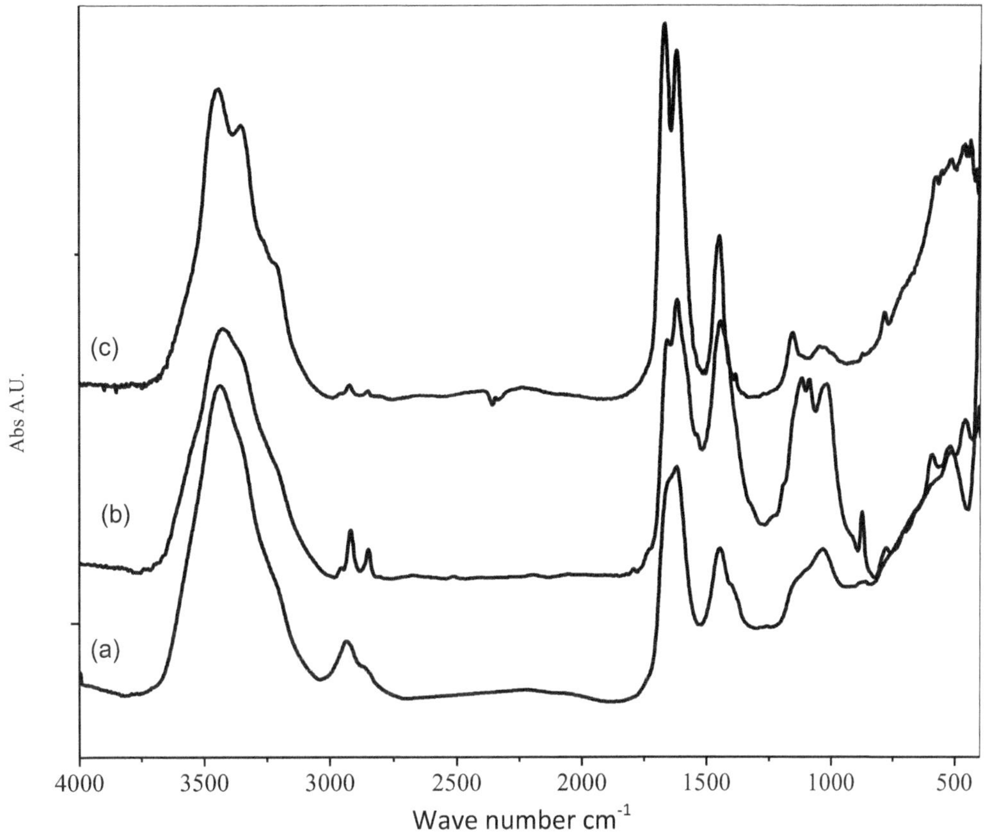

Figure 3. FTIR spectra of sample deposited with first (a) and third (c) concentrations under 1000 V for 6 h.

Based on the earlier mentioned observations, it could be concluded that increasing concentration from first to third concentration resulted in splitting the broad band of N-H stretching vibration into two peaks at 3450 and 3355 cm^{-1}, splitting the broad band at 1634 cm^{-1} into two sharp peaks at 1673 and 1625 cm^{-1} and shifting the peak appeared at 1431 cm^{-1} to 1454 cm^{-1}. It also appears that the latter peak became sharper. The decrease in the intensity of the C-H stretching vibrations and the increase in the number of N-H stretching vibrations may be caused by the increase in the nitrogen content in the film where NH_n group gradually replaces the CH_n group (Chengbing et al., 2008) and C=C becomes more active (Sreejith et al., 2005).

The appearance of the broad bands at 3436 and 1624 cm^{-1} in the spectra of the samples deposited under the first and the second concentrations suggests that the resulted carbon nitride material is in an amorphous state. While the splitting of these bands into doublets for the third concentration indicates that the amorphous state has been converted into crystalline state.

In this study the effect of the applied voltage has been taken into consideration. It is well known that increasing the electric field strength increases the concentration and energy of the precursor ions formed from the dissociation of the electrolyte, thereby affecting the incorporation of nitrogen in the amorphous carbon network.

The applied voltage was raised to 1000 V then to 1200 V. Figures 3 and 4 show the spectra of the samples deposited under various concentrations and applied voltages of 1000 and 1200 V for 6 h respectively. In case of the first and second concentrations, increasing the voltage from 800 to 1000 V resulted in an appearance of a shoulder at the higher frequency side of the band at 1666 cm^{-1}. While in case of the third concentration increasing the voltage to 1000 V resulted in an increase in the intensity of the 1666 cm^{-1} band relative to that of the 1625 cm^{-1} band.

In case of the first and the second concentrations, raising the voltage to 1200 V, resulted in no significant changes in the spectral features of the sample deposited at the applied voltage 1000 V. Raising the voltage to 1200 V with the third concentration produces the height of the band due to C=C observes equal to the height of the band due to C=N.

Furthermore the time of deposition was taken into account as a factor affecting the bond structure. The FTIR spectra of the samples deposited under different concentrations and 800 V for 10 h are shown in Figure 5. It is apparent from Figure 5a that increasing the time to 10 h produces no significant effects on the spectrum of the sample deposited under the first concentration

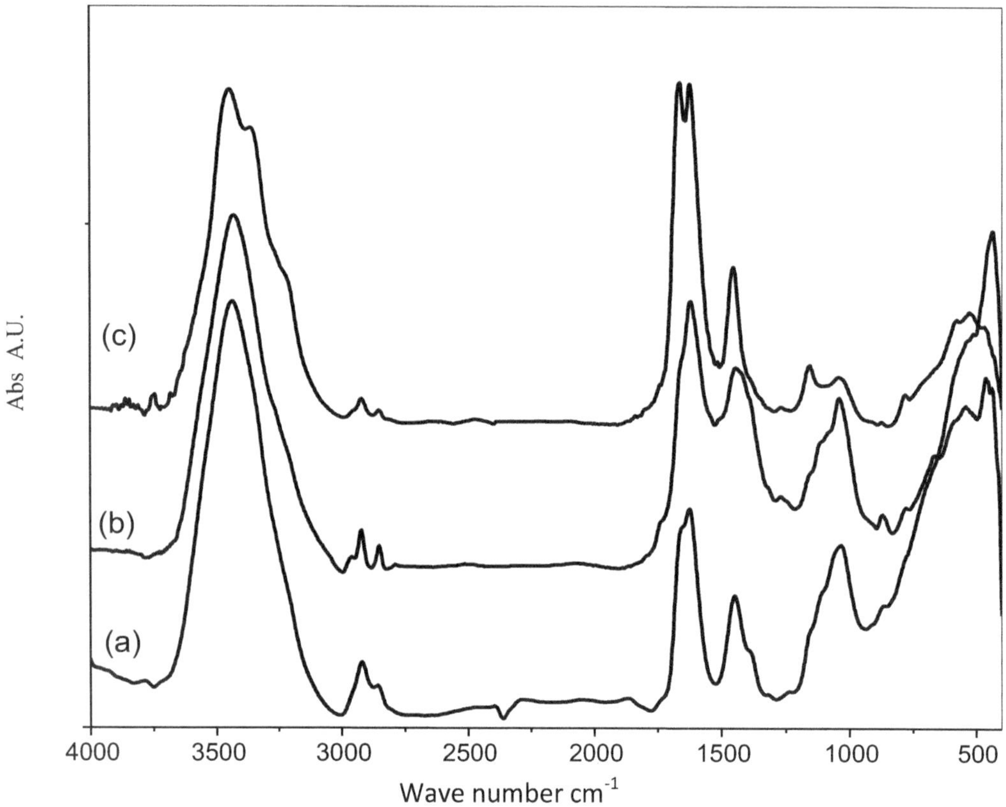

Figure 4. FTIR spectra of sample deposited with first (a), second (b) and third (c) concentrations under 1000 V for 6 h.

compared to that deposited for 6 h. For the case of the second concentration, Figure 5b shows that increasing the time to 10 h results in the appearance of two peaks at 3444 and 3336 cm^{-1}, and two bands at 1666 and 1623 cm^{-1} and a single sharp peak at 1450 cm^{-1}. In the case of the third concentration as shown in Figure 5c, the spectrum of the sample deposited for 10 h exhibits well defined peaks at 3443 and 3347 cm^{-1}, 1673 and 1625 cm^{-1} and sharp peak at 1463 cm^{-1}. It is evident from Figures 5b and 5c that the peak height of the band at 1625 cm^{-1} is lower than that of the band at 1666 cm^{-1}, in case of the second concentration; whereas in case of the third concentration, vice versa, that is, the peak height of the band at 1625 $cm^{-1}$is higher than that of the band at 1673 cm^{-1}. It is clear from the above mentioned observations that both time and concentration have significant effects on the concentration of C=C, C=N and C-N groups relative to each other. Thus increasing the time of deposition enhances the effect of the concentration; in addition, helps in the formation of crystalline phase.

Grazing incidence X-ray diffraction (GIXRD)

Figure 6 shows the GIXRD patterns of the films deposited by electrolysis of methanol urea organic solutions under various concentrations at applied voltage 800 V for 10 h. In case of the first concentration the deposited sample is mainly amorphous as shown in Figure 6a. This result confirms what was suggested by FTIR measurements. It is clear from Figure 6b that the GIXRD pattern of the sample deposited under the second concentration contains a number of diffraction peaks of different intensities at 2Θ= 31.64, 35.16, 38.08, 43.32, 48.04, 50.05, 52.56, 56.68, 62.96, 72.16, 74.24, and 79.48°. The pattern shows a number of diffraction peaks of different intensities at 2Θ= 22.25, 31.64, 35.16, 38.08, 43.32, 48.04, 50.05, 52.56, 56.68, 62.96, 72.16, 74.24 and 79.48°. The comparison between the experimentally measured lattice spacing of planes (*d*-values) and the theoretically predicted for the samples deposited under the second concentration assumes that, the peaks at 2Θ= 35.16, 48.04, 52.56, 56.68, 72.16 and 74.24° are matched to α–C_3N_4 while the peaks at 31.64, 43.32, 50.05, 62.96 and 79.48° are matched to β–C_3N_4 phase (Table 1). These peaks are similar to those reported in the literatures (GU et al., 1996; Zhang et al., 1999; 2000). The new additional peak at2Θ= 22.25° which appeared in the GIXRD pattern of the sample deposited under the third concentration may be due to a new phase of CN_x (Kundoo et al., 2003; Yan et al., 2004; GU et al., 1996). These findings give strong evidence that increasing the

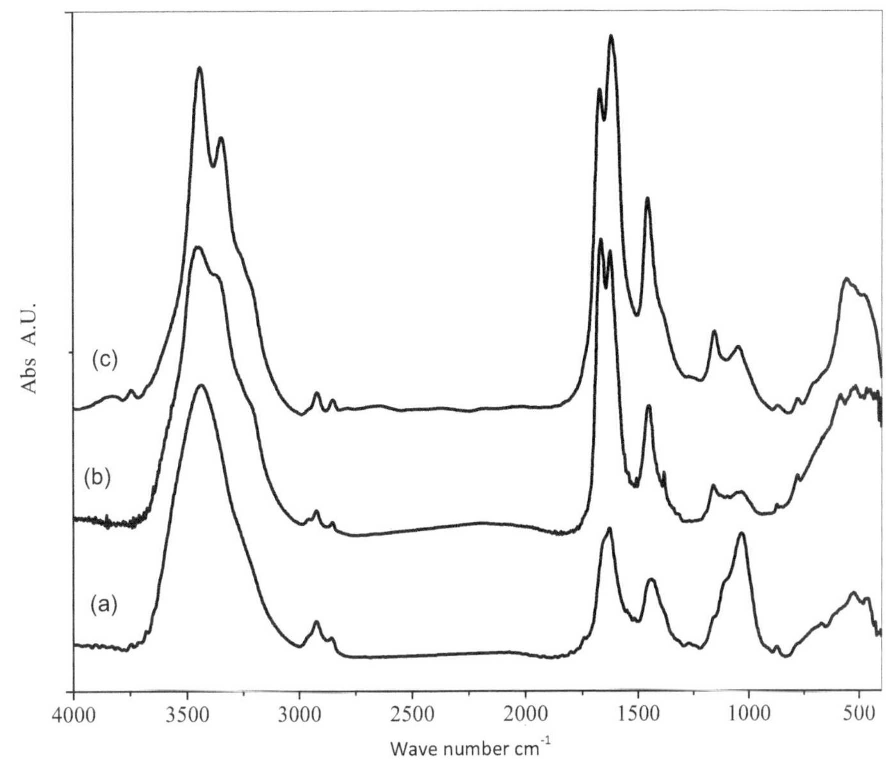

Figure 5. FTIR spectra of sample deposited with first (a), second (b) and third (c) concentrations under 800 V for 10 h.

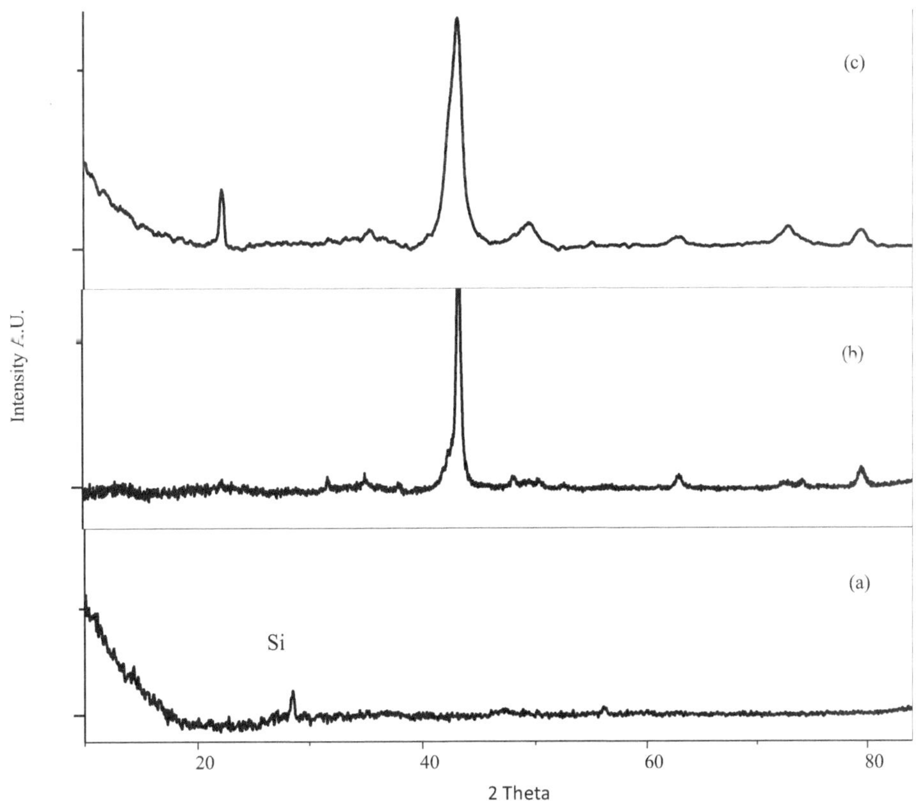

Figure 6. GIXRD of films deposited under first (a), second (b) and third (c) concentrations under applied voltage 800 V for 10 h.

Table 1. Theoretically predicted and experimentally measured d values for samples deposited under different concentrations and applied voltage 800 V for 10 h.

Experimental d (A°) concentration			Theoretical values phase		
First	Second	Third	α–C_3N_4	β–C_3N_4	CN_x
		3.99			4.09
	2.82	2.677		2.74	
	2.55	2.50	2.51		
	2.08	2.09		2.09	
	1.89		1.89		
	1.80	1.84		1.84	
	1.74		1.71		
	1.62		1.61		
	1.47	1.47		1.48	
	1.30	1.29	1.32		
	1.27		1.28		
	1.20	1.20		1.22	
		1.107			

concentration of urea from first to second concentration changes the structure from amorphous to crystalline one, that is, increasing the amount of urea in methanol results in an increase in the number of peaks belonging to some mixed polycrystalline phases of α- C_3N_4, and β- C_3N_4. Also, further increasing of the concentration of urea to the third concentration causes the appearance of a new phase of CN_x.

Figure 7a shows the pattern of the film deposited under the first concentration but at higher voltage (1000 V). Examination of the figure reveals that the deposited sample is mainly amorphous; while increasing urea concentration to the second and the third concentrations under the same applied voltage (1000 V) resulted in appearance of a number of diffraction peaks as is shown in Figures 7b and 7c respectively. For example increasing the voltage to 1000 V resulted in the formation of the new phase CN_x as well as the other two phases α-C_3N_4 and β- C_3N_4. It was also noted that, in case of the third concentration, the amount of the new phase becomes more intense relative to the two other phases α-C_3N_4, and β- C_3N_4.

Figure 8 illustrates the GIXRD diffractograms of the samples deposited under various concentrations and applied voltage 1200 V for 10 h. In case of the first concentration (Figure 8a), the diffractogram displays two small peaks at 2Θ= 22.15° and 2Θ = 35.48°. This result indicates that raising the applied voltage to 1200 V caused the appearance of tiny crystallite embedded with the amorphous phase of carbon nitride.

In case of the second concentration as is shown in Figure 8b, the recorded GIXRD shows a number of diffracted peaks at 2Θ= 22.13, 25.97, 29.22, 31.19, 33.21, 35.32, 43.29, 48.04, 49.38, 54.85, 62.99, 72.45 and 79.39°. Raising the voltage from 1000 to 1200 V caused no appreciable change in the features of the pattern apart from changes in the intensity of peaks belonging to the new phase of CN_x relative to the peaks of the two other phases.

In case of the third concentration as shown in Figure 8c, the recorded GIXRD pattern shows a number of diffraction peaks of different intensities at 2Θ= 22.25, 29.05, 31.69, 35.52, 40.60, 55.04, 67.4 and 86.60°. It is clear that the peak at 2Θ= 22.25° corresponds to the new phase becomes more predominant relative to those correspond to other phases.

Scanning electron microscope (SEM) and energy dispersive X-ray (EDX)

Figure 9 shows scanning electron micrograph of the carbon nitride films deposited on Si substrate under the second concentration at applied voltage of 800 V for 10 h. The figure shows the existence of the grains in the film, those are commonly observed in polycrystalline films. The average size of the spherical grains in the film can be measured, coming out at around 0.2 μm. The chemical composition of the deposited carbon nitride film was examined by EDX analysis. The atomic concentration of N and C are 0.60 and 76.86 at% respectively. Also there are amounts of O, Ni, Cu and Zn of atomic percent 7.46, 0.21, 4.59 and 10.29 respectively. The presence of oxygen is common in many of publications and the presence of Ni, Cu and Zn impurities come from the contacts which we used in the electrolytic cell.

For higher urea concentration, the same applied voltage and time of deposition more closely packed crystal-like structures are observed in Figure 10. This may be attributed to the presence of more nitrogen in the film (Kundoo et al., 2006). The atomic concentration of

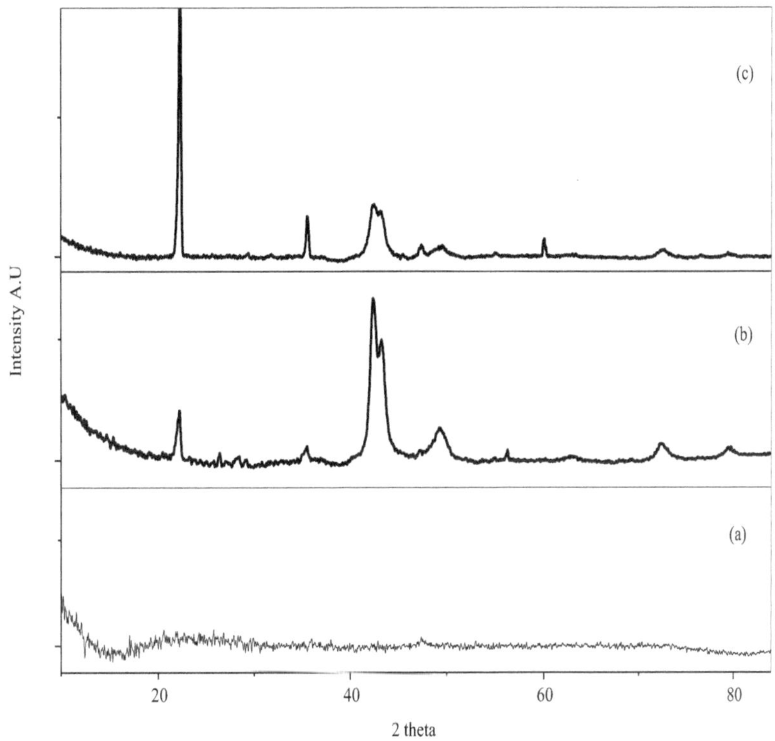

Figure 7. GIXRD of films deposited under first (a), second (b) and third (c) concentrations under applied voltage 1000 V for 10 h.

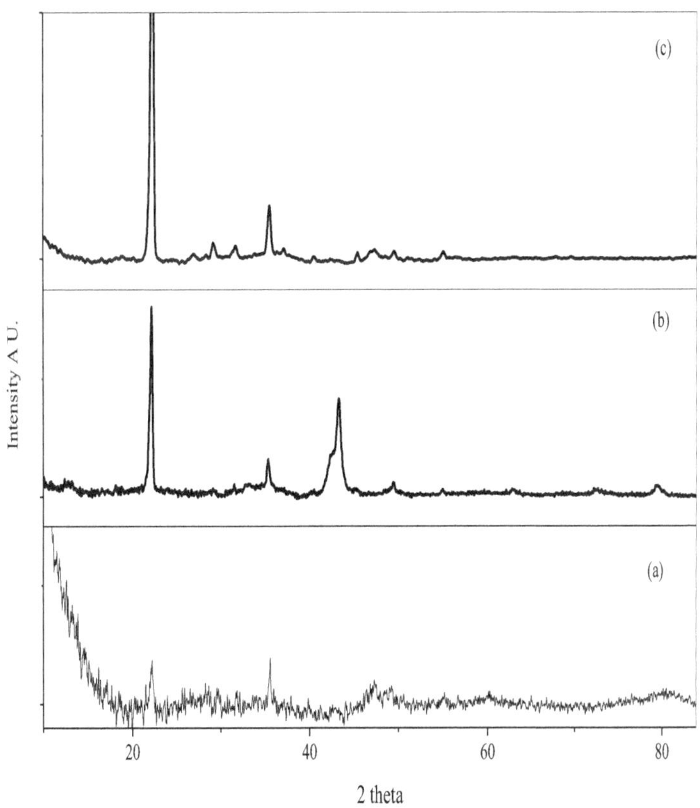

Figure 8. GIXRD of films deposited under first (a), second (b) and third (c) concentrations under applied voltage 1200 V for 10 h.

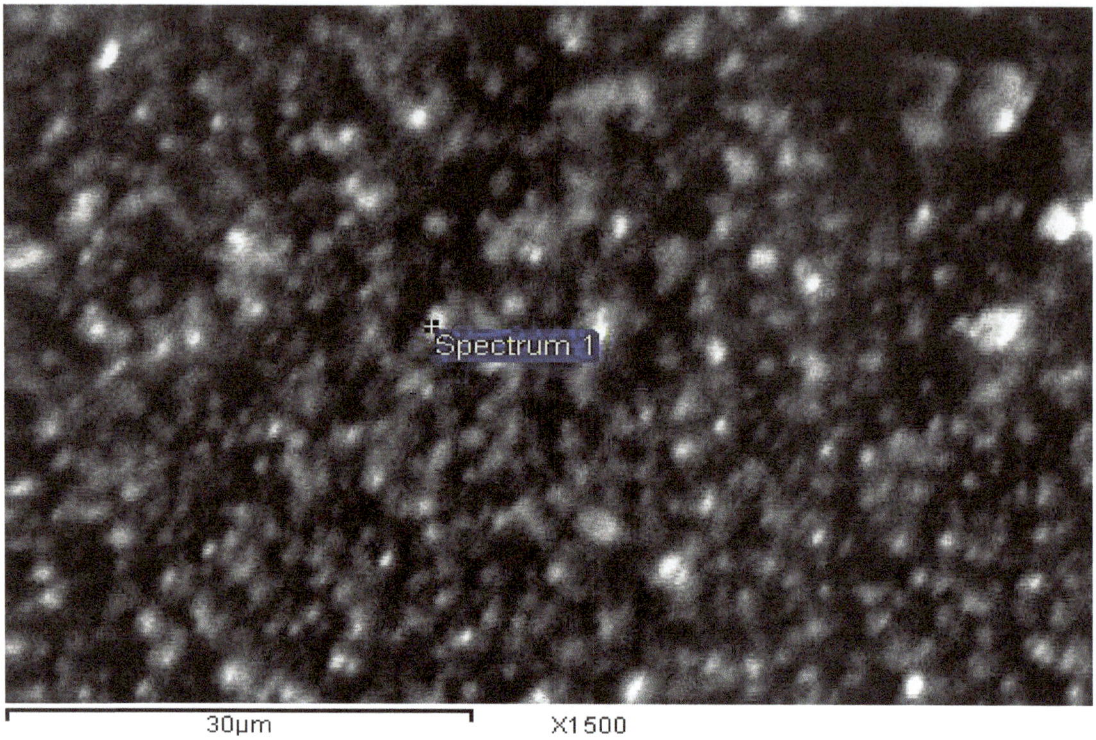

Figure 9. SEM micrograph of film deposited on Si substrate with second concentration and applied voltage of 800 V for 10 h.

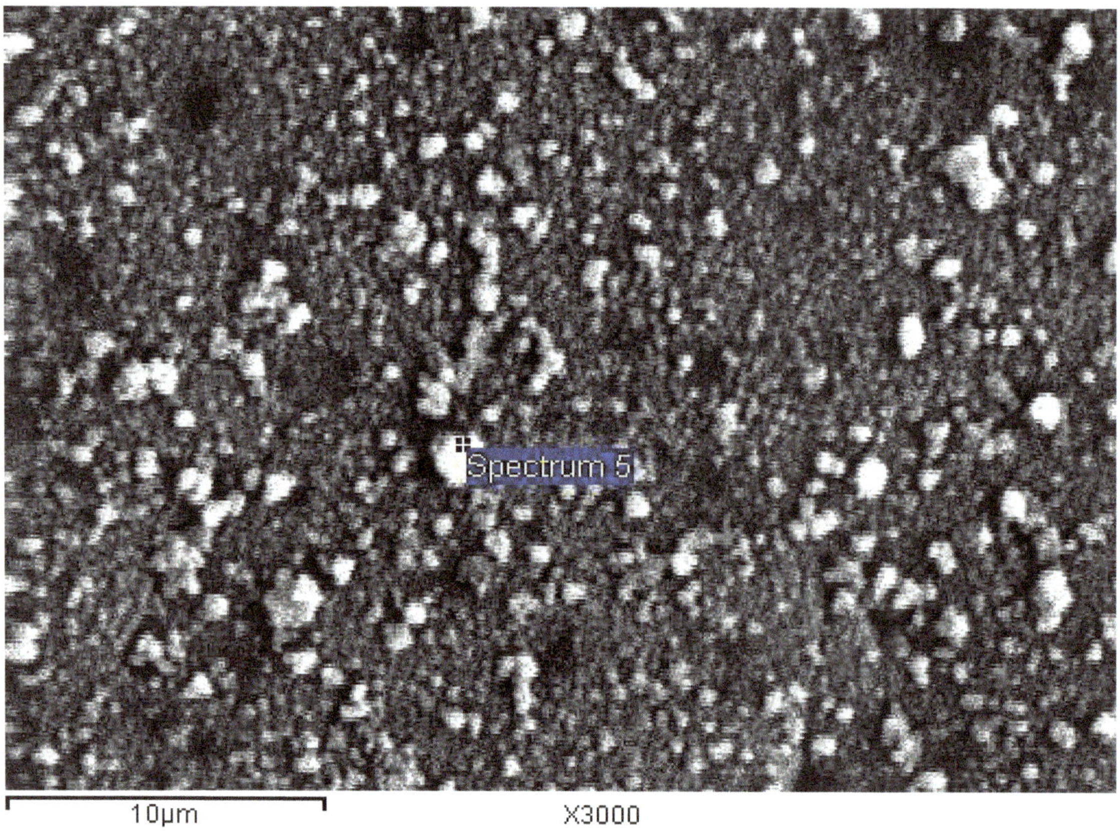

Figure 10. SEM micrograph of film deposited on Si substrate with the third concentration and applied voltage of 800 V for 10 h.

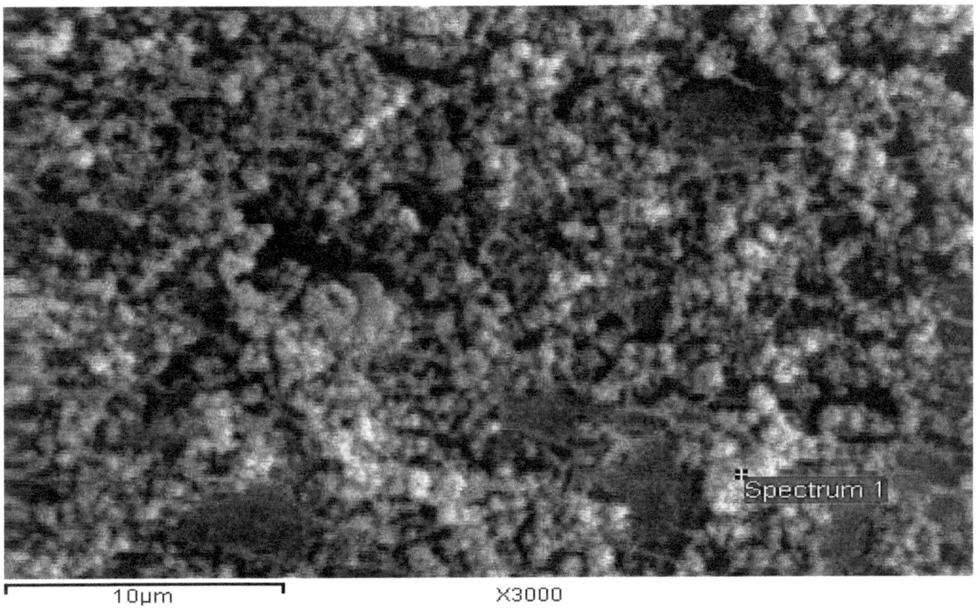

Figure 11. SEM micrograph of film deposited on Si substrate with the third concentration and applied voltage of 1200 V for 10 h.

N and C are 14.10 and 40.37% respectively. There are also O, Cu, and Zn of atomic percent which is 30.77, 8.04 and 6.73 respectively present in the film. It is noticed that the nitrogen concentration increased from 0.6 to 14.10% with increasing concentration of urea. Further, under the highest urea concentration (third) and highest applied voltage (1200 V) for 10 h the grains are observed more closely packed crystal-like structures (Figure 11). It means that nitrogen increased with increasing applied voltage. The atomic concentration of N and C are 29.54 and 29.90% respectively. There are also O and Cu of atomic percent of 33.12 and 4.45% respectively. It is noticed that under higher applied voltage (1200 V), the nitrogen and carbon concentrations are nearly equal.

Conclusions

It can be concluded that FTIR study of the deposited films revealed that carbon and nitrogen are bonded in C=N and C-N bonds as indicated by the existence of their absorption peaks at 1624 and 1450 cm^{-1} respectively. It is found that both the concentration and the applied voltage can influence the nitrogen fraction x incorporated in the CN_x films. When the applied voltage is increased, a unity of N/C ratio can be achieved with gradually increasing nitrogen content as verified by the FTIR results and confirmation by EDX data. Also it was found that the crystalline structure depends on the nitrogen content in the film where it changed from amorphous to poly-crystalline of α- C_3N_4, and β- C_3N_4 phases. With further increase in nitrogen content a new phase of CN_x appeared and at the highest nitrogen content, this phase became more predominates relative to the other phases.

REFERENCES

Chengbing W, Shengrong Y, Junyan Z (2008) The infrared characteristics investigation of carbon nitride films. Diamond Rel. Mater., 17: 174-179.

Cohen ML (1985). "Calculation of bulk moduli of diamond and zincblende solids" Phys Rev., B 32: 7988-7991.

Fu Q, Jiu JT, Cai K, Wang H, Cao CB, Zhu HS (1999)." Attempt to deposit carbon nitride films by electrodeposition from an organic liquid" Phys. Rev., B 59: 1693-1696.

GU YS, Pan LQ, Chang XR, Tian ZZ (1996)." Carbon nitride films prepared by ion implantation" J. Mater. Sci. Lett., 15: 1355-1357.

Hammer P, Victoria NM, Alvarez F (2000)" Effects of increasing nitrogen concentration on the structure of carbon nitride films deposited by ion beam assisted deposition" J. Vac. Sci. Technol., A 18: 2277–2287.

Kundoo S, Banerjee AN, Saha P, Chattopadhyay KK (2003)." Synthesis of crystalline carbon nitride thin films by electrolysis of methanol–urea solution" Mater. Lett., 57: 2193-2197.

Kundoo S, Saha P, Chattopadhyay KK (2006)." Structural characterisation of crystalline carbon nitride thin films synthesised by electrolysis of urea–methanol solution" Adv. Appl. Ceramics, 105:73

Liu AY, Cohen ML (1989)." Prediction of New Low Compressibility Solids" Science, 245: 841-842.

Lopez S, Dunlop HM, Benmalek M, Tourillon G, Wong MS, Sproul WD (1998). XPS, XANES and ToF-SIMS Characterization of Reactively Magnetron-sputtered Carbon Nitride Films" Surf. Interface Anal., 25(5): 315–323.

Majumdar A, Schröder K, Hippler R (2008)."Cytocompatibility of amorphous hydrogenated carbon nitride films deposited by CH4/N2 dielectric barrier discharge plasmas with respect to cell lines"J. Appl. Phys.,104: 074702.

Sreejith K, Nuwad J, Pillai CGS (2005). "Low voltage electrodeposition of CNx films and study of the effect of the deposition voltage on bonding configurations" Appl. Sur. Sci., 252: 321-329.

Teter DM, Hemley RJ (1996)." Low-Compressibility Carbon Nitrides" Science, 271: 53-55.

Yan X, Xu T, Chen G, Yang S, Xue HLQ (2004)." Preparation and

characterization of lectrochemically deposited carbon nitride films on silicon.
Zhang YP, Gu YS, Chang XR, Tian ZZ, Shi DX, Zhang XF, Yuan L. Characterization of carbon nitride thin films deposited by microwave plasma chemical vapor deposition. Sur. Coat. Tech., 127: 259-264.
Zhang Z, Guo H, Xu Y, Zhang W, Fan X (1999). "Corrosion resistance studies on α-C3N4 thin films deposited on pure iron by plasma enhanced chemical vapor deposition" J. Mater. Sci. Lett., 18: 685-687.
Zocco A, Perrone A, Anna ED, Leggieri G, Luches A, Klini A, Zergioti I, Fotakis C (1999). Deposition of carbon nitride films by reactive pulsed-laser ablation at high fluences. Diamond Relat. Mater., 8: 582–585

15

Emissions control catalysts: Links between Fecralloy preoxidation and coating adherence

Stephen A. Adegbite

School of Chemical Engineering, University of Birmingham, Edgbaston, Birmingham B15 2TT, United Kingdom.
E-mail: stephen_adegbite@hotmail.com.

Tightening emissions legislation for vehicles across the world has caused the use of monolith catalysts in automobile emission control to become ubiquitous. In this study, the physiochemical transformations caused by thermal pre-oxidation of the surface of Fecralloy® foil, an integral component of emissions control catalysts, have been studied to assess their role on coating film adherence. The foil coupons were pre-oxidised at 950°C for 0, 5, 10 and 30 h and the mass gain was found to fit a simplified hyperbolic model. Measurements were made of surface topography and microstructure using laser profiling interferometry (LPI) and scanning electron microscopy (SEM), with phase analysis being obtained from X-ray diffraction (XRD). The optimal surface roughness was obtained after 10 h of pre-oxidation, when the surface contained significant amounts of α-alumina arranged as randomly oriented whiskers. Upon coating of the treated foil coupons with γ–alumina slurry, the sample pre-oxidised for 10 h gave the best performance in terms of coating film loading of 7.94 mass % and adherence of less than 10 mass % loss.

Key words: Catalyst, Fecralloy®, pre-oxidation, alumina, coating, adherence.

INTRODUCTION

Tightening emissions legislation for vehicles across the world has caused the use of monolith catalysts in automobile emission control to become ubiquitous. Their impact has been considerable, with emission levels from exhaust systems of passenger cars falling by more than 90% since 1975, and with zero emissions now a future target. Environmentally harmful carbon monoxide and oxides of nitrogen in the exhaust stream are converted inside the monolith catalyst into carbon dioxide and nitrogen (Acres and Harrison, 2004; Twigg, 2006).

Monoliths are structures comprised of multiple parallel channels. One method of producing monolith catalysts is to coat the monolith surfaces with γ-alumina film containing active species (usually platinum group metals, PGMs, such as Pt, Pd and Rh). Monoliths are generally made from either metallic or ceramic materials, the former from thin metal sheet such as Fecralloy®, the focus of this study, and the latter typically from cordierite (Avila et al., 2005). Fecralloy® is comprised essentially of iron, Fe, chromium, Cr, and aluminium, Al. It is well suited for catalyst support applications because of its high mechanical strength, thermal conductivity and ductility which make it easy to process. It clearly outperforms cordierite as a monolith material in all these crucial properties (Hickman and Schmidt, 1992b; 1992a).

To ensure adherence of the γ-alumina coatings, Fecralloy® has to be pre-oxidised which changes its surface topography to provide a rough surface (Cybulski and Moulijn, 2006). Pre-oxidation also provides enrichment of the Fecralloy® surface with aluminium, therefore prolonging the product component life (Nicholls and Quadakkers, 2002). Some studies have assessed the effects of pre-oxidation on dip-coated foil coupons, by relating the surface microstructure obtained by scanning electron microscopy (SEM) to the coating adherence(Jia et al., 2007; Zhao et al., 2003). However, these comparisons are somewhat qualitative in nature and do not consider a quantitative measure of the surface roughness, nor do they address coating at a controlled shear rate. In this new study, a detailed assessment of Fecralloy® two-dimensional (2D) and three-dimensional (3D) surface topography by laser profiling interferometry (LPI) was performed as a function of pre-oxidation conditions. In addition, SEM and X-ray diffraction (XRD)

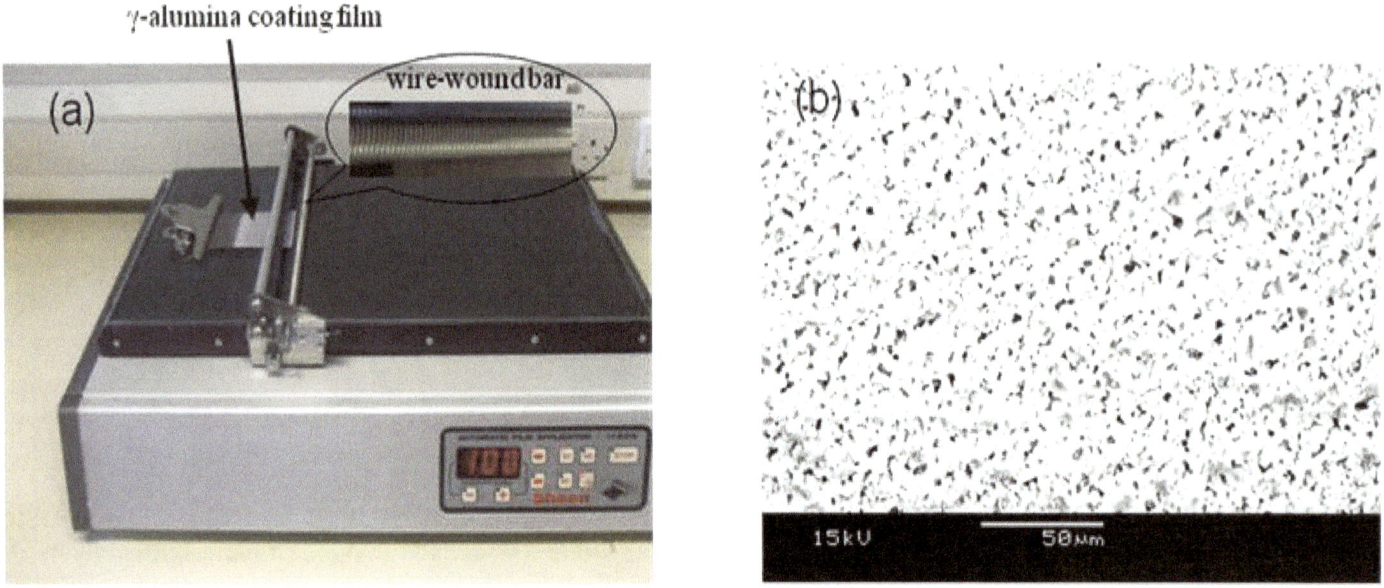

Figure 1. Coating of Fecralloy® supports: (a) γ-alumina coating film applied by 1132N applicator; (b) SEM micrograph of coating surface.

techniques have been applied to give a comprehensive description of the physiochemical transformations caused by pre-oxidation. To determine the effect of pre-oxidation on coating film adherence, γ-alumina slurry was deposited onto the Fecralloy® foil coupons at a controlled shear rate using an automatic film applicator, and the resulting composites were dried and calcined afterwards. The coating film loading and adherence were determined to reveal the optimal pre-oxidation conditions.

MATERIALS AND METHODS

Pre-treatment and assessment of Fecralloy® foil

Commercially available Fecralloy® foil (GoodFellow, UK) of 50 μm thickness with a composition of Fe (72.6 wt %), Cr (22.0 wt %), Al (4.80 wt %), Y (0.30 wt %) and Si (0.30 wt %) was used. The as-received foil was cut into 50 mm × 80 mm coupons without any prior mechanical polishing. The pre-treatment of the coupons consisted of the following steps (Jia et al., 2007):

1. Degreasing: ultrasonic cleaning at room temperature in acetone bath for 10 min, and then with deionised water at 80°C for 10 min
2. Pre-oxidation: thermal oxidation in a furnace (Lenton, UK) at 950°C for different durations: 0, 5, 10 and 30 h

The specific mass gain of the coupons as a function of pre-oxidation time was determined using a mass balance (Ohaus Corporation, USA). The surface topography (namely surface roughness and topography profiles) of the coupons was measured by LPI (Taylor Hobson, UK). The interferometer measured 1000 sampling lengths of 5 μm each, thus, covering a total evaluation length of 5 mm. The surface chemical composition of the coupons was determined by XRD using X' Pert Pro diffractometer (Phillips, The Netherlands) having Fe-filtered Co Kα radiation and a power of 45 kV × 30 mA. The surface microstructure of the coupons was assessed by using a JEOL 6060 SEM (Oxford Instruments, UK).

Coating of foils by an automatic film applicator

A γ-alumina slurry of average diameter $d_{0.5}$ = 4.36 μm and solids concentration of 40 wt% was prepared at a pH of 4 using ethanoic acid solution (Fischer Scientific, UK). The slurries were deposited onto the Fecralloy® coupons using an automatic film applicator (Sheen Instruments, UK) which provided close control of the traverse coating speed (Adegbite, 2010). The slurries were drawn across the coupons at a traverse speed of 100 mms^{-1} with a wire-wound bar of 100 μm gap (Figure 1a). The coated coupons were allowed to dry at room temperature, and were then oven dried at 110°C for 1 h and finally calcined at 500°C for 1 h (Jia et al., 2007). The surface view of a product coating obtained by SEM is shown in Figure 1b.

Assessment of coatings

The coating film loading on the Fecralloy® surface was then calculated by the percentage mass increase of the coupon. To determine coating film adherence, the coated foils were soaked into petroleum ether contained inside a sealed beaker for 30 min, then soaked in an ultrasonic bath (300 W and 60 kHz) for 1 h and later dried in an oven at 110°C for 2 h (Valentini et al., 2001). The percentage mass loss was then calculated, meaning that the lower the mass loss the better the coating adherence.

RESULTS AND DISCUSSION

Pre-oxidation profile

The pre-oxidation profile shown in Figure 2 represents the specific mass gain by foils as a function of pre-oxidation time. The increase in the specific mass of the foil with pre-oxidation time arises from the continuous formation of oxide layers on the foil surface. It is shown that the rate of oxide formation, which is the gradient of the curve, was highest at the start of pre-oxidation, and it

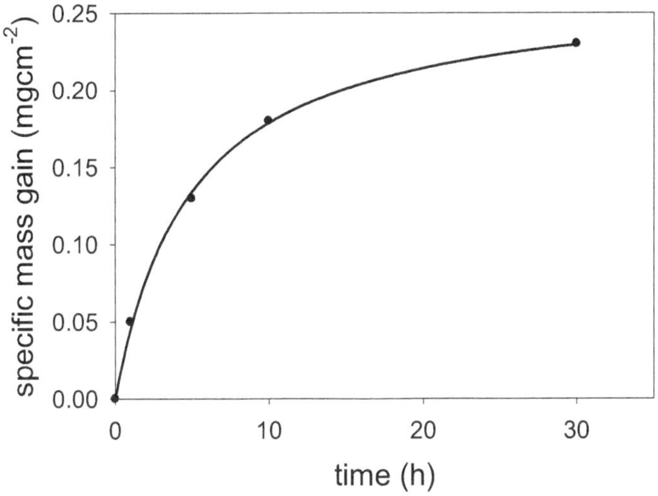

Figure 2. Pre-oxidation profile of Fecralloy® coupons for 0 – 30 h.

progressively diminished as pre-oxidation continued. A simplified hyperbolic model (Equation 1), derived from a previous study (Pragnell et al., 2005), properly describes the pre-oxidation profile of Fecralloy® for 0 to 30 h.

$$\overline{\Delta m} = \frac{Xt}{Y+t} \quad (1)$$

where, $\overline{\Delta m}$ is specific mass gain ($mgcm^{-2}$) and t is pre-oxidation time (h).
Constants X and Y were found to be 0.27 and 5.0 respectively.

Topography characterisation of Fecralloy® coupons by LPI

The surface topography of the coupons, represented by the 2D and 3D roughness parameters and the topography profiles, are shown in Table 1 and Figure 3. The roughness parameters for the untreated coupon were very low because the surface has not been transformed. The roughness parameters increased significantly after 5 h pre-oxidation, then reached the peak at 10 h pre-oxidation and later declined upon prolonged pre-oxidation for 30 h (Table 1). Clearly, the Fecralloy® coupon pre-oxidised at 950°C for 10 h had the roughest surface; this is supported by the textural profiles shown in Figure 3. Coating film adherence is promoted by rough surfaces which act as glues for coatings because of their irregularities (Jia et al., 2007; Zhao et al., 2003). The profiles shown in Figure 3 also exhibit order in their surface structure due to fabrication patterns, this is particularly evident in the sample shown in Figure 3c which is characterised by deep penetrating grooves.

XRD analysis of Fecralloy® surface

The XRD patterns in Figure 4 show the surface chemical composition of the coupons pre-oxidised at 950°C for 0 – 30 h. The untreated coupon comprises of a Fe (Cr) peak and some α-alumina peaks, which are present in small amounts (Figure 4a). Pre-oxidation for 5 h produced substantial amounts of α-alumina and a small amount of θ-alumina (Figure 4b). Increased pre-oxidation to 10 h produced sufficient transformation of θ-alumina into α-alumina, as the peaks of the latter are large and conspicuously visible (Figure 4c). Prolonged pre-oxidation however resulted into formation of α-alumina conglomerates as other alloy elements are spontaneously oxidised; this is characterized by diminishing peaks of α-alumina (Figure 4d). The Fecralloy® surface enrichment by aluminium during pre-treatment is caused by an elemental transport process (Jia et al., 2007; Nicholls and Quadakkers, 2002).

Assessment of Fecralloy® surface microstructure by SEM

The SEM micrographs displayed in Figure 5 clearly show the distinctions between the surface morphologies of the untreated and pre-oxidised coupons. The untreated coupon (Figure 5a) has a plain surface with no significant microstructure while the pre-oxidised coupons contain α–alumina whiskers (Figures 5b to d). After 5 h of pre-oxidation, there exists trace amounts of α-alumina. However, the coupon pre-oxidised for 10 h was characterized by uniform, conspicuous and randomly oriented α-alumina whiskers. It is shown that at this condition the α–alumina are properly formed and arranged in a manner which positions them for good adhering capabilities (Avila et al., 2005). The attendant effect of pre-oxidation for up to 30 h was the formation of non-uniform conglomerates of α-alumina.

Coating loading and adherence

The coating film loading and weight loss from adherence test are shown in Table 2. The values were averaged over 5 measurements with a standard deviation (SD) within 3.5%. Clearly, the duration of pre-oxidation has a major influence on the adherence of the coating film on the Fecralloy® coupons.

The results in Table 2 can be explained in the light of the Fecralloy® surface characterization afore mentioned. For the untreated coupon, the coating loading and adherence were very poor because the foil surface was the least rough and no wettability occurred. The coupons pre-oxidised for 5 and 30 h showed improved capabilities which are commensurate with their roughness characteristics. The optimal coating loading and adherence were obtained from the coupon with the

Table 1. Roughness parameters of coupons pre-oxidised at 950°C for 0 – 30 h.

Roughness parameters	Type	Definition	Values (µm)			
			0 h	5 h	10 h	30 h
R_a	2D	arithmetic mean	0.06	0.15	0.31	0.23
R_q	2D	root mean square	0.08	0.17	0.36	0.20
S_a	3D	arithmetic mean	0.20	0.35	0.83	0.52
S_q	3D	root mean square	0.24	0.42	1.01	0.69

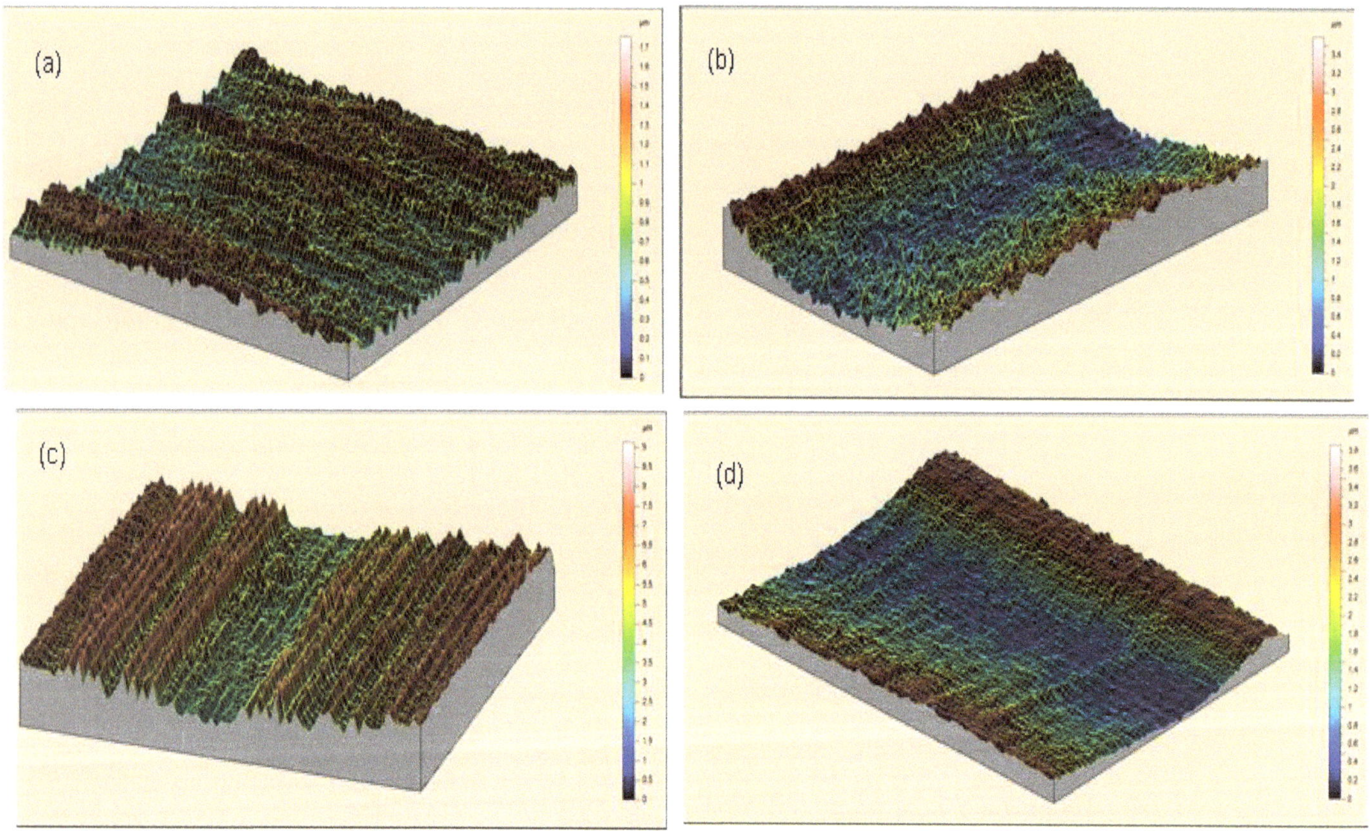

Figure 3. Topography profiles of Fecralloy® coupons pre-oxidised at 950°C for (a) 0 h; (b) 5 h; (c) 10 h; and (d) 30 h.

roughest surface i.e. the sample pre-oxidised for 10 h. This is because of the enhanced surface microstructure brought about by the randomly oriented α-alumina whiskers, thereby creating a suitable topography onto which coating film is firmly anchored. The images of the coatings after the adherence test are also given in Figure 6 which revealed the weakened areas where coatings had been lost.

Conclusions

Fecralloy® foils are materials from which monolith emissions catalysts are made. The Fecralloy® pre-oxidation conditions have been shown to be central to achieving increased performance, in terms of coating film loading and adherence. A simplified hyperbolic model provided the best fit for specific mass gain by the coupons as a function of pre-oxidation time from 0 – 30 h. The optimal coating film loading (7.94 mass %) and adherence (9.90 mass % loss) were obtained from coupons pre-oxidised at 950°C for 10 h, as these conditions produced the optimal surface topography and microstructure.

ACKNOWLEDGEMENT

This research was supported by the Engineering and Physical Sciences Research Council (EPSRC) in the UK.

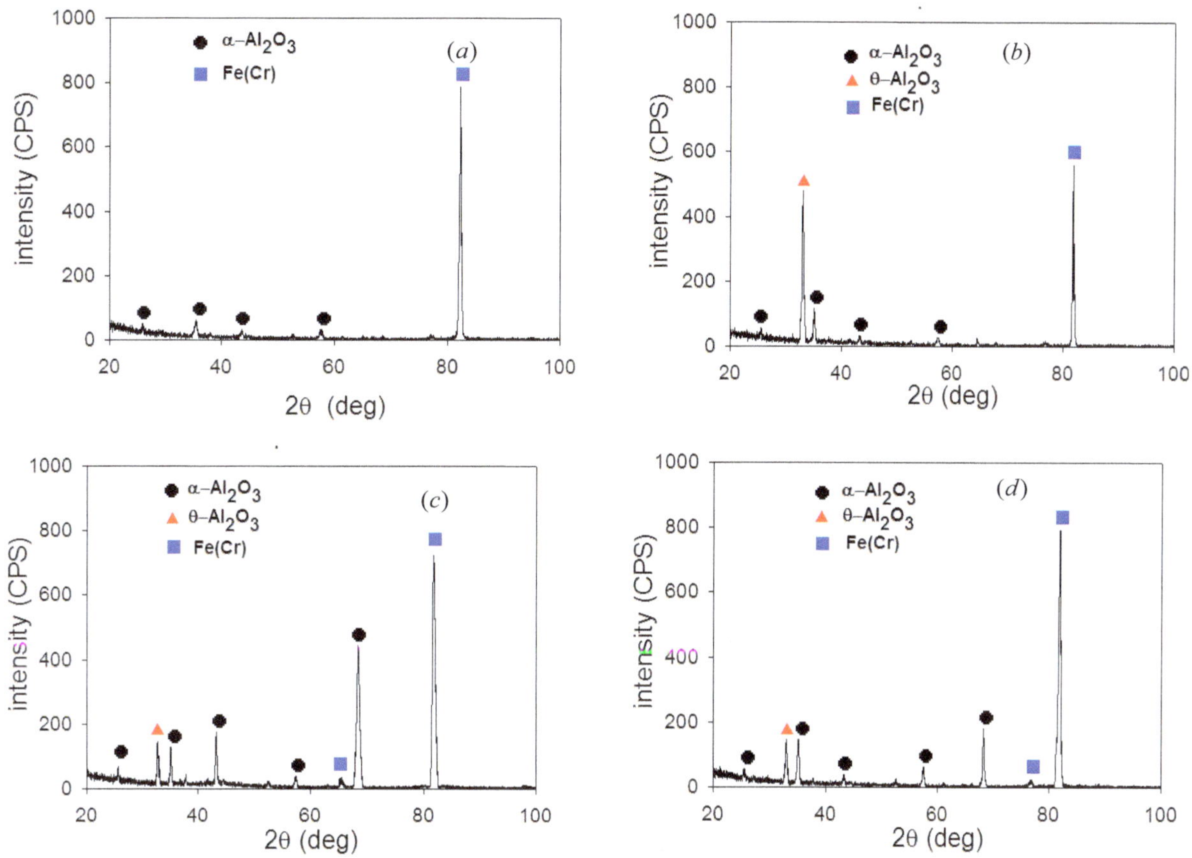

Figure 4. XRD patterns of Fecralloy® coupons pre-oxidised at 950°C for (a) 0 h; (b) 5 h; (c) 10 h; and (d) 30 h.

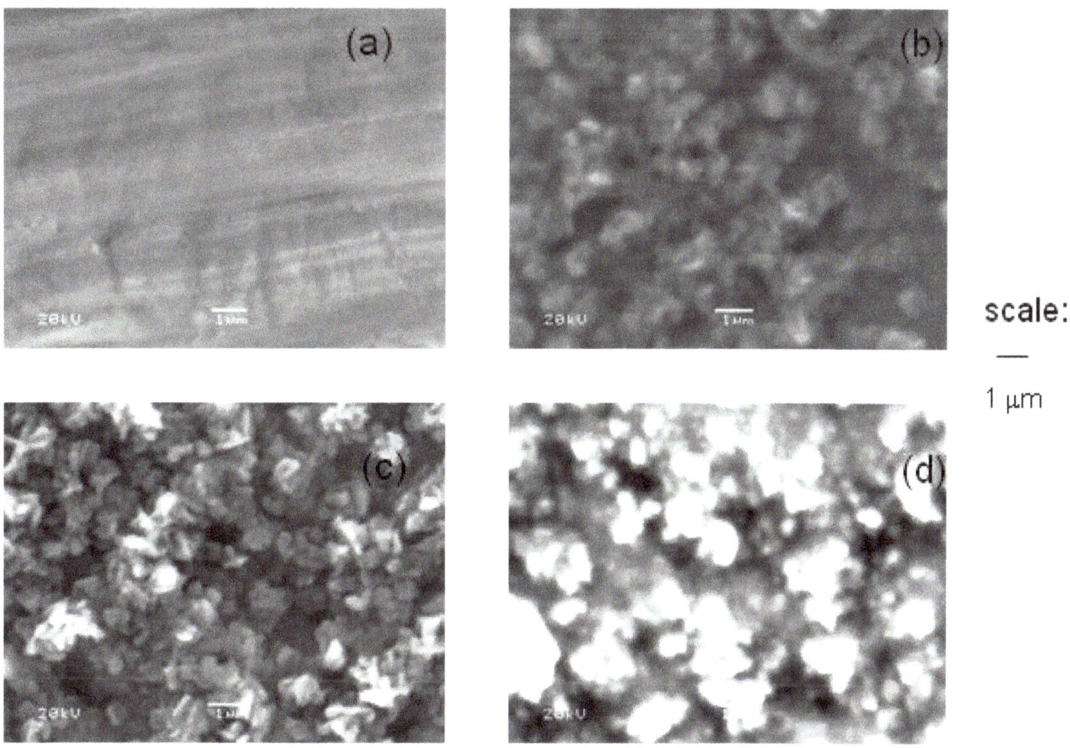

Figure 5. SEM micrographs of Fecralloy® coupons pre-oxidised at 950°C for (a) 0; (b) 5; (c) 10 and (d) 30 h.

Table 2. Coating properties for Fecralloy® coupons pre-oxidised at 950°C for 0 – 30 h.

Pre-oxidation time (h)	Coating loading (mass %) SD = ±3.1 % loading	Mass % loss from adherence test SD = ±3.5 % of loss
0	0.08	99.6
5	5.19	19.4
10	7.94	9.90
30	6.75	16.3

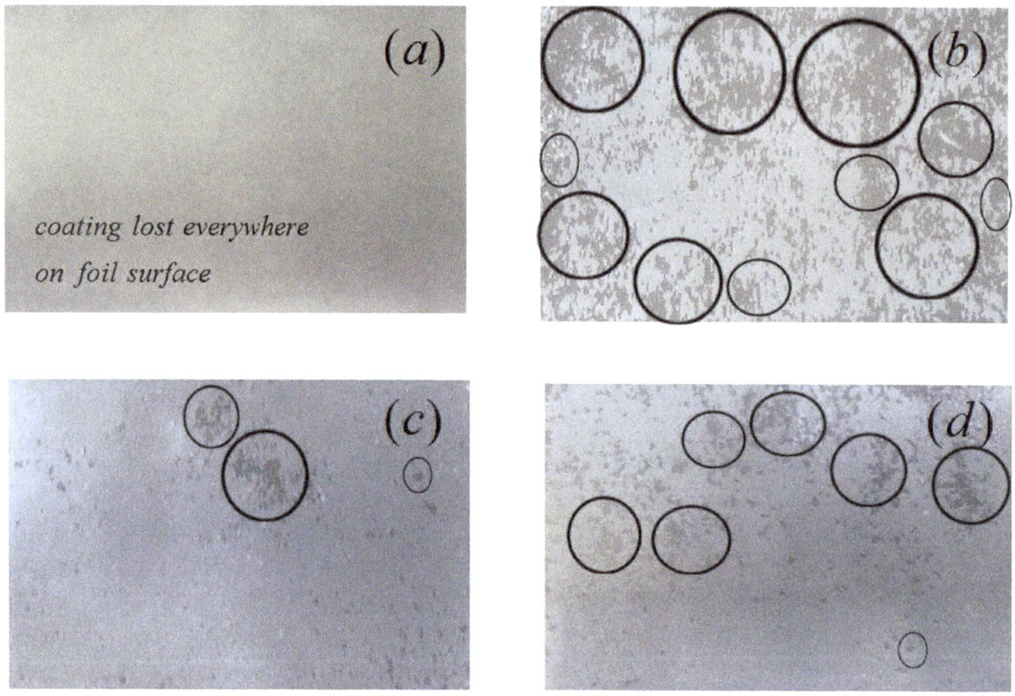

Figure 6. Coatings on Fecralloy® coupons after adherence test (circles indicating some areas of coating loss): (a) 0; (b) 5; (c) 10 and (d) 30 h.

REFERENCES

Acres GJK, Harrison B (2004). The development of catalysts for emission control from motor vehicles: early research at Johnson Matthey. Topics in Catalysis. 28: 3-11.

Adegbite SA (2010). Coating of catalyst supports: links between slurry characteristics, coating process and final coating quality, Ph.D. Thesis. University of Birmingham. pp. 131-132.

Avila P, Montes M, Miro EE (2005). Monolithic reactors for environmental applications - a review on preparation technologies. Chem. Eng. J., 109: 11-36.

Cybulski A, Moulijn JA (2006) Metal and coated metal catalysts. In: Structured catalysts and reactors. Taylor & Francis, London. pp. 71-105

Hickman DA, Schmidt LD (1992a). Synthesis gas-formation by direct oxidation of methane over Pt monoliths. J. Catal., 138: 267-282.

Hickman DA, Schmidt,L.D. (1992b). The role of boundary-layer mass-transfer in partial oxidation selectivity. J. Catal., 136: 300-308.

Jia LW, Shen MQ, Wang J (2007). Preparation and characterization of dip-coated γ-alumina based ceramic materials on FeCrAl foils. Surf. Coat. Technol., 201: 7159-7165.

Nicholls JR, Quadakkers WJ (2002). Materials issues relevant to the development of future metal foil automotive catalytic converters. In: Materials aspects in automotive catalytic converters. Bode H (ed), Wiley-VCH, Weinheim. pp. 31

Pragnell WM, Evans HE, Naumenko D, Quadakkers WJ (2005). Aluminium depletion in FeCrAlY steel during transitional alumina formation. Materials at High Temperatures. 22: 561-566.

Twigg MV (2006). Roles of catalytic oxidation in control of vehicle exhaust emissions. Catalysis Today. 117: 407-418.

Valentini M, Cristiani G, Levi M, Tronconi E, Forzatti P (2001). The deposition of γ-Al2O3 layers on ceramic and metallic supports for the preparation of structured catalysts. Catalysis Today. 69: 307-314.

Zhao S, Zhang JZ, Weng D, Wu XD (2003). A method to form well-adhered γ-Al2O3 layers on FeCrAl metallic supports. Surf. Coat. Technol., 167: 97-105.

Modified exponentially weighted moving average (EWMA) control chart for an analytical process data

Alpaben K. Patel[1]* and Jyoti Divecha[2]

[1]Directorate of Economics and Statistics, Sector-18, Gandhinagar-382009, India.
[2]Department of Statistics, Sardar Patel University, Vallabh Vidyanagar, 388120, India.

This article introduces modified exponentially weighted moving average (modified EWMA) control chart. The modified EWMA control chart is very effective in detecting small and abrupt shifts in monitoring process mean. This chart is based on modified EWMA control chart statistic, which is a correction of EWMA chart statistic and it is free from inertia problem. The advantage of using modified EWMA chart is its good performance for observations that are autocorrelated or independently normal distributed. The performances of Modified EWMA chart are illustrated along with EWMA chart for the two types of analytical process data. The average run length (ARL) properties of modified EWMA scheme are derived using Markov Chain approach.

Key words: Abrupt change, autocorrelated, average run length, exponentially weighted moving average (EWMA), modified.

INTRODUCTION

The control charts namely, Shewart chart (Shewhart, 1924), EWMA chart (Roberts, 1959) and CUSUM chart (Page, 1954) are often used for detecting shifts in a sequence of independent normal observations with common variance coming from a particular process. The usefulness of a chart is determined from the timely detection of shifts as they occur. The literature on evaluation of properties of these three charts indicates that Shewhart chart detect all single large shifts, CUSUM chart detects shifts by accumulating changes in a direction, while EWMA chart detects shifts through changes accumulated under exponential smoothing. In general, these three control charts will fail to detect shift in one case, either the large shift or the small shift in the process. Moreover, these charts cannot be used directly in chemical and pharmaceutical industries because the observations from processes in these industries are frequently autocorrelated. This autocorrelation has a large impact on the control charts developed under the independence assumption. A typical effect of autocorrelation is to decrease the in-control average run length (ARL) leading to a higher false alarm rate than for an independent process. Moreover, the autocorrelated processes may also have abrupt changes. Both the situations are bothersome for these industries.

In particular, in most of the industrial processes, temperature is an essential process variable and the process is sensitive to the cumulative small changes as well as some abrupt changes in temperature values. An abrupt change is the unpredictable large shift occurring in the process being monitored for small shifts. Recently, research efforts have been put for the process subject to abrupt changes. EWMA chart has received more attention as it also provides an estimate of forecast of process mean. Yaschin (1995) have discussed estimation of current process mean in a process subject to small changes, as well as abrupt changes on the basis of EWMA. Capizzi and Masarotto (2003) have provided a class of new control charts, AEWMA charts on the basis of EWMA to detect shifts of all types for independent data. Reynolds and Stoumbos (2006) suggested using two EWMA controls simultaneously, one for process mean and another for process variance to tackle detection of small shifts with abrupt changes. Woodall and Mahmoud (2005) have proposed a measure of inertia, the signal resistance, and shown that EWMA suffers from inertia problem more than the CUSUM. On the other hand, several authors have discussed process control methods for autocorrelated observations. Montgomery and Mastrangelo (1991) presented method

*Corresponding author. E-mail: alpakip@yahoo.co.in.

that consists of modeling the autocorrelative structure in the original data using EWMA and applying Shewhart control charts to the independent normal residuals. Their paper led to discussion, and one of the reviewers: Ryan (1991) stated that, residual chart for AR(1) will perform poorly unless autoregressive (AR) parameter is negative or extremely close to 1 and that the derivation of ARL properties was unclear because most of the time, the true state of autocorrelation in the process is not known. Faltin and Woodall (1991) appreciated their method for the virtue of simplicity in the interest of operators and engineers. Cox (1961) have shown that the EWMA is not a useful predictor for an AR(1) process having an autoregressive parameter less than 1/3. It is found that for ARMA (1,1) processes, the EWMA is useful as a predictor provided that, the AR term sufficiently dominates the moving average (MA) component. Harris and Ross (1991) studied the impact of traditional control charts based on residuals under several correlation structures of process and concluded that the traditional EWMA, CUSUM charting techniques perform poorly in the presence of serial correlation and the use of residuals in Shewhart chart is not effective for detecting small shifts when the observations are highly positively correlated. Lu and Reynolds (1999) considered the problem of monitoring the mean of a process in which the observations can be modeled as an AR(1) process plus a random error. They have suggested EWMA control chart based on the residuals from the forecast values of the model using an integral equation method. They showed that when the level of autocorrelation is low or moderate, the EWMA chart for residuals and the EWMA chart for original observations require about the same amount of time to detect various shifts; but for high levels of autocorrelation and large shifts, the EWMA chart of the residuals is faster a little. In all the efforts, an easy solution to control charting is missing. Control charts are effective tools for improving process quality and productivity and simplicity is always demanded by the users.

In this article, we introduce a Modified EWMA (Modified EWMA) chart that combines the features of a Shewhart chart and an EWMA chart in a simple way and has the ability to detect small, as well as large shift as soon as possible as required by some industrial processes with high level of first order autocorrelation. Modified EWMA control statistic gives weight to the past observations in slightly different way than EWMA and each current change is considered with full weight. This corrects EWMA statistic for suffering from inertia problem. The underlying idea is to adapt the weight of the past observations, past changes, the current observation and the current change. Unlike AEWMA control chart which corrects EWMA statistic by considering the difference between process value and control statistic value as the "error"$=Y_n-X_{n-1}$ to detect shifts of different sizes, Modified EWMA control chart does it by measuring the "current fluctuation" contained in EWMA control statistics, as $Y_n - Y_{n-1}$, the difference between two consecutive process values to detect abrupt changes. This is a new approach and does not require consideration of score function (error function) and separate detection limits; hence it is simpler to operate in comparison to AEWMA chart. We feel that just like EWMA, AEWMA should be applicable to autocorrelated processes, whenever EWMA is apt. The Modified EWMA control statistic is defined as:

$$X_n = (1-\lambda)X_{n-1} + \lambda Y_n + (Y_n - Y_{n-1}),\ 0< \lambda \leq 1.$$

Note that X_n is the geometric sum of past observations, past changes, current observation and the current change in the process. This make Modified EWMA scheme perform like an EWMA scheme for small shifts and at the same time detect abrupt changes alike Shewhart scheme.

EXPONENTIAL WEIGHT BASED QUALITY CONTROL CHARTS

Let Y_n, n =1, 2,... be a sequence of first order autocorrelated or independent normal observations with the process target mean as μ_0 and a common variance σ^2, such as single measurements of the process.

EWMA control chart

An EWMA control chart is based on the statistic:

$$X_n = (1-\lambda)X_{n-1} + \lambda Y_n, \quad (1)$$

where λ is a suitable constant, such that $0< \lambda \leq 1$, X_n is n^{th} statistic and Y_n is n^{th} observation . The quantity X_0 represents the starting value, often the target value μ_0.

This schemes signals when X_n value exceeds specified control limit. In general upper and lower control limits of EWMA Chart are:

$$UCL = \mu_0 + L\sigma\sqrt{\frac{\lambda}{(2-\lambda)}}$$

$$CL = \mu_0$$

$$LCL = \mu_0 - L\sigma\sqrt{\frac{\lambda}{(2-\lambda)}}. \quad (2)$$

L is suitable in control width limit, and σ is the process standard deviation. The detailed discussion on EWMA control schemes can be found in Montgomery (1996).

Adaptive EWMA control chart

EWMA control chart cannot detect abrupt change/s, and thus fail in worst-case situation. To overcome this problem, Capizzi and Masarotto (2003) designed Adaptive EWMA (AEWMA) control charts. Their class of control charts considers the following statistic:

$$X_n = X_{n-1} + \phi(e_n),\ X_0 = \mu_0 \quad (3)$$

Where $e_n = Y_n - X_{n-1}$ and $\phi(e_n)$ is a "score" function. An alarm is raised when $|X_n - \mu_0| > h$, where μ_0 denotes the target value of the process mean and h is a suitable threshold. The h value is mainly determined by ensuring that the desired mean time between false alarms is large. They have designed AEWMA schemes for three score functions, each based on exponential weight parameter λ and specified values of one or two constants. The AEWMA scheme is effective in detecting shifts of different sizes, but the monitoring scheme is not simple.

Modified EWMA (Modified EWMA) control chart

The desirable properties of a control chart are that it is easy to implement and is effective for detecting shifts of all sizes as per technical specifications. The Modified EWMA chart that we introduce considers past observations similar to EWMA scheme and additionally considers past changes, as well as latest change in the process. A Modified EWMA control chart is based on the statistic:

$$X_n = (1-\lambda)X_{n-1} + \lambda Y_n + (Y_n - Y_{n-1}), \quad (4)$$

where, n = 0,1,... $0 < \lambda \leq 1$ is a constant and the starting value is, $X_0 = \mu_0 = Y_0$.

The mean and variance of control statistic X_n are respectively, the process mean μ and $\left[\frac{\lambda}{(2-\lambda)} + \frac{2\lambda(1-\lambda)}{(2-\lambda)}\right]\sigma^2$ for autocorrelated process (Appendix 1), based on $\rho \to 1$, the first order autocorrelation coefficient, and correlation structure of the process observations and process fluctuations.

Therefore, the upper and lower control limits of Modified EWMA chart are:

$$UCL = \mu_0 + L\sigma\sqrt{\frac{\lambda}{2-\lambda} + \frac{2\lambda(1-\lambda)}{2-\lambda}}$$

$$CL = \mu_0$$

$$LCL = \mu_0 - L\sigma\sqrt{\frac{\lambda}{2-\lambda} + \frac{2\lambda(1-\lambda)}{2-\lambda}} \quad (5)$$

Where μ_0 is the target mean, σ^2 is the process variance, λ and L are suitable Modified EWMA scheme constants representing exponential weight and in control width limit. This chart reduces to Shewhart chart for highly autocorrelated process for λ=1, L=3. Modified EWMA chart is capable for capturing signals of small shift in the process, as well as abrupt changes in an autocorrelated process. The process is in control, as long as the values are plotted within the control limits. A point plots outside the control limits is interpreted as evidence that the process is out of control, and action are required to find and eliminate the assignable causes responsible for this behavior.

We know that EWMA chart suffer from inertia problem because of 'error' in EWMA statistic. Modified EWMA statistic is a corrected EWMA statistic and EWMA is the best predictor in the class of linear predictors (Yashchin, 1993). Hence Modified EWMA is the best predictor of process mean; its mean square error (MSE) is nil for autocorrelated process with nearly one autocorrelation (Appendix 1). It forecasts the process mean state accurately. This makes Modified EWMA chart free from inertia problem. Now we apply Modified EWMA control scheme monitoring along with EWMA scheme on temperature data and capsule weight data from an analytical process.

Application 1: Modified EWMA chart for highly autocorrelated normal process

Table 1 displays the part of measurements on temperature column taken every minute from a chemical process that is working in control and out of control situations, abrupt changes and small shifts occur. The target mean is 110.468°C, and the process standard deviation is σ = 1.131°C (σ^2 =1.279) (Table 1).

Here, EWMA (λ=0.1, L=2.814, UCL= 111.198, LCL=109.738, CL=110.468) control scheme could not detect two abrupt changes at 21st and 28th observations, and the small shifts were detected from 71st observation. Modified EWMA (λ=0.1, L=1.683, UCL = 111.199, LCL=109.737) detects those two abrupt changes and desired small shifts early, timely in 61st run. In Modified EWMA statistic, the smaller the value of λ, the larger is the effect of past history of the process. Therefore, if we select λ=0.1, then L=1.683 is automatically fixed through ARL.

Application 2: Modified EWMA chart for independent normal process

The data in Table 2 are part of the measurements (in grams) taken every 30 s from a manufacturing process that is working in control and an out of control situation,

Table 1. Monitoring performance of EWMA and modified EWMA for the chemical process temperature data.

No. n	Y_n AR(1)≈1	EWMA	Modified EWMA
		λ = 0.1	
		L = 2.814	L = 1.683
0		110.468	110.468
1	110.004	110.422	109.958
…	…	…	…
20	110.379	110.282	110.390
21	108.750	110.123	108.597*
22	110.419	110.129	110.448
…	…	…	…
27	110.518	110.291	110.543
28	107.800	110.042	107.551*
29	110.596	110.097	110.651
…	…	…	…
60	111.169	110.969	111.191
61	111.189	110.991	111.211*
62	111.208	11.013	111.230*
…	…	…	…
70	111.366	111.181	111.387*
71	111.386	111.201*	111.407*
…	…	…	…
780	113.610	113.628*	113.608*

*Shift.

Table 2. Monitoring performance of EWMA and Modified EWMA for the pharmaceutical process capsule weight data.

No. n	Y_n	EWMA	Modified EWMA
		λ = 0.04	
		L = 2.477	L = 1.423
0	5	5	5
1	5.22	5.009	5.229*
2	4.95	5.007	4.948
3	5.2	5.014	5.208*
4	5.41	5.030	5.426*
5	5.2	5.037	5.207*
6	5.02	5.036	5.019
7	5.11	5.039	5.113*
8	5.26	5.048	5.269*
9	5.27	5.057	5.279*
10	3.83	5.009	3.780*

*Shift.

abrupt change occurs. The target mean is 5 g, and the process standard deviation is $\sigma = 0.3$ g. Data were adapted from Capazzi and Masarotto (2003).

Here, EWMA (λ=0.04, L=2.477, UCL= 5.106, LCL=4.894, CL=5) control scheme could not detect any changes in observations. Modified EWMA (λ = 0.04, L=1.423, UCL = 5.104, LCL=4.896) detects upper shifts at 1st , 3rd to 5th , 7th to 9th, and lower shift at 10th observations.

From Figure 1, we can see that EWMA chart could not detect any Shift. On the other hand Modified EWMA chart detects all type of shifts as shown in Figure 2.

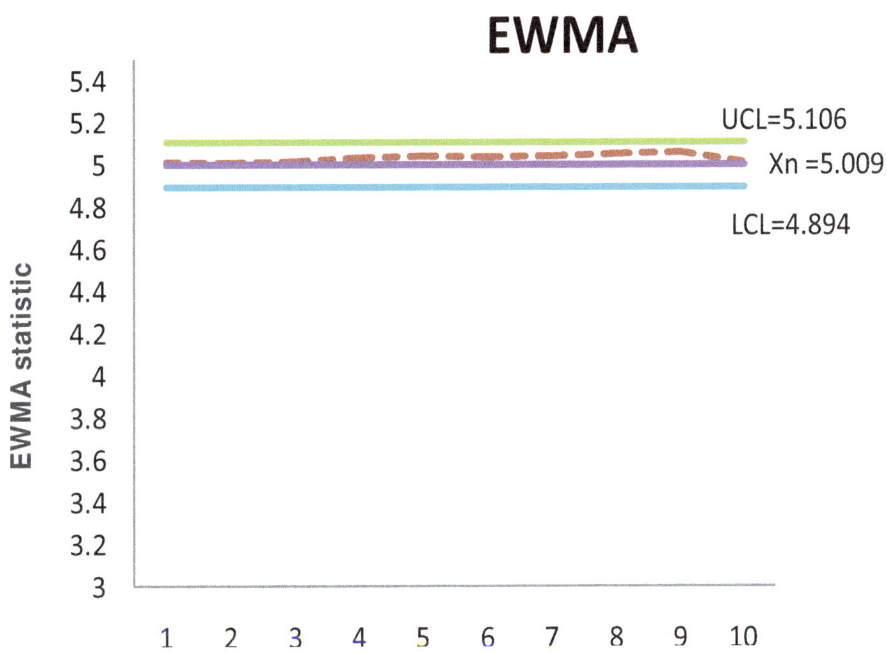

Figure 1. Plot of EWMA control chart.

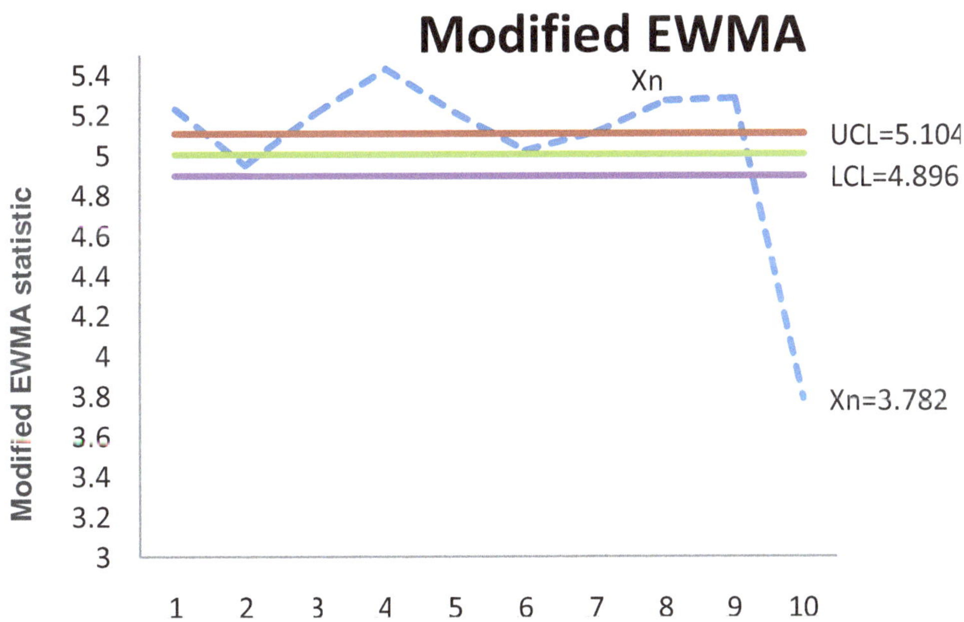

Figure 2. Plot of modified EWMA control chart.

PROPERTIES OF MODIFIED EWMA SCHEME AND COMPARISONS WITH EWMA SCHEME

ARL properties

All the ARL computations were carried out using Markov-chain approach described in Appendix 2. ARL properties of Modified EWMA scheme have been derived for different choices of {λ, L} and in control ARL to be 500 for detection of small shift only, because the abrupt changes are detected as they occur.

The Modified EWMA scheme introduced in this article is a modification of EWMA scheme for detecting small shifts along with abrupt shifts. Here, we have compared the ARL values for Modified EWMA scheme with ARL values of EWMA scheme over a wide range of parameter

Table 3. Average run lengths of modified EWMA schemes.

	Modified EWMA $ARL_0 = 500$							
L	**2.5**	**2.155**	**2.048**	**1.948**	**1.9**	**1.843**	**1.683**	**1.423**
λ	**0.75**	**0.5**	**0.4**	**0.3**	**0.25**	**0.2**	**0.1**	**0.043**
Shift \ Abrupt	**> 2 σ**	**> 2 σ**	**> 2 σ**	**>2 σ**	**>1 σ**	**>1 σ**	**>0.5σ**	**>0.5σ**
0.0	501	505	502	497	500	500	498	500
0.25	293	232	209	183	170	150	104	73.4
0.50	131	83.1	67.7	53.7	47.7	41.2	30.3	27.3
0.75	58.3	33.3	26.73	21.3	19.24	17.3	14.9	16.6
1.0	28.4	15.87	13.01	10.9	10.2	9.6	9.35	11.6
1.50	8.99	5.52	4.87	4.53	4.49	4.53	5.09	7.08
2.0	3.96	2.8	2.61	2.58	2.64	2.76	3.37	4.99
2.50	2.29	1.84	1.75	1.75	1.8	1.91	2.45	3.8
3.0	1.65	1.47	1.42	1.39	1.4	1.44	1.87	3.03
3.50	1.36	1.34	1.32	1.27	1.24	1.23	1.47	2.33
4.0	1.21	1.28	1.29	1.23	1.23	1.17	1.2	1.98
5.0	1.04	1.07	1.08	1.1	1.13	1.06	1.07	1.15

Table 4. Average run lengths of EWMA scheme (Lucas and Saccucci, 1990).

	EWMA $ARL_0 = 500$											
Shift	**0**	**0.25**	**0.50**	**0.75**	**1**	**1.50**	**2**	**2.5**	**3**	**3.5**	**4**	**5**
λ = 0.75, L = 3.087	500	321	140	62.4	30.5	9.86	4.52	2.67	1.87	1.46	1.23	1.04
λ = 0.5, L = 3.071	499	254	88.4	35.7	17.3	6.44	3.58	2.47	1.91	1.58	1.36	1.10
λ = 0.1, L = 2.814	492	104	30.6	15.5	10.1	5.99	4.31	3.41	2.85	2.47	2.20	1.83
λ = 0.04, L = 2.477	487	78.4	28.0	16	11.2	7.03	5.18	4.14	3.48	3.02	2.68	2.22

values. In Table 3, the ARL values of Modified EWMA for all choices of λ are smaller than those for EWMA scheme. ARL values for EWMA schemes as per Markov Chain approach are shown in Table 4 for the sake of easy comparisons.

Thus, detection of small, as well as large shifts by a single Modified EWMA chart is possible with proper choice of λ and knowledge of AR(1) parameter and/or correlation structure of the process observations and the process fluctuations. It should be clear that, the process fluctuation variance would be smaller for process being controlled for small shifts and few large abrupt changes.

Inertia (Signal resistance) properties

Woodall and Mahmoud (2005) proposed a measure of inertia, the signal resistance, to be the largest standard deviation from target not leading to an immediate out-of-control signal. They define a simple measure of inertia that they called "signal resistance". In physics, "inertia" refers to the state of resistance an object has to a change in its state of motion. Similarly, in statistical process control, "inertia" can refer to a measure of the resistance of a chart to signaling a particular process shift. For the EWMA chart, the signal resistance is:

$$SR(EWMA) = [\, h - (1-\lambda)\, w \,] / \lambda \qquad (6)$$

Where w is the value of the EWMA statistic, $h = L\,\sigma \sqrt{\frac{\lambda}{(2-\lambda)}}$. The asymptotic control limits for the EWMA chart are $\pm h$. The signal resistance for an EWMA control chart in conjunction with Shewhart control limits is:

$$SR\,(EWMA + Shewhart) = \begin{cases} L & \text{if } -h \le w < (h-\lambda L)/(1-\lambda) \\ [h-(1-\lambda)w]/\lambda & \text{if } (h-\lambda L)/(1-\lambda) \le w \le h \end{cases} \qquad (7)$$

Where L is the value of the multiplier used to obtain the Shewhart limit and w is the value of the EWMA statistic.

Obviously, in this case, the signal resistance cannot exceed the value of the multiplier used to obtain the Shewhart limit, that is, L. The signal resistance for Modified EWMA control chart is:

SR(Modified EWMA) = $[h_1 - (1-\lambda)w_1]/\lambda \quad if -h_1 \leq w_1 \leq h_1$ (8)

Where w_1 is the value of the Modified EWMA statistic,

$h_1 = L\sigma\sqrt{\frac{\lambda}{2-\lambda}+\frac{2\lambda(1-\lambda)}{2-\lambda}}$ is the Modified EWMA decision limit.

Conclusion

A simple control chart for monitoring small, as well as large shifts in highly autocorrelated or independent normally distributed observations, such as analytical processes are given. In addition, the Modified EWMA chart can also be used to forecast the observation in the next period, which can help analysts take preventive actions before process departures to the out-of-control state.

REFERENCES

Brook D, Evans DA (1972). An Approach to the Probability Distribution of CUSUM Run Lengths, Biometrika, 59: 539-549.

Capizzi G, Masarotto G (2003). An Adaptive Exponentially Weighted Moving Average Control Chart. Technometrics, 45: 199-207.

Cox DR (1961). Prediction by Exponentially Weighted Moving Averages and Related Methods. J. Roy. Stat. Soc., 23: 414-442.

Crosier RB (1986). A New Two-Sided Cumulative Sum Quality Control Scheme. Technometrics, 28: 187-194.

Harris TJ, Ross WH (1991). Statistical Process Control Procedures for Correlated Observations. The Can. J. Chem. Eng., 69: 48-57.

Lu CW, Reynolds MR Jr. (1999). EWMA Control charts for Monitoring the Mean of Autocorrelated Processes. J. Qual. Tech., 32(2): 166-188.

Lucas JM, Saccucci MS (1990). Exponentially Weighted Moving Average Control Schemes: Properties and Enhancements. Technometrics, 32: 1-12.

Montgomery DC, Mastranglo CM (1991). Some Statistical Process Control Methods for Autocorrelated Data. J. Qual. Tech., 23(3): 179-204.

Montgomery DC (1996). Introduction to Statistical Quality Control., 3rd Ed, Wiley, p. 9

Reynolds MR Jr, Stoumbos ZG (2006). Comparisons of Some Exponentially weighted Moving Average Control Charts for Monitoring the Process Mean and Variance. Technometrics, 48(4): 550-567.

Page ES (1954). Continuous Inspection Schemes. Biometrika, (41): 100-115.

Roberts SW (1959). Control chart tests Based on Geometric Moving Averages. Technometrics, 1: 239-250.

R Language: Copyright 2004, The R Foundation for Statistical Computing Version 2.0.1 (2004-11-15), ISBN 3-900051-07-0

Wardell DG, Moskowitz H, Plante RD (1994). Run Length Distributions of Special Cause Control Charts for Correlated Processes, Technometrics, (36): 3-17.

Woodall WH, Mahmoud (2005). The Inertial Properties of Quality Control Charts. Technometrics, 47: 4.

Yashchin E (1995). Estimating the Current Mean of a Process Subject to Abrupt Changes. Technometrics, 37: 311-323.

APPENDIX

Appendix 1: Mean and variance of modified EWMA control statistic

Let $\{Y_n, n \geq 1\}$ are first order auto correlated $(\rho \to 1)$ process observations following normal distribution with mean μ and variance σ^2. For mean and variance, consider expanding Modified EWMA statistic:

$X_n = \lambda Y_n + (1-\lambda) X_{n-1} + (Y_n - Y_{n-1})$
$= \lambda Y_n + (1-\lambda) [\lambda Y_{n-1} + (1-\lambda) X_{n-2} + (Y_{n-1} - Y_{n-2})] + (Y_n - Y_{n-1})$
$= \lambda Y_n + \lambda(1-\lambda)Y_{n-1} + (1-\lambda)^2 X_{n-2} + (1-\lambda)(Y_{n-1} - Y_{n-2}) + (Y_n - Y_{n1})$
$= \lambda Y_n + \lambda(1-\lambda)Y_{n-1} + (1-\lambda)^2 [\lambda Y_{n-2} + (1-\lambda) X_{n-3} + (Y_{n-2} - Y_{n-3})] + (1-\lambda)(Y_{n-1} - Y_{n-2}) + (Y_n - Y_{n-1})$
$= \lambda Y_n + \lambda(1-\lambda)Y_{n-1} + \lambda(1-\lambda)^2 Y_{n-2} + (1-\lambda)^3 X_{n-3} + (1-\lambda)^2 (Y_{n-2} - Y_{n-3}) + (1-\lambda)^1 (Y_{n-1} - Y_{n-2}) + (1-\lambda)^0 (Y_n - Y_{n-1})$

And continuing like this recursively for X_{n-j}, j = 2, 3, . . . ,n, we obtain:

$$X_n = \lambda \sum_{j=0}^{n-1} (1-\lambda)^j Y_{n-j} + (1-\lambda)^n X_0 + \sum_{j=0}^{n-1} (1-\lambda)^j (Y_{n-j} - Y_{n-j-1}).$$

Here, $\sum_{j=0}^{n-1} (1-\lambda)^j (Y_{n-j} - Y_{n-j-1})$ accounts for sum of the past and latest change/fluctuations in the process; the unaccounted current fluctuations accumulated to time n in EWMA statistic:

$$\text{Let } X_n = \lambda \sum_{j=0}^{n-1} (1-\lambda)^j Y_{n-j} + (1-\lambda)^n Y_0 + \sum_{j=0}^{n-1} (1-\lambda)^j (Y_{n-j} - Y_{n-j-1})$$

Taking expectation on both side:

$$E(X_n) = \lambda \sum_{j=0}^{n-1} (1-\lambda)^j E(Y_{n-j}) + (1-\lambda)^n E(Y_0) + \sum_{j=0}^{n-1} (1-\lambda)^j E(Y_{n-j} - Y_{n-j-1})$$

$$\text{But } \lambda \sum_{j=0}^{n} (1-\lambda)^j = \frac{\lambda\,[1-(1-\lambda)^n]}{[1-(1-\lambda)]} = [1-(1-\lambda)^n]$$

$$\therefore E(X_n) = [1-(1-\lambda)^n]\mu + (1-\lambda)^n \mu + 0 \qquad \text{(i)}$$
$$= \mu$$

The starting value of process is, $X_0 = \mu_0 = Y_0$ and $0 < \lambda \leq 1$ is a constant. The mean is:

$E(X_n) = E((1-\lambda)X_{n-1} + \lambda Y_n + (Y_n - Y_{n-1})) = \mu_0$.

The variance of Modified EWMA control statistic X_n is:

$$V(X_n) = V(\lambda \sum_{j=0}^{n-1} (1-\lambda)^j Y_{n-j}) + V((1-\lambda)^n Y_0) + V(\sum_{j=0}^{n-1} (1-\lambda)^j (Y_{n-j} - Y_{n-j-1}))$$

$$V(X_n) = (1-\lambda)^{2n} V(Y_0) + \sum_{j=0}^{n-1} \lambda^2 (1-\lambda)^{2j} V(Y_j) + 2\sum_{j=0}^{n-1} \lambda^2 (1-\lambda)^{2j+1} Cov(Y_{n-j}, Y_{n-j-1})$$
$$+ \sum_{j=0}^{n-1} (1-\lambda)^{2j} V(Y_{n-j} - Y_{n-j-1}) + 2\sum_{j=0}^{n-1} (1-\lambda)^{2j+1} Cov[(Y_{n-j} - Y_{n-j-1}), (Y_{n-j-1} - Y_{n-j-2})]$$
$$+ \sum_{j=0}^{n-1} \lambda(1-\lambda)^{2j} Cov(Y_{n-j}, (Y_{n-j} - Y_{n-j-1})) + \sum_{j=0}^{n-1} \lambda(1-\lambda)^{2j+1} Cov(Y_{n-j-1}, (Y_{n-j} - Y_{n-j-1}))$$

Since Y_n's are autocorrelated normal with variance σ^2, the variance of $(Y_n - Y_{n-1})$ $(n \geq 1)$ is:

$$\sigma_1^2 = 2\sigma^2 - 2\rho\sigma^2 = 2(1-\rho)\sigma^2 \text{ (small when } \rho \to 1).$$

The weights $\lambda(1-\lambda)^{2j}$ decrease geometrically with the age of sample mean. Suppose Y_n's are correlated to the forward fluctuation $(Y_n - Y_{n-1})$ $(n \geq 1)$ with common correlation ρ_1 and correlated to the backward fluctuation $(Y_{n+1} - Y_n)$, $(n \geq 0)$ with common correlation ρ_2, and forward fluctuation $(Y_n - Y_{n-1})$ are correlated to the backward fluctuation $(Y_{n+1} - Y_n)$, $(n \geq 1)$ with common correlation ρ_3, then asymptotic variance for large n is given as:

$$V(X_n) = \frac{\lambda}{(2-\lambda)}\sigma^2 + \frac{2\lambda(1-\lambda)}{(2-\lambda)}\rho\sigma^2 + \frac{2(1-\rho)\sigma^2}{\lambda(2-\lambda)} + \frac{4\rho_3(1-\rho)(1-\lambda)\sigma^2}{\lambda(2-\lambda)} + \frac{2\sqrt{2}\rho_1\sqrt{(1-\rho)}\sigma^2}{(2-\lambda)} + \frac{(1-\lambda)2\sqrt{2}\rho_2\sqrt{1-\rho}\,\sigma^2}{(2-\lambda)} \qquad \text{(ii)}$$

In normal autocorrelated process (a) with ρ_3 nearly negative half and ρ_1 , ρ_2 nearly equal and opposite in sign and being monitored for small shifts, (b) with autocorrelation ρ nearly one, the aforementioned expression (ii), reduces to:

$$V(X_n) = \frac{\lambda}{(2-\lambda)}\sigma^2 + \frac{2\lambda(1-\lambda)}{(2-\lambda)}\rho\sigma^2 \qquad \text{(iii)}$$

Let $\frac{2\lambda(1-\lambda)}{(2-\lambda)}\rho\sigma^2$ be a small value for high value of $\rho \to 1$ and small λ, sometimes even negligibly small such that

Modified EWMA limits equal EWMA limits.

Therefore:

$$V(X_n) = \left[\frac{\lambda}{(2-\lambda)} + \frac{2\lambda(1-\lambda)}{(2-\lambda)} \right] \sigma^2. \quad \text{(iv)}$$

The upper-lower control limits for Modified EWMA chart are:

$$\mu_0 \pm L\,\sigma \sqrt{\frac{\lambda}{(2-\lambda)} + \frac{2\lambda(1-\lambda)}{(2-\lambda)}}. \quad \text{(v)}$$

Appendix 2: Cyclical steady state average run lengths of modified EWMA scheme using Markov-chain approach

The use of Markov chains to approximate ARL of CUSUM schemes was discussed by Brook and Evans (1972). Crosier (1986) modified their Markov chain approach to compute conditional and cyclical steady state ARLs for CUSUM scheme. Further, Lucas and Saccucci (1990) customized their approach for ARL properties of EWMA scheme. They evaluated the properties of the continuous state Markov chain by discretizing the infinite state transition probability matrix. This procedure involves dividing the interval between the upper and lower control limits into t = 2m+1 subintervals of width 2δ. The control statistic, X_n, is said to be in transient state i at time n, if $S_i - \delta < X_n \le S_i + \delta$ for i = -m,-m+1, . . . , m, where S_i represent the midpoint of the ith interval. For example, when the LCL = -0.6455759 to UCL = 0.6455759 distance is divided into 5 sub-intervals each of width 2δ (δ=0.1291), -0.6455759, -0.3873455 ,-0.1291152, 0.1291152 , 0.3873455 , 0.6455759; the S_i 's are -0.5165 ,-0.2582 , 0.0000, 0.2582, 0.5165.

The transition probability matrix (t.p.m.), represented in partitioned matrix form, is given by:

$$P=\begin{pmatrix} \mathbf{R} & (\mathbf{I}-\mathbf{R})\mathbf{1} \\ \mathbf{0}^T & 1 \end{pmatrix},$$

where the sub matrix R contains the probabilities of going from one transient state to another, **I** is the identity matrix, and **1** is a column vector of ones. Hence P_{ij} represents the probability that the control statistic goes from state (i) to state (j) in one step. We have derived algorithm for calculating cyclical steady state ARL of Modified EWMA scheme using the Markov chain approach following Crosier (1986) and Lucas and Saccuscci (1990). The transition probabilities in **R** are approximated by assuming that the control statistic is equal to S_i whenever it is in state (i), i = -m, -m+1, … , m.

Let Y_n, n = 1, 2 … denote the sequence of process observations, which are first order autocorrelated normal with mean μ and variance σ^2. Then the transient change occurring in state of Modified EWMA control statistic is $X_n-(1-\lambda)X_{n-1}$, which is equivalent to $\lambda Y_n + (Y_n - Y_{n-1})$. The transition in Modified EWMA control chart statistic has the following mean and variance:

$$E(S_n|S_{n-1}) = E[\lambda Y_n + (Y_n-Y_{n-1})] = \lambda\mu, \text{ and}$$
$$\begin{aligned} V(S_n|S_{n-1}) &= V[\lambda Y_n + (Y_n-Y_{n-1})] \\ &= \lambda^2 V(Y_n) + V(Y_n-Y_{n-1}) + 2\text{cov}\{Y_n, (Y_n-Y_{n-1})\} \\ &= \lambda^2\sigma^2 + 2(1-\rho)\sigma^2 + 2\sqrt{2(1-\rho)}\,\rho_1\,\sigma^2 \\ &= \lambda^2\sigma^2 \end{aligned}$$

where variance, is a small value for nearly 1 values ρ.

Then the in control transition probabilities are approximated as:

P_{ij} = Pr(going to S_j | in S_i)
≈ Pr $[(\lambda\sigma)^{-1}\{(S_j-\delta)-(1-\lambda)S_i\} < Y_n \le (\lambda\sigma)^{-1}\{(S_j+\delta)-(1-\lambda)S_i\}]$,

i, j = -m, -m+1, …, m.

(m =12 imply t = 25), where Φ represents the standard normal distribution function.

Appendix 3: Algorithm and R program for ARL computation of modified EWMA

Step-1: Choose the parameter constant λ, target mean μ, standard deviation σ, Limit L.

Step -2: Compute UCL=$\mu_0 + L\,\sigma \sqrt{\frac{\lambda}{(2-\lambda)} + \frac{2\lambda(1-\lambda)}{(2-\lambda)}}$,

LCL=$\mu_0 - L\sigma \sqrt{\frac{\lambda}{(2-\lambda)} + \frac{2\lambda(1-\lambda)}{(2-\lambda)}}$.

Step-3: Choose number of sub-intervals (states), t = 2 m + 1.
Step-4: Compute width w = (UCL - LCL)/t, shift δ = w/2.
Step-5: Compute the values of S_i i = -m, -m+1, …, m for each sub-interval.
Step-6: Compute the transition probability matrix (t. p. m.). Compute R,
$R = \Phi[(\lambda\sigma)^{-1}\{(S_j+\delta)-(1-\lambda)S_i-\lambda\mu\}] - \Phi[(\lambda\sigma)^{-1}\{(S_j-\delta)-(1-\lambda)S_i-\lambda\mu\}]$
Step-7: Adjust the t. p.m. such that row sums are unity.
Step-8: Compute u = $[I - R]^{-1}$ **1**
Step-9: Compute q = R'* I
Step-10: ARL = q' * u, OR ARL = $q'[I - R]^{-1}$ **1**

Enhanced photoluminescence from ZnO/ZnS core-shell structure

Sarmila Dutta[1]*, Shibabrata Basak[2] and Pijus Kanti Samanta[3]

[1]Department of Physics, IMPS College of Engineering and Technology Malda, India.
[2]School of Applied Sciences, Haldia Institute of Technology, Haldia, India.
[3]Department of Physics, Ghatal Rabindra Satabarsiki Mahavidyalaya, Ghatal, Paschim Medinipur, India.

The effect of ZnS coating over ZnO core is discussed in this present report. Very simple two step wet chemical method was used for the fabrication of this core-shell structure. XRD and HRTEM data reveal the formation of good shell coating over the ZnO core. Photoluminescence spectra shows that emission intensity is at least four times higher when coated with ZnS compare to bare ZnO. Band gap is also calculated using UV-vis spectroscopy. This study will be very useful towards tuning the emission properties and developing intense violet light-emitting and nanophotonic devices.

Key words: Core-shell, photoluminescence, violet emission, green emission.

INTRODUCTION

Semiconductor nanostructures have gained considerable interest during the past decade. These materials offer advanced applications in the various optoelectronic (Chengyao et al., 2009), biological (Manish et al., 2010) areas due to their size dependant physical and optical properties. ZnO being a wide direct band gap (~3.37 eV) material with very stable emission properties (exciton binding energy 60 meV) at near UV to visible region (Jadwisienczak et al., 2002; Xiaosheng et al., 2010; Mingsong et al., 2010) is one of the most investigated nanomatrials. Different synthesis routes (Ye et al., 2008; Hong et al., 2006; Nause and Nemeth, 2005; Kaidashev et al., 2003; Lim et al., 2008; Pai-Chun et al., 2004; Walter and Sheng-Yuan, 2002; Pijus et al., 2011; Hua et al., 2011) with variation of synthesis parameters (that is, precursor concentration, reaction time, pH level, temperature, doping) (Bin et al., 2006; Rai et al., 2011; Vernardou et al., 2007) has been employed to tune the luminescent property of the fabricated structures. In this context surface coating with different band-gap materials also leads to dramatic change of the luminescence properties (Iqbal et al., 2009; Tang et al., 2010) of the host material.

*Corresponding author. E-mail: sarmila.dutta09@gmail.com.

ZnS with wide band-gap energy of ~3.66 eV at room temperature is a well-known luminescent material, having prominent applications in flat-panel displays, electroluminescent devices, sensors and lasers (Peng et al., 2006; Murugadoss and Arun, 2008; Zengxing et al., 2005). In this paper we analyze the coating effect of higher band gap ZnS over ZnO prepared using simple, low cost two step chemical route synthesis of ZnO/ZnS with high quantum yield.

METHODOLOGY

Material preparation

In the typical preparation of ZnO/ZnS core–shell nanostructure, 1 M 50 ml of zinc acetate stirred vigorously for 25-30 min over a magnetic stirrer. Oxide particles were precipitated by slowly addition of 50 ml of 1 M NaOH, after 4 h 50 ml of 1 M Na_2S solution was added to the above mixture and stirring was continued for an hour. The same procedure was used for the synthesis of ZnO, except for the addition of Na_2S solution. The precipitate was filtered and washed several times with ethanol. Finally the precipitate was dried at 40°C for further characterization.

Material characterizations

The grazing incidence X-ray diffraction was carried in Rigaku X-ray diffractometer system over an angular range of 30° < 2θ < 60° using

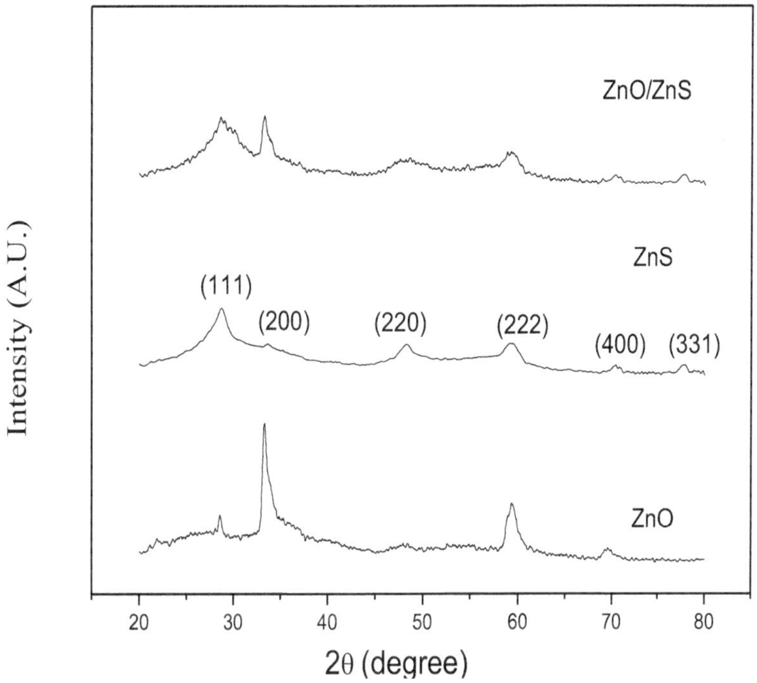

Figure 1. XRD pattern of the ZnO, ZnS and ZnO/ZnS core-shell structures.

Cu-k_α radiation of wavelength λ = 1.54Å. For further structural characterization transmission electron microscopy images were taken in a JEOL transmission electron microscope. Room temperature photoluminescence data were recorded using a PERKIN ELMER LS-55 with a xenon lamp with excitation of 325 nm. UV-visible spectroscopy was used to calculate the band gap.

RESULTS AND DISCUSSION

X-ray diffraction

Figure 1 shows the XRD patterns of the ZnO nanocrystals, ZnS nanoparticles and the ZnO/ZnS core–shell structures. All the peaks of ZnO were consistent with the literature data of JCPDS 21-1486. After the introduction of Na_2S, both ZnO and cubic ZnS (JCPDS card no.80-0020) were found to coexist in the synthesized products. Because the peaks of ZnS were almost at same position of ZnO, the peaks were overlapped together in ZnO/ZnS core-shell structures.

HRTEM analysis

To study the structural information in details, HRTEM was done. Figure 2a shows the structures are of 50 nm in diameter with almost uniform shell coating. Figure 2b shows the SAED pattern of the core-shell structures. The SAED of an individual ZnO/ZnS core/shell structures displays a spotted pattern that corresponded to the single crystal of ZnO and a set of diffraction rings that fitted to the polycrystalline ZnS nanoparticles.

Photoluminescence study

Figure 3 shows the PL spectra of as grown ZnO/ZnS core-shell structures. To compare the luminescence intensity of the ZnO/ZnS core-shell PL spectra of bare ZnO is use as a reference. Two emitting bands, including a strong violet emission centered at around 422 nm and a weak green band centered at around 530 nm, were observed in ZnO nanocrystals. The origin of the violet and green emission could be attributed to the defect-related emissions between Zn_i and VB; O_i and VB respectively. The PL spectra of ZnO/ZnS core–shell structures show an enhanced broad violet emission and a disappearing green emission compared to that of ZnO nanocrystals. The enhancement in violet emission was because ZnS nanoparticles has higher band gap than ZnO and it suppressed the tunneling of the charge carriers from the core to the ZnS nanoshell. As a result, more photo generated electrons and holes are confined inside the ZnO core, giving rise to a high quantum yield about 4 times more that of nano ZnO alone. However, the green emission is weakened to a great extent.

As we know, the green emissions originate from oxygen interstitials. During the process of the formation of ZnO/ZnS core–structures, the hydrolysis of Na_2S can produce H_2S and Sulphur atom can replace and squeeze

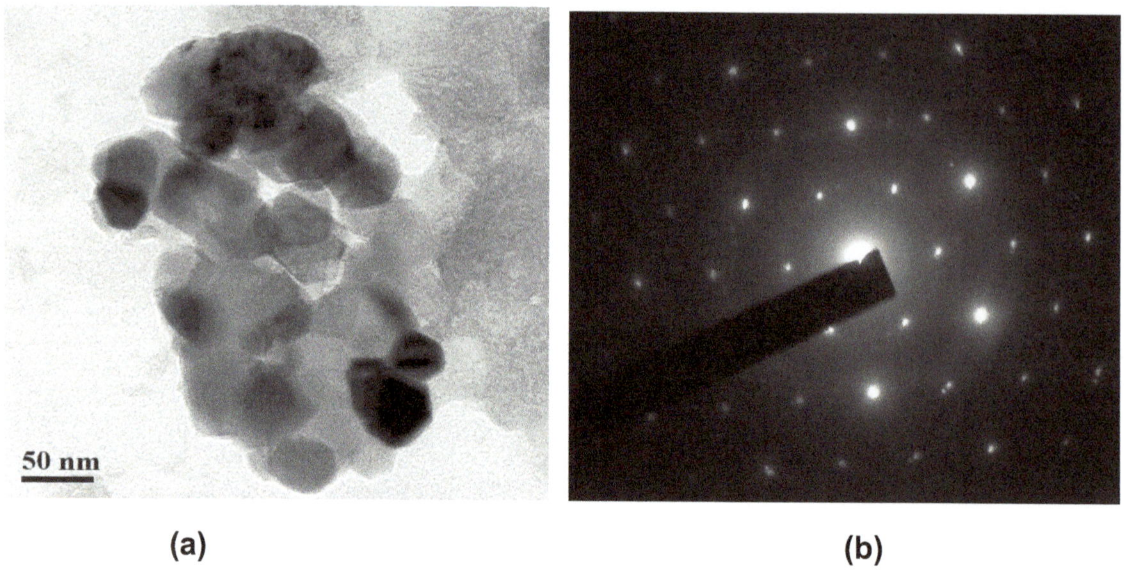

Figure 2. (a) HRTEM image of ZnO/ZnS core-shell structure and (b) its SAED pattern.

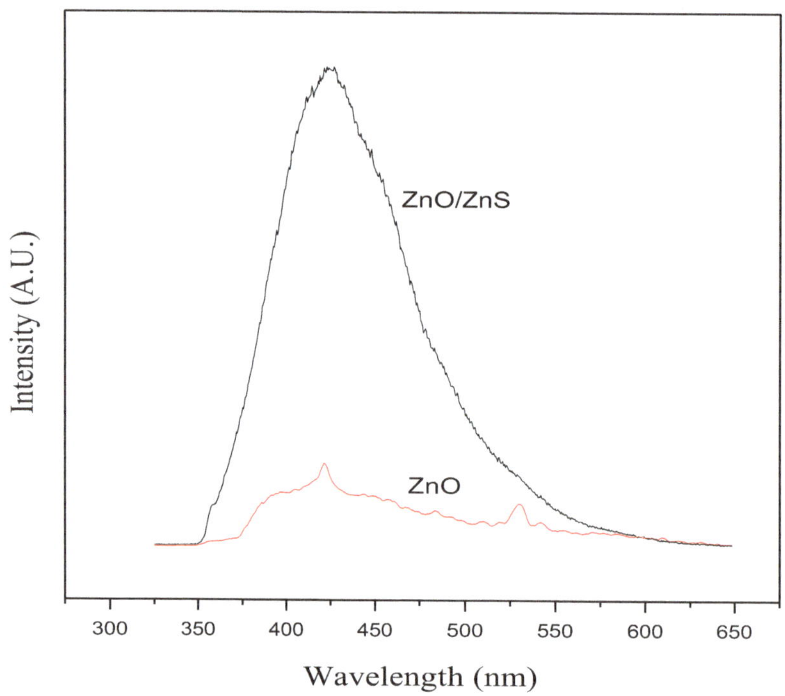

Figure 3. Room temperature PL spectra of ZnO and ZnO/ZnS core-shell structure.

out the oxygen interstitials from the surface of ZnO nanocrystals. As a result, the concentration of oxygen interstitials is reduced greatly and the green emission is weakened. The above results indicated that the sulfidation process has a great effect on the relative intensity and position of typical PL properties of ZnO nanocrystals. Therefore, the PL properties of ZnO nanocrystals could be tuned by this approach.

UV-Vis Spectroscopy

UV-Visible spectroscopy is frequently employed for band gap calculation (Samanta and Roy, 2011). For this purpose, dilute solution of the sample is prepared mixing deionized water in ultrasonic bath and in UV spectrometer transmittance (T) with respect to wavelength (λ) is measured. Absorbance (α) at the

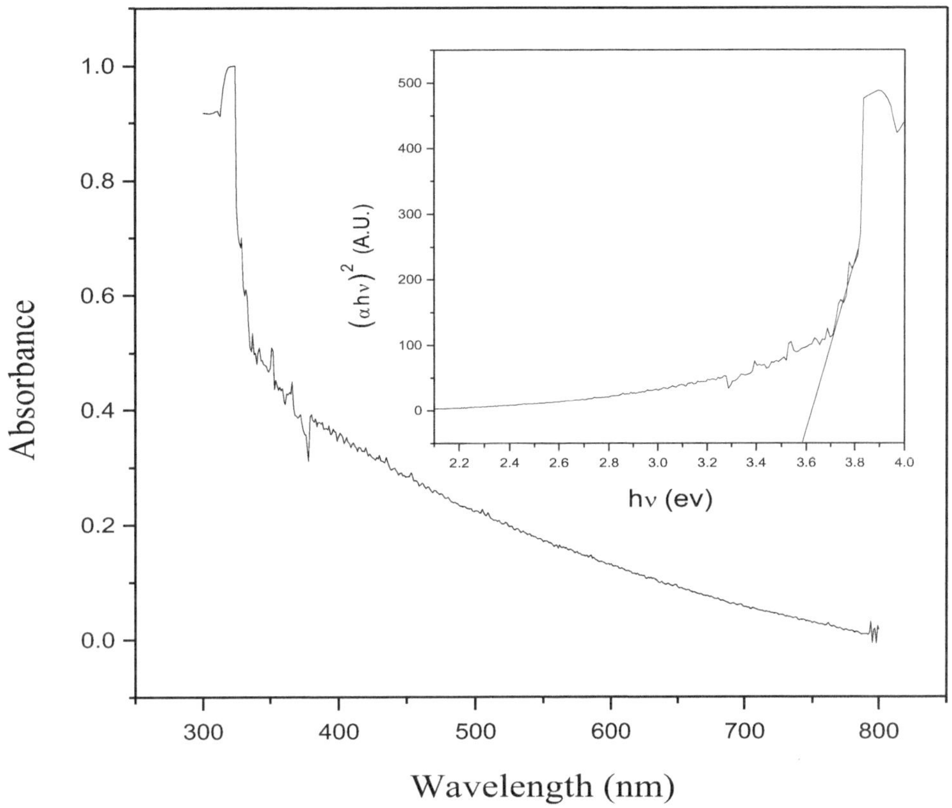

Figure 4. Absorbance spectra of ZnO/ZnS and inset show the band gap.

corresponding wavelengths λ obtained using Beer-Lambert's relation, $\alpha = \frac{1}{d}\ln\left(\frac{1}{T}\right)$, where, *d* is the path traversed by the light inside the solution. For direct band gap ZnO, $(h\nu) \propto \frac{\sqrt{h\nu - E_g}}{h\nu}$, where E_g is the band gap and hv is the incident energy of the photon. Thus band gap energy (E_g) obtained from the intercept of the straight plot of $(\alpha h\nu)^2$ vs hv. Figure 4 shows the UV absorbance spectra of ZnO/ZnS core-shell structures which reveal the highest absorbance at ~ 325 nm. Band gap of 3.58 eV is calculated from the $\alpha(h\nu) \propto \frac{\sqrt{h\nu - E_g}}{h\nu}$ as discussed above, inset of Figure 4.

Conclusion

This simple two-step, solution-based route to synthesize ZnO/ZnS core-shell structures have a wide potential application in large-scale conversion from metal oxides to metal sulfide. The study on the PL spectra of ZnO/ZnS core–shell nanostructures reveals enhanced violet emissions by a factor of four and also passivated green emission. Thus the present study shows a way to tune the visible emission of the ZnO by the formation of ZnO/ZnS core–shell structure. Also high intensity violet emission leads to potential fabrication of nanoscale optical devices.

REFERENCES

Bin C, Wensheng S, Joette MR-T, Lei Z, Edward TS (2006). Synthesis of Variable-Aspect-Ratio, Single-Crystalline ZnO Nanostructures. Inorg. Chem., 45. 1208.

Chengyao L, Min G, Chen D, Xiaoxian Z, Lihuan Z, Qing C, Lian-Mao P (2009). *In situ* comprehensive characterization of optoelectronic nanomaterials for device purposes. Nanotechnology, 20(17): 175703.

Hong JF, Bodo F, Roland S, Cameliu H, Andreas B, Hartmut L, Armin D, Alois K, Silke C, Ulrich G, Margit Z (2006). Vapour-transport-deposition growth of ZnO nanostructures: switch between c-axial wires and a-axial belts by indium doping. Nanotechnology, 17(11): S231.

Hua G, Sun L, Tian Y, Chen X, Zhang L, Dai J, Hu L, Dai S(2011). A Facile Chemical Route to Synthesize ZnO Nanoarrays: One-Dimensional Nanowire Arrays and Two-Dimensional Porous Nanosheet Arrays. J. Nanosci. Nanotechnol., 11(4): 3488.

Iqbal J, Baiqi W, Liu X, Huichao Z, Dapeng Y, Ronghai YJ (2009). Role of sp-d Exchange Interactions in Room-Temperature Photoluminescence and Ferromagnetism of CuCo Co-Doped ZnO Nanorods. Nanosci. Nanotechnol., 9: 6823.

Jadwisienczak WM , Lozykowski HJ , Xu A, Patel B (2002). Visible emission from ZnO doped with rare-earth ions. J. Electronic Mat., 31 (7): 776.

Kaidashev EM, Lorenz M, von Wenckstern H, Rahm A, Semmelhack H-C, Han K-H, Benndorf G, Bundesmann C, Hochmuth H, Grundmann M (2003). High electron mobility of epitaxial ZnO thin films on c-plane sapphire grown by multistep pulsed-laser deposition. Appl. Phys. Lett., 82: 3901.

Lim SJ, Soonju Kwon, Kim H (2008). ZnO thin films prepared by atomic layer deposition and rf sputtering as an active layer for thin film transistor. Thin Solid Films, 516(7): 1523

Manish S, Dennis BP, Marc RK (2010). Biological Surface Effects of Metallic Nanomaterials for Applications in Assembly and Catalysis.Langmuir, 26 (19): 15121.

Mingsong W, Eun WS, Jin SC, Seung HH, Eui JK, Sung HH, Kee-Kahb K (2010). Tunable visible emission and warm white photoluminescence of lithium-doped zinc oxide thin films. J. Mater. Sci., 45(15): 4111.

Murugadoss A, Arun C(2008). Tuning photoluminescence of ZnS nanoparticles by silver. Bull. Mater. Sci., 31(3): 533.

Nause J, Nemeth B (2005). Pressurized melt growth of ZnO boules. Semicond. Sci. Technol., 20(4): S45

Pai-Chun C, Zhiyong F, Dawei W, Wei-Yu T, Wen-An C, Juan H, Jia GL(2004). ZnO Nanowires Synthesized by Vapor Trapping CVD Method. Chem. Mater., 16(24): 5133.

Peng WQ, Cong GW, Qu SC, Wang ZG (2006). Synthesis and photoluminescence of ZnS:Cu nanoparticles. Opt. Mater., 29: 313.

Pijus KS, Shibabrata B, Partha RC (2011). Fern leaves: The secret life of zinc oxide. Mater. Today, 14(6): 295.

Rai P, Jo JN, Wu XF, Yoon JM, Yu YT (2011). Synthesis of well dispersed, regular shape ZnO nanorods: effect of pH, time and temperature, J. Nanosci. Nanotechnol., 11(1): 64.

Samanta PK, Roy CP (2011). Growth and Optical Properties of Chemically Grown ZnO Nanobelts. Sci. Adv. Mater., 3: 112.

Tang B, Deng H, Shui ZW, Zhang Q (2010). Synthesis and Optical Properties of Vertically Aligned ZnO Nanorods. J. Nanosci. Nanotechnol., 10: 1842.

Vernardou D, Kenanakis G, Couris SEK, Kymakis E, Katsarakis N (2007). pH effect on the morphology of ZnO nanostructures grown with aqueous chemical growth. Thin Solid Films, 515: 8764.

Walter W, Sheng-Yuan C (2002). Physical and structural properties of ZnO sputtered films. Mater. Lett., 55: 67.

Xiaosheng T, Eugene SGC, Ling L, Jun D, Junmin X (2010). Synthesis of ZnO Nanoparticles with Tunable Emission Colors and Their Cell Labeling Applications. Chem.Mat., 22(11): 3383.

Ye S, Neil AF, Jason RD, Michael NRA (2008). Hydrothermal Growth of ZnO Nanorods Aligned Parallel to the Substrate Surface. J. Phys. Chem. C, 112(25): 9234-9239.

Zengxing Z, Jianxiong W, Huajun Y, Yan G, Dongfang L, Li S, Yanjuan X, Xiaowei Z, Lifeng L, Shudong L, Xinyuan D, Shicheng M, Weiya Z, Sishen X(2005). Low-Temperature Growth and Photoluminescence Property of ZnS Nanoribbons. J. Phys. Chem. B, 109(39): 18352.

Chlorine adsorption kinetics of activated carbon from selected local raw materials

Yusufu, Mohammed I.[1]*, Ariahu, C. C.[2], Nkpa, N. N.[3] and Igbabul, B. D.[2]

[1]Department of Food Science and Technology, Federal Polytechnic, Idah- Nigeria.
[2]Department of Food Science and Technology, University of Agriculture, Makurdi, Nigeria.
[3]Department of Chemistry, University of Agriculture, Makurdi, Nigeria.

Chlorine adsorption kinetics of activated carbons from Cattle bone, coconut-shell and wood as influenced by mode of activation (heat and acid) were studied using a commercial activated carbon (CAC) as control. Batch chlorine adsorption by each activated carbon was investigated by contacting with fixed amount of chlorine solution at 10, 15, 20, 25, and 30°C respectively at 10 min intervals. The residual chlorine data were fitted with Freundlinch, Langmuir isotherms, and first order kinetic model. The data were also fitted to Fick's unsteady state mass transfer equation and chlorine diffusivities were calculated from the slopes. Heat (H) activation was at 950°C for 3 h to produce heat activated bone carbon (HCB), and 850°C for 2.5 h for heat activated coconut-shell and wood carbon (HCS and HWC) while concentrated phosphoric acid (A) (1:3-1:4) carbon: acid (w/w) was at 600°C for 2 h to produce acid activated wood (AWC), acid activated coconut-shell (ACS) and acid activated bone (ACB) samples. The rates of chlorine adsorption were faster within the first 10-30 min; follow first order kinetics and Freundlich models. Chlorine diffusivities ranged from 3.674 × 10^{-6} m^2/s for Acid activated cattle bone carbon to 5.017 × 10^{-6} m^2/s for Commercial activated carbon. Porosity characterization of the carbons shows a wide range of pore sizes for both gas and liquid phase applications. This study indicated that efficient activated carbons could be obtained from local raw materials by controlled activation with acid or heat for home and industrial purposes.

Key words: Adsorption, kinetics, activations, isotherms, diffusivities, porosity.

INTRODUCTION

The importance and relevance of activated carbon to an ever growing society cannot be overemphasized considering its enormous uses. Its uses range from liquid phase to gaseous phase applications. One of the important applications of activated carbon is in water de chlorination. Chlorination is the process of adding the element chlorine to water as a method of water purification. Chlorine is widely popular for this application because of its ability to kill bacteria and other disease-causing organisms at relatively low concentrations. The chlorine binds and destroys the outer surface of bacteria and viruses thereby preventing waterborne diseases. The chlorine that is not used is called free residual chlorine. There are many health concerns regarding residual chlorine in water. This is mainly because chlorine is not naturally needed by the body and as a halogen, it tends to inactivate enzymes (Strands, 2008). Chlorinated water contains chemical compounds called trihalomethanes which are carcinogens. Bathing with chlorinated water could have the same effect as drinking because of dermal absorption of chlorine (Nedalee and Abdullah, 2009). Other concerns are aesthetic such as chlorine taste and odour. These concerns inform the need for an

*Corresponding author. E-mail: yusufu_ikagu@yahoo.com.

Abbreviations: ACB, Acid activated bone carbon; **ACS,** Acid activated coconut shell carbon; **AWC**, Acid activated wood carbon; **CAC,** Commercial activated carbon; **HCB**, Heat activated bone carbon; **HCS**, Heat activated coconut shell carbon; **HWC,** Heat activated wood carbon.

effective means of water de chlorination after disinfection at low cost. In Europe, the most common means of destroying microorganisms in water system is the use of ozone treatment. In Nigeria, chlorine is extensively used in water treatment because it is cheaper than ozone. Activated carbons are good at removing chlorine residual from water but the actual rates of chlorine removal by activated carbons obtained from different local raw materials need to be determined. Fast adsorption rate and high adsorption capacity are important and must be considered in the selection of an activated carbon for a given purpose. The performance of activated carbon is indicated by its adsorptive characteristics, which is derived from the specific surface area, pore size and pore volume of the carbon (Akinyemi and Taiwo, 2004). In this study, adsorption isotherms and chlorine removal kinetics from water using activated carbon from selected local raw materials were considered.

MATERIALS AND METHODS

Source/sample preparation

Coconut shells, wastes pieces of hard wood Prosopis *Africana* were respectively collected from coconut and fire wood sellers in Idah, Kogi state while cattle bones were collected from an abattoir opposite Federal Polytechnic, Idah, kogi state -Nigeria. Each material was crushed manually to small chips of approximately equal sizes (1.0 - 0.8 cm). Analytical grade- Calcium hypochlorite, free chlorine reagent, and a commercial activated carbon (Chemivivon, Sussex, UK) were purchased from Charlitech Laboratories, Makurdi, Nigeria.

Carbonization and activation

Carbonization was carried out using the method described by Gimba et al. (2001). The carbonization time was varied between 0.5 and 3 h at 600°C to determine optimum carbonization for each raw material by fixed carbon measurement without ash formation. Acid activation was carried out using the method described by Xu and Liu (2008). However, the carbon: acid ratio was varied between 1:1 and 1:6 to determine the optimum activation for each of the raw material by iodine value measurement. Heat activation was carried out using the method described by Ajayi and Olawale (2009) and the activation temperature was varied between 700 – 950°C to determine optimum activation by iodine value measurement.

Chlorine adsorption experiment and order of adsorption

The experimental procedure basically involves batch adsorption performed by shaking specified carbon sample in 100 ml of chlorine solution in a glass beaker using an SMZ Stuart magnetic stirrer (Stuart Scientific Co Ltd; Britain) with an intensity of agitation of 6 rps. The experiments were performed by adding a specified amount of carbon to each of the 5 beakers, the beakers were labelled 1-5. Beaker one was placed in a thermostat set to 30°C, while beakers 2, 3, 4 and 5 were subjected to 25, 20, 15 and 10°C respectively for temperature variation. Under all the varied experimental conditions, concentrations of free chlorine remaining in solution after shaking and decanted at the specified time were determined photometric ally using multipara meter bench photometer (manufacturer-Hanna Co. Ltd, Hungary).

The chlorine adsorption was modelled to 1st order adsorption using the equation:

$$dC/dt=KC; \quad (1)$$

Integrating equation (1) gives:

$$\ln C=\ln C_0-Kt \quad (2)$$

Where C is chlorine concentration at time t, C_0 is the concentration at time zero and K is the rate constant.

Adsorption isotherm and determination of chlorine diffusivity (rate of diffusion)

The chlorine adsorption data at the temperature of (30°C) were fitted to two most familiar isotherm equations which characterizes most adsorption processes rather well under conditions present in most water treatment systems (Ademoroti, 2003). The Langmuir and Freundlich constants were calculated from the slopes and intercepts of the isotherms respectively (Okuo et al., 2008). Chlorine diffusivities were calculated from the slopes of Fick's unsteady state mass transfer plots.

The Langmuir equation can be written as:

$$C/N=C/K_2+1/K_1K_2 \quad (3)$$

C is the concentration; K1 and K_2 are Langmuir constants

The Freundlich equation takes the form:

$$N=KC^n \quad (4)$$

The linear form of the equation can be written as:

$$LogN=LogK+nLogC \quad (5)$$

Where N is the amount of chlorine adsorbed by a unit mass of activated carbon, C is the concentration, K and n are Freundlich constants.

Fick's unsteady state mass transfer plot of $\ln C-Ce / C_0-C_e$ against time (t)

Where C = instantaneous chlorine concentration, C_o = the initial chlorine concentration, Ce = the equilibrium chlorine concentration.

The slope K is given as:

$$K=D\pi^2/4L^2 \quad (6)$$

D is the diffusivity (m^2s^{-1}) and L is the thickness of the material (m)

Porosity characterization

This was performed by the adsorption of N_2 at 77 k using micrometric surface area analyser (ASAP2010) Micrometric Inc. USA. Before the adsorption of N_2, the samples were subjected to degasing for 2 h at final pressure of 133.32 x10^{-4} pa. The total pore volume (V_T) was estimated from the volume of N_2 (as liquid) adsorbed at relative pressure of (p/p^0) of 0.95 (Xu and Liu, 2008). The microspore volumes were calculated from adsorption isotherms and mesopore was calculated as the difference between total pore volume and microspore volume.

Statistical analysis

Experimental results were analysed statistically. Means were

Table 1. Regression parameters for first-order chlorine adsorption.

Carbon material	Treatment		Temperature (°C)				
			30	25	20	15	10
Wood	Heat	N	5	5	5	5	5
		r^2	0.999	0.989	0.999	0.999	0.999
		A	1.078	1.081	1.096	1.128	1.137
		K	0.056	0.058	0.062	0.068	0.072
	Acid	N	5	5	5	5	5
		r^2	0.999	0.999	0.999	0.998	0.994
		A	1.086	1.075	1.072	1.097	1.085
		K	0.047	0.048	0.052	0.055	0.058
Coconut-shell	Heat	N	5	5	5	5	5
		r^2	0.997	0.998	0.999	0.999	0.999
		A	1.135	1.000	1.126	1.136	1.137
		K	0.042	0.044	0.048	0.050	0.051
	Acid	N	5	5	5	5	5
		r^2	0.999	0.998	0.999	0.998	0.999
		A	1.116	1.089	1.102	1.110	1.110
		K	0.042	0.044	0.048	0.051	0.053
Bone	Heat	N	5	5	5	5	5
		r^2	0.999	0.999	0.999	0.999	0.989
		A	1.135	1.137	1.115	1.123	1.130
		K	0.041	0.042	0.046	0.048	0.050
	Acid	N	5	5	5	5	5
		r^2	0.999	0.999	0.0999	0.998	0.994
		A	1.126	1.117	1.087	1.095	1.107
		K	0.041	0.043	0.046	0.048	0.050
Commercial carbon		N	5	5	5	5	5
		r^2	0.985	0.999	0.999	0.999	0.989
		A	1.107	1.124	1.161	1.162	1.052
		K	0.060	0.065	0.073	0.077	0.082

N = Number of entries, r = Correlation coefficient, A = Intercept, and k = Rate constant per minute.

compared using test of significant difference (Steel and Torrie, 1980). The least square linear regression was employed for calculating the gradients and intercepts of the linearized forms of the kinetic models.

RESULTS

The results of regression parameters for the first order adsorption are shown in Table 1. The r^2 values for 1st order ranged from 0.998 - 0.999. The rate constants (per minute) for heat activated wood carbon ranged from 0.056 - 0.072, coconut-shell carbon 0.042 – 0.051, and bone 0.041 – 0.050. For acid activated carbons, the rate constants ranged from 0.042 to 0.052 (wood), 0.042 to 0.050 (coconut-shell), and 0.047 to 0.058 (bone) as compared to 0.060 to 0.082 for the commercial activated carbon. The regression parameters for Freundlich and Langmuir adsorption isotherm is shown in Table 2. The r^2 values for Freundlich isotherm ranged from 0.982 for CAC – 0.999 for HWC and ACS while for Langmuir isotherm it ranged from 0.834 for AWC – 0.994 for HWC.

The Freundlich adsorption exponents (n) ranged from 1.218 for AWC to 1.678 for CAC while the adsorption coefficient ranged from 1.062 for HCB to 2.244 for CAC. There was no significant ($P \geq 0.05$) difference between the Freundlich adsorption exponent of CAC and those of HWC, HCS, ACB and HCB samples. On the other hand, there is no Significant ($P \geq 0.05$) difference between the Freundlich adsorption coefficient (K) of CAC and that of HWC. The equilibrium parameter (R_L) indicating adsorptions were 0.394, 0.099, 0.335, 0.214, 0.249, 0.241 and 0.063 for AWC, HWC, ACS, HCS, ACB, HCB and CAC respectively. There was no significant ($P \geq 0.05$) difference between R_L values of CAC and those of HWC

Table 2. Regression parameters for Freundlich and Langmuir isotherm.

Model	Regression parameter	Activated carbon sample							
		AWC	HWC	ACS	HCS	ACB	HCB	CAC	LSD (0.05)
Freundlich	N	4	4	4	4	4	4	4	-
	r^2	0.995	0.999	0.999	0.983	0.988	0.996	0.985	-
	n	1.218	1.527	1.252	1.538	1.486	1.522	1.678	0.221
	K	1.875	2.178	1.493	1.138	1.112	1.062	2.244	0.339
Langmuir	N	4	4	4	4	4	4	4	-
	r^2	0.834	0.994	0.991	0.916	0.921	0.932	0.983	-
	K_1	3.637	2.012	3.493	1.949	2.114	1.992	1.751	-
	K_2	0.804	3.049	0.662	1.221	1.004	1.050	4.966	-
	R_L	0.394	0.099	0.335	0.214	0.249	0.241	0.063	0.163

N= Number of entries, r = correlation coefficient, n = Adsorption exponent and K is a measure of adsorption capacity, K_1 = Langmuir adsorption capacity, K_2 the Langmuir constant related to energy of adsorption and R_L is an indicator of adsorption

and HCS while AWC, ACS, ACB and HCB had statistically similar R_L values. The R_L values ranged from 0.099 for CAC - 0.394 for HAW. The regression parameters for Fick's expression of chlorine adsorption are presented in Table 3. Variation of chlorine diffusivities with temperatures is presented in Figure 1. Chlorine diffusivities increase with decreased in temperature for all the carbons. Diffusivities (m^2/s) decreased from 3.559×10^{-6} - 3.968×10^{-6} for AWC, 3.909×10^{-6} - 4.667×10^{-6}, for HWC, 3.209×10^{-6} - 3.784×10^{-6} for ACS, 3.092×10^{-6} - 3.617×10^{-6} for HCS, 3.209×10^{-6} - 3.675×10^{-6} for ACB, 3.267×10^{-6} - 3.675×10^{-6} for HCB and 4.084×10^{-6} - 5.017×10^{-6} for CAC respectively. The pore size distribution curves of the carbons are presented in Figure 2.

DISCUSSION

The high correlation coefficients in addition to linearity of the curves (Table1) indicated that chlorine removal from water by the activated carbons can be modelled using first order kinetics. Earlier studies indicated that first order kinetic model adequately described chlorine desorption by activated woody pod carbon and silica (Ejikeme and Ochonogor, 2008), (Sakamota et al., 1994). The results presented in Table 2 shows Langmuir isotherm is essential in assessing the adsorption efficiency of adsorbent. This model was based on assumption of formation of its initial chlorine monolayer of adsorb ate species onto the surface of adsorbent. This study is useful in optimizing the operating condition for effective adsorption. Equilibrium parameter R_L which is considered as a more reliable indicator of adsorption (Omonmhenle et al., 2008), (McKay and McConvey, 1985) was computed from the expression represented as follows:

$$R_L = \frac{1}{1+K_2} \qquad (7)$$

Where C initial chlorine concentration (mg/l), K_2 is the Langmuir constant values between 0&1 indicate favourable adsorption for all initial concentration and temperature studied. R_L indicates isotherm shape and whether it is favourable or not as per the criteria given subsequently (Inamullah et al., 2008).

Adsorption isotherm is the relationship between the amount of substance adsorbed by an adsorbent and the equilibrium pressure or concentration at constant temperature. The isotherms show that correlation coefficients (r) for Freundlich and Langmuir were high. The high correlations suggest that both isotherms describe chlorine sorption on the activated carbons with comparable success, with the Freundlich model being superior. An indicator of adsorption RL from Langmuir constants indicates favourable adsorption for all carbons at temperatures studied (Table 4). The lower the R_L values, the more favourable the adsorption process. R_L indicates isotherm shape and whether it is favourable or not as per the criteria given subsequently (Geetha et al., 2008). In a related study, favourable adsorption of 2-4-Dichlorophenol on soil activated samples with R_L values between 0.649 to 0.719 for different treated clay samples were reported (Omonmhenle et al., 2008). Several researchers have reported the applicability of Freundlich and Langmuir isotherms in carbon (Okuo et al., 2008), (Onogwu et al., 2001), (Ejimeke and Ochonogo, 2008), (Ufuah et al., 2004). The K and n from Freundlich isotherm and the K_1 and K_2 from Langmuir isotherm incorporates all the factors affecting adsorption capacity and intensity. It has been suggested that n values of between 1 and 10 represent favourable adsorption (Ufuah et al., 2004).

The increase in diffusivities with decrease in temperature (Figure 1) could be attributed to exothermicity of adsorption (Okuo and Ozioko, 2001).Adsorption is an exothermic process and is favoured by lower energy. In addition, at lower temperature, more of the chlorine exists in adsorb able

Table 3. Regression paramers for ficks unsteady state mass transfer.

Carbon sample	Treatment	Parameter	Temperature(°C)				
			30	25	20	15	10
Wood	Heat	Entries	6	6	6	6	6
		SD Error	0.016	0.016	0.015	0.015	0.014
		Slope	0.067	0.070	0.072	0.072	0.080
		Diffusivity(m^2/s)	3.909×10^{-6}	4.111×10^{-6}	4.200×10^{-6}	4.434×10^{-6}	4.667×10^{-6}
	Acid	Entries	6	6	6	6	6
		SD Error	0.015	0.015	0.016	0.014	0.014
		Slope	0.061	0.062	0.065	0.065	0.066
		Diffusivity(m^2/s)	3.559×10^{-6}	3.617×10^{-6}	3.792×10^{-6}	3.852×10^{-6}	3.968×10^{-6}
Coconut-shell	Heat	Entries	6	6	6	6	6
		SD Error	0.015	0.016	0.016	0.015	0.015
		Slope	0.053	0.056	0.060	0.061	0.063
		Diffusivity(m^2/s)	3.092×10^{-6}	3.267×10^{-6}	3.500×10^{-6}	3.559×10^{-6}	3.617×10^{-6}
	Acid	Entries	6	6	6	6	6
		SD Error	0.015	0.015	0.016	0.016	0.017
		Slope	0.055	0.058	0.060	0.062	0.064
		Diffusivity(m^2/s)	3.209×10^{-6}	3.384×10^{-6}	3.500×10^{-6}	3.639×10^{-6}	3.734×10^{-6}
Bone	Heat	Entries	6	6	6	6	6
		SD Error	0.017	0.016	0.016	0.015	0.015
		Slope	0.056	0.057	0.060	0.062	0.063
		Diffusivity(m^2/s)	3.267×10^{-6}	3.326×10^{-6}	3.403×10^{-6}	3.617×10^{-6}	3.675×10^{-6}
	Acid	Entries	6	6	6	6	6
		SD Error	0.015	0.016	0.016	0.016	0.015
		Slope	0.055	0.056	0.060	0.062	0.063
		Diffusivity (m^2/s)	3.209×10^{-6}	3.267×10^{-6}	3.403×10^{-6}	3.617×10^{-6}	3.675×10^{-6}
Commercial carbon		Entries	6	6	6	6	6
		SD Error	0.014	0.014	0.013	0.014	0.014
		Slope	0.070	0.073	0.080	0.081	0.086
		Diffusivity(m^2/s)	4.084×10^{-6}	4.259×10^{-6}	4.667×10^{-6}	4.726×10^{-6}	5.017×10^{-6}

form and more adsorption rather than desorption occurs. The diffusivities (m^2/s) ranged from 5.017×10^{-6} to 4.667×10^{-6}.The variation in diffusivities of chlorine for the different carbons could be attributed to the differences in structural properties of the carbons (Carolina et al., 1995).

The pore size distribution curves (Figure 2); showed a broader range of pore sizes for ease of movement of adsorb ate (chlorine) through the carbon pores and for the adsorption. The BET surface area of the carbons ranged from 1015 for HWC to $782 m^2g^{-1}$ for ACB while that of control (CAC) was $1022\ m^2g^{-1}$. The surface area for the activated carbons fall within the range of 600 to $1200\ m^2g^{-1}$ reported for most widely used commercial activated carbons (Ng et al., 2002). The high surface area of the carbons is important since it limits the amount of adsorb ate which can be adsorbed.

The BET surface areas were higher for acid activated carbons and this is expected because chemical activation normally develops more porosity and gives high surface area when compared with thermal activation (Ajayi and Olawale, 2009). The increase BET surface areas and microspore (0.746 to $0.641 cm^3g^{-1}$) structure of acid activated carbons was brought about by oxidation of the carbons by phosphoric acid. The macrospore volume of the carbons ranged from 0.779 for HWC- $0.656 cm^3g^{-1}$ for HCB and the mesopore volume ranged from 0.008 for HWC – $0.040\ cm^3g^{-1}$ for HCS. The broader pore size for these carbons is important for a wide application since pore volume limits the size of molecules to be adsorbed.

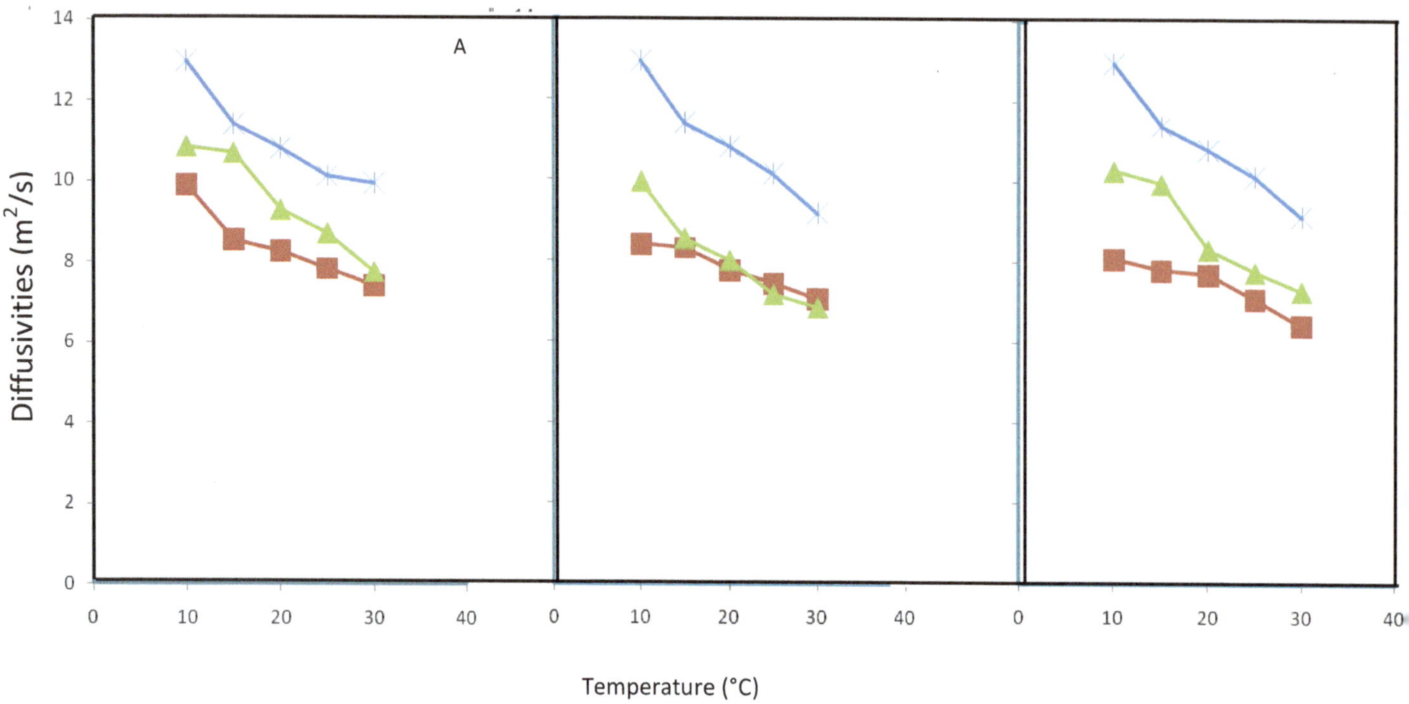

Figure 1. Effect of temperature on chlorine diffusivities by commercial carbon (*), heat activated (▲) and acid activated (■) carbons from wood (plot A), coconut-shell (plot B) and bone (plot C) respectively.

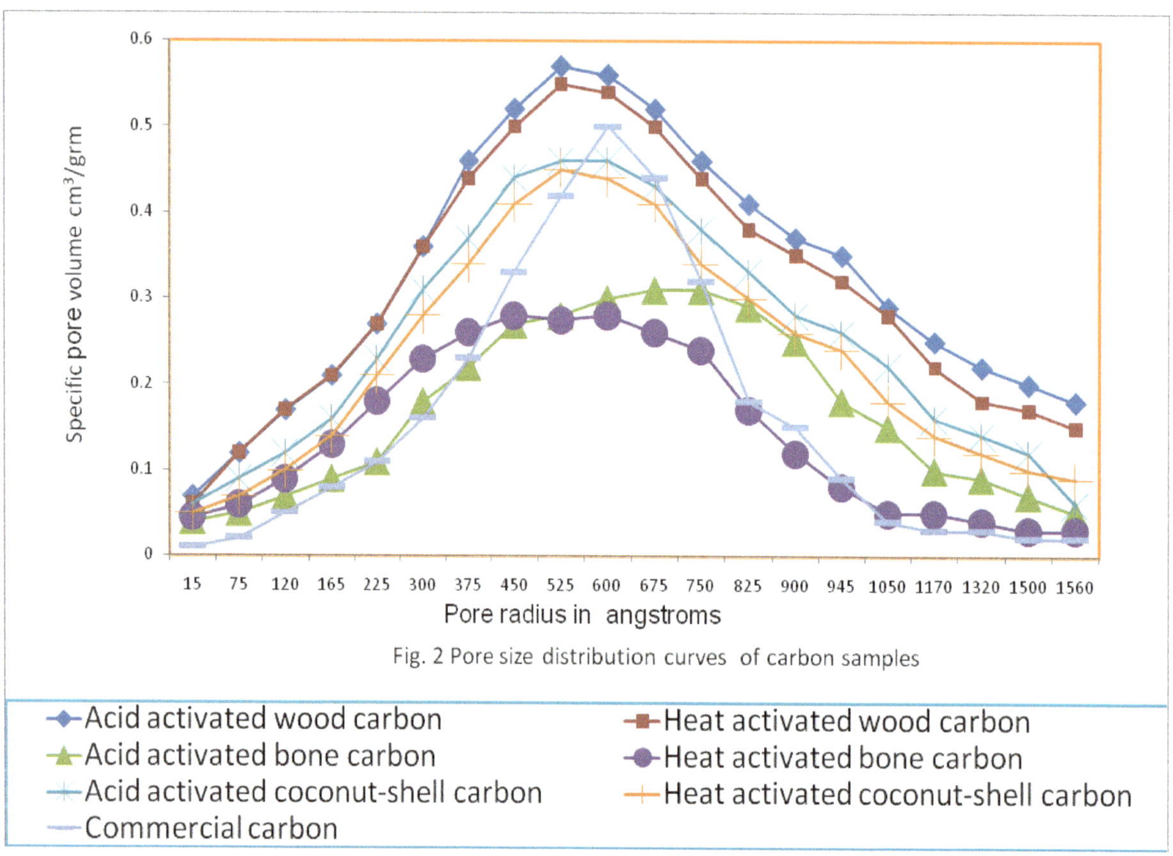

Figure 2. Pore size distribution curves of carbon samples.

Table 4. Adsorption isotherm.

RL values	Adsorption
RL> 1	Unfavourable
RL =1	Linear
0<RL<1	Favourable
RL =0	Irreversible

Conclusion

This study indicates that efficient activated carbons can be obtained from bone, wood and coconut shell by controlled activation with phosphoric acid or heat for water de chlorination and for other uses. The adsorption of chlorine by the carbons compared favourably with a commercial carbon and from chlorine diffusivity (D) values, phosphoric acid activation is the best for coconut shell while thermal (heat) activation gives the best results for wood and bone carbon.

REFERENCES

Ademoroti CMA (2003). Chemical thermodynamics. 2nd edn. Foludex press ltd. Ibadan, Nigeria. pp. 283-286.

Ajayi OA, Olawale AS (2009): A comparative study of thermal and chemical activation of *Canarium Schweinfurthii* Nutshell. J. Appl. Sci. Res., 5(12): 2148-2152.

Akinyemi OP, Taiwo EA (2004). Production of activated carbon from agricultural wastes. Proceedings of Nigerian society of chemical engineers held in Warri- Nigeria.

Carolina RL, Mirta LT, Alfredo (1995). Effects of zinc on the structure and growth dynamics of a natural freshwater phytoplankton assemblage reared in the laboratory. Environ. Pollut., 88: 275-281.

Ejikeme, P, M, and Ochonogor, A. E (2008).Adsorption kinetics of basic violet-1 removal by *penthaclethra mychrophylla* woody pod activated carbon. J. Chem. Soc. Nigeria. 33(1): 132-137.

Geetha A, Sujatha M, Sivakuma P, Palanisani PN (2008). Orthophosphoric Acid Activated Babul Seed Carbon as an Adsorbent for the Removal of Methylene Blue. J. Chem., 5(4):742-753, http://www.e-journals.ne

Gimba CC, Olayemi JY, Siyaka ZM (2001). Comparative studies on the adsorption of dyes using activated carbon derived from *Arachis hypogee and Cocos nucifera* shells. Proc. Chem. Soc. Niger.

Inamullah B, Khadija Q, Kazi RA, Abdul KA (2008). Preparation and Characterisation of Chemically Activated Almond Shells by Optimization of Adsorption Parameters for Removal of Chromium VI from Aqueous Solutions. World Acad. Sci. Eng. Technol., 34: 199-204.

McKay G, McConvey IF (1985). Adsorption of dye onto wood meal by solid diffusional mass. Chemical Eng. Process, 19: 287-295.

Nedalee R, Abdullah S (2009). Chlorinated water- The hidden danger. Articlebase. Aeticlebase.com.

Ng C, Marshal R, Rao M, Rishipal RB, Jack NL, Ralph JA(2002). Granulated activated carbons from agricultural by- products: Process description and estimated cost of production. LSU Ag center. Bulletin Number 881.

Okuo JM, Akapaja EO, Eguono WO (2008). Kinetics of cd^{2+} and Se^{4+} ions removal from solution by the fruiting bodies of white rof fungi (plenrotus tuberregium) J. Chem. Soc. Nig., 33(1): 56-64.

Okuo JM, Ozioko AC (2001). Adsorption of lead and mercury ions on chemically treated periwinkle shell. J. Chem. Soc. Nig., 26(1): 60-65.

Omonmhenle S, Ofomaja AE, Okieimen FE (2008). Effect of H2SO4 treatment on the distribution of 2, 4 didichlorophenol in soil and I. J. Chem. Soc. Niger., 33(1): 70-78.

Onogwu JO, Oketunde FO, Sha'ato R (2001).Sorption of O,O-Dimethyl-O-O(3,5,6- Trichloro-2-pyridylphosphothionate on Benue Alluvial soil. Proc. Annu. Conf. Chem. Soc. Nig. Book Abstr., pp. 105.

Sakamoto H, Toyokazu H, Yuji T, Horie T, Miyamota N (1994). First-order isothermal desorption kinetics of chlorine on SiH_2Cl_2-adsorbed Si (100) surface (100). J. Appl. Surf. Sci., 75(1-4): 27-32.

Steel RGD, Torri JW (1980). Principles and Procedures of Statistics. Biometric approach. 2nd edn. Auckland New Zealand. McGraw Hill.

Xu T, Liu X (2008). Peanut shell activated carbon: Characterization, surface modification and adsorption of, Pb2+ from Aqueous Solution. Chin. J. Chem. Eng., 19(3): 401-406.

Synthesis and characterization of (Thallium-Tin) doped Bismuth based superconducting materials

Rashid Mahmood[1]* and Muhammad Javed Iqbal[2]

[1]Division of Science and Technology, University of Education, Township Lahore-54000, Pakistan.
[2]Centre of Nano-Science and Technology, Preston University, H-8 Islamabad.

Doping with combinations of $Tl_{0.3-x}Sn_x$ in a certain concentration range enhances the zero resistivity temperature $Tc_{(0)}$ from 104 to 108K. Both high-Tc (2223) and Low-Tc (2212) phases were detected by the X-ray diffraction analysis. However the low-Tc phase is found to dominate over the high-Tc phase with the increase in Sn content. Critical temperatures were determined by the electrical resistivity and ac magnetic susceptibility measurements. Amongst the doped samples studied here a sample containing Sn content of x = 0.06 exhibits comparatively the lowest value of ΔT = 5K, the lowest room temperature resistivity of ρ_{RT} = 0.01494 ohm-cm and the highest value of $Tc_{(0)}$ = 108K.

Key words: Bismuth based high-Tc superconductors, X-ray diffraction, resistivity, susceptibility.

INTRODUCTION

The superconducting bismuth system $Bi_2Sr_2Ca_nCu_{n+1}O_y$ comprises multiple phases corresponding to n = 0, 1 and 2 with critical temperatures of 10, 85 and 110K, respectively (Maeda et al., 1988). It is one of the two main high temperature superconducting (HTS) materials from which the flexible wire or tape carrying high density of current is being manufactured for practical applications in power generation (Yi et al., 2004; Arndt et al., 2002; Fujikami et al., 2002; Balachandran et al., 1998). Doping of aliovalent impurities or oxygen non-stoichiometry can change the charge carrier concentration considerably in Bi-based system (Thamizhavel et al., 1997; Prabhakaran and Subaramanian, 1997; Prabitha et al., 2005). The chemical inhomogeneity and disorder due to the doping vis-à-vis its location (site) strongly affect the critical temperature of Bi-based superconductors system (Aisaki et al., 2004). Doping of impurity atoms into HTS produces atomic level crystal defects, lattice strains, non-superconducting inclusions or precipitates and other structural defects. These inhomogeneties can improve Tc and self-field Jc which act as effective flux pinner in the system and thus enhance the Jc in applied fields (Wang et al., 2001; Wakata et al., 1992).

Partial replacement of Bi by doping with Pb and Sb produces materials of extremely high phase purity, with Antimony playing an important role in accelerating the formation of the phase with higher superconductive transition in the Bi-Sr-Ca-Cu-O system (Cava, 2000). The role of Antimony appeared to be the enhancement of low-Tc to high Tc phase reaction beyond that achievable by Lead incorporation alone (Iqbal and Mehmood, 2006). The presence of Antimony, as Sb_2O_3 enhances the reactivity and the kinetic of reaction (especially decarbonation) as well as the promotion of the high-Tc phase (Fruth et al., 2004). By the addition of low concentrations of Sn into Bi (2223) system results in increment of Tc because it improves the weak link of grain boundary (Kim et al., 1998). Doping with (Pb-Sn) is reported to enhance Jc in materials used for practical conductors (Li et al., 2001). The enhancement of $Tc_{(0)}$ by Sb-content in $Bi_{1.7}Tl_{0.3-x}Sb_xPb_{0.4}Sr_2Ca_2Cu_3O_y$ system is also reported (Iqbal and Mehmood, 2009). T_c and T_0 were found to decrease and the normal state resistance and hole concentration were observed to increase by increasing the Mo level in $Bi_2Sr_2Ca_2Cu_{3-x}Mo_xO_{10+y}$ system (Aksan and Yakinci, 2007). The presence of rare earth elements like Y, Er and Lu in bismuth based system, decreases the $Tc_{(0)}$ from 109 to 104K (Pop et al., 2000).

*Corresponding author. E-mail: rashid_meh786@yahoo.com.

Pr at Ca-site helps in the formation of low-Tc Bi (2212) phase (Celebi et al., 2004). $Tc^{(onset)}$ and $Tc_{(0)}$ decrease by doping Ni at Ca-site (Terzioglu et al., 2005). The replacement of Ca^{2+} by rare earth Sm^{3+} provides an additional electron, which, in turn decreases the hole carrier concentration which leads to decrease of $Tc_{(0)}$ from = 107 to 70 K suggesting the transformation of the Bi (2223) phase into Bi (2212) phase (Mishra et al., 2000). It was found that addition of MgO in Bi (2223) system influenced the microstructure of the samples and decreased the mean size of grains (Guilmeau et al., 2003). It has been recently suggested that the substitution of Bi^{+3} by Pb^{+2} draws electronic charges out of the CuO_2 layers and thus increases the number of oxygen hole (charge carrier) in these layers (Rahaman, 2003). In the present study, binary mixtures of Tin and Thallium are added to B_iPb (2223) system that enhances the value of $Tc_{(0)}$ by approximately 4K. However, it was observed that Bi (2212) phase dominated the Bi (2223) phase in almost all the samples investigated here.

EXPERIMENTAL

The samples with the nominal composition $Bi_{1.7}Tl_{0.3-x}Sn_xPb_{0.4}Sr_2Ca_2Cu_3O_y$ (where x = 0.00, 0.02, 0.04, 0.06, 0.08, 0.10 and 0.20) were prepared by the conventional solid-state reaction method (Maeda et al., 1988). The stoichiometric amounts of high purity Bi_2O_3 (99.9%) PbO (99.9%) Sn_2O_3 (99.9%) and $SrCO_3$ (99.9%), supplied by Aldrich, and CuO (>99%), $CaCO_3$ (99.9%) and Tl_2O_3 (>99%) supplied by Fluka, were used as starting materials. The required quantities of the materials were weighed, mixed and ground in an Agate mortar and pestle for 1 to 2 h using an electrical grinder. The mixtures were first calcined in air using a temperature programmed muffle furnace initially at 800°C for 48 h at optimized heating rate of $14°Cmin^{-1}$.The resulting powders were ground and pellets of 13 mm were prepared at 95 kN. m^{-1} pressure. These pellets were sintered in air at 850°C for 96 h, 860°C for 24 h and then at 870°C for 24 h in the muffle furnace at the heating rate of $14°C.min^{-1}$. The samples were allowed to cool at natural furnace cooling. The ac magnetic susceptibility of the samples was measured with laboratory-built ac susceptometer using a lock-in amplifier (Stanford SR-830). The electrical resistivity was measured with a conventional dc four-point probe method employing Janis Research Module (VPF -100) combined with a temperature controller (Lakeshore-321). High conductivity silver paint (Agar Scientific) was used for resistivity measurements. Powder X-ray diffraction patterns of the samples were obtained with (Philips X'pert PRO 3040/60) diffractometer which uses Cu Kα radiation source of 40 kV and 30 mA with a step of 0.02 over the range of 7 to 70° and with a scan speed of 8° min^{-1}. The scanning electron microscopy was performed by using Joel Jsem- 6460 scanning electron microscope.

RESULTS AND DISCUSSION

Figure (a–f) shows the room temperature X-ray diffraction patterns of bismuth based superconducting materials containing different Tl:Sn ratios, that is, $Bi_{1.7}Tl_{0.3-x}Sn_xPb_{0.4}Sr_2Ca_2Cu_3O_y$ as a result of variation of Sn content in the range of x = 0.00 to 0.10. The intensity of reflection peak (115) at 2θ ~ 27.6° corresponding to the low-Tc 2212 phase initially increases gradually from 885.99 to 1569.63 and to 1795.31 cps respectively, on the addition of Sn contents of x = 0.00, 0.04 and 0.06 and then decreases to 1694.66, 1538.39 and to 1123.94 cps, respectively for x = 0.08, 0.10 and 0.20. In our previous studies where combinations of Tl: Sb were doped [26], the reflection peak (115) at 2θ ~ 27.6° was observed to decrease for Sb contents of x = 0.00 to 0.20. The X-ray diffraction data of Sn-doped samples also reveals that the amount of low-Tc (2212) phase regularly increases from x = 0.00 to 0.10, however for x = 0.20 a prominent peak at 2θ ~ 33.2211° is clearly seen for high-Tc (2223) phase which indicates that Sn content in a certain concentration range in Tl and Sn binary mixture helps in the formation of low-Tc phase while beyond certain limits it contributes to the conversion of a low Tc (2212) phase to high-Tc (2223) phase. As shown in Table 1, the other characteristic reflection peaks at 2θ ~ 29, 31 and 33° also correspond to the low-Tc phase and their intensity increases at a certain lower concentration range and then decreases for x = 0.10 and 0.20, respectively.

Figure 2 shows the temperature versus resistivity curves for $Bi_{1.7}Tl_{0.3-x}Sn_xPb_{0.4}Sr_2Ca_2Cu_3O_y$ samples containing different Sn contents (x = 0.00, 0.04, 0.06, 0.08, 0.10 and 0.20). All the samples studied in this work exhibit a metal-like behavior in the normal state region before transition to the superconducting phase. The values of the room temperature resistivity (ρ_{RT}) vary from 0.01326 to 0.02125 ohm-cm at different temperatures in the range of (77 to 300K) and evidently the sample with Sn content of x = 0.00 has the lowest value of 0.01326 ohm-cm. There is almost regular increase in the value of ρ_{RT} for samples containing x = 0.00 to x = 0.2 with the exception of x = 0.06 (Table 2). It is evident from the resistivity versus temperature plots that almost all the synthesized samples have the residual resistivity (ρ_o) of less than 0.006 ohm-cm. The observed values of $Tc_{(0)}$ are in the range of (104 to 108)K as shown in Table 2. Amongst the synthesized samples, samples containing Sn contents of x = 0.06 and x = 0.20 show highest value of $Tc_{(0)}$ = 108K whereas a sample with x = 0.00 exhibits lowest value of 104K. The values $Tc^{(onset)}$ are noted in the range of (112 to 117)K as given in Table 2. The sample with x = 0.06 show the lowest value of the transition width, ΔT = 5K which varies from 5 to 11K.

Figure 3 shows temperature versus magnetic susceptibility curves for $Bi_{1.7}Tl_{0.3-x}Sn_xPb_{0.4}Sr_2Ca_2Cu_3O_y$ samples containing different Sn contents. AC magnetic susceptibility of the samples is measured with a frequency of 270 Hz in a temperature range of 77 to 300K. All the samples exhibit diamagnetic character and show Messiner effect. The samples show diamagnetic transition in the range of (105 to 114)K with the variation of Sn concentration.The presence of high-Tc and low-Tc phases is clearly indicated in the susceptibility curves by the deviation from the sudden and sharp decreasing

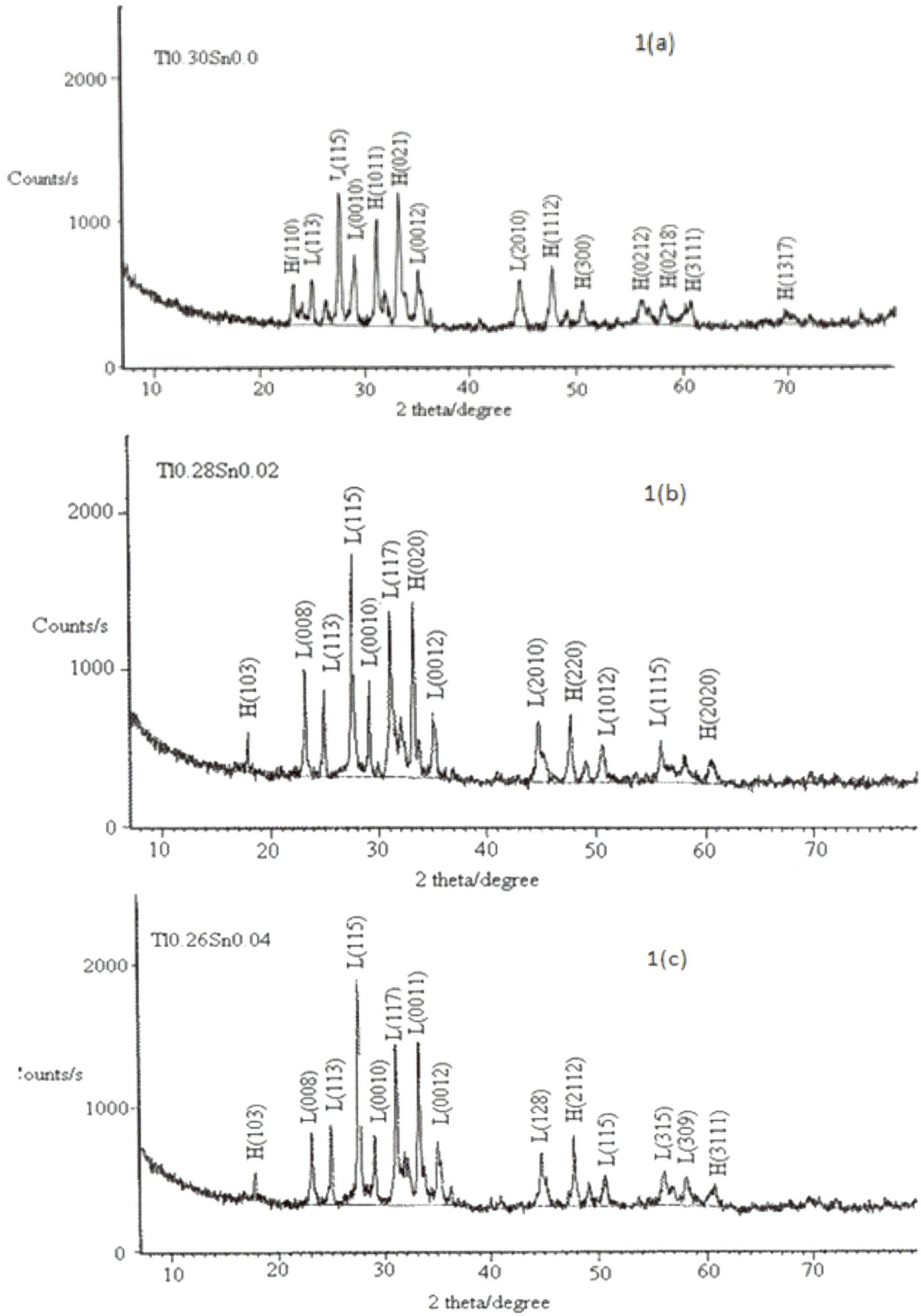

Figure 1. X-ray diffraction patterns of Bi1.7Tl0.3-xSnxSr2Ca2Cu3Oy samples containing Sn contents, x = (a) 0.00, (b) 0.02, (c) 0.04, (d) 0.06, (e) 0.08 and (f) 0.10.

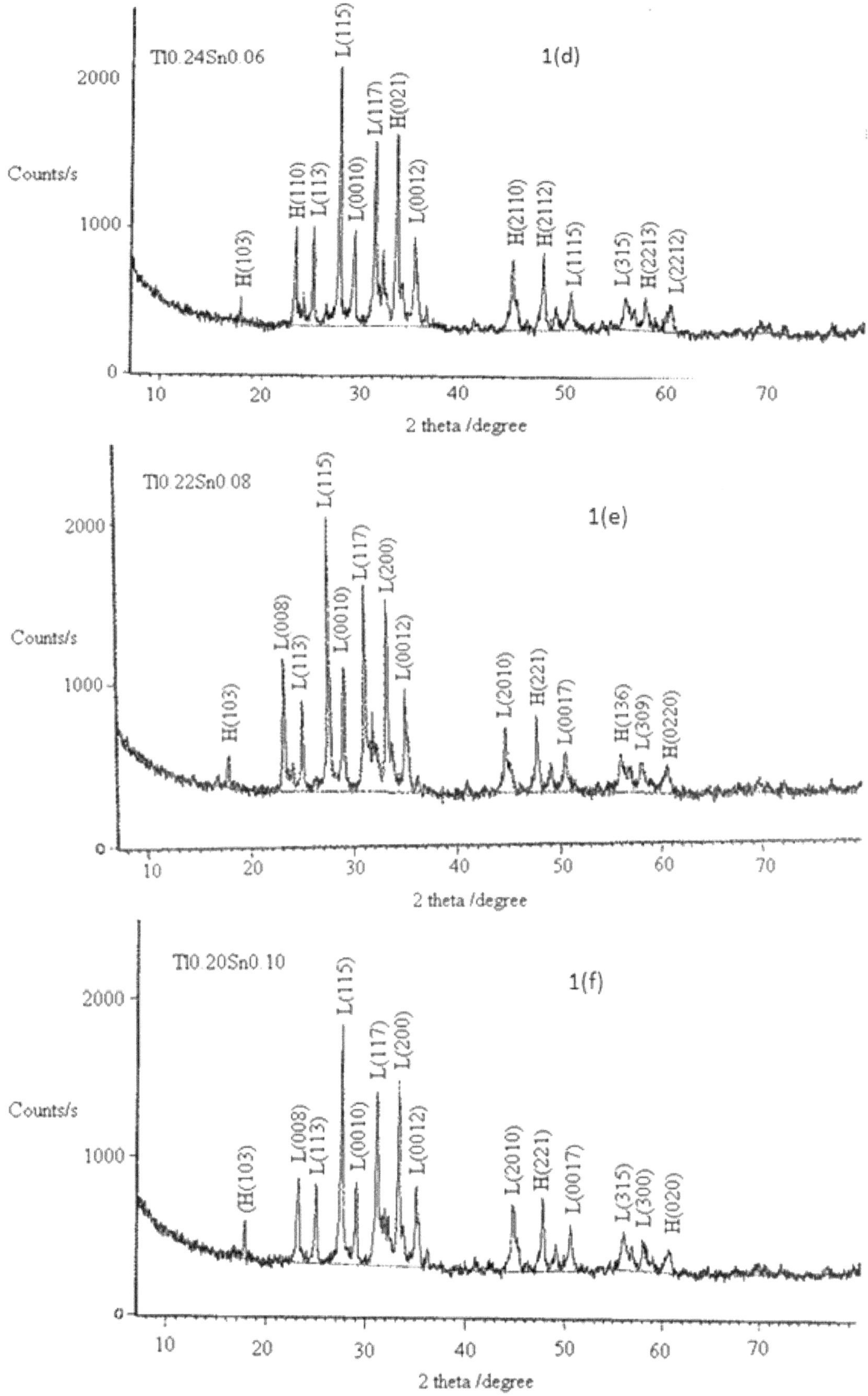

Figure 1. Contd.

Table 1. Comparison of 2θ, d- value, characteristic peak intensities, H (hkl) and L (hkl) of Tl-Sn doped samples $Bi_{1.7}(Tl_{0.3-x}Sn_{x})Pb_{0.4}Sr_2Ca_2Cu_3O_y$ where x = 0.0, 0.04, 0.06, 0.08, 0.10 and 0.20.

Samples	2θ	d-values	Characteristic peak intensities	Phase (*hkl*)	Relative intensities of highest peak intensities of low and high-Tc phases
$Bi_{1.7}Tl_{0.3}Sn_{0.0}Pb_{0.4}Sr_2Ca_2Cu_3O_y$	27.6669	3.22433	885.99	L(1 1 5)	
	29.0865	3.07011	405.31	L(0 010)	
	31.2040	2.86644	748.33	H(011)	
	3.2931	2.69119	928.04	H(0 21)	
					0.955
$Bi_{1.7}Tl_{0.26}Sn_{0.04}Pb_{0.4}Sr_2Ca_2Cu_3O_y$	27.5969	3.23234	1596.63	L(1 1 5)	
	29.0739	3.07085	483.60	L(0010)	
	31.1252	2.87351	1156.29	L(1 1 7)	
	33.2959	2.69097	1143.12	L (0011)	
$Bi_{1.7}Tl_{0.24}Sn_{0.06}Pb_{0.4}Sr_2Ca_2Cu_3O_y$	27.6480	3.22648	1795.31	L(1 1 5)	
	29.1478	3.06379	651.05	L(0010)	
	31.1760	2.86894	1273.19	L(1 1 7)	
	3.3319	2.68815	1296.78	H(0 21)	
					1.38
$Bi_{1.7}Tl_{0.22}Sn_{0.08}Pb_{0.4}Sr_2Ca_2Cu_3O_y$	27.5856	3.23364	1694.66	L(1 1 5)	
	29.0718	3.07163	778.25	L(0 010)	
	31.1100	2.87488	1286.24	L(1 1 7)	
	3.2767	2.69248	1199.57	L(2 0 0)	
$Bi_{1.7}Tl_{0.20}Sn_{0.10}Pb_{0.4}Sr_2Ca_2Cu_3O_y$	27.5760	3.23474	1538.39	L(1 1 5)	
	29.0494	3.07394	530.48	L(0010)	
	31.1001	2.87577	1101.79	L(1 1 7)	
	33.2861	2.69209	1158.61	L(2 0 0)	
					1.05
$Bi_{1.7}Tl_{0.10}Sn_{0.20}Pb_{0.4}Sr_2Ca_2Cu_3O_y$	27.5559	3.23706	1123.94	L(1 1 5)	
	28.9573	3.08351	396.72	H(0012)	
	31.0848	2.87715	747.65	L(1 1 7)	
	33.2211	2.69686	1069.47	H(0 2 0)	

Table 2. Comparison of room temperature resistivity ρ_{RT}, critical transition temperature $T_c^{(onset)}$, zero resistivity temperature $Tc_{(0)}$ and transition width ΔT values calculated for different samples of $Bi_{1.7}Tl_{0.3-x}Sn_xPb_{0.4}Sr_2Ca_2Cu_3O_y$ (where x = 0- 0.20).

Nominal composition	ρ_{RT} (ohm-cm)	$Tc_{(0)}$ (K)	$Tc^{(onset)}$ (K)	ΔT (K)
$Bi_{1.7}Tl_{0.3}Sn_{0.0}Sr_2Ca_2Cu_3O_y$	0.01326	104	114	10
$Bi_{1.7}Tl_{0.26}Sn_{0.04}Sr_2Ca_2Cu_3O_y$	0.01578	105	112	07
$Bi_{1.7}Tl_{0.24}Sn_{0.06}Sr_2Ca_2Cu_3O_y$	0.01494	108	113	05
$Bi_{1.7}Tl_{0.22}Sn_{0.08}Sr_2Ca_2Cu_3O_y$	0.01825	107	116	09
$Bi_{1.7}Tl_{0.20}Sn_{0.10}Sr_2Ca_2Cu_3O_y$	0.02125	104	117	11
$Bi_{1.7}Tl_{0.10}Sn_{0.20}Sr_2Ca_2Cu_3O_y$	0.01575	108	114	06

trend. The diamagnetic transition shifts towards lower temperature with increase in Sn content, however not in a regular order which indicates the presence of low-Tc (2212) phase in excess. The difference in sharpness and saturation level of the diamagnetic signals is due to the interconnectivity of the grains.

Figure 4 shows the scanning electron micrographs of $Bi_{1.7}Tl_{0.3-x}Sn_xPb_{0.4}Sr_2Ca_2Cu_3O_y$ samples containing different Sn contents. A predominantly uniform phase is clearly evident in the concentration range of x = 0.00 to 0.08 however there is some indication of some impurity phase in the cases of the samples with x = 0.10 to 0.20.

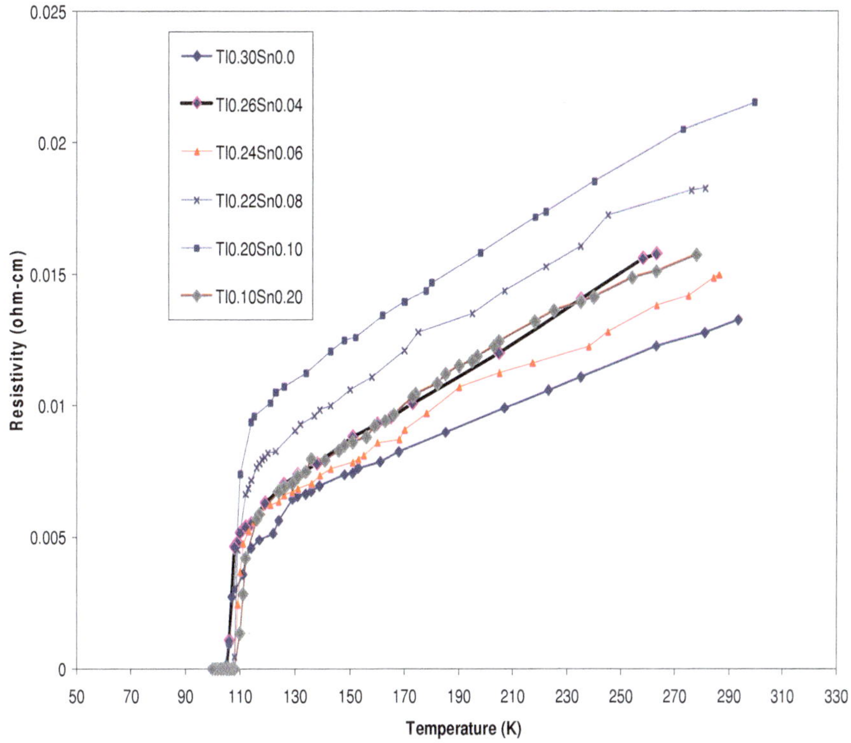

Figure 2. The dc electrical resistivity versus temperature curves of Bi1.7Tl0.3-xSnxSr2Ca2Cu3Oy samples containing Sn contents, x = 0.00, 0.04, 0.06, 0.08, 0.10 and 0.20.

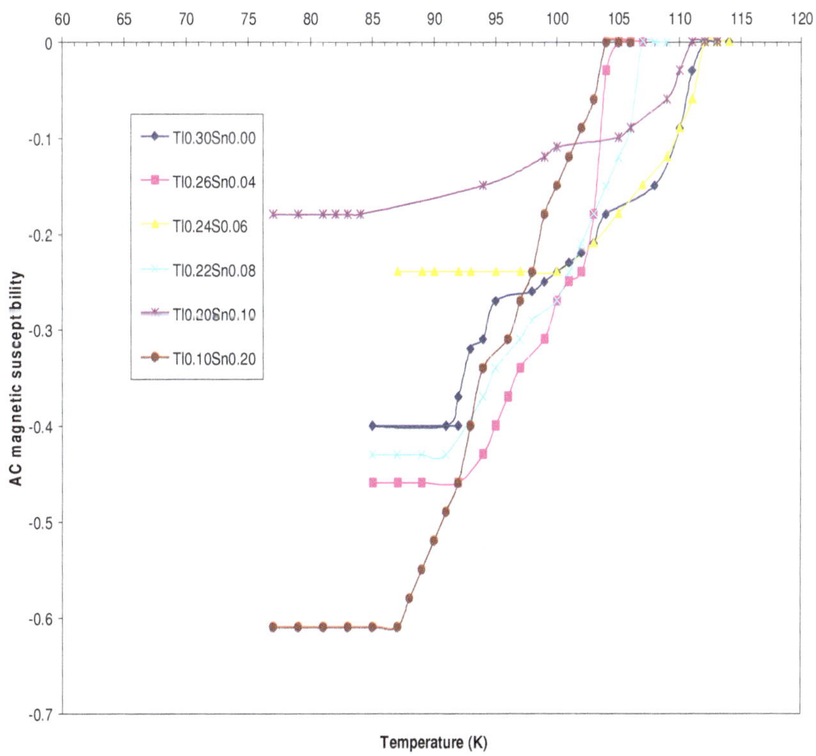

Figure 3. Ac magnetic susceptibility versus temperature curves of Bi1.7Tl0.3-xSnxSr2Ca2Cu3Oy samples containing Sn contents, x = 0.00, 0.04, 0.06, 0.08, 0.10 and 0.20.

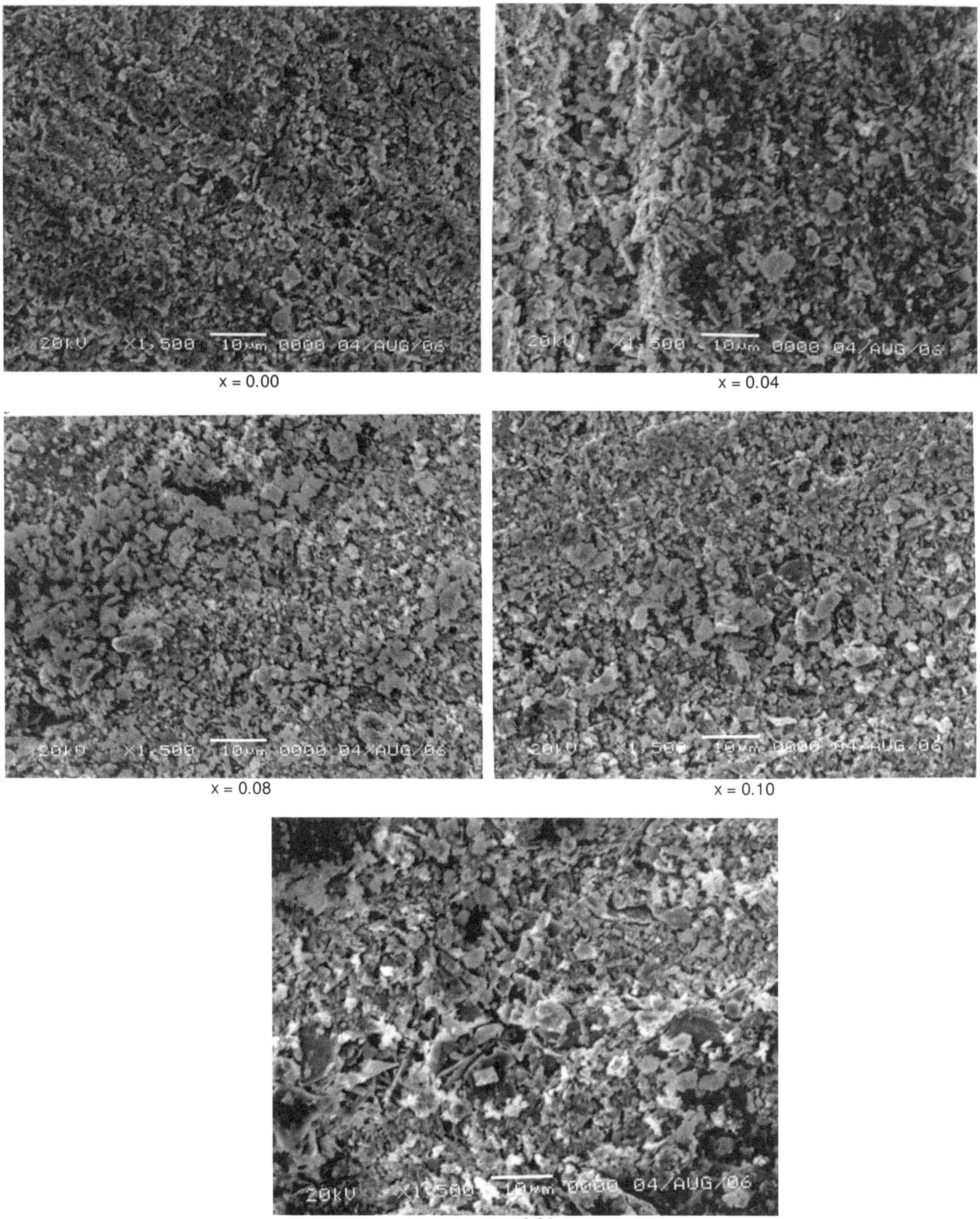

Figure 4. Scanning electron micrographs of $Bi_{1.7}Tl_{0.3-x}Sn_xSr_2Ca_2Cu_3O_y$ samples containing Sn contents, x = 0.00, 0.04, 0.08, 0.10 and 0.20.

Conclusion

Both Low-Tc (2212) and high Tc (2223) were found to be present in all the synthesized samples. Increase in Sn-concentration caused an increase in the relevant amount of Low-Tc phase. A sample with x = 0.06 exhibits comparatively the lowest value of ΔT = 5K, the lowest room temperature resistivity of ρ_{RT} = 0.01494 ohm-cm and the highest value of $Tc_{(0)}$ = 108K.

ACKNOWLEDGEMENT

Higher Education Commission (HEC) of Pakistan is acknowledged for the financial support through indigenous 300-scholarship scheme.

REFERENCES

Aisaki H, Kancko N, Fung DL, Damascelli A, Mang PK, Shen KM, Shen ZX, Greven M (2004). Effect of chemical in homogeneity in bismuth-based copper oxide superconductors, Phys. Rev. B. 69: 064512-064519.

Aksan MA, Yakinci ME (2007). Effect of Mo substitution on the structural and transport properties of $Bi_2Sr_2Ca_2Cu_{3-x}Mo_xO_{10+y}$ system, J. Alloys. Comp. 433: 22-32.

Arndt TJ, Aubele A, Fisher B, Krauth H, Sailer B, Szulczyk A (2002). Bi-2223 tapes for applications at high temperatures and/or high fields—designs, long length processing and properties, Physica C, 372-376: 887- 890.

Balachandran U, Salvamanickam V, Haldar Eror NG (1998). Development of Ag-clad Bi-2223 superconductors for electric power applications Supercond. Sci. Technol. 11: 978-981.

Cava RJ (2000) Oxide Superconductors, J. Am. Ceram. Soc. 83: 5-28.

Celebi S, Malik AI, Inanir F, Halim SA (2004). ac losses in Ni-doped $Bi_{1.6}Pb_{0.4}Sr_2(Ca_{1-x}Ni_x)_2Cu_3O_\delta$ ceramic superconductors, , Supercond. Sci. Technol. 17: 1121-1125

Fruth V, Popa M, Berger D, Ionica CM, Jitianu M (2004). Phases investigation in the antimony doped Bi_2O_3 system, J. Eur. Ceram. Soc. 24: 1295-1299.

Fujikami J, Kaneko T, Ayai N, Kobayashi S, Hayashi K, Takei H, Sato K (2002). Development of the industrial scale Ag/Bi 2223 tape, Physica C, 378-381: 1061-1064.

Guilmeau E, Andrzejewski B, Noudem JG (2003). The effect of MgO addition on the formation and the superconducting properties of Bi(2223) phase, Physica C, 387: 382-390

Iqbal MJ, Mehmood R (2006), Synthesis and characterization of antimony-doped Bi-based superconducting materials", Mat. Sci. Eng. B. 135: 166-171.

Iqbal MJ, Mehmood R (2009). Improvement in high-Tc phase formation in (thallium–antimony) doped bismuth-based superconducting materials, J. Alloys. Comp. 477: 386-390.

Kim HS, Jeng RA, Lee GJ, Lee DH, Ahn K, Kim KH (1998). Formation of high-T_c phase in Sn-incorporated (Bi,Pb)-Sr-Ca-Cu-O superconducting system, Mater. Chem. Phys. 56:147-152.

Li Y, Kaviraj S, Berenov A, Perkins GK, Driscoll J, Caplin JAD, Cao GH, Ma QZ, Wang B, Wei L, Zaho ZX (2001). Enhancement of critical current density of (Pb,Sn)-doped Bi-2212 superconductors at high temper*ature*, Physica C, 355: 51-58.

Maeda H, Tanaka Y, Fikutomi M, Asano T (1988). A New High-T_c Oxide Superconductor without a Rare Earth Element, Jpn. J. Appl. Phys. 27: L209- L210.

Mishra DR, Sharma SV, Sharma RG (2000). Role of vanadium in Bi(2223) ceramics, PRAM. J. Phys. 54: 317

Pop AA, Deac IG, Ilonca G, Ciurchea Pop V (2000). *Physica B* 284-288: 1101-1102.

Prabhakaran D, Subaramanian C (1997). Metal-insulator transition in the Pr substituted Bi-2212 bulk textured crystals Physica C, 291: 73-78.

Prabitha YG, Biju A, Kumar RGA, Sarun PM, Aloysius RP, Syamaprasad U (2005). Effect of Sm addition on (Bi,Pb)-2212 superconductor Physica C, 433: 28-36.

Rahaman MN (2003). Ceramic Processing and Sintering, Marcel Dekker, Second. p.787.

Terzioglu C, Yilmazlar M, Ozturk O, Yanmaz E (2005). Structural and physical properties of Sm-doped $Bi_{1.6}Pb_{0.4}Sr_2Ca_{2-x}Sm_xCu_3O_y$ superconductors, Physica C, 423: 119-126

Thamizhavel D, Prabhakaran A, Jayavel R, Subaramanian C (1997). Growth and characterization of superconducting $Bi_2Sr_2Ca_{1-x}Ce_xCu_2O_{8+\delta}$ single crystals, Physica C, 275: 279-283.

Wakata M, Takano S, Munakata F, Yamauchi H (1992) Effects of cation substitution on flux pinning in Bi-2212 superconductors Cryogenic, 32: 1046.-1051.

Wang XL, Liu HK, Dou SX, Horvat J, Millikon D, Heine G, Lang W, Luo HM, Ding SY (2001). Superconductivity and flux pinning in Y and heavily Pb codoped Bi-2212 single crystals, J. Appl. Phys. 89: 7669.

Yi HP, Han Z, Zhang JS, Liu T, Li MY, Fang J, Liu Q, Zheng YK (2004). Research status of the manufacturing technology and application properties of Bi-2223/Ag tapes at Innost, Physica C, 412-414: 1073-1078.

Kinetic and thermodynamic studies on the adsorption behavior of Rhodamine B dye onto animal bone meal

Mohammadine El Haddad

Equipe de Chimie Analytique and Environnement, Faculté Poly-disciplinaire, Université Cadi Ayyad, BP 4162, 46000 Safi, Maroc. E-mail: elhaddad71@gmail.com.

Removal of Rhodamine B from aquoeus media was achieved onto animal bone meal (ABM) as a new low cost adsorbent. The latter was characterized by Infra-Red and X-ray diffraction. The adsorption of Basic Red 46 occurred by studying the effect of adsorbent amount, dye concentration, contact time and temperature. The adsorption rate data were analyzed using the pseudo-first order and the pseudo second order kinetics models to determine adsorption rate constants. Thermodynamic parameters were also evaluated for the dye-adsorbent system and revealed that the adsorption process is endothermic nature and spontaneous process at the temperatures under investigation. All results found concluded that animal bone meal could be effectively employed as effective new low cost adsorbent for the removal textile dyes from aquoeus solutions.

Key words: Rhodamine B, adsorption, kinetics, thermodynamic, animal bone meal, animal bone meal (ABM).

INTRODUCTION

Dyes are one of the most important hazardous species found in textile industry produces in wastewater. Its presence in water bodies reduces light penetration, precluding the photosynthesis of aquoeus flora (Boyer et al., 2010; Al-Degs et al., 2008), besides being aesthetically objectionable for drinking water. Also, dyes can cause allergy, dermatitis, skin irritation and also provoke cancer and mutation in humans (de Lima et al., 2007; Rosenkranz et al., 2007). The color and the non-biodegradable nature of the spent dye baths cintitue can cause serious environmental problems.

Many treatment methods have been investigated to remove dyes from wastewater. These methods can be classified as chemical coagulation/flocculation, ozonation, oxidation processes, chemical precipitation, ion exchange, reverse osmosis and ultra filtration etc. The removal of dyes from dye containing wastewater has serious restrictions such as high cost, formation of hazardous by-products or intensive energy requirements (Banat et al., 1996). Therefore, the development of efficient, low-cost and environmental friendly technologies to reduce dye content in wastewater is extremely necessary. Among treatment technologies, adsorption is rapidly gaining prominenece. Activated carbon is the most widely used adsorbent for dye removal, but it is too expensive (Malik, 2003), consequently, numerous low-cost alternative adsorbents have been proposed including: chemically modified sugarcane bagasse lignin (da Silva et al., 2011), pistachio hull waste (Moussavi and Khosravi, 2011), coffee husk-based activated carbon (Ahmad and Rahman, 2011), pine cone (Mahmoodi et al., 2011), rice husk (Safa and Bhatti, 2011), synthetic calcium phosphates (El Boujaady et al., 2011), natural untreated clay (Errais et al., 2011), pillared clays (Gil et al., 2011), swelling clays (Li et al., 2011).

The purpose of this work; adsorption of Rhodamine B dye onto animal bone meal (ABM) has been investigated. The effect adsorbent amount and initial dye concentration has been studied. Kinetic experiments have been also conducted to determine the rate of Rhodamine B adsorption onto ABM. The obtained experimental data were analyzed using isotherm models namely, Langmuir and Freundlich.

MATERIALS AND METHODS

Preparation of adsorbent

Animal bones were collected from nearby butcher shops. All the attached meat and fat were removed and cleaned from the bones.

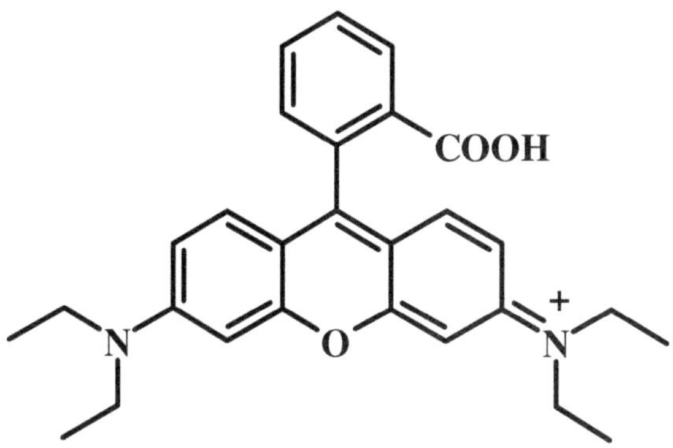

Figure 1. Chemical structure of Rhodamine B.

The bones were then washed several times with tap water and left in open air for several days to get rid of odors. Later, they were transferred to the oven at 80°C for drying. The dried bones were crushed and milled into different particle sizes in the range 45 to 200 µm then calcined for 2 h at 800°C. The residue was washed with water and was used after drying for 24 h at 80°C. The residue was washed with water and was dried overnight at 100°C in a conventional drying oven, and then calcined at a heating rate of 2°C/min to 400°C and kept at this temperature for 4 h. Identification of ABM was carried out by X-Ray diffraction (Philips X'Pert PRO) and IR spectroscopy (spectrometer Bruker-Tensor 27).

Adsorbate

Rhodamine B is the cationic dye used in this study, which was supplied from Fluka and was used without purification. The chemical structure of this dye (Rhodamine B) is shown in Figure 1. Colored solutions were prepared by dissolving requisite quantity of Rhodamine B in distilled water. The final volume prepared was 500 ml. Adsorption studies for the evaluation of ABM adsorbent for the removal of Rhodamine B dye from aquoeus solutions were carried out in triplicate using a batch contact adsorption method.

Adsorption procedure

The Adsorption experiments were carried out in batch. Proliminary experiments demonstrated that the equilibrium was established in 60 min. A 40 mg sample of ABM was mixed with 100 ml dye solution of 20 mg/L in batch. Samples of 5 ml of mixture were withdrawn from the batch at predetermined time intervals and the supernatant was centrifuged for 15 min at 3600 rpm. All dye solutions prepared were filtered by Millipore membrane type 0.45 µm HA, and the concentrations of dyes were determined from its UV-Vis absorbance characteristic with the calibration method. A BioMate 6, England UV/Visible spectrophotometer was used. A linear correlation was established between the dye concentration and the absorbance at λ_{max} = 554 nm, in the dye concentration range 0 to 30 mg/L with a correlation coefficient r^2 = 0.99.

The adsorption capacity of Rhodamine B was calculated as follows:

$$q_t = \frac{(C_0 - C_t)}{w} V \qquad (1)$$

Where q_t (mg/g) is the amount of Rhodamine B adsorbed at contact time t (min), C_0 (mol/L) is the initial dye concentration, C_t (mol/L) is the dye concentration at time (t) and w (g) is the ABM amount in the solution, V is the volume of the solution (L).

The effects of varying the ABM amount on dye adsorption were carried out by adding 10, 20, 40, 50 and 60 mg samples of ABM to 100 ml solution of Rhodamine B aquoeus solution 20 mg/L as initial concentration. The effect of the initial dye concentration was investigated as follows: 40 mg sample of ABM was added to 100 ml solution of Rhodamine B with initial concentrations varying from 5 to 20 mg/L.

RESULTS AND DISCUSSION

Characterization of animal bone meal (ABM) adsorbent

In order to investigate the surface characteristic of ABM, elemental analysis, IR and X-Ray diffraction spectrums were studied. Elemental analysis of ABM shows a high yield of Ca (49.62%) and P (42.36%) with a Ca/P ratio of 1.17. Small amounts of Si (3.88%), Mg (1.32%), Na (0.77%), Al (0.35%), Fe (0.24%), Cl (0.24%), S (0.11%), K (0.07%), Sr (0.03%), Cu (0,03%) and Zn (0.02%) are found. The IR absorption spectrum of ABM in Figure 2 shows bands characteristics of hydroxyapatite and more particularly a carbonated fluorapatite type B. We noted that the bands appeared near 1455 and 1430 cm^{-1}. These wave numbers are comparable with those of carbonated fluorapatites Type B prepared according to the procedure used by Bonel (1972). Moreover, the IR shows independently of the bands of phosphates, bands located between 780 and 800 cm^{-1} which could appear from the vibration of silicates groups. X-ray diffraction analysis confirms the presence of hydroxyapatite as shown in Figure 3. The specific surface area of ABM was determined by BET method from adsorption-desorption isotherm of nitrogen at its liquid temperature (77°K) and was found to be S_p = 85 m^2/g. We have already used this support as catalyst for organic compounds synthesis (Riadi et al., 2011, 2010; Mamouni et al., 2010).

Effect of animal bone meal (ABM) adsorbent amount onto dye removal

Equilibrium experiments were carried out by contacting different amounts of ABM with 100 ml of Rhodamine B dye solution (20 mg/L). The agitation was made for 24 h, which was found to be sufficient time to achieve equilibrium. The removal percent of Rhodamine B dye was increased with increasing adsorbent dosage. A maximum of 92.4% removal of the dye was observed by 20 mg of ABM as shown in Figure 4.

It was observed that the uptake of the dye increased by the amount of ABM added and that the maximum dye removal was achieved within the amount 40 mg. This implied that the number of adsorption sites increased as adsorbent mass increases as shown in Figure 4. From

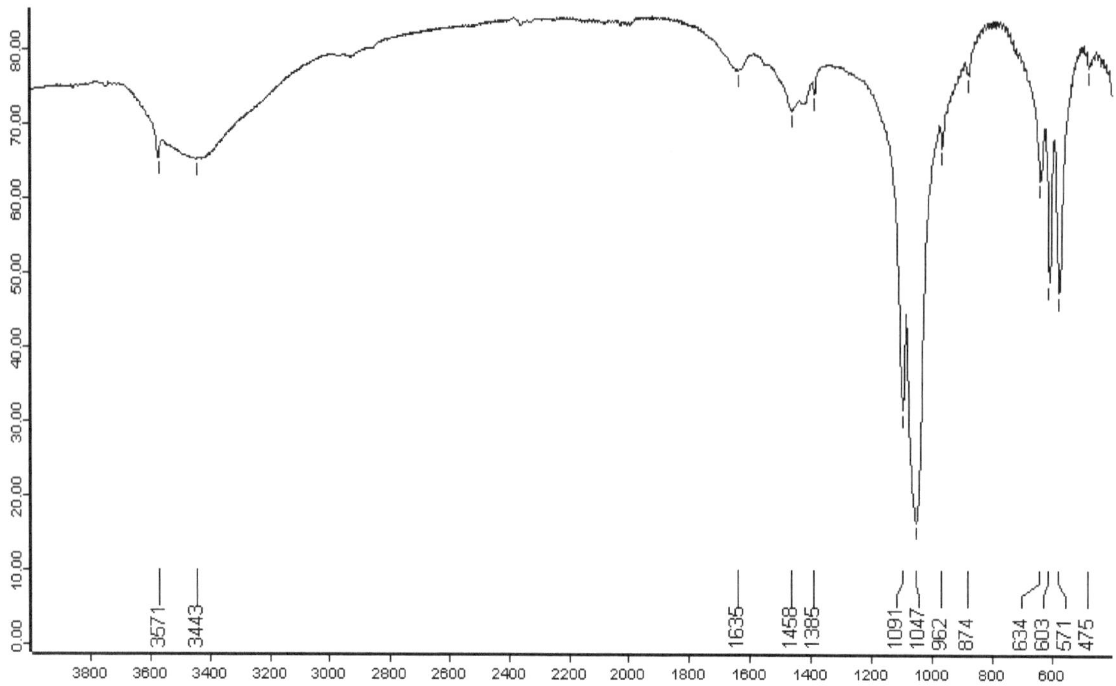

Figure 2. IR spectra of ABM.

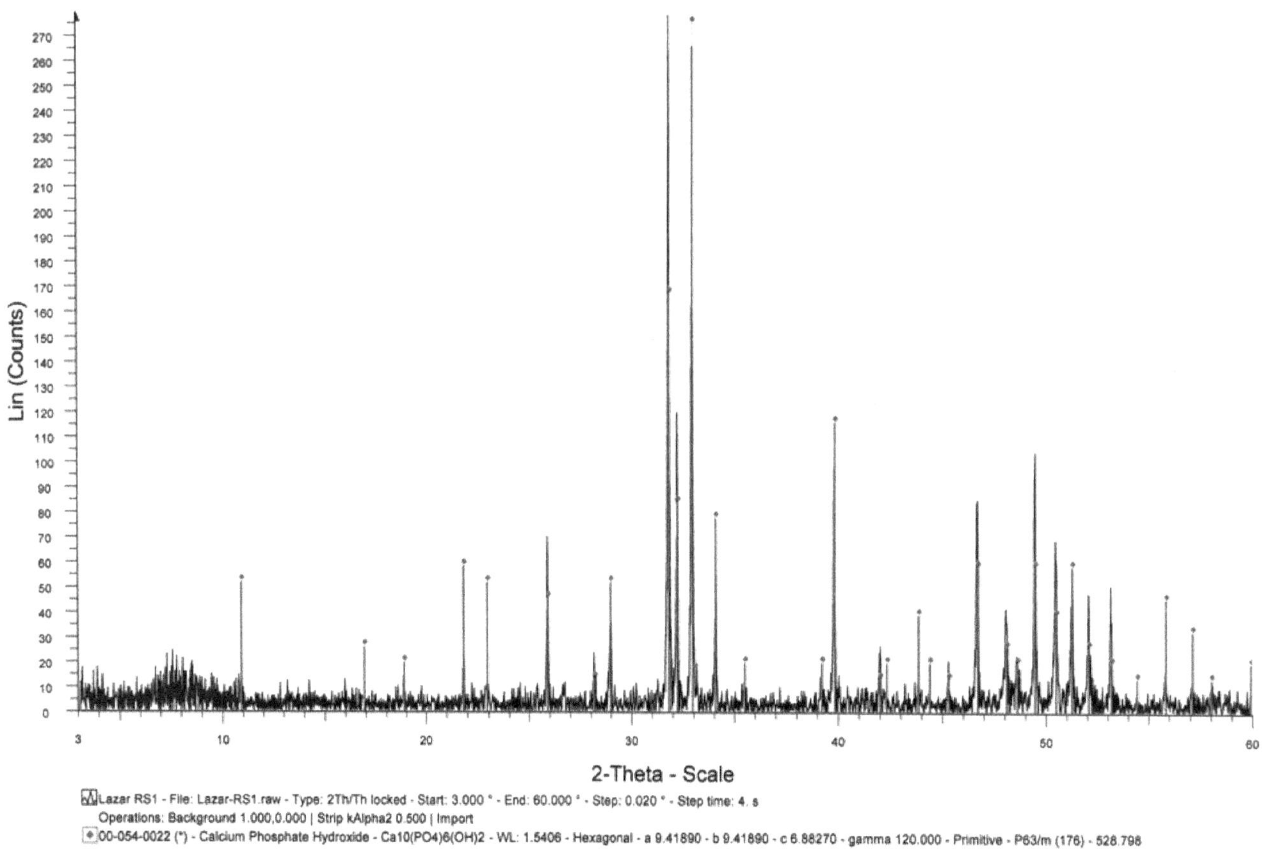

Figure 3. X-ray diffraction of ABM.

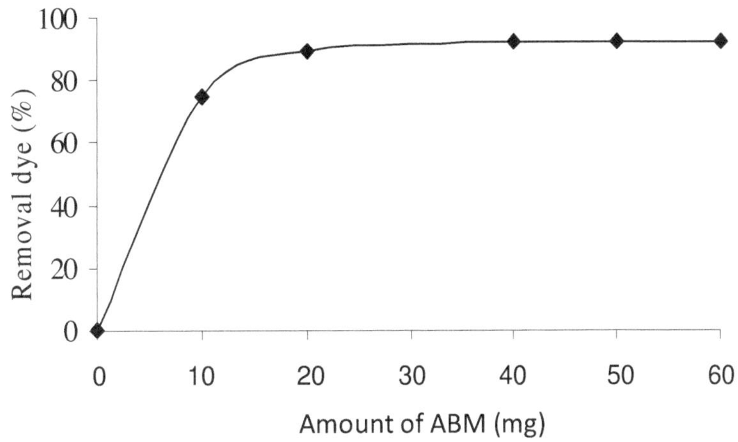

Figure 4. Effect of adsorbent amount of ABM on Rhodamine B removal from 20 mg/L dye solution at 25°C.

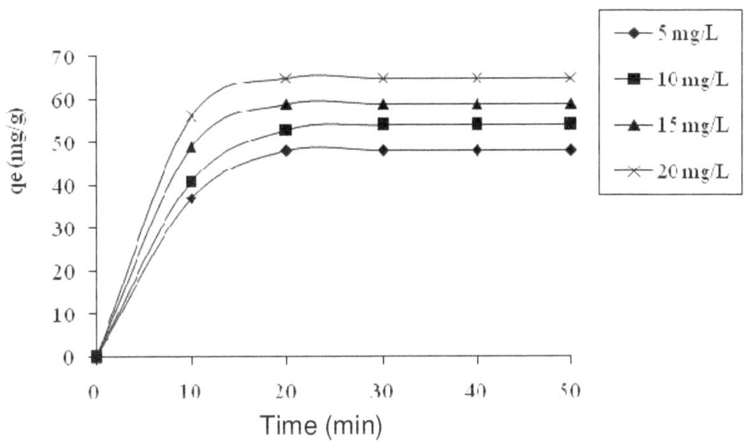

Figure 5. Effect of contact time and dye concentration on the adsorption amount of Rhodamine B onto 20 mg of ABM at 25°C.

the curve, the amounts of dye adsorbed after 90 min of contact time are illustrated in Figure 4. This isotherm belongs to type L of the Giles et al. classification (Giles et al., 1960) which indicates that, as more sites in the substrate are filled, it becomes increasingly difficult for the solute molecules to find an available vacant site. This could be either because the adsorbed molecules are more likely to be adsorbed on monolayer on a surface containing a finite number of identical sites and there is no strong competition from the solvent.

Effect of contact time of animal bone meal (ABM) and Rhodamine B dye concentration

The dye adsorption behavior onto ABM was studied by the variation of the equilibrium time in the range of 0 to 60 min. The adsorption capacity of the dye as a function of contact time plotted in Figure 5. The initial Rhodamine B concentrations used are 5, 10, 15 and 20 mg/L. This figure shows the effect of initial Rhodamine B concentration on the adsorption rate of the dye 25°C. An increase in initial dye concentration leads to an increase in the adsorption capacity. As the initial dye concentration increases from 5 to 20 mg/L, the adsorption capacity of Rhodamine B onto ABM changes from 48.32 to 65.76 mg/g. Thus, this indicates that the total amount of dye uptake was found to occur in the first rapide stage. The higher adsorption rate at the initail period may be due to an increased number of vacant sites available at the initial stage, which is because of the existed increase in the concentration gradients between adsorbate in solution and adsorbate on the adsorbent surface. As time proceeds, the dye concentration is reduced to the

Table 1. Kinetic models parameters for adsorption of Rhodamine B onto ABM.

Dye (mg/L)	$q_{e\ exp}$(mg/g)	Pseudo-first-order kinetic			Pseudo-second-order kinetic		
		q_e (mg/g)	k_1 (g/mg.min)	r^2	q_e (mg/g)	k_2 (g/mg.min)	r^2
5	47.14	15.67	0.23	0.71	48.18	0.05	0.998
10	53.21	17.87	0.31	0.75	54.78	0.02	0.999
15	59.34	21.65	0.34	0.66	60.55	0.01	0.999
20	65.23	29.76	0.36	0.71	66.78	0.01	0.998

accumulation of dye particles in the vacant sites, leading to a decrease in the adsorption rate at later stages. The obtained curves show single, smooth and continuous, indicating monolayer coverage of dye on the surface of adsorbent.

Kinetic studies

The kinetic models of pseudo-first order model and pseudo-second order were used to examine the adsorption mechanism. Pseudo-first order reaction model is described as follows (Lagergren, 1898):

$$\frac{dq_t}{dt} = k_1(q_e - q_t) \tag{2}$$

Where q_e and k_1 are the amount of dye adsorbed at equilibrium (mg/g) and the equilibrium rate constant of pseudo-first order kinetics (mn^{-1}), respectively. After integration by applying conditions, $q_t = 0$ at $t = 0$ and $q_t = q_t$ at $t = t$, then Equation (2) becomes:

$$\log(q_e - q_t) = \log q_e - \frac{k_1}{2.303}t \tag{3}$$

The rate constant k_1 and q_e were obtained from the slope and intercept of the linear plots of log ($q_e - q_t$) against t, respectively. Data were applied to be the pseudo-second order kinetic which is expressed as Ho (1995, 2001).

$$\frac{dq_t}{dt} = k_2(q_e - q_t)^2 \tag{4}$$

Where k_2 is the equilibrium rate constant of pseudo-second order (g/mg.mn). On integrating the Equation (4), Equation (5) is obtained:

$$\frac{t}{q_t} = \frac{1}{k_2 q_e^2} + \frac{1}{q_e}t \tag{5}$$

The rate constant k_2 and q_e were obtained from the slope and intercept of the linear plots of t/q_t against t, respectively.

Pseudo-first order and pseudo-second order are the two kinetic models that were tested to explain the experimental data found. The agreement between experimental data and model calculated values is expressed by the correlation coefficient r^2. The results are presented in Table 1. Kinetic adsoprtion of Rhodamine B onto ABM occurs with pseudo-second order model. In this fact, the higher values of ($r^2 > 0.99$) and the good agreement between the experimental and calculated equilibrium describes correctly the adsorption kinetics. So, the lower values of r^2 and the difference of experimental and calculated equilibrium with pseudo-first order model shows that the pseudo-first-order model failed to describe the adsorption kinetics.

Adsorption isotherms and thermodynamic parameters

Isotherms correlate the equilibrium data with different mathematical models to describe the behavior of the adsorption process; an optimized design of adsorption system provides valuable information. Accordingly, we evaluated the fitness of the equilibrium data obtained from the experiments with the Langmuir and Freundlich models. The best fitted model was selected based on the determination of correlation coefficient r^2. A mathematical expression of Langmuir model (Langmuir, 1918) can be written as:

$$q_e = \frac{Q_0 K_L C_e}{1 + K_L C_e} \tag{6}$$

Where q_e (mg/g) is the adsorbed amount at equilibrium, C_e is the equilibrium concentration of the adsorbate (mg/L), K_L is Langmuir equilibrium constant (L/mg) and Q_0 the maximum adsorption capacity (mg/g). The linear form of Langmuir equation is:

$$\frac{C_e}{q_e} = \frac{1}{K_L Q_0} + \frac{C_e}{Q_0} \tag{7}$$

The essential characteristic of Langmuir isotherm can

Table 2. Isotherm modeling (Langmuir and Freundlich) of Rhodamine B onto ABM.

Isotherm model	Equilibrium temperature (°C)			
	25	35	45	55
Langmuir				
K_L	0.23	0.27	0.33	0.34
q_{max}	60.56	62.31	63.78	64.54
R_L	0.08	0.07	0.06	0.09
r^2	0.999	0.998	0.999	0.998
Freundlich				
K_F	19.36	20.44	23.78	25.76
n	3.31	3.38	3.73	3.90
r^2	0.899	0.897	0.914	0.889

Table 3. Thermodynamic parameters of Rhodamine B adsorption onto ABM.

Temperature °C	Thermodynamic parametres		
	ΔG^0 (kj/mol)	ΔH^0 (kj/mol)	ΔS^0 (j/mol°K)
25	- 1.31		
35	- 2.08	11.35	42.56
45	- 2.50		
55	- 2.83		

be expressed by the dimensionless constant called equilibrium parameter, R_L, defined by:

$$R_L = \frac{1}{1+K_L C_0} \tag{8}$$

Where C0 is is the initial dye concentration (mg/L). R_L values indicate the type of isotherm to be irreversible ($R_L = 0$), favorable ($0 < R_L < 1$), unfavorable ($R_L > 1$) Mahmoodi and Arami (2008).

The Freundlich isotherm endorses the heterogeneity of the surface and assumes that the adsorption occurs at sites with different energy of adsorption. The energy of adsorption varies as a function of the surface coverage Freundlich (1906). A mathematical expression of Freundlich isotherm was as follows:

$$q_e = K_F C_e^{1/n} \tag{9}$$

Where K_F (L/mg) is Freundlich constant and *n* is the heterogeneity factor. The K_F value is related to the adsorption capacity; while the *1/n* value is related to the adsorption intensity. *1/n* values indicate the type of isotherm to be irreversible ($1/n = 0$), favorable ($0 < 1/n < 1$), unfavorable ($1/n > 1$) Mahmoodi and Arami (2008).

Equation (9) can be rearranged to the following linear form:

$$\log q_e = \log K_F + \frac{1}{n}\log C_e \tag{10}$$

For this study, we use a sample of 40 mg of ABM and were added to 100 ml of dye solution at 20 mg/L. The experiments were carried out at 25, 35, 45 and 55°C in a constant temperature shaker bath which controlled to within ±1°C. The data from the isotherm evaluation is summarized in Table 2. The correlation coefficient r^2 of the Langmuir isotherm model for all tested temperatures was higer than the Freundlich model, showing that the experimental equilibrium data was better explained by the Langmuir model. This finding supports the assumption that Rhodamine B is adsorbed as a homogeneous monolayer onto ABM particles sites. The thermodynamic data reflect the feasibility and favourability of the adsorption. The parameters such as free energy change indicated that the adsorption kinetics of dye on ABM followed the pseudo-second order at different dye ΔG^0, enthalpy change ΔH^0 and entropy change ΔS^0 can be estimated by the change of equilibrium constants with temperature. The free energy change of the adsorption reaction is given by:

$$\Delta G^0 = -RTLnK_C \tag{11}$$

Where ΔG^0 is the free energy change (kj/mol), R is the universal gas constant (8.314 j/mol°K), T is the absolute temperature (°K) and K_C states the equilibrium constant (q_e/C_e). The values of ΔH^0 and ΔS^0 can be calculated from the Van't Hoff equation (Freundlich, 1906).

$$LnK_C = -\frac{\Delta H^0}{RT} + \frac{\Delta S^0}{R} \tag{12}$$

When LnK_C is plotted against 1/T, a straight line with slope ($-\Delta H^0/R$) and intercept ($\Delta S^0/R$) are found.

The calculated thermodynamic parameters ΔG^0, ΔH^0, and ΔS^0 are depicted in Table 3. The positive values of ΔH^0 shows that the adsorption is endothermic process while positive ΔS^0 values reflects the increasing randomness at the solid/solution interface during the adsorption. The changes in free energy for physical and chemical reactions are between -20 and 0 kj/mol and -80 and -400 kj/mol respectively (Ozcan et al., 2006).

Conclusion

The results indicated that ABM is a promising new low cost adsorbent for removal of Rhodamine B from aquoeus solutions. The kinetics studies of dyes on ABM concentration values. The equilibrium data have been analyzed. The results showed that the Rhodamine B flollowed Langmuir isotherm model. Thermodynamic studies indicated that the dye adsorption onto ABM was a spontaneous, endothermic and physical reaction.

REFERENCES

Ahmad MA, Rahman NK (2011). Equilibrium, kinetics and thermodynamic of remazol brillant orange 3R dye adsorption on coffee husk-based activated carbon. Chem. Eng. J. 170:154-161.

Al-Degs YS, El-Barghouthi MI, El-Sheikh AH, Walker GM (2008). Effect of solution pH, ionic strength, and temperature on adsorption behavior of reactive dyes on activated carbon, Dyes Pigm. 77:16-23.

Banat IM, Nigam P, Singh D, Marchant R (1996). Microbial decolorization of textile dye-containing effluents: A review. Bioresour. Technol. 58: 217-227.

Bonel G (1972). Paper Contribution à l'étude de la carbonation des apatites I. physico-chimiques des apatites carbonates de type A, Annales Chimie. 7:127-144.

Boujaady HEI, El Rhilassi A, Ziatni MB, El Hamri R, Taitai A, Lacout JL (2011). Removal of a textile dye by adsorption on synthetic calcium phosphates. Desalination in press.

Boyer B, Cardoso NF, Lima EC, Macedo TR (2010). A useful organofunctionalized layered silicate for textile dye removal. J. Hazard. Mater. 181:366-374.

da Silva LG, Ruggiero R, Gontijo PM, Pinto RB, Royer B, Lima EC, Fernandes THM, Calvete T (2011). Adsorption of brillant red 2BE dye from water solutions by chemically modified sugarcane bagasse lignin. Chem. Eng. J. 168:620-628.

de Lima ROA, Bazo AP, Salvadori DMF, Rech CM, Oliveira DP, Umbuzeiro GA (2007). Matagenic and carcinogenic potential of a textile azo dye processing plant effluent that impacts a drinking water source. Mutat. Res. Genet. Toxicol. Environ. Mutagen. 626:53-60.

Errais E, Duplay J, Darragi F, M'Rabet I, Aubert A, Huber F, Morvan G (2011). Efficient anionic dye adsorption on natural untreated clay: Kinetic study and thermodynamic parameters. Desalination 275:74-81.

Freundlich HMF (1906). Uber die adsorption in lasugen, Z. Phys. Chem. (Leipzig). 57A:385-470.

Gil A, Assis FCC, Albeniz S, Korili SA (2011). Removal of dyes from wastewaters by adsorption on pillared clays. Chem. Eng. J. 168:1032-1040.

Giles CH, Macewan TH, Smith DJ (1960). Studies in adsorption. Part XI. A system of classification of solution adsorption isotherms, and its use in diagnosis of adsorption mechanisms and in measurement of speci.c surface areas of solids. Chem. Soc. XI:3973-3993.

Ho YS (1995). Adsorption of heavy metals from waste streams by peat, PhD. Thesis. The University of Birmingham, Birmingham, U. K.

Ho YS (2001). Sorption studies of acid dye by mixed sorbents. Adsorption 7:139-147.

Lagergren S (1898). Zur theorie der soggenannten adsorption gel ster stoffe, K. Sven. Vetenskapsakad. Handl. 24:1-39.

Langmuir I (1918). The adsorption of gases on plane surfaces of glass, mica and platinium, J. Am. Chem. Soc. 40:1361-1403.

Li Z, Change PH, Jiang WT, Jean JS, Hong H (2011). Mechanism of methylene blue removal from water by swelling calys. Chem. Eng. J. 168:1193-1200.

Mahmoodi NM, Arami M (2008). Modelling and sensitivity analysis of dyes adsorption onto natural adsorbent from colored textile wastewater. J. Appl. Polym. Sci. 109:4043-4048.

Mahmoodi NM, Hayati B, Arami M, Lan C (2011). Adsorption of textile dyes on pine cone from colored wastewater:kinetic, equilibrium and thermodynamic studies. Desalination 268:117-125.

Malik PK (2003). Use of activated carbons prepared from sawdust and rice-husk for adsorption of acid dyes:a case study of acid yellow 36. Dyes Pigm. 56:239-249.

Mamouni R, Riadi Y, Abrouki Y, El Haddad M, Saffaj N, El Antri S, Routier S, Guillaumet G, Lazar S (2010). First Animal Bone Meal (ABM) as a new catalyst in thia-Michael addition. Lett. Org. Chem. 7 (3):269-271.

Moussavi G, Khosravi R (2011). The removal of cationic dyes from aquoeus solutions by adsorption onto pistachio hull waste. Chem. Eng. Res. Des. in press.

Ozcan A, Oncu EM, Ozcan AS (2006). Kinetics, isotherm and thermodynamic studies of adsorption of Acid Blue 193 from aquoeus solutions onto natural sepiolite. Colloids Surf. A277:90-97.

Riadi Y, Mamouni R, Azzalou R, El Haddad M, Routier S, Guillaumet G, Lazar S (2011). An efficient and reusable heterogeneous catalyst Animal Bone Meal for facile synthesis of benzimidazoles, benzoxazoles and benzothiazoles. Tetrahedron Lett. 52: 3492-3495.

Riadi Y, Mamouni R, Boulahjar R, El Haddad M, Routier S, Guillaumet G, Lazar S (2010). Animal bone meal as an efficient catalyst for crossed-aldol condensation. Tetrahedron Lett. 51:6715-6717.

Rosenkranz HS, Cunningham SL, Mermelstein R, Cunningham AR (2007). The challenge of testing chemicals for potential carcinogenicity using multiple short-term assays:an analysis of a proposed test battery for hair dyes. Mutat. Res. Genet. Toxicol. Environ. Mutagen 633:55-66.

Safa Y, Bhatti HN (2011). Kinetic and thermodynamic modeling for the removal of direct red 31 and direct red 26 dyes from aquoeus solutions by rice husk. Desalination 272:313-322.

Autocatalysed oxidation of etophylline by permanganate in aqueous sulphuric acid medium-kinetics and mechanistic study

H. V. Rajeshwari, K. S. Byadagi, S. T. Nandibewoor and S. A. Chimatadar*

Department of Studies in Chemistry, Karnatak University, Pavate Nagar, Dharwad 580003, India.

The reaction kinetics of oxidation of etophylline by permanganate ions has been investigated in sulphuric acid medium using spectrophotometric techniques at 25°C and at constant ionic strength, 1.60 mol dm^{-3}, under pseudo-first order conditions. An autocatalysis reaction is observed due to the formation of one of the products, Mn^{II}. The order with respect to etophylline and Mn^{VII} concentrations are found to be unity in each, whereas fractional order with respect to autocatalyst, Mn^{II}. As the concentration of acid increases the rate of the reaction increases. The order with respect to acid concentration is less than unity. Based on the experimental results a suitable mechanism is proposed. The influence of temperature on the rate of reaction is studied. The activation parameters and thermodynamic quantities have been determined with respect to slow step of the mechanism.

Key words: Autocatalysis, oxidation, etophylline, permanganate.

INTRODUCTION

Permanganate is a unique oxidizing agent in neutral, alkaline and acid medium. In general reduction of the permanganate, in acid medium goes to either Mn^{IV} or Mn^{II} having the reduction potential of the couple Mn^{VII}/ Mn^{IV}: 1.695 V and Mn^{VII}/ Mn^{II}: 1.51 V (Day and Selbin, 1985). During the oxidation by permanganate, it is evident that permanganate is reduced to various oxidation states in acidic, alkaline and neutral media. The manganese chemistry involved in these multistep redox reactions is an important source of information as the manganese intermediates are relatively easy to identify when they have a sufficiently long lifetimes, and the oxidation states of the intermediate permit useful conclusion as to the possible reaction mechanisms, including the nature of the intermediates. In acid medium it exists in different forms such as $HMnO_4$, $H_2MnO_4^+$, $HMnO_3$, Mn_2O_7 and depending on the nature of reductant, the oxidant has been assigned both inner sphere and outer sphere mechanism path ways in their redox reactions (Joaquin and Perenz-Benito, 2009; Babatunde, 2008). Etophylline, (3, 7-dihydro-7-(2-hydroxyethyl)-1, 3-dimethyl-1H-purine-2, 6-Dione) (ETO) is a bronchodilator and is normally applied in combination with theophylline. Etophylline also acts as a proconvolusant (Girish et al., 1996). Although the pharmacological actions of the etophylline are generally considered like those of theophylline, Merkus and Zuidema pointed out different aspects of the two drugs (Merkus and Zuidema, 1980). In view of potential pharmaceutical importance of etophylline and lack of literature on the oxidation of this drug by any oxidant and the complexity of the reaction, a detailed study of the reaction becomes important.

In autocatalytic pathways in which Mn^{II} ions have been determined responsible for the effect, a mechanism with

*Corresponding author. E-mail: schimatadar@gmail.com.

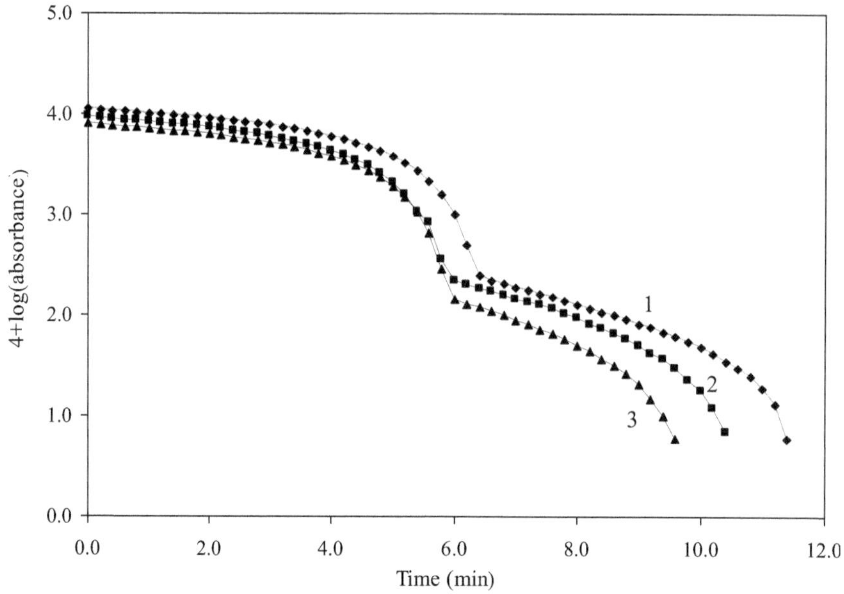

Figure 1. Plot of log [MnO_4^-] versus time [ETO] = 2.0 x 10^{-3}; [H_2SO_4] = 0.50; [MnO_4^-] x 10^4 mol dm^{-3} = (1) 3.0, (2) 4.0, (3) 5.0.

no free radicals has been suggested (Arrizbalaga et al., 1996; Arrizbalaga et al., 1997). Though, permanganate is used as an oxidant in the oxidation of several organic and inorganic substrates (Wiberg, 1965; Moghadasi et al., 2001; Surendra Rao, 1979; Ernst and Cyfert, 1987; Purohit et al., 1981; Solanki et al., 1982; Ameta et al., 1980; Tiwari et al., 1979), literature survey shows that the kinetic study of etophylline oxidation by Mn^{VII} is lacking. Hence we have investigated the oxidation of etophylline by acid permanganate. In this paper, distinct evidence of an autocatalytic influence of Mn^{II} for permanganate oxidation of etophylline in strong acid medium has been presented and free radical mechanism has been proposed.

MATERIALS AND METHODS

Throughout the work the double distilled water and reagent grade chemicals have been used. A stock solution of etophylline (AR, Arti industries) was prepared by dissolving in water. Permanganate stock solution was obtained by dissolving potassium permanganate (Glaxo, Analar) in water and standardized by titrating against oxalic acid (Vogel, 1967). The manganese (II) solution was made by dissolving manganese sulphate (AR, Analar) in water. Na_2SO_4 (AR, Rankem) and H_2SO_4 (AR, Rankem) were used to provide required ionic strength and acidity respectively.

All kinetic measurements were performed under the pseudo-first order condition with etophylline concentration greater than permanganate concentration at constant ionic strength of 1.60 mol dm^{-3}.

The reaction was initiated by mixing thermally equilibrated (25 ± 0.1°C) solution of etophylline and permanganate which contained the required quantities of H_2SO_4 and Na_2SO_4. Since the initial reaction was too fast to be monitored by usual methods, the title reaction was followed by measuring the absorbance of permanganate in the reaction mixture at 525 nm in a 1cm cell placed in the thermostatted compartment of Varian Cary 50 Bio UV-vis spectrophotometer connected to a rapid kinetic accessory (HI-TECH SFA-12, U.K) at its absorption maximum of 525 nm as a function of time. Application of Beer's law was verified between 1.0 x 10^{-4} to 1.0 x 10^{-3} mol dm^{-3} of permanganate at 525 nm under the reaction conditions and the molar extinction coefficient was found to be ε = 2200 ± 50 dm^3 mol^{-1} cm^{-1} (Literature value ε =2400 dm^3 mol^{-1} cm^{-1} (Timmanagoudar et al., 1996)). The kinetic runs were followed more than 95% completion of the reaction. The pseudo first order rate constants, k_{obs} were calculated from the plots of log (absorbance) versus time. The plots were linear over 60% completion of the reaction (Figure 1) The k_{obs} values were reproducible within ± 5% and are the average of at least three independent kinetic runs (Tables 1 and 2).

RESULTS AND DISCUSSION

Stoichiometry and product analysis

Different sets of concentration of reactants in 0.50 mol dm^{-3} sulphuric acid at constant ionic strength, 1.60 mol

Table 1. Effect of variation of Manganese (VII) and etophylline concentrations on the oxidation of etophylline by acidic permanganate at 25°C and I = 1.60 mol dm^{-3}.

$[Mn^{VII}] \times 10^4$ (mol dm^{-3})	[ETO] x 10^2 (mol dm^{-3})	$[H^+]$ (mol dm^{-3})	$k_{obs} \times 10^2$ (s^{-1})	$k_{sec\ order} = k_{obs}/[ETO]$ (dm^3 mol^{-1} s^{-1})
0.5	2.0	0.50	0.43	-
1.0	2.0	0.50	0.46	-
2.0	2.0	0.50	0.45	-
3.0	2.0	0.50	0.42	-
4.0	2.0	0.50	0.41	-
5.0	2.0	0.50	0.40	-
2.0	0.5	0.50	0.11	0.21
2.0	1.0	0.50	0.21	0.21
2.0	2.0	0.50	0.42	0.21
2.0	3.0	0.50	0.63	0.21
2.0	4.0	0.50	0.84	0.21
2.0	5.0	0.50	1.05	0.21

Table 2. Effect of variation of H^+ ions concentrations on the oxidation of etophylline by acidic permanganate at 25°C and I = 1.60 mol dm^{-3}.

$[H_2SO_4]$ (mol dm^{-3})	$[HSO_4^-]$ (mol dm^{-3})	$[SO_4^{2-}]$ (mol dm^{-3})	$[H^+]$ (mol dm^{-3})	$k_{obs} \times 10^2$ (s^{-1})
0.1	0.175	0.857	0.0247	0.04
0.3	0.490	0.542	0.109	0.18
0.5	0.721	0.311	0.278	0.42
0.7	0.847	0.185	0.552	2.73
0.9	0.905	0.127	0.894	1.03
1.0	0.928	0.104	1.071	1.15

dm^{-3}, were kept for over 5 h at 25°C in a closed container. When [permanganate] > [etophylline], the remaining permanganate concentration was assayed by measuring the absorbance at 526 nm. The reaction products were identified as Mn^{II} and 1, 3-dimethyl-2, 6-dioxo-1, 2, 3, 6, tetrahydro-purin-7yl)-acetaldehyde. The results indicate that 5 moles of etophylline consumes two moles of permanganate according to Appendix 1.

The stoichiometric ratio of oxidation of etophylline by permanganate suggests that the main product as 1, 3-dimethyl-2, 6-dioxo-1, 2, 3, 6, tetrahydro-purin-7yl)-acetaldehyde (A). The product was extracted with ether and recrystallized from aqueous alcohol. The obtained product was further purified by column chromatography (Harihar et al., 2000) using the mixture of benzene (65%) and ethyl acetate (35%) as mobile phase and fractions are subjected to spectral investigations. From the spot test (Hosahalli et al., 2010), IR (Figure 2), GCMS (Figure 3) spectra, the main oxidation product was identified as A. The IR spectroscopy showed an aldehydic, C=O stretching at 1707 cm^{-1} indicating the presence of aldehydic C=O functional group. C=O stretching at 1657 and 1604 cm^{-1} indicating the presence of two amide group in the product. The presence of A was also confirmed by GC-MS analysis (Figure 3), obtained on a Shimadzu 17A gas chromatograph with a Shimadzu XP-5000A mass spectrometer using EI ionization technique. The mass spectrum showed the molecular ion peak at 224 amu consistent with the molecular mass of 224 amu. All other peaks are observed in the GC-MS can be interpreted in accordance with the structure of A.

The reaction orders were determined from the slopes of log k_{obs} versus log (concentrations) plots by varying the concentration of etophylline, permanganate and sulphuric acid and in turn while keeping all other concentrations and conditions constant. The oxidant permanganate

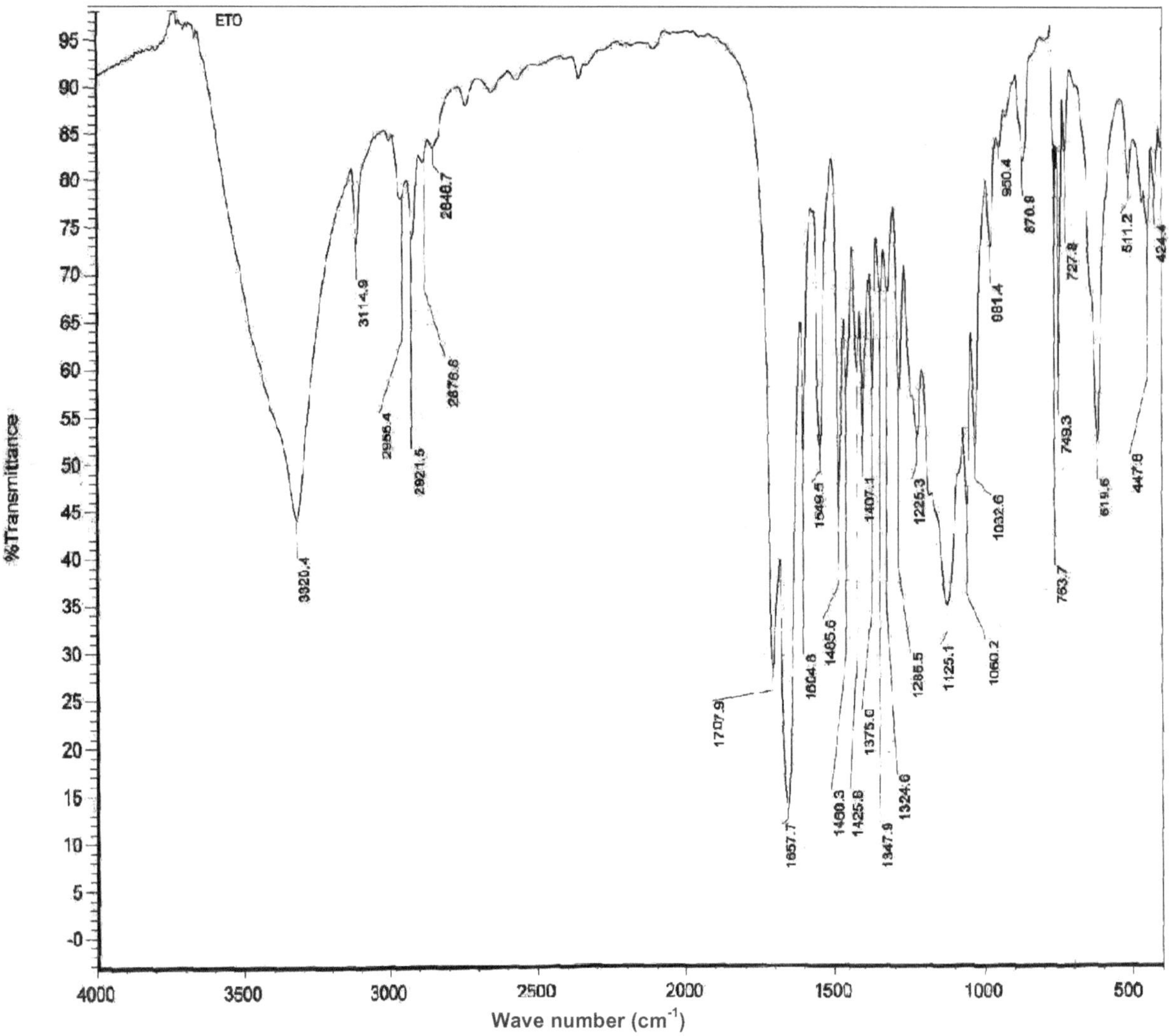

Figure 2. FT-IR spectra of the product 1, 3-dimethyl-2, 6-dioxo-1, 2, 3, 6, tetrahydro-purin-7yl)-acetaldehyde of oxidation of etophylline by permanganate.

(MnO_4^-), concentration was varied in the range 5.0 × 10^{-5} - 5.0 × 10^{-4} mol dm^{-3}. The observed pseudo-first order rate constants k_{obs} were almost constant (Table 1) indicates first order with respect to permanganate concentrations. The effect of variation of etophylline on the rate of reaction was studied in the concentration range, 5.0 x 10^{-3} - 5.0 x 10^{-2} mol dm^{-3}, at constant concentration of permanganate, acid and at constant ionic strength. It was observed that, as the etophylline concentration increases, k_{obs} also increase (Table 1). The value of the slope of the plot of log k_{obs} versus log [etophylline] was found to be unity, which indicates first order with respect to etophylline concentrations. This was also confirmed by the plot of k_{obs} versus ETO concentration (Figure 4), which is a straight line passing through the origin. The effect of variation of sulphuric acid on the rate of reaction was studied in the concentration range, 0.1 to 1.0 mol dm^{-3} at fixed concentration of oxidant, [MnO_4^-] = 3.0 × 10^{-4} mol dm^{-3}, reductant, [etophylline] = 3.0 × 10^{-2} mol dm^{-3} and with other conditions remaining constant, The *in situ* H^+ ion concentration in the sulphuric acid – sulphate media was calculated using the known ionization constant of acid sulphate as in an earlier study (Patil et al., 2008; Ya et al., 1965). The k_{obs} was found to be increased with increase in H^+ concentration (Table 2). The order with

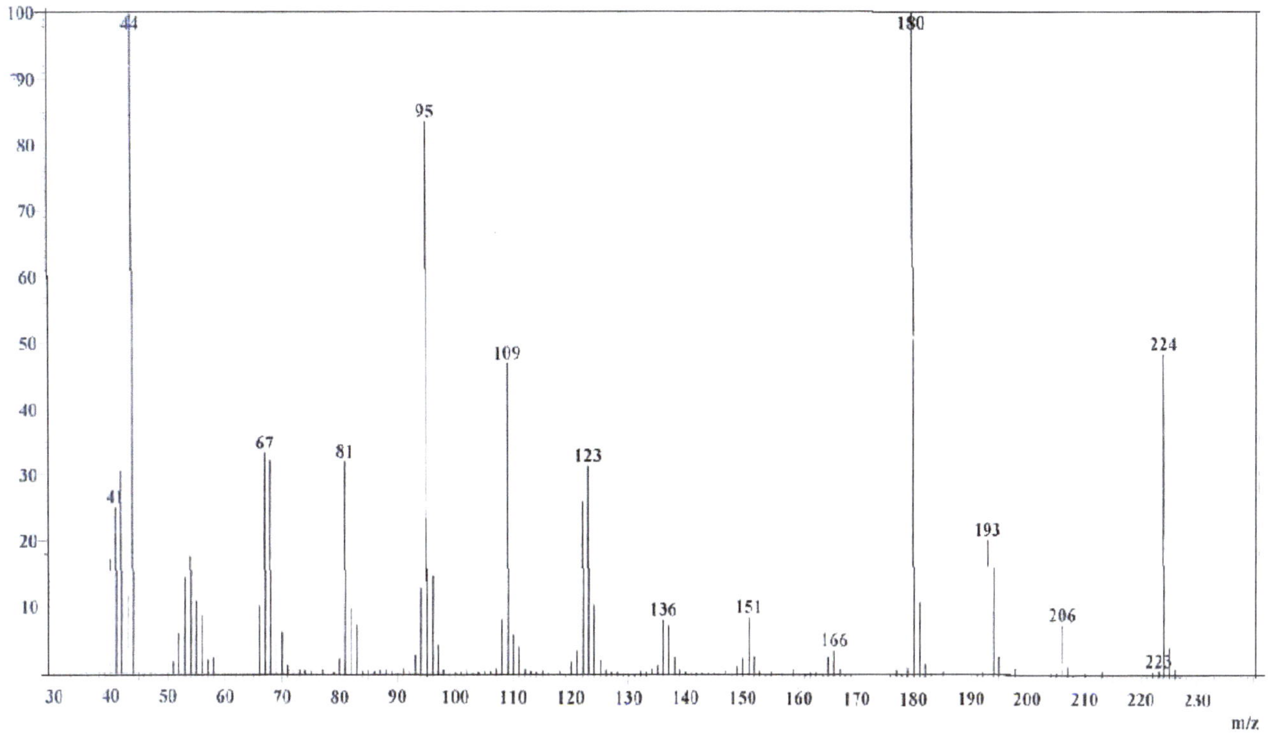

Figure 3. GC-MS spectra of the product 1, 3-dimethyl-2, 6-dioxo-1, 2, 3, 6, tetrahydro-purin-7yl)-acetaldehyde, showed molecular ion peak at m/z 224 amu.

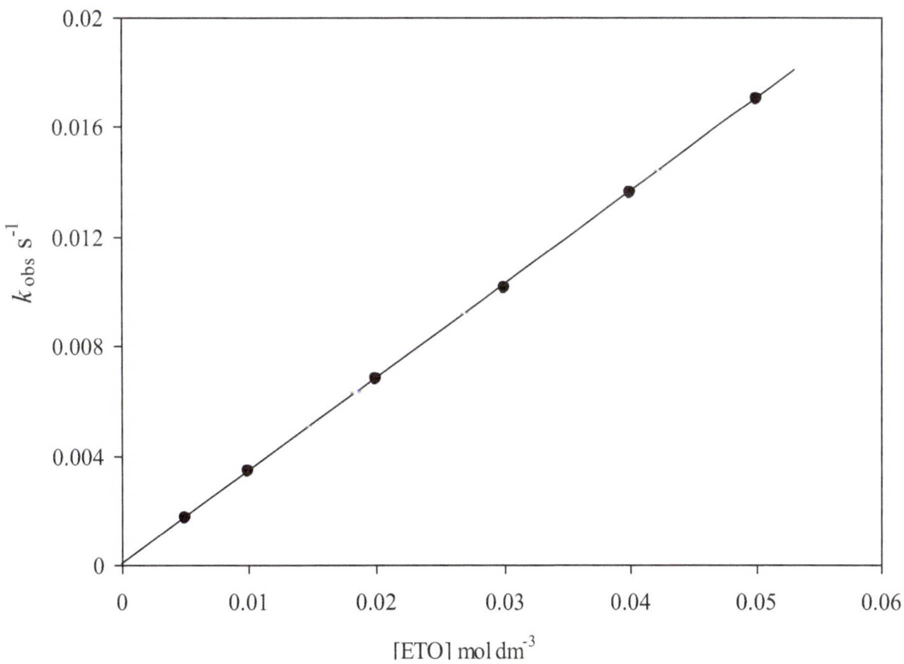

Figure 4. Plot of k_{obs} versus [ETO] (conditions as in Table 1).

Table 3. Effect of added of Manganese(II) concentration on the oxidation of etophylline by aqueous acidic permanganate at 25°C.

$[Mn^{II}] \times 10^4$ (mol dm^{-3})	$k_{autocat} \times 10^2$ (s^{-1})
0.5	0.66
1.0	0.95
2.0	1.18
3.0	1.37
4.0	1.53
5.0	1.61

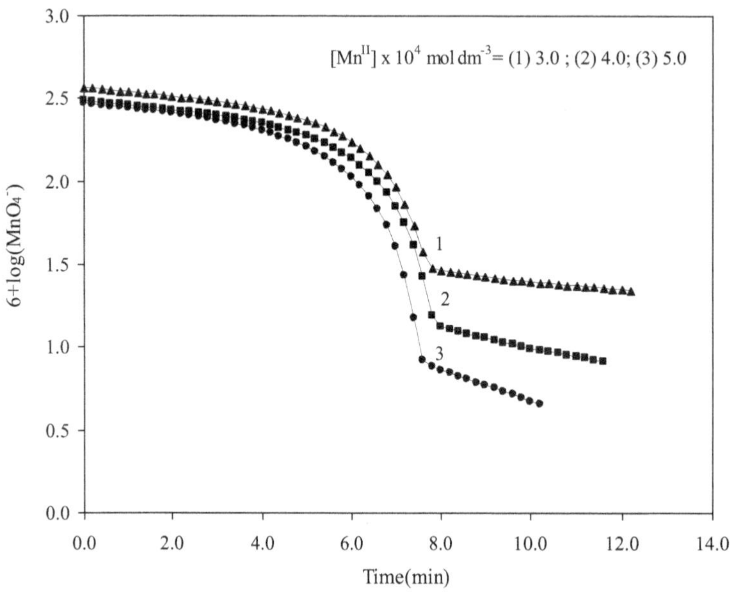

Figure 5. Effect of added product, Manganese (II) on the oxidation of etophylline by permanganate.

respect to [H^+] was found to be less than unity (0.87).

The initially added products, manganese (II) were studied in the range 0.5×10^{-5} to 5.0×10^{-4} mol dm^{-3} respectively, while keeping the reactant concentrations and all other conditions constant. As the initial concentration of Mn^{II} ions increased, a notable increase in the rate (Table 3) of reaction was seen (Figure 5). In all the experiments performed, curves with the sigmoid profile (Figures 1 and 5) indicate the autocatalytic nature of the product manganese (II). The added product 1, 3-dimethyl-2, 6-dioxo-1, 2, 3, 6, tetrahydro-purin-7yl)-acetaldehyde does not change the rate appreciably.

At constant concentrations of reactants and other conditions constant, the ionic strength was varied by varying the concentration of sodium sulphate between 1.60 and 3.20 mol dm^{-3}. Increasing ionic strength had negligible effect on the rate of reaction. At constant acidity and other constant conditions, as the acetic acid content increases from 0 to 50% (v/v) in the reaction mixture, change in dielectric constant had negligible effect on the rate of reaction.

The intervention of the free radicals in the reaction was examined as follows. The reaction mixture, to which a known quantity of acrylonitrile monomer was initially added, was kept for 2 h in an inert atmosphere. On diluting the reaction mixture with methanol, a white precipitate was formed, indicating the intervention of free radicals in the reaction. The blank experiments of either etophylline or permanganate alone with acrylonitrile did not induce any polymerization under the same condition

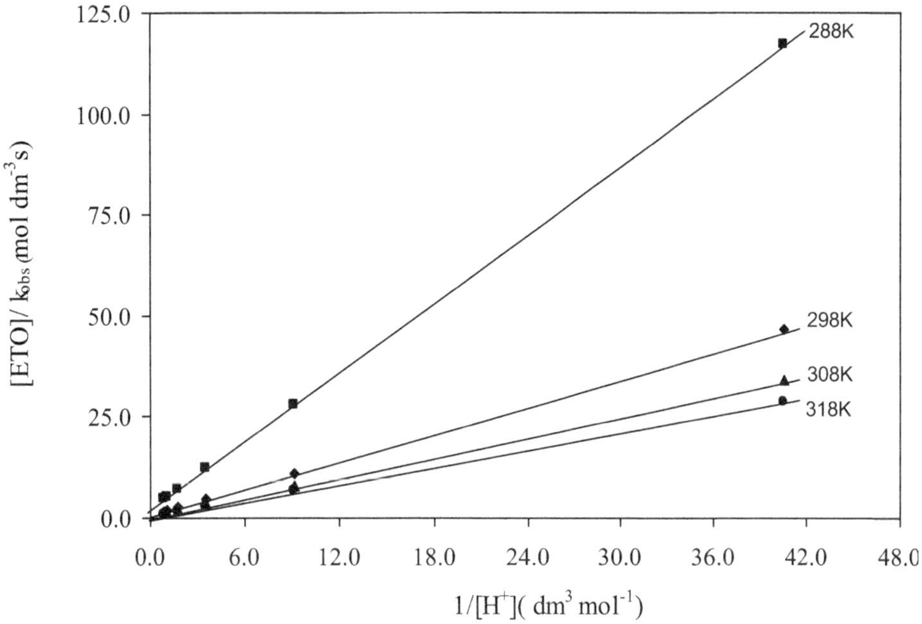

Figure 6. Verification of rate law (7) in the form of equation (8) at different temperatures.(Conditions as in Tables 1 and 2).

Table 4. Effect of temperature on the oxidation of etophylline by permanganate in sulphuric acid, I = 1.60 mol dm^{-3}.

(A) Activation parameters with respect to slow step of Appendix 2		
Temp (K)	**k_1 ($dm^3 mol^{-1} s^{-1}$)**	**Activation parameters**
288	0.52	Ea = 46 ± 3 kJ/mol
298	1.49	$\Delta H^{\neq}$ = 43 ± 3 kJ/mol
308	2.38	$\Delta S^{\neq}$ = -94 ± 4 J/K/mol
318	3.31	$\Delta G^{\neq}$ = 71 ± 2 kJ/mol
		log A = 8± 0.4
(B) Thermodynamic quantities with respect to first step of Appendix 2		
Temp (K)	**K_1 ($dm^3 mol^{-1}$)**	**Thermodynamic quantities**
288	0.68	ΔH = -11.6 ± 0.5 kJ/mol
298	0.59	ΔS = -43.3 ± 3 J/K/mol
308	0.50	ΔG = -2.20 ± 0.03 kJ/mol
318	0.42	

as those induced for reaction mixture. Initially added acrylonitrile decreased the rate of reaction indicating free radical intervention (Farokhi and Nandibewoor, 2004).

The rate of the reaction was measured at four different temperatures 15, 25, 35 and 45°C at varying sulphuric acid concentration. The rate of the reaction increased with the increase in temperature. The rate constant, k_1 of the slow step of Appendix 2 were obtained from the intercept of the plots [ETO]/ k_{obs} versus 1/ [H^+] (Figure 6) at four different temperatures (Table 4). The energy of activation was obtained by the plot of log k_1 versus 1/T, from which the activation parameters were calculated (Table 4A). The formation constant of $HMnO_4$ (K_1) was evaluated from the intercept and slope of the plot [ETO]/

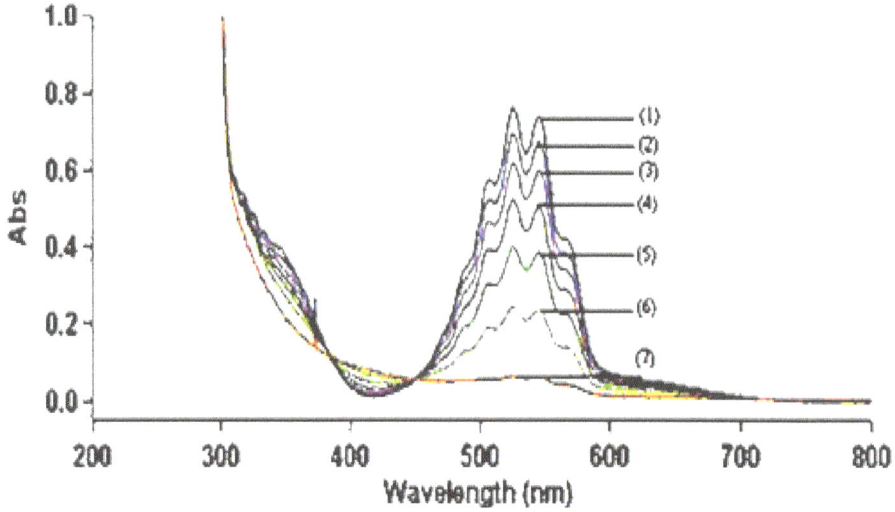

Figure 7. Spectroscopic changes occurring in the oxidation of etophylline by acidic permanganate at 25°C. $[MnO_4^-]$ = 3.0 x 10^{-4} ; [ETO] = 2.0 x 10^{-2} ; $[H^+]$ = 0.5 and *I* = 1.60 mol dm^{-3} with scanning interval of 1 min.

k_{obs} versus 1/ $[H^+]$ (Figure 6) at four different temperatures (Table 4B). The *van't* Hoff plot (log K_1 versus 1/*T*) was drawn and the thermodynamic quantities were calculated (Table 4B).

The expected oxidizing species of permanganate in acid media are MnO_4^-, $H_2MnO_4^+$, $HMnO_3$, Mn_2O_7. Among them MnO_4^-, ion is powerful oxidizing agent in aqueous alkaline as well as in acidic medium. The stable reduction product of MnO_4^- in acid medium is Mn^{II}. Figure 7 illustrates the spectroscopic changes occurring in the oxidation of etophylline by acidic permanganate at 25°C with scanning interval of 1 min. The literature survey reveals that (Joaquin and Perenz, 2009) Mn^{IV} ion absorbs in the region 400 - 650 nm. Figure 7 shows no features in this wavelength area indicating that MnO_2 is not a reaction product. The reaction between etophylline and MnO_4^- has a stoichiometry 5:2 with first order dependence on $[MnO_4^-]$ and [ETO] and fractional order dependence on both $[H^+]$ and $[Mn^{II}]$ in acid medium. Mn^{II} is the reduced product of Mn^{VII} in the reaction may be that etophylline shows strong reducing character in sulphuric acid medium.

The active species of permanganate in aqueous acid medium may be deduced from the dependence of the rate on $[H^+]$ in the reaction medium. The apparent order of less than unity in $[H^+]$ may be an indication of permanganate species as permanganic acid. In acid medium, permanganic acid, $HMnO_4$ is more efficient oxidant species of manganese than permanganate ion (Chimatadar et al., 1991; Wiberg, 1965). In addition it has been observed that the rate of reaction increased with increase in $[H^+]$ and was tending to attain a limiting value at high acidities (Bailar et al., 1975). The plot of k_{obs} versus $[H^+]$ is a curve of increasing slope (convex to the rate axis) (Figure 8) from which it can be inferred (Farokhi and Nandibewoor, 2004) that, a rapid equilibrium between the protonated form is involved. At higher acidities protonation is almost complete, leading to the limiting rate, this indicates that only the protonated form is active. The ionic strength on the rate of reaction also confirms that $HMnO_4$ as the active species of MnO_4^-. Thus the acid permanganate equilibrium is represented as:

$$H^+ + MnO_4^- \underset{}{\overset{K_1}{\rightleftharpoons}} HMnO_4 \quad (1)$$

where K_1 is the formation constant of $HMnO_4$. The reaction between the permanganate and etophylline in alkaline medium has stiochiometry 5:2 (ETO: MnO_4^-) with first order dependence on both substrate and oxidant and apparent less than unit order with respect to acid concentration. Based on the experimental results, a mechanism is proposed for which all the observed orders in each constituent such as [oxidant], [reductant] and $[H^+]$ may be well accommodated. The results suggest the formation of protonated form of permanganate is the active species of MnO_4^- in a prior equilibrium step, which

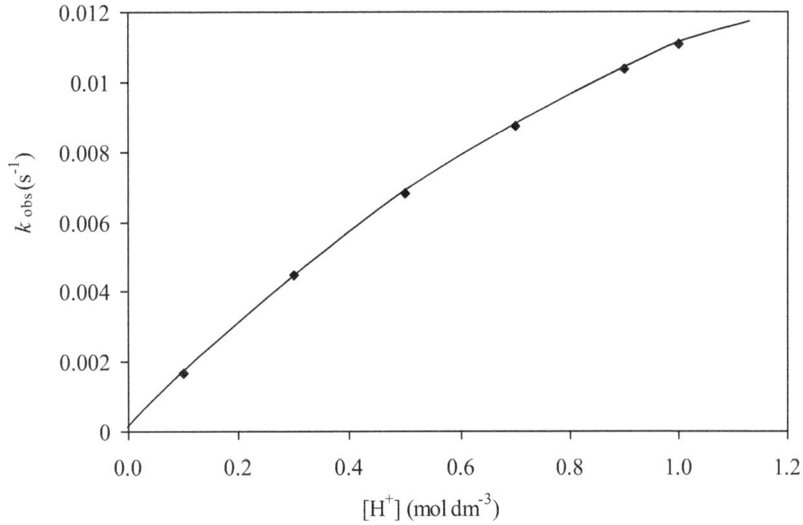

Figure 8. The plot of k_{obs} versus [H$^+$], a curve of increasing slope (convex to the rate axis) (Conditions as in Table 2).

reacts with the one mole of etophylline in the rate determining step to give a free radical derived from etophylline and an intermediate Mn^{VI}. In further fast steps the intermediate Mn^{VI} reacts with a free radical derived from etophylline to produce the product, 1,3-dimethyl-2,6-dioxo-1,2,3,6,tetrahydro-purin-7yl)-acetaldehyde and an intermediate Mn^{V}. In further fast steps Mn^{V} subsequently reduced to the end product Mn^{II} satisfying stoichiometric observations. The mechanism is proposed as in Appendix 2.

Since none of the intermediate could be detected, Appendix 2 is only one of the possible mechanisms for the reaction in the presence of free radical. Attempts were made to allow spectroscopic detection of intermediate Mn^{V} and Mn^{III} as the reaction proceeded in the oxidation of etophylline by MnO_4^-. Unfortunately, the low concentration of Mn^{III} and Mn^{V} intermediates obtained under our experimental conditions made the detection unsuccessful. However, there are reports in the literature for the existence of Mn^{III} and Mn^{V} (Bailar et al., 1975). Appendix 2 leads to the rate law (8); according to Appendix 2:

$$\text{Rate} = k_1[\text{HMnO}_4]\,[\text{ETO}]$$

$$= k_1K_1[\text{MnO}_4^-]_f[\text{ETO}]_f[\text{H}^+]_f \qquad (2)$$

The total concentration of MnO_4^- is given by, (where 't' and 'f' stands for total and free)

$$[\text{MnO}_4]_t = [\text{MnO}_4^-]_f + [\text{HMnO}_4]_f$$

$$= [\text{MnO}_4^-]_f + K_1[\text{H}^+][\text{MnO}_4^-]$$

$$= [\text{MnO}_4^-]_f(1 + K_1[\text{H}^+])$$

Therefore,

$$[\text{MnO}_4^-]_f = \frac{[\text{MnO}_4^-]_t}{1 + K_1[\text{MnO}_4^-]} \qquad (3)$$

Similarly,

$$[\text{H}^+]_f = \frac{[\text{H}^+]_t}{1 + K_1[\text{MnO}_4^-]} \qquad (4)$$

Substituting the Equations (3) and (4) in Equations (2) (and omitting 't' and 'f' subscripts),

$$\text{Rate} = \frac{-d[\text{MnO}_4^-]}{dt} = \frac{k_1K_1[\text{MnO}_4^-][\text{ETO}][\text{H}^+]}{(1 + K_1[\text{H}^+])\,(1 + K_1[\text{MnO}_4^-])} \qquad (5)$$

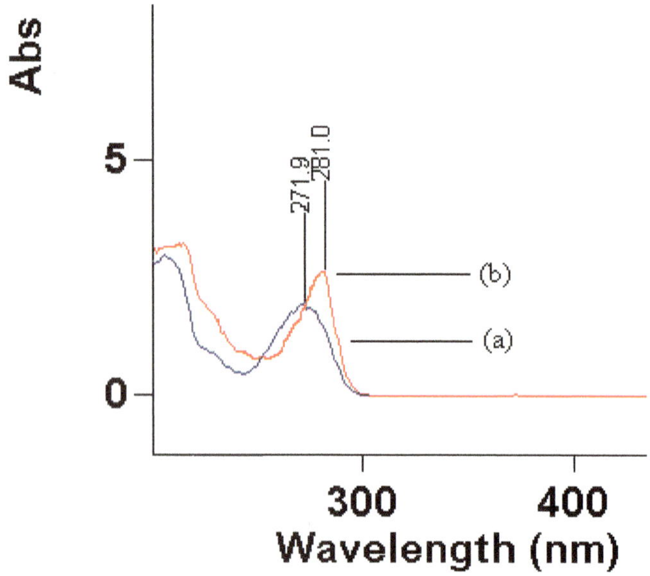

Figure 9. Spectroscopic evidence for the complex formation between etophylline and Manganese (II). (a) UV-vis spectra of etophylline (271 nm); (b) UV-vis spectra of mixture of etophylline and Manganese(II) (281 nm).

$$= \frac{k_1 K_1 [MnO_4^-][ETO][H^+]}{1 + K_1[H^+] + K_1[MnO_4^-] + K_1^2[H^+][MnO_4^-]} \quad (6)$$

The terms, such as $K_1[MnO_4^-]$ and $K_1^2[H^+][MnO_4^-]$ in the denominator of Equation (6) are neglected in the view of the low concentration of MnO_4^- used in the experiments. Hence the Equation (6) becomes (7)

$$k_{obs} = \frac{Rate}{[MnO_4^-]} = \frac{k_1 K_1 [ETO][H^+]}{1 + K_1[H^+]} \quad (7)$$

The rate law (7) may be verified by rearranging in the form:

$$\frac{[ETO]}{k_{obs}} = \frac{1}{k_1 K_1 [H^+]} + \frac{1}{k_1} \quad (8)$$

where k_{obs} is first order rate constant. According to Equation (8), the plot of $[ETO]/k_{obs}$ versus $1/[H^+]$ is expected to be linear, and is found to be so (Figure 6). The slope and the intercept of such plot lead to the values of K_1 and k_1 at 25°C as 0.59 $dm^3\ mol^{-1}$ and 1.49 $dm^3\ mol^{-1}s^{-1}$. The negligible effect of ionic strength and relative permittivity is consistent with reaction between two neutral molecules as in Appendix 2.

The Mn^{II} acts as an auto catalyst in case of oxidation of permanganate is well known (Zahedi et al., 2004). In the present case Mn^{II} was found to auto catalyst the rate of oxidation of etophylline by permanganate. In all the experiments performed, curves with a sigmoid profile (Figures 1 and 5) have been obtained. This characteristic suggests the existence of an autocatalytic mechanism. In order to demonstrate that Mn^{II} species is responsible for the autocatalytic effect, kinetic runs were carried out in the presence of Mn^{II}. As the concentration of Mn^{2+} ions increased, a notable increase in the rate of reaction was observed (Figure 5 and Table 3).

The catalytic effect of Mn^{II} can be interpreted in either of two ways, (a) Mn^{II} may form a complex with the substrate which is then oxidized by $HMnO_4$ or (b) Mn^{II} first reacts with Mn^{VII} to produce Mn^{III} which accelerates the rate of reaction. The apparent order is less than unity in Mn^{II} ions may be attributed to the complex formation between etophylline and Mn^{II}. The complex (C_1) is then subsequently involved in the interaction with MnO_4^-. These steps in Appendix 3 will form part of Appendix 2. The probable structure of the complex (C_1) is given in Appendix 4.

The evidence for the complex (C_1) formation was obtained from UV-vis spectra etophylline and mixture of etophylline and Mn^{II}, which indicates the bathochromic shift of 10 nm from 271 to 281 nm (Figure 9). Indeed such complex formation is observed in the literature (Kim et al., 1997; Farokhi and Nandibewoor, 2004). MnO_4^- ions are primarily responsible for the oxidation of substrate etophylline. The intermediate complex subsequently undergoes decomposition to give the reaction products.

According to Appendix 3:

$$Rate = k_2 [C_1]$$

$$= k_2 K_2 [Mn^{II}]_f [ETO]_f$$

The total concentration of Mn^{II} is given as:

$$[Mn^{II}]_t = [Mn^{II}]_f + [C_1]$$

$$= [Mn^{II}]_f \{1 + K_2[ETO]\} \quad (9)$$

$$[Mn^{II}]_f = \frac{[Mn^{II}]_t}{1 + K_2[ETO]} \quad (10)$$

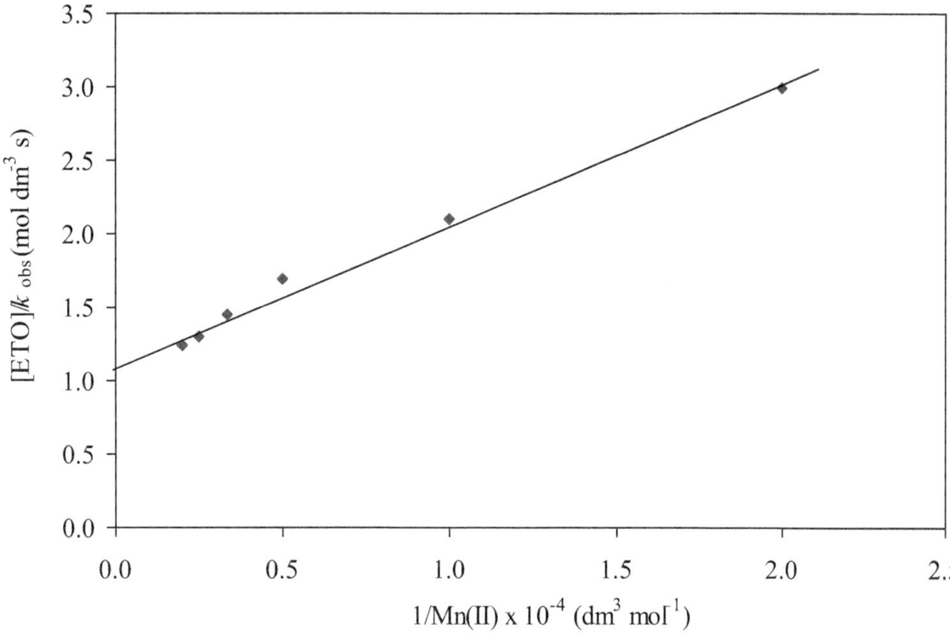

Figure 10. Verification of rate law (16) (Conditions as in Table 3).

Similarly,

$$[ETO]_f = \frac{[ETO]_t}{1 + K_2 [Mn^{II}]} \quad (11)$$

Substituting Equations (10) and (11) in Equation (9) and omitting the subscripts 't' and 'f'

$$k_{autocat} = \frac{k_2 K_2 [Mn^{II}] [ETO]}{\{1 + K_2[ETO]\} \{1 + K_2[Mn^{II}]\}} \quad (12)$$

Thus, when Mn^{II} was initially present, a composite Scheme involving all the steps of the Appendix 2 and 3 operates and rate law becomes

$k_{gross} = k_{obs} + k_{autocat}$
$k_{autocat} = k_{gross} - k_{obs}$

$$k_{gross} = \frac{k_1 K_1 [ETO][H^+]}{1 + K_1[H^+]} + \frac{k_2 K_2 [Mn^{II}] [ETO]}{\{1 + K_2[ETO]\} \{1 + K_2[Mn^{II}]\}} \quad (13)$$

$$k_{autocat} = \frac{k_2 K_2 [Mn^{II}][ETO]}{\{1 + K_2[ETO]\} \{1 + K_2[Mn^{II}]\}} \quad (14)$$

$$k_{autocat} = \frac{k_2 K_2 [Mn^{II}][ETO]}{1 + K_2[ETO] + K_2[Mn^{II}] + K_2^2[ETO][Mn^{II}]} \quad (15)$$

The term K_2^2[ETO][Mn(II)] can be neglected. The rate law (15) may be verified by rearranging in the form:

$$\frac{[ETO]}{k_{autocat}} = \frac{1}{k_2[Mn(II)]} + \frac{1}{k_2 K_2[Mn(II)]} + \frac{1}{k_2} \quad (16)$$

where $k_{autocat}$ is first order rate constant. According to Equation (16), the plot of [ETO]/$k_{auotocat}$ versus 1/ [Mn(II)] (Figure 10) is expected to be linear, and is found to be so (Figure 10). The slope and the intercept of such plot lead to the values of K_1 and k_1 at 25°C as 0.65 dm^3 mol^{-1} and 8.11 dm^3 $mol^{-1}s^{-1}$. The value of K_2 is in agreement with the literature (Farokhi and Nandibewoor, 2004). Indeed it is to be noted that the plot shows an intercept which is in agreement with complex formation as in Appendix 3.

Conclusion

In acid medium, the title reaction exhibits autocatalytic nature by one of the products, Mn^{II}. The sigmoid profile observed for permanganate absorption variation versus time at 525 nm was completely transformed to a line under the condition $[Mn^{II}]_0 = [KMnO_4]$. Thus we report the first conclusive evidence of autocatalytic oxidation process of etophylline in acid medium in which Mn^{II} species are responsible for such effect. Mn^{II} is the reduced product of Mn^{VII} in the reaction may be that, etophylline shows a strong reducing character in H_2SO_4 solution. The interventions of free radicals were observed, and the mechanisms satisfying the experimental observations for the autocatalysed pathways were presented.

REFERENCES

Ameta CS, Pande PN, Gupta HL, Chowdhry HC (1980). Kinetic study of oxidative decarboxylation of L-β-phenylalanine by potassium permanganate. Z. Phys. Chem. 261(6):1222.

Arrizbalaga A, Andres Ordax FJ, Fernandez Aranguiz MY, Pche R (1996). Kinetics and mechanism of oxidation of L-α-amino-n-butyric acid by permanganate in acid medium. Int. J. Chem. Kinet. 29:799.

Arrizbalaga A, Andres Ordax FJ, Fernandez Aranguiz MY, Pche R (1997). Kinetic studies on oxidation of amino acids, effect of length of amino acid carbon chain. Int. J. Chem. Kinet. 29:18.

Authur Vogel I (1967). Vogel's Text Book of Macro and Semi Micro Qualitative Inorganic Analysis. p. 291.

Babatunde OA (2008). A study of the kinetics and mechanism of oxidation of L-ascorbic acid by permanganate ion in acid medium. World J. Chem. 3:27.

Bailar JC, Emeleus HJ, Nyholm R, Dickenson AFT (1975). Comprehensive Inorganic Chemistry, Pergamon Press Ltd, New York. p. 771.

Chimatadar SA, Hiremath SC, Raju JR (1991). Oxidation of Thallium(I) by permanganate in aqueous perchloric acid medium. Ind. J. Chem. 30A:190.

Day MC, Selbin J (1985). Theoretical Inorganic Chemistry, Reinhold Publishing Corporation, New York. p. 344.

Ernst T, Cyfert M (1987). Kinetics and mechanism of sulphite oxidation by permanganate in acid medium. Z. Phys. Chem. 268:175.

Farokhi SA, Nandibewoor ST (2004). The kinetic and mechanism of the oxidative decarboxylation of benzilic acid by acid permanganate (stopped flow technique) - Autocatalytic study. Can. J. Chem. 82:1372.

Girish K, Jayaprakash MS, Ganagadhar BN, Janakiramaiah N (1996). Etophylline as a proconvulsant. Convuls. Ther. 12:196.

Harihar AL, Kembhavi MR, Nandibewoor ST (2000). Kinetic and mechanistic study of oxidative deamination and decarboxylation of L-valine by alkaline permanganate. Montash fur.Chem. 131:739.

Hosahalli RV, Savanur AP, Nandibewoor ST, Chimatadar SA (2010). Kinetics and mechanism of uncatalysed and Ruthenium(III)-catalysed oxidation of D-panthenol by alkaline permanganate. Transit. Met. Chem. 35:237.

Joaquin F, Perenz-Benito (2009). Autocatalytic reaction pathway on manganese dioxide colloidal particles in the permanganate oxidation of glycine. J. Phys. Chem. 113:15982.

Kim S, Zhang W, Pinnavaia JT (1997). Catalytic Oxidation of Styrene by Manganese(II) Bipyridine Complex Cations Immobilized in Mesoporous Al-MCM-41. Catal. Lett. 43:149.

Merkus FW, Zuidema J (1980). Theophylline and hydroxyethyltheophylline are different drugs. Int. J. Clin. Phamacol. Ther. Toxicol. 18:97.

Moghadasi M, Rashidi N, Iloukhani H (2001). Oxidation of L-arginine by manganese(VII) in concentrated sulphuric acid medium. Phys. Chem. Liq. 39:267.

Patil RK, Chimatadar SA, Nandibewoor ST (2008). Mechanistic study of cerium(IV) oxidation of antimony(III) in aqueous sulphuric acid in the presence of micro amounts of manganese(II) by stopped flow technique. Transit. Met. Chem. 33:625.

Purohit R, Ameta CS, Cowdhary HC (1981). Kinetic study of oxidation of tartaric acid by acid permanganate. Z. Phys. Chem. 262:737.

Solanki SK, Khadikar PV, Nandwana RK (1982). Kinetic studies in redox reaction of EDTA complex. I. Evidence for free groups in the complex. Z. Phy. Chem. 263(5):1069.

Surendra Rao VB, Sethuram BS, Navaneeth Rao T (1979). Kinetics of oxidation of some amino acids by $KMnO_4$ in moderately concentrated H_2SO_4 medium in presence and absence of Ag^+. Int. J. Chem. Kinet. 11:165.

Timmanagoudar PL, Hiremath GA, Nandibewoor ST (1996). Permanganate oxidation of Thallium(I) in sulphuric acid: A kinetic study by stopped flow technique. Pol. J. Chem. 70:1459.

Tiwari N, Purushottam CS, Ameta Shastry, Vijayendra R (1979). Kinetics of oxidation of acetic acid by potassium permanganate. Z. Phys. Chem. 260(1):167.

Wiberg KB (1965). Oxidation in Organic Chemistry (Part A), New York, Academic Press, 2:48.

Zahedi M, Bahrami H (2004). Kinetic and mechanism of autocatalytic oxidation of L-asperagine in moderately concentrated sulphuric acid medium. Kinet. Catal. 45:351.

APPENDIX

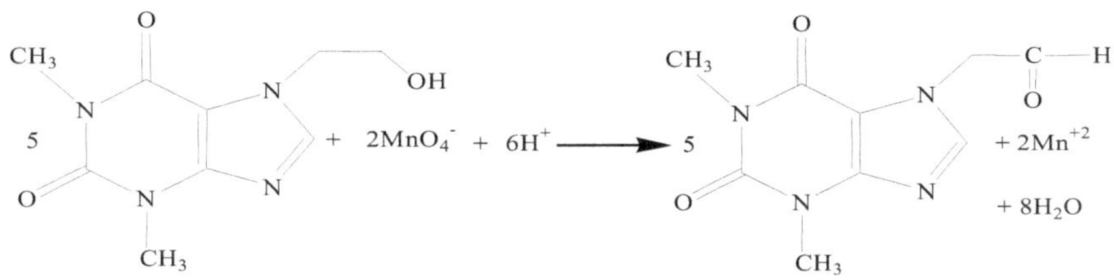

Appendix 1.

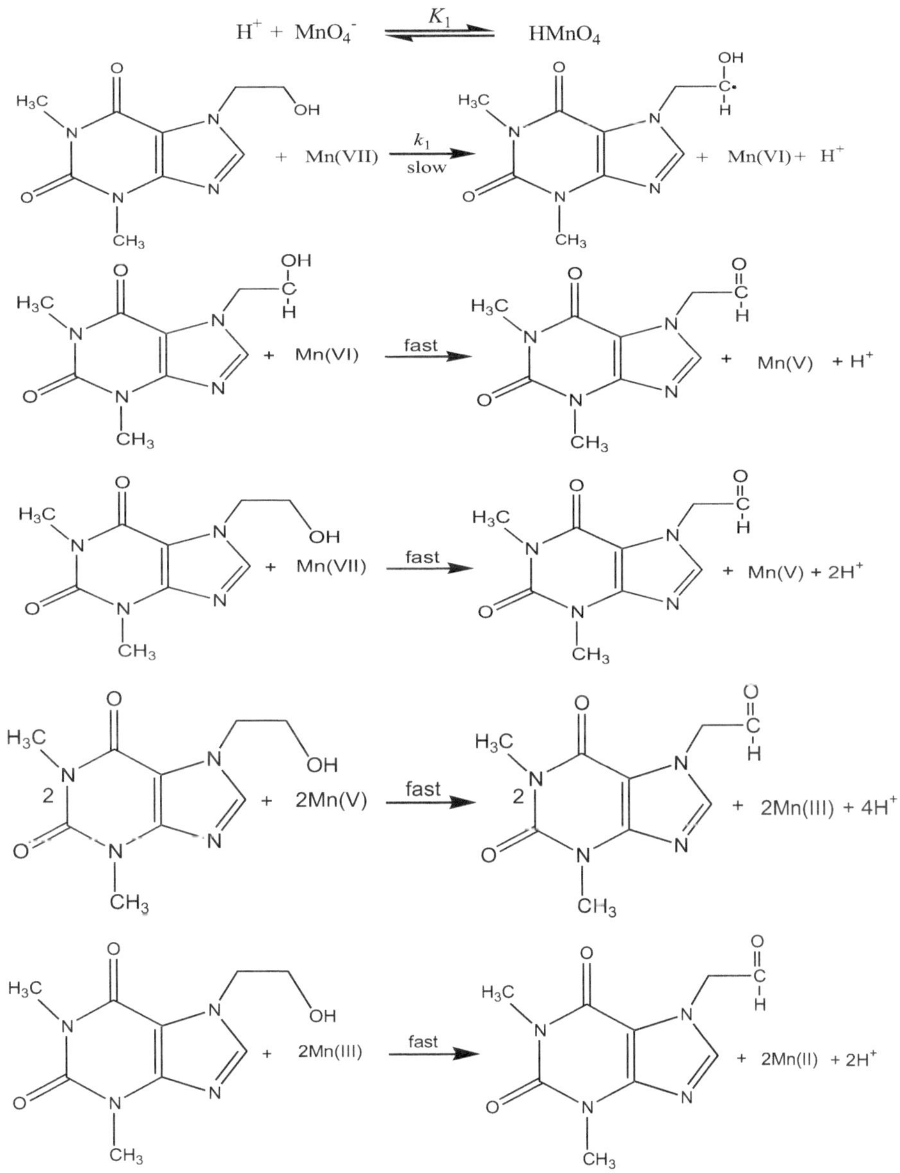

Appendix 2.

$$\mathrm{Mn^{II}} + \text{Ettophylline} \underset{}{\overset{K_2}{\rightleftharpoons}} \text{Complex } (\mathrm{C_1})$$

$$\text{Complex}(\mathrm{C_1}) \xrightarrow[\text{slow}]{k_2} \text{Products}$$

Appendix 3.

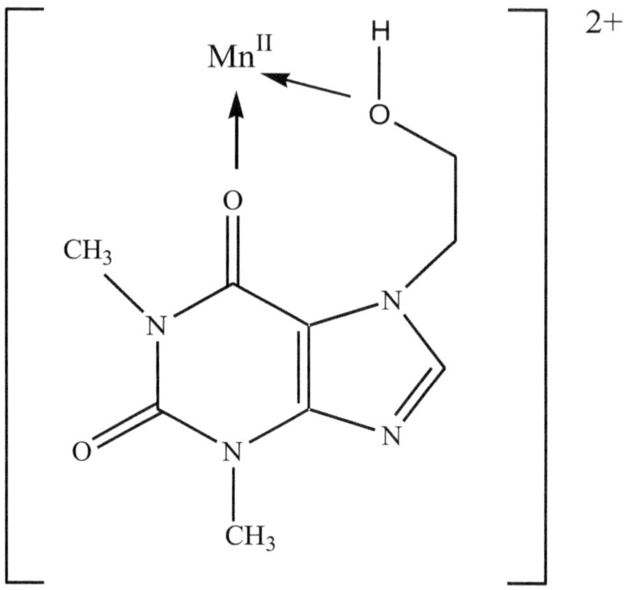

Appendix 4.

Removal of ammonium and phosphates from aqueous solutions by activated and modified Bulgarian clinoptilolite

Nadejda Taneva

University of Architecture, Civil Engineering and Geodesy, "Christo Smirnenski"blvd, Sofia 1046, Bulgaria.
E-mail: nadejda_taneva@yahoo.com or nadejda.taneva@abv.bg.

Results from a laboratory investigation on the capability of activated and modified Bulgarian clinoptilolite samples from Beli plast deposit in Eastern Rhodopes for removing ammonium and phosphates from aqueous solutions are presented in this report. A series of ion exchange batch experiments were carried out using model water composed of NH_4Cl and $Na_2HPO_4.12H_2O$. Total ion exchange capacity was determined by batch experiments as a function of initial ammonium and phosphates concentration, solution pH and temperature. The influence of the presence of Ca^{2+} on the removal of ammonia was investigated. It was found that the adsorption and ion exchange activity of natural clinoptilolite towards phosphate ions could be increased by chemical activation and modification.

Key words: Clinoptilolite, ammonium, phosphates, ion-exchange, sorption.

INTRODUCTION

Sources of ammonium and phosphates in wastewater are municipal, agricultural and industrial wastewater. Nitrogen is presented in the form of ammonium and organic nitrogen. Ammonium nitrogen decreases the dissolved oxygen required for the aquatic life and also accelerates the corrosion of metals and construction materials. Biological nitrification and denitrification, air sripping and ion exchange are the most preferred methods of removal of NH_4^+. The ion exchange method is preferred over the other methods since it is stable, suits automation and quality control and is easy to maintain. Phosphorus can exists like orthophosphates, three polyphosphates, pyrophosphates and organic phosphorus. Orthophosphates are the prevailing form in municipal wastewater (Hultman et al., 2000). Basic sources of contamination of groundwater and surface water with nitrogen and phosphorus are infiltration of rainfall in fertilised agricultural areas and leakage from municipal sewage. Traditional methods for removing phosphorus are chemical precipitation (with aluminium, iron and calcium salts), biological process that rely on biomass growth (bacteria, algae, plants) or intracellural bacterial polyphosphate accumulation (De-Bashan and Bashan, 2004) and sorption (natural apatite, activated carbon, etc.) (Degs et al., 2001). In recent years considerable attention has been paid to the investigation of different types of low-cost sorbents, such as alum sludge (Galarneau and Gehr, 1997), red mud (Shiao and Akashi, 1977), fly ash (Oguz, 2005), calcite, goetite, apatite and zeolite (clinoptilolites) (Tillman et al., 2005).

The study aims to determine the optimal parameters of ion-exchange of ammonium ions and sorption of phosphate ions using activated and modified Bulgarian clinoptilolite.

MATERIALS AND METHODS

Pre-treatment procedure of clinoptilolite

Clinoptilolite is composed of rings of 8 and 10 elements or 5.6 Å unit window show an ideal molecular size for ammonium ion. The zeolite that was used for the experiments is clinoptilolite and comes from Bulgaria. It has been extracted from the site of Beli Plast in the town of Kardjali, in the Rhodopi Mountains. The cation selectivity row of the used zeolite decreases in the sequence:

$Cs^+ > Tl^+ > Rb^+ > K^+ > NH_4^+ > Ag^+ > Na^+ > Ba^+ > Pb^{2+} > Sr^{2+} > Ca^{2+} > Mg^{2+}$

Practically the result of any pre-treatment operation is the increase of the content of a single cation what is called homoionic form. Therefore, pre-treatment aims to remove certain ions from the structure of the material and locate more easily removable ones, prior to any ion-exchange application of it (Semmens and Martin, 1998). Due to the high content of potassium ions in the structure of clinoptilolite, which block the capacity for NH_4^+ fixation, the natural material was previously exchanged with 1M solution of NaCl.

Ammonium ion-exchange studies were carried out using activated (modified) clinoptilolite samples. For the activation 100 g of a natural sample 1 M NaCl solution was used. The natural sample of clinoptilolite was treated 3 days at magnetic stir, then dried at 378 K for 5 h, washed with distilled water, dried at 378 K for 5 h and stored in a desiccator. Figure 1 shows two X-ray diffraction diagrams of natural and activated with 1M NaCl solution sample clinoptilolite.

Phospates ions adsorption studies were carried out using three types of modified clinoptilolite. Sample 1 (Cl -Al): 100 g of a natural sample of clinoptilolite was treated 3 days at magnetic stir with 1 M NaCl solution, then dried at 378 K for 5 h, washed with distilled water, treated with 6 g/L$Al_2(SO_4)_3.18H_2O$ (500 mg/L Al^{3+}) in a batch system in a plexiglass column, dried at 378 K for 5 h, washed with 0.002 M NaOH pH=10.5 and 0.1 M HCl pH=3.5 dried at 378 K for 5 h and stored in a desiccator.

Sample 2 (Cl - Fe): 100 g of a natural sample of clinoptilolite was treated 3 days at magnetic stir with 1 M NaCl solution, then dried at 378 K for 5 h, washed with distilled water, treated with 1.5 g/L $FeCl_3$ (500 mg/L Fe^{3+}) in a batch system in a plexiglass column, dried at 378 K for 5 h, washed with 0.002 M NaOH pH=10.5 and 0.1 M HCl pH=3.5 dried at 378 K for 5 h and stored in a desiccator.

Sample 3 (Cl - Fe/Mn) 100 g of a natural sample of clinoptilolite was treated 3 days at magnetic stir with 1 M NaCl solution, then dried at 378 K for 5 h, washed with distilled water, treated with 1.5 g/L $FeCl_3$ (500 mg/L Fe^{3+}) and Mn Cl_2 (500 mg/L Mn^{2+}) in a batch system in a plexiglass column, dried at 378 K for 5 h, washed with 0.002 M NaOH pH=10.5 and 0.1 M HCl pH=3.5 dried at 378 K for 5 h and stored in a desiccator.

For the measurements of concentration of ammonium and phosphates the Hach-Lange cuvette tests were used.

Adsorption studies

The experiments were carried out at T=25℃ with an activated sample of clinoptilolite and then with the three types of modified clinoptilolite. The clinoptilolite fraction grain size was 0.5 to 1.2 mm. The goal was to determine the total theoretical exchange capacity (TEC) of the zeolite towards ammonia ions and to determine the influence of calcium ions on this capacity.

RESULTS AND DISCUSSION

Kinetic study

The rate of ammonium removal process is a very important factor for exchanger application, since low exchange rates often pose serious limitations to the applicability of the process. The rate of reaching the ion exchange equilibrium depends on a number of technological parameters such as the temperature, size and distribution of aluminosilicate particles, hydrodynamics of the reaction system, concentration and the hydrated radii of exchangeable ions. 100 g of activated clinoptilolite were put in Plexiglas column and saturated with ammonium in a batch system. The amount of adsorbed ammonium ions for three initial concentrations was calculated and the effect of contact time on the ion exchange process was studied. Figure 2 indicates that ammonium removal increased with increasing contact time. The adsorption rate by all sets occurred rapidly at the beginning of the contact and gradually decreased with increasing contact time. The equilibrium in the batch system is reached for 90 min for all concentrations. When using concentration of 100 mg/L all present in the solution ammonium cations are ion-exchanged. When using concentration of 500 mg/L equilibrium is reached without using the full ion-exchange capacity of clinoptilolite. When using concentration of 1000 mg/L the full ion-exchange capacity of clinoptilolite (8 meq/g) is reached.

Effect of initial ammonium and phosphates concentrations

When determining the sorption capacity of a filter material with an agitation experiments the initial concentration of the solution greatly impacts the sorption results. In general, the sorption increases with a higher initial concentration of the species. This phenomenon was seen in the conducted experiments.

Sorption isotherm studies were conducted by adding 1 g of Cl-Na to a 250 ml solution containing various ammonia and calcium ion concentrations in the range of 20 to 550 mg/L. The clinoptilolite sample was in contact with the solutions for 48 h at T=25℃.

The total theoretical capacity for Cl-Na is 8.5 meq NH_4^+/gCl, but decreases to 6 meq NH_4^+/gCl in the presence of calcium ions. A competition between the two kinds of cations (Na^+ and Ca^{2+}) for the available ion-exchange sites in the clinoptilolite structure takes place and the removal of ammonium is influenced negatively. This data lead to the conclusion that using clinoptilolite for ammonium removal in hard and very hard water (high calcium and magnesium concentrations) is restricted. Water hardness can vary a lot depending on the water source (surface or ground waters), and is a point of concern since the concentration of calcium and magnesium cations is often higher than the one of ammonium in wastewaters, then the selectivity row is not valid anymore.

Table 1 summarizes the experimental results obtained at various initial concentrations of ammonium and calcium.

Figure 3 and Table 1 illustrate that the presence of calcium ions in solution decreases the ammonia

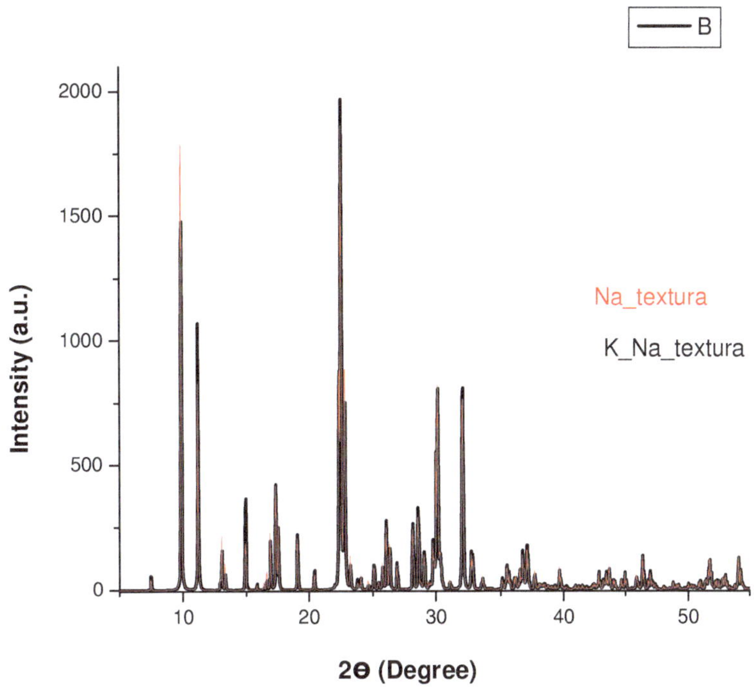

Figure 1. X-ray diffraction diagram of a sample activated clinoptilolite.

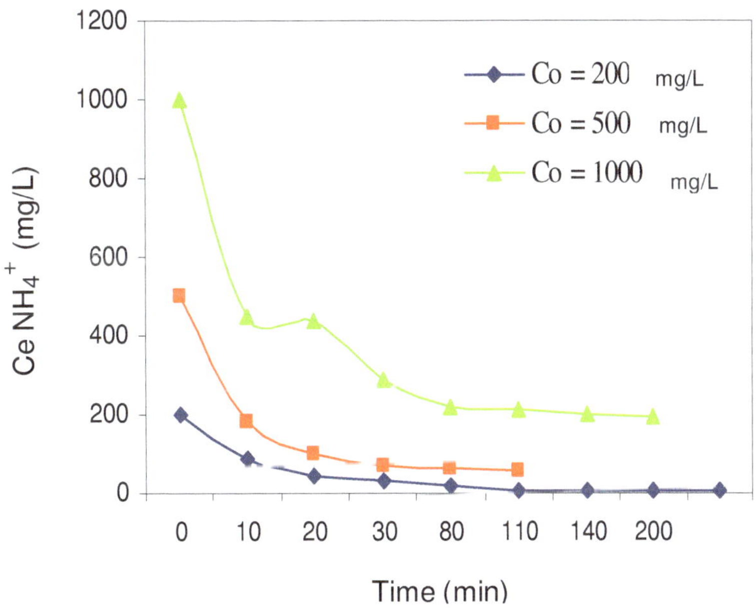

Figure 2. Kinetic study for ammonia ion-exchange process.

Table 1. Ion-exchange capacity of activated clinoptilolite reached as function of Co NH_4^+.

Co NH_4^+ (mg/L)	Capacity (meq NH_4^+/gCl) without Ca^{2+}	Capacity (meqNH_4^+/gCl) with Ca^{2+}
100	6	4
200	8	4.5
350	8.5	6
550	8.5	6

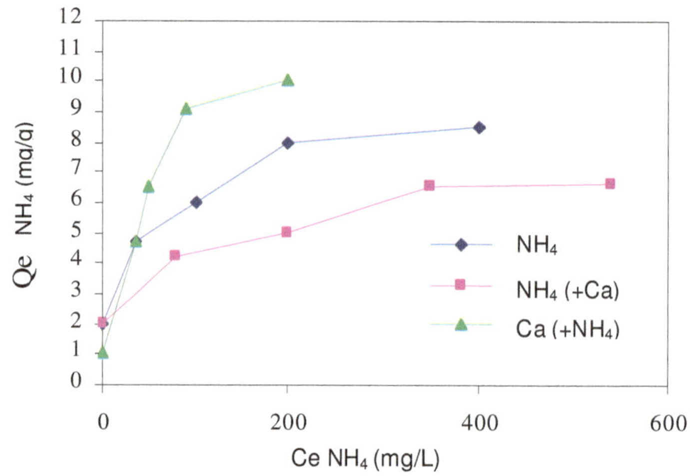

Figure 3. Ammonium adsorption isotherm for activated clinoptilolite.

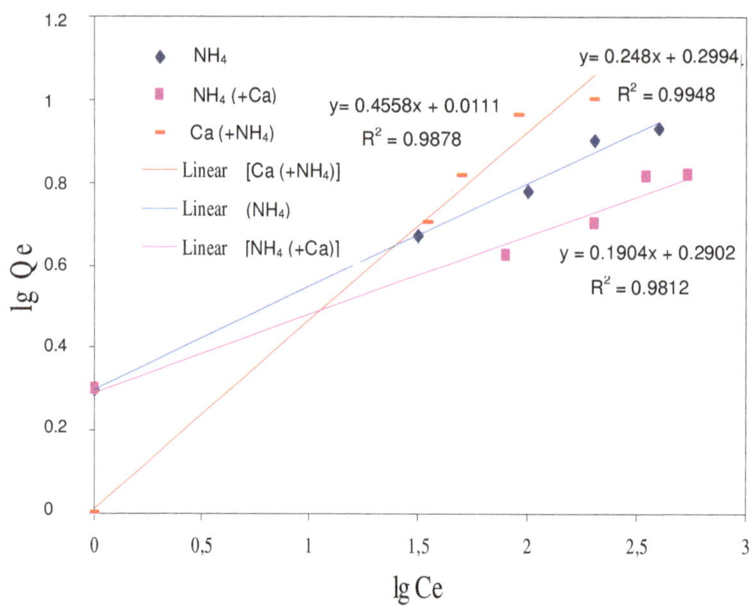

Figure 4. Freundlich adsorption isotherm applied to the equilibrium relationship for ammonia and calcium adsorption onto activated clinoptilolite.

adsorption capacity. Calcium adsorption is influenced by the presence of ammonium ions in less degree.

$$Qe = (Co - Ce) \times V/m \qquad (1)$$

Where C_O and Ce are:

For the measurements of calcium, titration with ethelyenediaminetetraacetic acid (EDTA) using a coloured indicator, Eriochrome Black T, at pH = 10, was the most suitable.

Adsorption isotherms are essential for the description of how NH_4^+ concentration will interact with zeolites and are useful to optimize the use of zeolites as adsorbents. Therefore, empirical equations (Langmuir and Freundlich isotherm model) are important for adsorption data interpretation and prediction. The data obtained in these experiments were fitted to both Fruendlich and Langmuir adsorption isotherms, two most commonly employed models in describing the distribution of ammonium between the solid phase and solution phase. The data fitted better with Freundlich isotherm Figure 4. The Langmuir model assumes only one solute molecule per site, and also assumes a fixed number of sites. The Freundlich model describes a logarithmic relationship between the solid and solution concentrations, and is

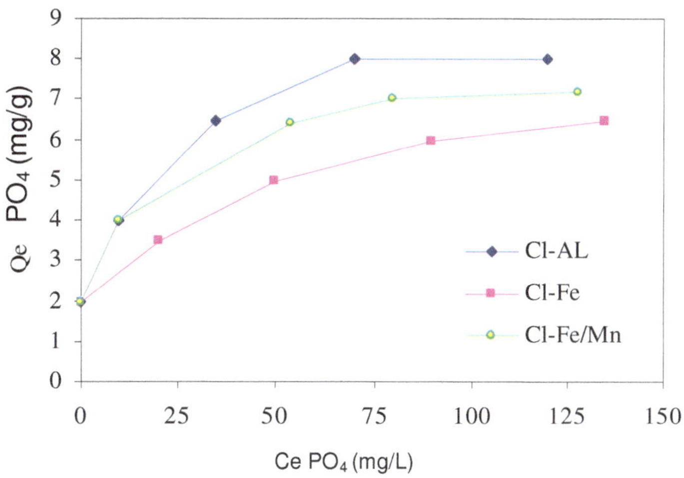

Figure 5. Phosphates adsorption isotherm for modified clinoptilolite

6. Effect of initial pH.

useful in correlating isotherm data obtained over a wide range of concentrations. The Freundlich isotherm relates the uptake of solute onto the exchanger to the solution equilibrium concentration, as shown in Equation (2).

$$Q_e = kC_e^{1/n} \qquad (2)$$

This isotherm can also be presented by the following equation:

$$\log Q_e = \log k + \frac{1}{n}\log C_e \qquad (3)$$

When log Q_e is plotted against log C_e the coefficients k and n, which are generally temperature dependent, can be calculated. Knowing all coefficients and using experimental date for C_e, Freundlich theoretical Q_e can be calculated according to Equation (3).

Sorption isotherm studies were conducted by adding 1 g of C Al, C Fe and C- Fe/Mn to a 250 ml solution containing various phosphate ion concentrations in the range of 10 to 140 mg/L. The clinoptilolite samples were in contact with the solutions for 48 h. The proposed mechanism of removal of phosphates is the formation of low soluble phosphate salts on the surface of the clinoptilolite sample. Comparison between the adsorption capacities of three modified clinoptilolite samples can be seen on Figure 5. The aluminium modification has the best adsorption capacity in static conditions.

To investigate the pH effect on ammonium removal, a number of equilibrium batch processes were conducted. The experiments were carried out at room temperature and Co NH_4^+ = 50 mg/L. 1 g of activated sample was in contact with 250 ml of solution containing 12.5 mg ammonium. Initial pH values of solutions were adjusted by addition of 1 M HCl and 1 M NaOH solutions. The results are shown on Figure 6. The obtained results indicate that ammonium exchange increases with increasing the pH up to pH = 7. At pH > 7.5, ammonium exchange shows a rapid decrease. It is assumed that NH_4^+ is converted into NH_3 specimen in alkaline media and is removed from the solution as ammonia. In very acid media (pH =1 to 3), the exchange of ammonium ions is considerably reduced compared to their exchange in slightly acid, neutral and slightly alkaline media (pH = 6 to 7.5). According to reaction (1):

$$MA(s) + x\ H_2O(l) \leftrightarrow HA(s) + (x-1)\ H_2O(l) + M^{z+}(l) + OH-(l) \qquad (1)$$

where M^{z+} is the exchangeable cation and z is the charge number of cations in the aluminosilicate structure A, with the decrease of the solution pH, the equilibrium is shifted to the right, that is, the concentration of exchangeable ions in solution increases. Thus, the conditions of ion exchange are more favourable in a more acid medium, however a stronger attraction of H^+ ions to exchangeable sites and competition between H^+ and NH_4^+ is observed. For this reason, the experiments were performed in a stable pH range, optimal to study the binding of ammonium ions.

When the initial pH of the solution increases from 2.5 to 10 the removal of phosphates ions increases till pH=6 and then decreases. The decrease in the phosphate ion uptake, occurring beyond pH=6 (Figure 7), implies probably a competition between phosphate and hydroxyl ions for the sorption on the surface of the sorbents. The experiments were carried out at room temperature and Co PO_4^{3-} = 10 mg/L. 1 g of each of the modified samples was in contact with 250 ml solution containing 2.5 mg phosphates.

The increase of pH value from 6 to 10 implies a decrease in the development of positive charges in the

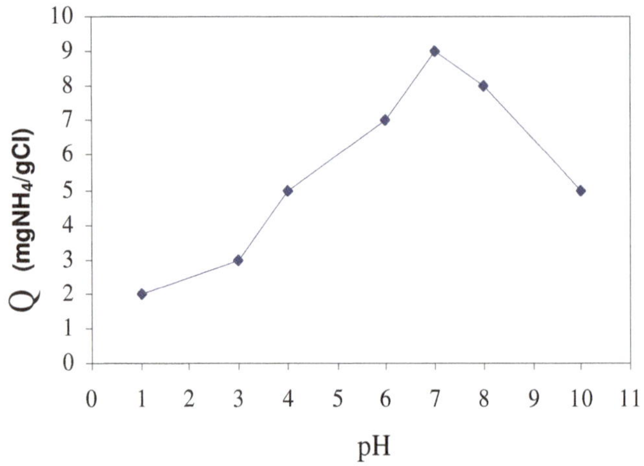

Figure 6. Removal of ammonia as a function of pH.

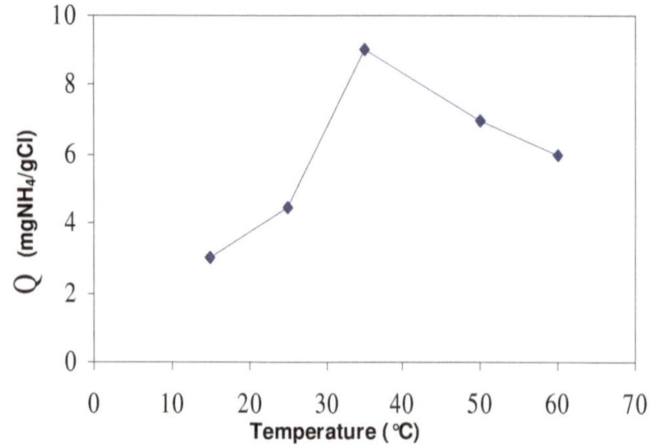

Figure 8. Removal of ammonia as a function of temperature.

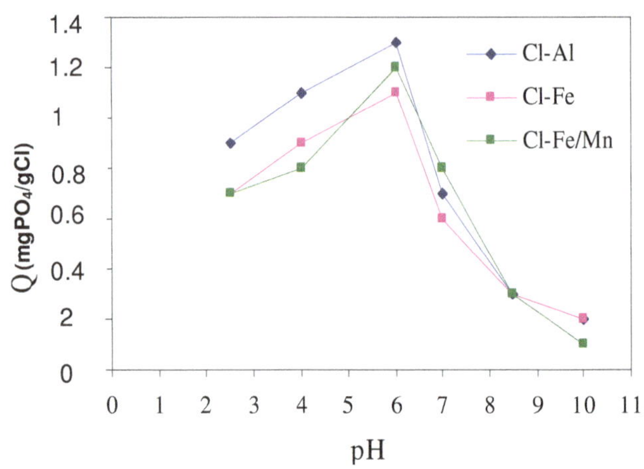

Figure 7. Removal of phosphates as a function of pH.

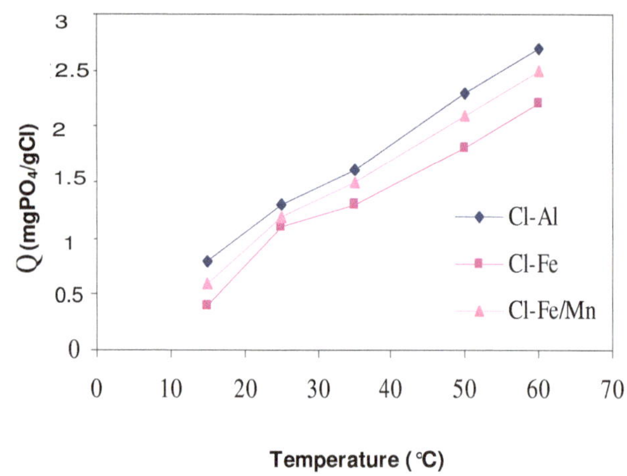

Figure 9. Removal of phosphates as a function of temperature.

particles surface, which means a lower electrostatic interaction between the adsorbent and the adsorbate. This decrease is more signficant for Cl-Fe particles as adsorbent. The large variations observed for Cl-Al as compared with the Cl-Fe supports the assumption that the interaction between adsorbent and adsorbate are mainly electrostatic for Cl-Fe/PO_4 systems and are both electrostatic and hydrogen-bonding for Cl-Al/PO_4 systems.

Effect of temperature

The experiments with activated clinoptilolite sample were conducted at pH=7 in the range of values of temperature between 15 and 60°C. Figure 8 shows that the removal efficiency of ammonia increases till 35°C and above this value the ion-exchange process decreases. A decrease in the removal efficiency of NH_4^+ ion with the rise in temperature was due to the increasing tendency to desorb from the interface to the solution. This result also indicated that the adsorption process of NH_4^+ onto natural zeolite was exothermic in nature.

The experiments with modified clinoptilolite samples were conducted at pH=6. The removal of phosphates increases with increasing temperature (Figure 9). The results indicate that surface adsorption of phosphates is an endothermic process.

Conclusion

1. The presence of calcium ions decreases the ammonia adsorption capacity. Calcium adsorption is influenced by the presence of ammonium ions in less degree.
2. The aluminium modification has the best adsorption capacity in static conditions.
3. The optimal studied parameters for ammonium ion-exchange is pH = 7 and temperature of 35°C.

4. The optimal pH for phosphates adsorption is 6.
5. Surface adsorption of phosphates is an endothermic process and is favoured by higher values of the temperature.

REFERENCES

De-Bashan, LE Bashan Y (2004). Recent advances in removing phosphorus from wastewater and its future use as fertilizer. Water Res. 38:4222-4246.

Degs YA, Khraisheh MAM, Tutunji MF (2001). Sorption of lead ions on diatomite and manganese oxides modified diatomite. Water Res. 35:3724-3728.

Galarneau E, Gehr R (1997). Phosphorus removal from wastewaters: Experimental and theoretical support for alternative mechanisms. Water Res. 31(2):328–338.

Hultman B, Levlin E, Johansson L, Al-Najar N, Li P, Plaza E, Minter R (2000). Nutrient removal and sludge handling. In LC Lundin (Ed.) Sustainable water management in the Baltic Sea Basin. Book 2 Water use and management .The Baltic University Programme. pp. 171-178.

Oguz E (2005). Sorption of phosphate from solid/liquid interface by fly ash. Coll. Surf.A Physicochem. Eng. Aspects 262(1-3):113-117.

Semmens MJ, Martin WP (1998). The influence of Pretreatment on the capacity and Selectivity of Clinoptilolite for Metal Ions. Water Res. 22(5):537-542.

Shiao SJ, Akashi K (1977). Phosphate removal from aqueous solutions using neutralised bauxite. J. WPCF 49(2):280–285.

Tillman FD, Barlet-Hunt SL, Craver VA (2005). Relative metal ion sorption on natural and engineered sorbent batch and column studies. Environ. Eng. Sci. 22:400-409.

Electrochemical behaviour of some copper alloys in sodium chloride solutions containing different inorganic additives

H. E. El-Feky*, N. H. Helal and M. R. Negem

Department of Chemistry, Faculty of Science, University of Fayoum, Fayoum, Egypt.

The corrosion behavior of Cu-Fe and Cu-Fe-Al was studied in NaCl solutions without and with different concentrations of some inorganic additives, Na_2MoO_4, $K_2Cr_2O_7$, and KIO_3. The measuring techniques used were, open-circuit potential, potentiodynamic polarization and electrochemical impedance spectroscopy (EIS). The potentiodynamic polarization measurements indicated that the corrosion current, I_{corr}, of Cu–Fe is less than that of Cu–Fe–Al. The investigated additives in 3.5% NaCl shift corrosion potential Ecorr to more positive values (except in $K_2Cr_2O_7$ reverses the action for Cu-Fe alloy) and increase Icorr. Impedance measurements indicated that Nyquist plots for Cu–Fe and Cu–Fe–Al may be regarded as part of semicircle (arc), the values of R_p for Cu–Fe are higher than those for Cu–Fe–Al except in $K_2Cr_2O_7$. On the other hand, the values of C_{dl} for Cu–Fe are lower than those obtained for Cu–Fe–Al. The presence of these additives in NaCl solutions increases the values of polarization resistance (R_p) and decreases the values of (C_{dl}).

Key words: Copper alloys, corrosion, passivators, NaCl.

INTRODUCTION

Copper is one of the oldest metals known to man, so the corrosion of this material has been studied for many years (Bengough et al., 1920; Bonfiglio et al., 1973; Faita et al., 1975; Ohsawa et al., 1979; Lee et al., 1985; Barcia et al., 1993). Cu–Fe alloys are materials that are more and more extensively applied in industry, they are utilized as master alloys to produce new copper alloys for very special purposes. They are used as materials for electrical device components, for example, semiconductor lead frames, electrical connectors and electrical fuses. Also, copper and its alloys are widely used as preferred materials for the transport of water in domestic and industrial water utilities, in plumbing within buildings and heat exchangers. They are also used in the petrochemical, marine, food, paper, and textile industries. This is related to their corrosion resistance, mechanical workability, excellent electrical and thermal conductivity and resistance to macrofouling (Nomoto et al., 1999). It has been reported that iron-alloying addition to copper alloys may be used to improve their corrosion resistance (Cairs et al., 1990). The corrosion resistance of copper alloys has been attributed to the formation of a protective film of cuprous oxide (Cu_2O) and to the doping of Cu_2O layer with ions such as iron (Blundy et al., 1972; North et al., 1970). This doping reduces the ionic or electronic conductivity of the film, improving the corrosion resistance. In practice, alloying additions of Al, Zn, Sn, Fe and Ni are used to dope the corrosion product films. The films retard both the anodic dissolution of the alloy and the rate of O_2 reduction (De Sanchez et al., 1982). Therefore, it is not surprising that failure is always associated with the deterioration of the protective film (Gudas et al., 1979; El-Feky et al., 2009), the corrosion behavior of Cu–Fe and Cu–Fe–Al was studied. Sury (1970) showed that the thickness of the corrosion layers of Cu-Al alloys decreases with increasing Al-content of the corrosion layer. The protection nature of these films was ascribed to the formation of a Cu_2O layer which was detected on binary Cu- Al alloys in NaCl solution by X-ray technique. The asymptotic drop of the corrosion rate during the layer formation process was explained with the simultaneous decrease of the anodic reaction inside growing Cu_2O

*Corresponding author. E-mail: heshamelfiky@yahoo.com.

layer, and the cathodic reaction on its surface. The role of Al in passivation was assumed to be due to its incorporation in the Cu_2O lattice, reducing the rate of O_2 reduction on the Cu_2O surface. The protection of metals against corrosion in neutral electrolytes can be achieved either with inhibitors or with passivators (Rozenfeld, 1981). Ateya et al. (1994) studied the corrosion behaviour of α-Al bronze (Cu-7% Al) in 3.4% NaCl solution for extended periods and stated that corrosive attack occurred very early leading to general corrosion and pit nucleation, coalescence, growth and eventual merging of neighboring pits together to give multiple tough-like cavities. In the present work, the kinetic studies on the stability of the passive layer formed on copper-iron and copper-aluminium-iron alloys in different media are aimed. These are performed by studying the effect of some inorganic additives (passivators) on the corrosion and passivation behaviours on the investigated alloys in neutral media. The inorganic additives, used in the present work are: sodium molybdate, potassium iodate and potassium dichormate.

MATERIALS AND METHODS

Materials

Two types of the copper alloys are used in this study having the chemical composition (%wt): Alloy (I) Cu-4.47%Fe and Alloy (II) Cu-10.67%Al-5%Fe. The electrodes were jammed in a small copper mould with about 30 cm wire tail which was fixed in a glass tube with Araldite (Ciba- production, Switzerland) epoxy adhesive resin. The apparent cross sectional area of all electrodes were 0.2826 cm2.

Test medium

The test solution of 3.5% NaCl used in this work was prepared from Analar grade reagents. The inorganic additives were: potassium iodate, potassium dichromate, and sodium molybdate. These additives are chemically pure grades and obtained from Aldrich Chemical Co Ltd. All measurements were carried out at temperature of 20℃.The measurements were carried out in an EA876G- 20 Jacket Cell Bottom. The thermostat used is Compact Thermostat Model PC-Italy. The reference electrode (RE) was Ag/AgCl/ Cl- electrode and the counter electrode was Pt electrode. The electrodes were mechanically polished with successive grades of 600, 800, 1000 and 1200 emery papers till the surface appears as a bright mirror. The electrodes were then rinsed several times with deionized water and degreased with acetone.

Electrochemical tests

The potential of the working electrode under open circuit conditions was measured against Ag/AgCl/Cl- (3M KCl) reference electrode (eo = -197 mV vs NHE) and Pt- electrode as a counter electrode. The potential were recorded by means of multimeter (Keithley, Model 130 A, USA). Electrochemical impedance spectroscopy (EIS) measurements were carried out using impedance measurements system (IM6 Zahner electric, Meßtechnik, Germany). Impedance measurements were conducted, with excitation amplitude of 10 mV peak to peak, and the frequency domain from 0.1 Hz to 100 kHz. Potentiodynamic polarization technique experiments were performed after obtaining a more or less steady state at 35 min open circuit potential using the same instrument and software with a scan rate 10 mV /sec.

RESULTS

Open-circuit conditions

In these series of experiments, the potential of Cu-Fe and Cu–Fe–Al was recorded as a function of time in 3.5% NaCl solution in absence and in presence of different concentrations of the inorganic additives: Na_2MoO_4, $K_2Cr_2O_7$ and KIO_3 under open-circuit conditions. Figures 1a, b, and c shows open circuit potential (OCP) for Cu-Fe in 3.5% NaCl solutions without and with different concentrations of Na_2MoO_4, $K_2Cr_2O_7$ and KIO_3. The potential of the electrode was measured up to 45 min or more exposure time. The values of the electrode potential are listed in Table 1 for the different concentrations of the inorganic additives. Figures 2a, b, and c represent the potential-time plots under open-circuit conditions for the Cu–Fe–Al in 3.5% NaCl solutions without and with different concentrations of Na_2MoO_4. Similar plots are obtained in presence of different concentrations of $K_2Cr_2O_7$ and KIO_3. The values of the electrode potential are listed in Table 2 for the different concentrations of the inorganic additives. Generally, the potential of Cu-Fe is shifted to more positive values with time except in case of presence of $K_2Cr_2O_7$, it shifts the potential to more negative direction. The potential of Cu–Fe–Al is shifted to more negative values with time in all solutions except in presence of KIO_3, the potential is shifted to positive direction. The addition of inorganic compounds shifts the steady state potential to positive and negative direction. The ranging in potential with time is due to heterogeneity of layer formed.

Potentiodynamic polarization

Figures 3a, b and c shows cathodic and anodic polarization curves recorded for Cu-Fe in 3.5% NaCl solutions without and with different concentrations of Na_2MoO_4, $K_2Cr_2O_7$ and KIO_3 respectively. The values of different electrochemical parameters for corrosion (corrosion potential Ecorr, corrosion current density icorr, and anodic and cathodic Tafel slopes βa, βc) are listed in Table 3 for Cu-Fe in 3.5% NaCl solutions without and with the inorganic additives. The similar curves are showed for Cu–Fe–Al in Figures 4a, b and c in 3.5% NaCl containing different concentrations of $K_2Cr_2O_7$, Na_2MoO_4 and KIO_3. The corrosion parameters for the alloy in different media are listed in Table 4. Generally, the addition of inorganic compound shifts the Ecorr to more positive values except in $K_2Cr_2O_7$, where it shifts Ecorr to more negative direction for Cu-Fe alloy. The presence of inorganic additives causes shifting of E_{corr} to more positive value

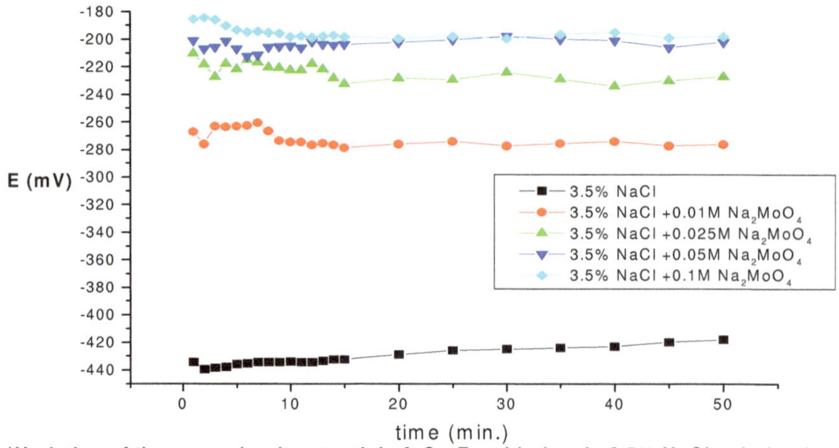

Figure 1a. Varation of the open-circuit potential of Cu-Fe with time in 3.5% NaCl solution in the presence of different concentration of Na_2MoO_4 at 20℃.

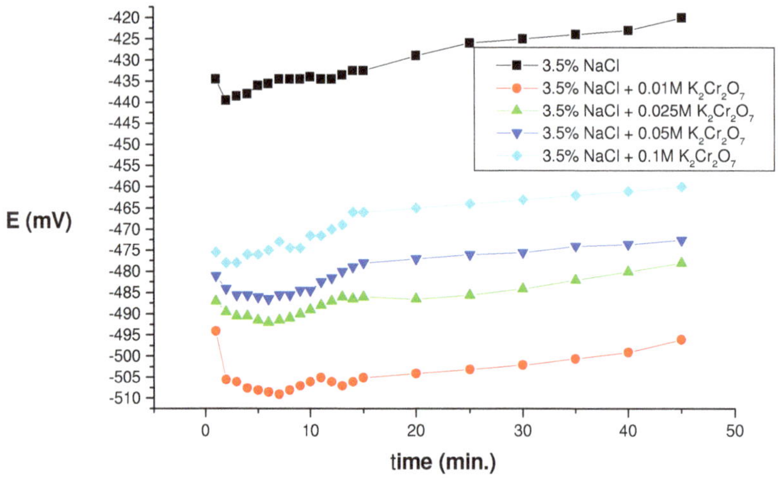

Figure 1b. Varation of the open-circuit potential of Cu-Fe with time in 3.5% NaCl solution in the presence of different concentration of $K_2Cr_2O_7$ at 20℃.

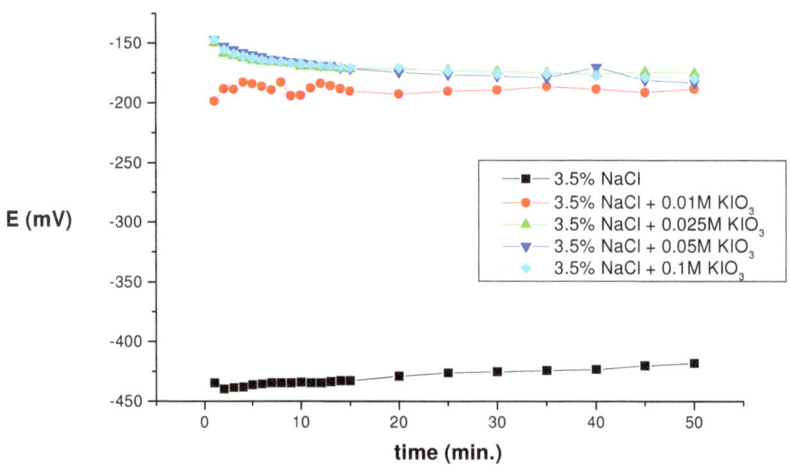

Figure 1c. Varation of the open-circuit potential of Cu-Fe with time in 3.5% NaCl solution in the presence of different concentration of KIO_3 at 20℃.

Table 1. Data of open circuit potential measurements for Cu-Fe immersed in 3.5% NaCl solutions without and with different concentrations of inorganic additives at 20°C.

Additive conc. (M)	Na_2MoO_4	$K_2Cr_2O_7$	KIO_3
	Potential E (mV)		
0.00	-430	-430	-430
0.01	-370	-510	-180
0.025	-310	-485	-170
0.05	-290	-480	-170
0.1	-270	-460	-170

Electrochemical impedance spectroscopy

The impedance measurements is used to study the corrosion behavior of Cu-Fe and Cu–Fe–Al alloys in 3.5% NaCl solutions with presence of different concentrations of the above mentioned inorganic additives. In principle, the method involves direct measurements of impedance of the electrochemical system in the frequency domain from 0.1 to 105 Hz. The electrode impedance, Z and the phase shift (q) are presented in Bode and Nyquist plots. These experiments are carried out at steady state under open circuit conditions. Figure 5a - c, 6a - c and 7a - c show the Bode, Nyquist and phase plots for Cu-Fe alloy in presence of different inorganic additive (0.01 M KIO_3, 0.01 M $K_2Cr_2O_7$ and 0.01 M Na_2MoO_4, respectively) at different immersion time. In the presence of 0.01 M $K_2Cr_2O_7$, the resistance of passive films that formed on the alloy surface increases with time and displays one phase maxima. This means that one electrode processes occur on the alloy surface. But with 0.01 M Na_2MoO_4 and 0.01M KIO_3, the film resistance decreases with increasing of immersion time, in addition to the electrode display one phase maxima. The values of polarization resistance R_p and capacitance of double layer Cdl for the electrode are listed in Table 5. Figures 8a - c, 9a - c and 10a - c shows the Bode, Nyquist and phase plots for Cu–Fe–Al alloy, in presence of different inorganic additive (0.01 M Na_2MoO_4, 0.01 M $K_2Cr_2O_7$ and 0.01 M KIO_3, respectively. The film resistance and reciprocal of charge transfer increases with elapsing of time in presence of inorganic additives. The addition of different concentrations of dichromate, malybdate and iodate, lead to increasing of the film resistance than that was recorded with 3.5% NaCl inorganic compound-free.

The values of polarization resistance R_p and capacitance of double layer Cdl are listed in Table 6.

DISCUSSION

The potential of Cu-Fe is shifted to more positive values with time, while in the presence of $K_2Cr_2O_7$, the potential is shifted to negative direction. The potential of Cu–Fe–Al shifted to more negative values with time, while the presence of KIO_3 shifts the potential to positive direction. The addition of inorganic compounds shifts the steady state potential to positive and negative direction. Also, there are ranging in potential with time in most cases due to heterogeneity of layer formed (Rezenfeld et al., 1961). In pure chloride solution, the electrode is in active-passive state and the observed oscillations of the potential, which is associated with the fact that two concurrent processes are taking place at the electrode: activation and passivation. If passivating ions are added to the electrolyte, the electrode potential becomes stable and oscillations cease. Such an electrode can be polarized without activation up to the positive potentials. A concurrent adsorption of the passive state of the metal depends on the ratio between the numbers of activating and passivating ions, the charge, the potential of the electrode and the composition of the alloy. The role of anions as accelerators or inhibitors is directly related to their polarizability, is a measure of their deformability (Piontelli et al., 1951). Polarization determines the ability of the ion to inter the double layer and influence the electrode reaction. Mayre (1960) found that the attack on the protective film on the metal is sufficiently low in the presence of ions having polarizability equal to or lower than OH-, In the presence of ions of larger polarizability than OH-, film formation does not take place. e mechanism of action of inhibitors of the non-oxidizing type can be assumed to be as follows: after retarding the anodic reaction of ionization of the metal by forming sparingly soluble compounds or by direct electrochemical oxidation of the metal, these inhibitors shift the potential of the metal to values at which oxidation of the metal by oxygen of water becomes possible. Covering a considerable part of the electrode with sparingly soluble compound reduces the passivation current. Thus the current densities which are produced by the ordinary reaction of oxygen reduction are sufficient to keep the potential at the level necessary for passivation. Here a unique internal synergistic effect is involved, whereby the inhibitor facilitates passivation of the metal by oxygen of water, without participating in the cathodic reaction.

The inhibitors with strong properties act primarily as stimulators of the cathodic process and indirectly retarding the anodic reaction. Based on the effect of anions of the type M on iron, Cartledge (1955) and Lee (1955) advanced their electrostatic polarization hypothesis. The M anions do not exhibit certain behaviour. Thus, Mn reacts readily with iron to give non-protective film, S does not inhibit, P is only effective in the presence of oxygen and Cr causes passivity even in absence of oxygen. According to this hypothesis, an inhibitor should have suitable geometry for adsorption, a high for central charge, a high degree of internal polarity and low enough redox potential to prevent its reduction. The action of KIO_3 may be attributed to its oxidizing properties even in neutral solutions according to the following half reaction: This oxidizing property primarily stimulates the cathodic

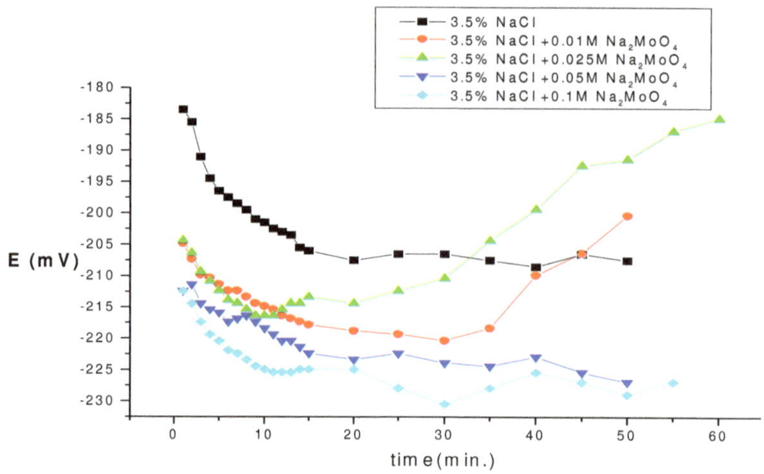

Figure 2a. Varation of the open-circuit potential of Cu–Fe–Al with time in 3.5% NaCl solution in the presence of different concentration of Na_2MoO_4 at 20℃.

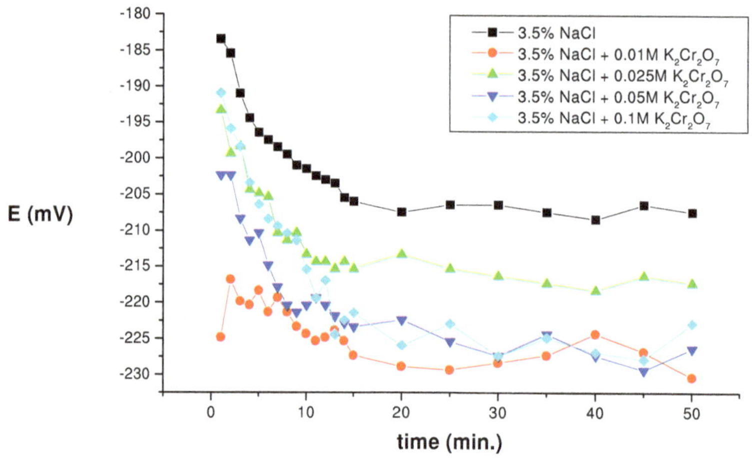

Figure 2b. Varation of the open-circuit potential of Cu–Fe–Al with time in 3.5% NaCl solution in the presence of different concentration of $K_2Cr_2O_7$at 20℃.

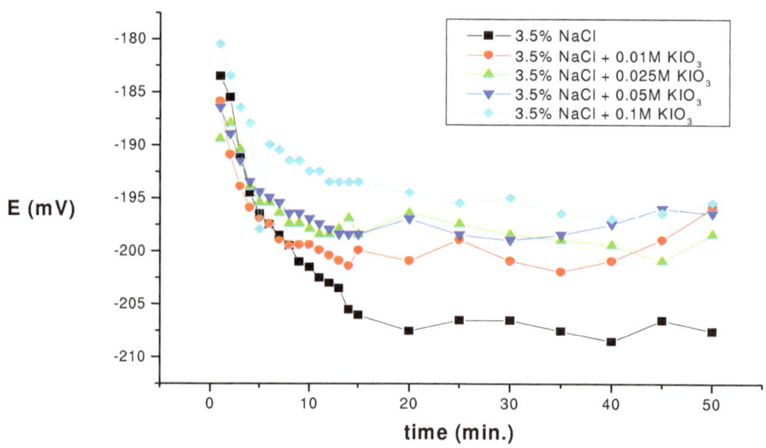

Figure 2c. Varation of the open-circuit potential of Cu–Fe–Al with time in 3.5% NaCl solution in the presence of different concentration of KIO_3 at 20℃.

Table 2. Data of open circuit potential measurements for Cu–Fe–Al immersed in 3.5% NaCl solutions without and with different concentrations of inorganic additives at 20°C.

Additive conc. M	Na_2MoO_4	$K_2Cr_2O_7$ Potential E (mV)	KIO_3
0.00	-206	-206	-206
0.01	-216	-227	-202
0.025	-212	-216	-199
0.05	-221	-226	-199
0.1	-225	-225	-195

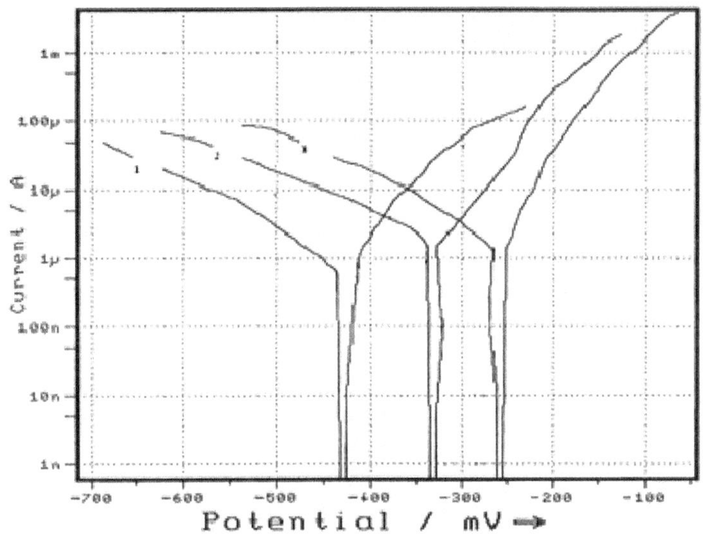

Figure 3a. Potentiodynamic polarization plots for Cu-Fe alloy 3.5 %NaCl at 35 min of immersion time at 20°C. 1=3.5% NaCl, 2= 3.5% NaCl+ 0.01M Na_2MoO_4, 3= 3.5% NaCl +0.1M Na_2MoO_4.

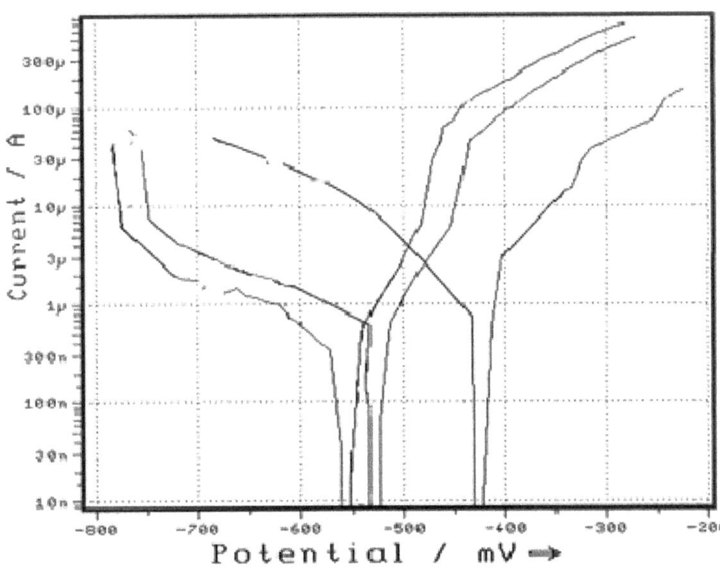

Figure 3b. Potentiodynamic polarization plots for Cu-Fe alloy 3.5 %NaCl at 35 min of immersion time at 20°C. 1=3.5% NaCl, 2= 3.5% NaCl+ 0.01M $K_2Cr_2O_7$, 3=3.5% NaCl + 0.1 M $K_2Cr_2O_7$.

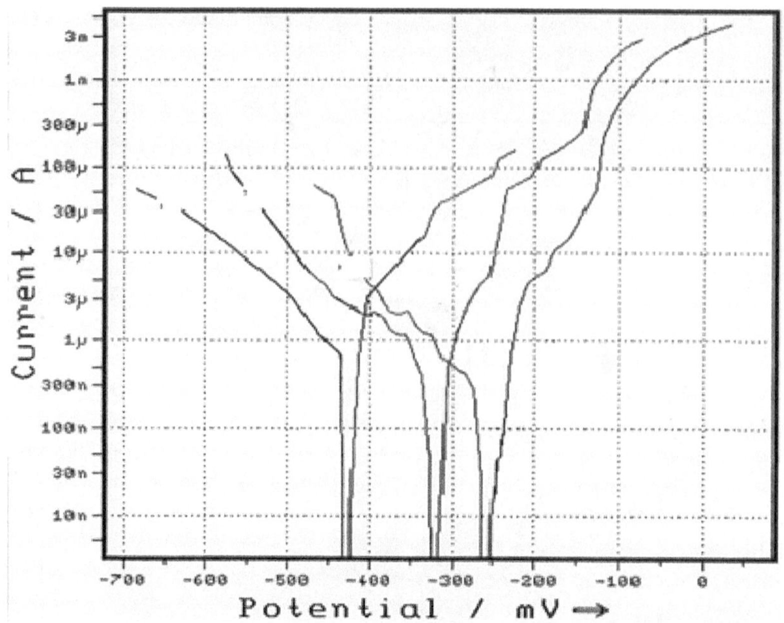

Figure 3c. Potentiodynamic polarization plots for Cu-Fe alloy 3.5 %NaCl at 35 min of immersion time at 20℃. 1 = 3.5% NaCl, 2 = 3.5% NaCl + 0.01 M KIO_3, 3= 3.5% NaCl +0.05 M KIO_3.

Table 3. Corrosion parameters for Cu-Fe in 3.5% NaCl Solutions without and with different concentrations of inorganic additives at 20℃.

Additive	Conc. mol/dm^3	-β_c (mV/d)	β_a (mV/d)	$i_{corr.}$ (μA/cm^2)	$E_{corr.}$ (mV)
NaCl	3.5%	118	66.3	0.241	-453.4
	0.01	75.5	38.7	0.488	-323.4
Na_2MoO_4	0.025	108	62.9	0.968	-317.2
	0.05	83.3	45.3	0.509	-263
	0.1	615	184	1.31	-274.1
$K_2Cr_2O_7$	0.01	217	50.4	0.497	-515.3
	0.025	133	65.8	0.453	-536.7
	0.05	159	52.8	0.428	-546
	0.1	102	49.8	0.238	-558.9
KIO_3	0.01	136	51.1	0. 539	-313.9
	0.025	105	81.7	1.91	-237.7
	0.05	104	55.3	0.126	-226.6
	0.1	72.3	47.4	0.242	-228.3

process and indirectly retarding the anodic reaction. In addition, the above half-reaction shows that the hydroxyl ions are produced. For a sufficiently high concentration of hydroxyl ions in the electrolyte, it becomes possible for a passivating oxide to form at the expense of hydroxyl oxygen according to the reaction: without the anodic polarization (Rozenfeld, 1981). Also Cu_2O and $Cu_2(OH)_3Cl$, may be formed by the following reactions: Also such a behaviour of the electrode is undoubtly associated with concurrent adsorption of chloride ions, and the produced iodide ions, and the hydroxyl ions. The amount of adsorbed chloride ions decreases when is introduced into

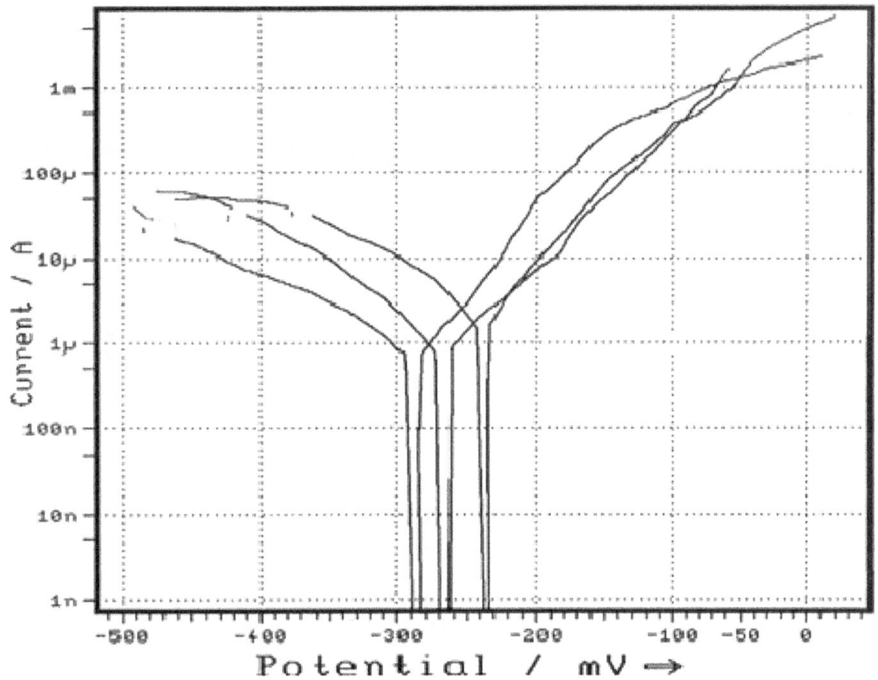

Figure 4a. Potentiodynamic polarization plots for Cu–Fe–Al alloy 3.5 %NaCl at 35 min of immersion time at 20℃. 1 = 3.5% NaCl, 2 = 3.5% NaCl+ 0.01 M Na_2MoO_4, 3 = 3.5% NaCl +0.1 M Na_2MoO_4.

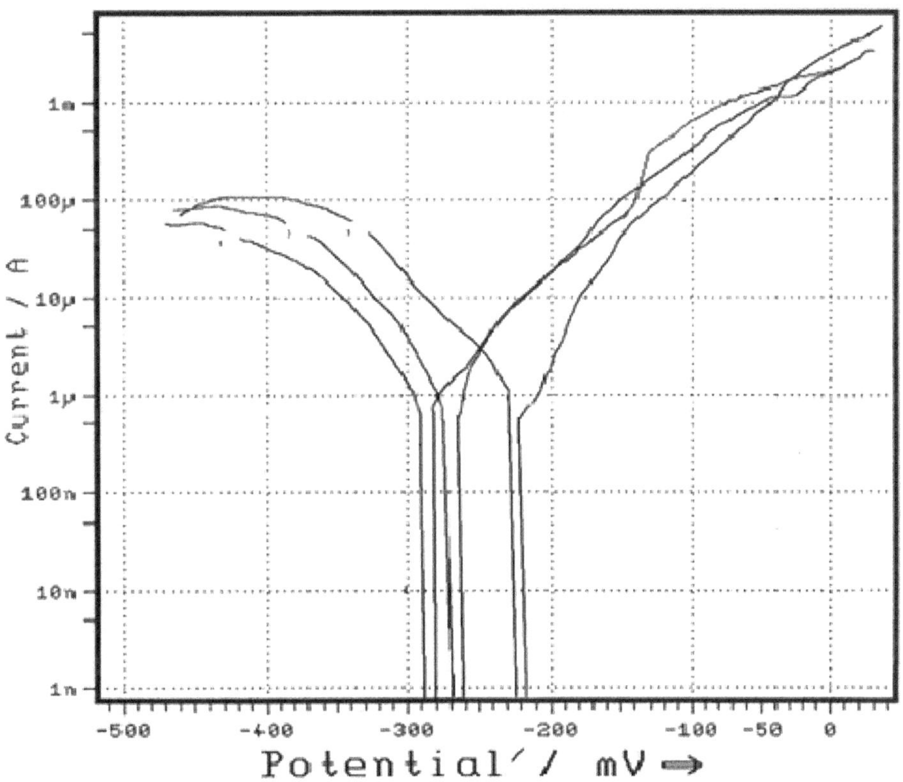

Figure 4b. Potentiodynamic polarization plots for Cu–Fe–Al alloy 3.5 %NaCl at 35 min of immersion time at 20℃. 1 = 3.5% NaCl, 2 = 3.5% NaCl + 0.01 M $K_2Cr_2O_4$, 3 = 3.5% NaCl + 0.1 M $K_2Cr_2O_4$.

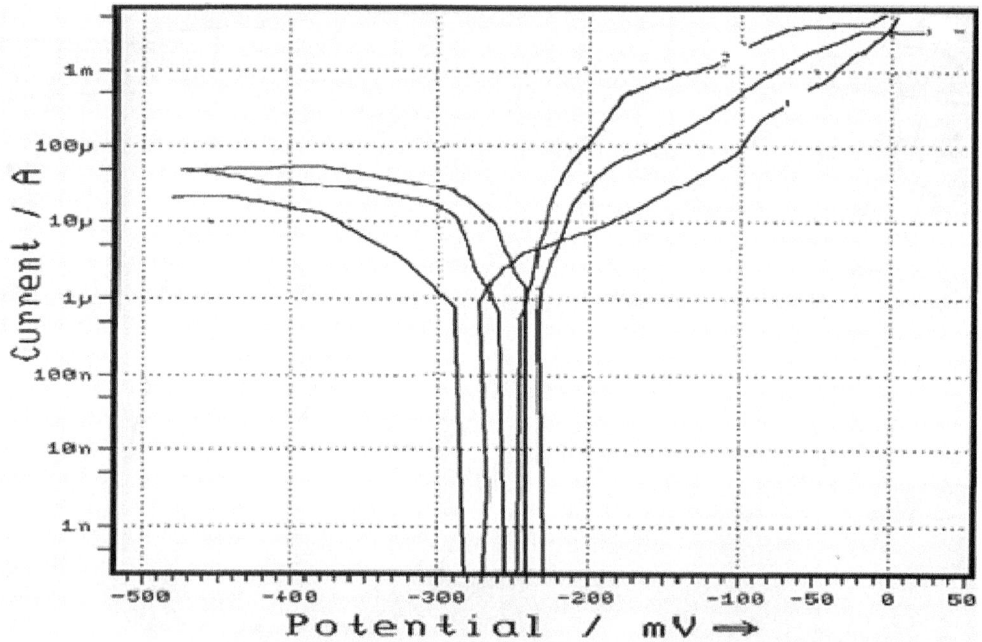

Figure 4c. Potentiodynamic polarization plots for Cu-Fe alloy 3.5 %NaCl at 35 min of immersion time at 20℃. 1 = 3.5% NaCl, 2 = 3.5% NaCl + 0.01 M KIO_3, 3 = 3.5% NaCl + 0.1 M KIO_3.

Table 4. Corrosion parameters for Cu–Fe–Al in 3.5% NaCl solutions without and with different concentrations of inorganic additives at 20℃.

Additive	Conc. mol/dm^3	-βc (mV/d)	βa (mV/d)	$i_{corr.}$ (μA/cm^2)	$E_{corr.}$ (mV)
NaCl	3.5%	1350	149	0.993	-286.1
	0.01	116	10.7	0.771	-263.5
Na_2MoO_4	0.025	142	69.1	0.795	-241.5
	0.05	166	71	0.889	-256
	0.1	136	43.6	1.152	-231.5
$K_2Cr_2O_7$	0.01	35.2	79.8	0.98	-266.6
	0.025	32.9	14.5	0.574	-260
	0.05	112	36.9	0.955	-217.7
	0.1	53.8	170	0.986	-217.8
KIO_3	0.01	31	40	0.439	-262.1
	0.025	105	90.9	0.747	-286
	0.05	63.5	161	1.19	-294.9
	0.1	106	110	2.12	-254.8

the electrolyte. For higher concentrations of passivating ions, the adsorption of activating ions can be excluded completely.

Compounds such as Na_2MoO_4 can protect the alloy against corrosion. Upon hydrolysis of these compounds, which are salts of weak acid and strong base, hydroxyl ions appear which are capable of forming a passivating oxide. Salts of a weak acid and strong base take preference over ordinary alkali, since the hydroxyl ions used in the formation of the passivating oxide are automatically

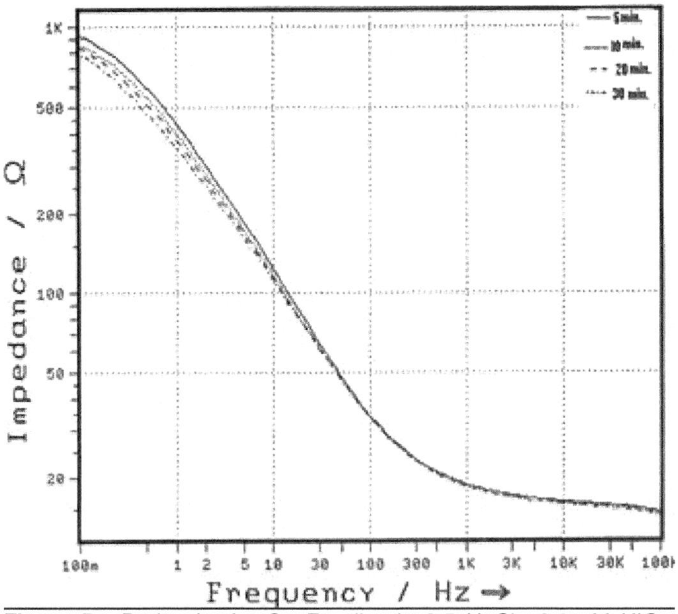

Figure 5a. Bode plot for Cu–Fe alloy in 3.5 NaCl +0.01 M KIO_3 solution as a function of immersion time at 20℃.

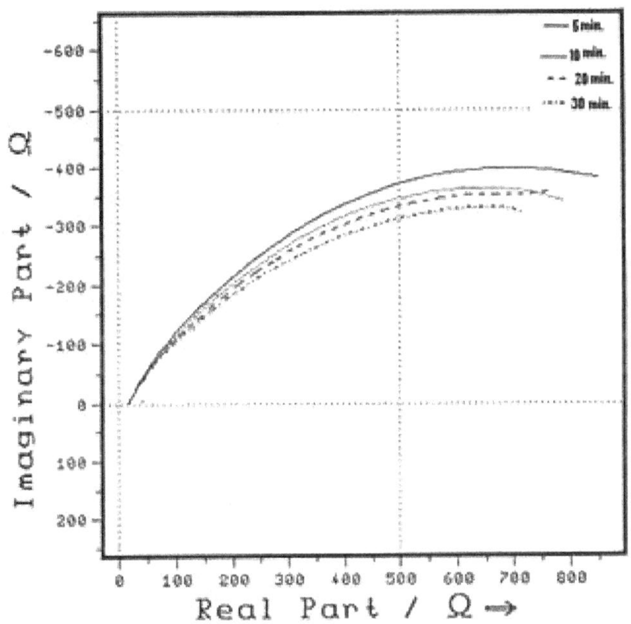

Figure 5b. Nyquist plot for Cu–Fe alloy in 3.5 NaCl + 0.01 M KIO_3 solution as a function of immersion time at 20℃.

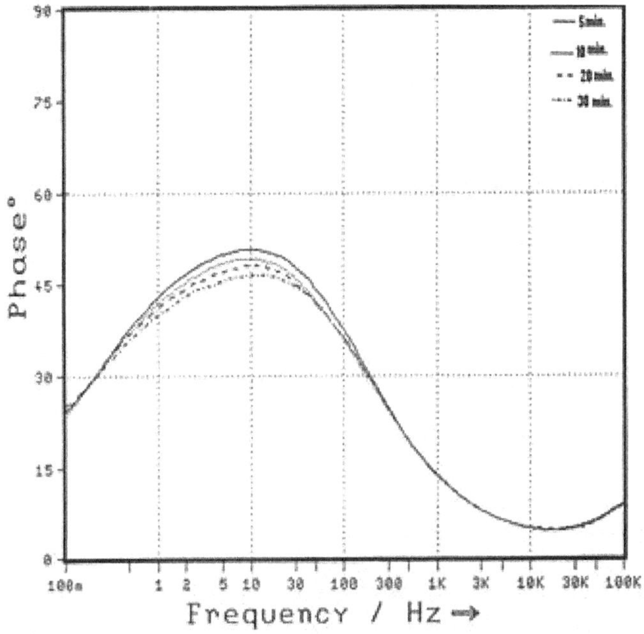

Figure 5c. Phase plot for Cu–Fe alloy in 3.5 NaCl +0.01 M KIO_3 solution as a function of immersion time at 20℃.

Table 5. Impedance parameters for Cu-Fe in 3.5% NaCl solutions without and with different concentrations of inorganic additives at 20℃.

Additive	Conc. (mol /dm^3)	C_{dl} (μF. cm^{-2})	R_p (Ω-cm)
NaCl	3.5%	383.6	1343.29
Na_2MoO_4	0.01	802.5	1684.73
	0.025	283.6	1733.7
	0.05	322.8	2094.4
	0.1	246	2443.63
$K_2Cr_2O_7$	0.01	227	2269.25
	0.025	240.6	2644.45
	0.05	141	2740.92
	0.1	63.4	2091.4
KIO_3	0.01	471.6	917.56
	0.025	259.5	2874.7
	0.05	321.6	3433.08
	0.1	433.6	3199.08

replenished by hydrolysis. For a sufficiently high concentration of hydroxyl ions in the electrolyte, it becomes possible for a passivating oxide to form at the expense of the hydroxyl oxygen according to the above mentioned reactions. Also, the formed hydroxyl ions may compete the activating chloride ions, for the adsorption on the electrode surface. Compound such as $K_2Cr_2O_7$, the dichromate ions, can exist only in the acidic media and in neutral media these ions transform to chromate ions according to the equilibrium: and the dichromate ions can not behave as an oxidizing agent in this neutral medium. For this reason the presence of in low concentrations in NaCl solution is accompanied by shifting of the

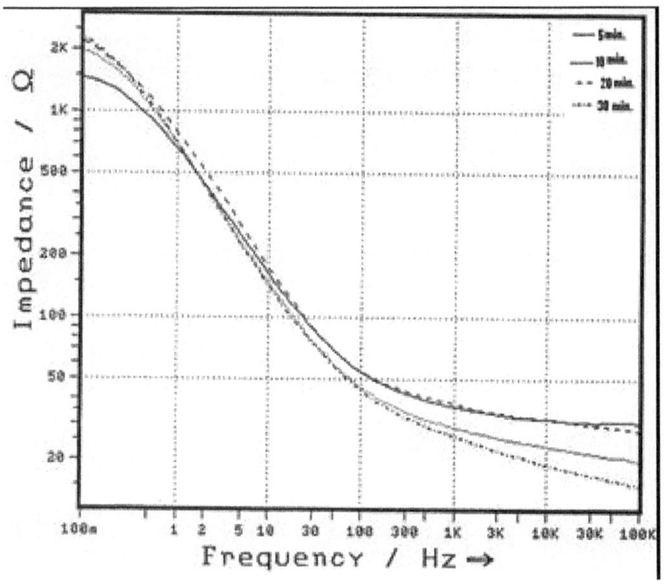

Figure 6a. Bode plot for Cu–Fe alloy in 3.5 NaCl +0.01 M $K_2Cr_2O_7$ solution as a function of immersion time at 20℃.

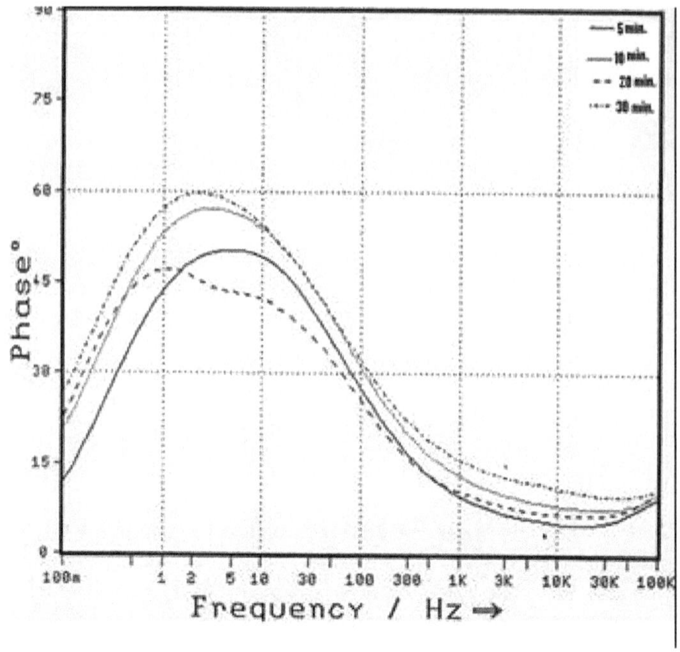

Figure 6c. Phase plots for Cu–Fe alloy in 3.5 NaCl +0.01 M $K_2Cr_2O_7$ solution as a function of immersion time at 20℃.

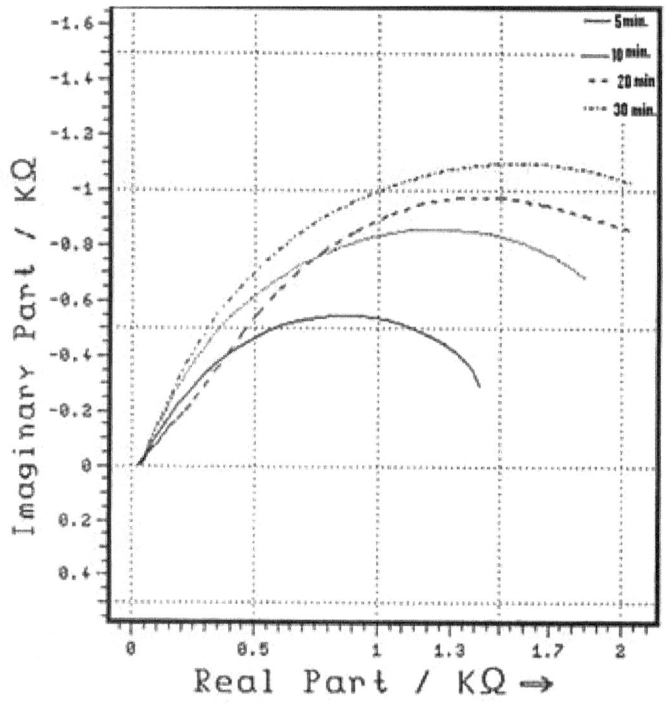

Figure 6b. Nyquist plots for Cu–Fe alloy in 3.5 NaCl +0.01 M $K_2Cr_2O_7$ solution as a function of immersion time at 20℃.

Table 6. Impedance parameters for Cu–Fe–Al in 3.5% NaCl solutions without and with different concentrations of inorganic additives at 20℃.

Additive	Conc. (mol /dm^3)	C_{dl} (μF. cm^{-2})	R_p (Ω-cm^2)
NaCl	3.5%	978.3	789.87
Na_2MoO_4	0.01	901.9	749.73
	0.025	489.2	1300.87
	0.05	313	1727.98
	0.1	549.3	1641.21
$K_2Cr_2O_7$	0.01	63.5	10019.33
	0.025	105.1	4554.92
	0.05	92.9	4410.38
	0.1	304.9	3225.95
KIO_3	0.01	157.2	2294.36
	0.025	184.97	2340.61
	0.05	153.1	2208.58
	0.1	184.5	2021.89

electrode potential to the more negative value than that obtained in sodium chloride solution. This means that the low concentration of these additives may accelerate the corrosion of the alloy in sodium chloride solution. The negative shift of electrode potential decreases with the increase of the additive concentration and may become positive at higher concentrations. This behaviour occurs at a certain concentration, depending on the type of additive. The ennoblement of the electrode potential in the presence of higher concentration of $K_2Cr_2O_7$ means that it can behave as an inhibitor and can retard the corrosion effect of sodium chloride. The retarding effect of $K_2Cr_2O_7$ can be explained by adsorption of the ions of inhibitor which reduce the free energy of the system and

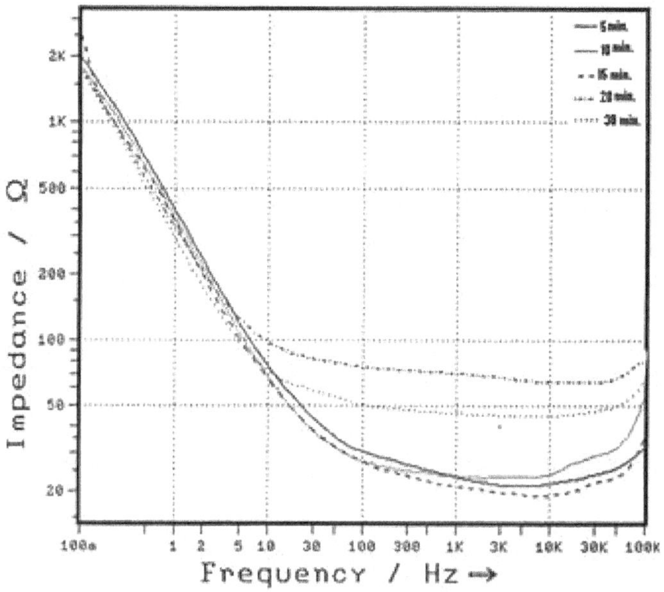

Figure 7a. Bode plot for Cu–Fe alloy in 3.5 NaCl +0.01 M Na_2MoO_4 solution as a function of immersion time at 20 °C.

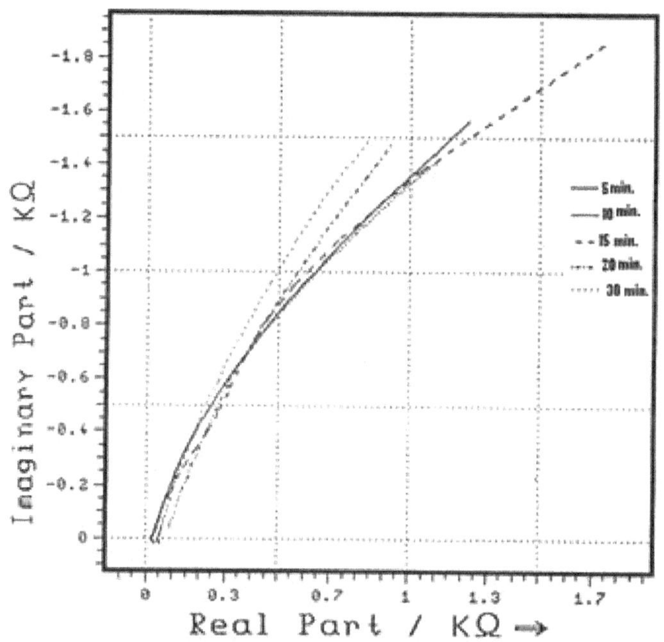

Figure 7b. Nyquist plot for Cu–Fe alloy in 3.5 NaCl +0.01 M Na_2MoO_4 solution as a function of immersion time at 20 °C.

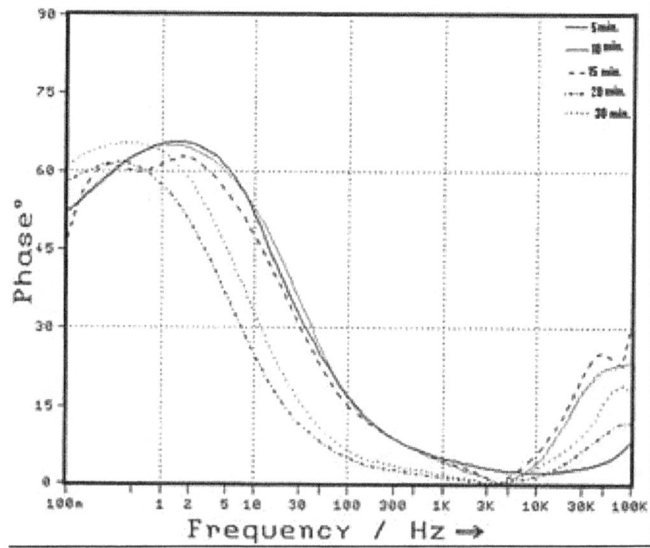

Figure 7c. Phase plot for Cu–Fe alloy in 3.5 NaCl +0.01 M Na_2MoO_4 solution as a function of immersion time at 20 °C.

impedes the passage of ion-atoms of metal from the lattice into the solution. It must, however, be kept in mind that the passivating properties of these additives depend greatly on the presence of activating ions (e.g. Cl-) in solution and the ratio between them. Also, such behaviour of the electrode is undoubtly associated with concurrent adsorption. In NaCl solutions and under the anodic polarization Cu_2O, CuCl and CuO can from respectively (De chialov et al., 1985; Brossard, 1983) in the potential region from 0.2 to + 0. 1 V (SCE).

For copper alloys in the potential range between -0.2 and 0.0 V (SCE), the corrosion reactions should yield Cu_2O, CuCl, Cu (I), and Cu (II) such as , and were found in solution depending on the chloride concentration (Lee et al., 1984). Paratacamite ($Cu_2(OH)_3Cl$) is stable at 0.12 V (SCE) and pH = 7.5 (Bianchi et al., 1978).

In the case of alloy (II) the presence of aluminum in the alloy leads to the formation of heterogeneous phases which have a potential difference. The phases of $CuAl_2$ and $FeAl_3$ may precipitate at the grain boundaries, where these phases are cathodic to the matrix. The passivation of this alloy occurs on the account of the formation of iron oxides, copper oxides and copper oxchlorides or also aluminum oxides.

The results of Sury (1970) showed that the thickness of the corrosion layers of Cu-Al alloys decreases with increasing Al-content of the corrosion layer. The protection nature of these films was ascribed to the formation of Cu_2O layer which was detected on binary Cu-Al alloys in NaCl solution by X-ray technique. The asymptotic drop of corrosion rate during the layer formation process was explained with the simultaneous decrease both of the anodic reaction inside growing Cu_2O layer and the cathodic reaction on its surface. The role of Al in passivation was assumed to be due to its incorporation in the Cu_2O lattice, reducing the rate of O_2 reduction on the Cu_2O surface.

Ateya et al., 1994 on studying the corrosion behaviour of α-Al bronze (Cu-7% Al) in NaCl solution stated that corrosive attack occurred very early leading to general corrosion and pit nucleation. The corrosion products formed under free corrosion conditions appeared as a

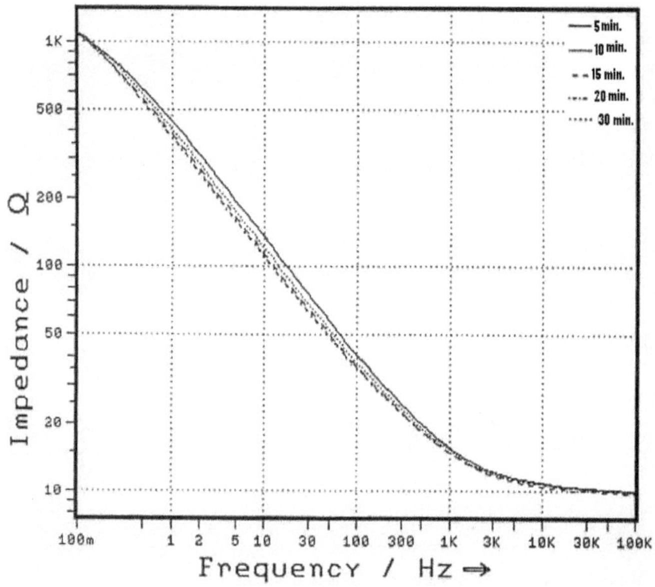

Figure 8a. Bode plot for Cu–Al-Fe alloy in 3.5 NaCl + 0.01 M Na_2MoO_4 solution as a function of immersion time at 20°C.

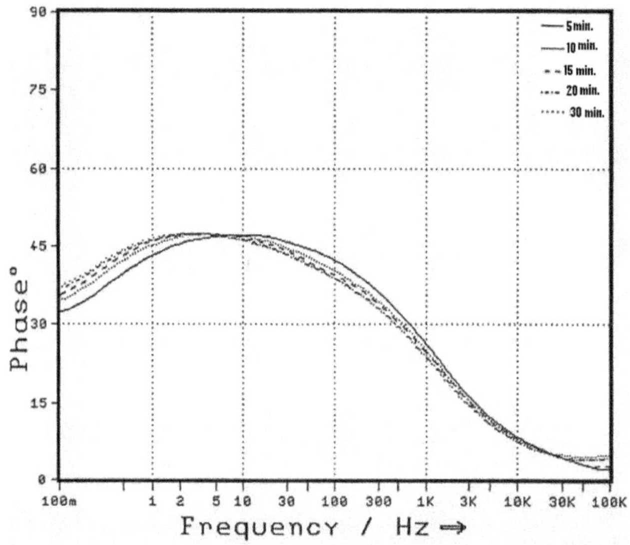

Figure 8c. Phase plot for Cu–Al-Fe alloy in 3.5 NaCl + 0.01 M Na_2MoO_4 solution as a function of immersion time at 20°C.

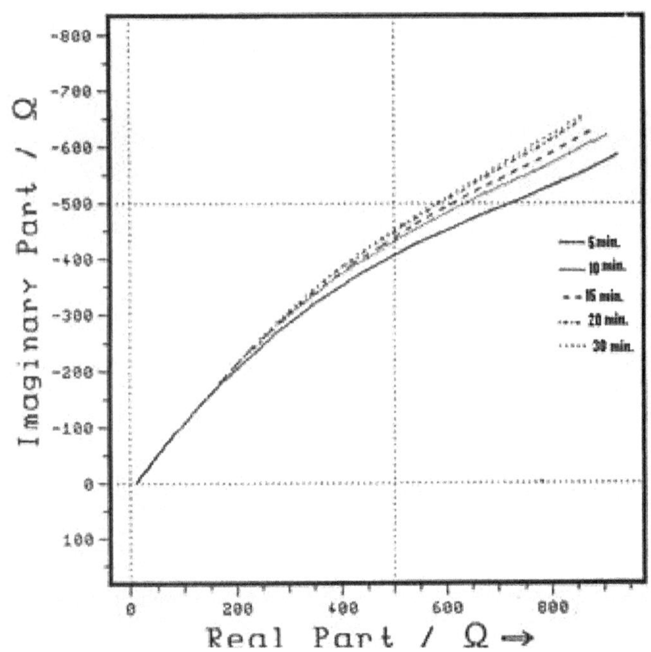

Figure 8b. Nyquist plot for Cu–Al-Fe alloy in 3.5 NaCl + 0.01 M Na_2MoO_4 solution as a function of immersion time at 20°C.

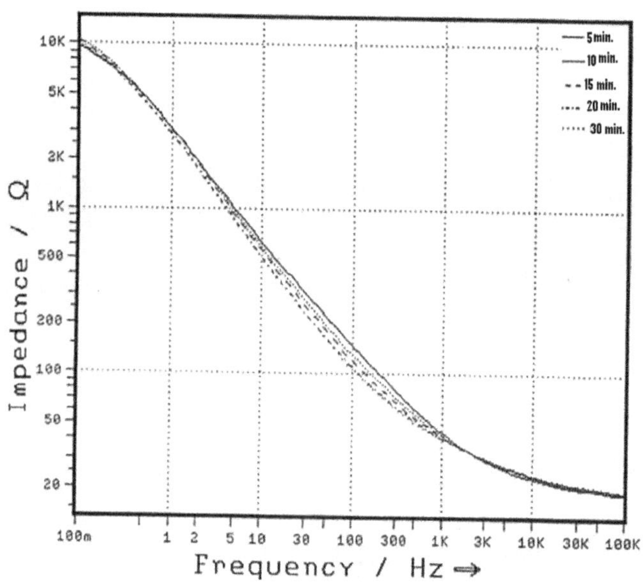

Figure 9a. Bode plot for Cu–Fe–Al alloy in 3.5 NaCl + 0.01 M $K_2Cr_2O_7$ solution as a function of immersion time at 20°C.

duplex film with an inner-adherent layer of Al_2O_3 and an outer layer of Cu_2O, $Cu_2(OH)_3Cl$ and Cu (OH) Cl. The polarization measurements carried out for Cu–Fe and Cu–Fe–Al alloys in NaCl solution indicated that the corrosion current of Cu–Fe is less than that of Cu–Fe–Al. Addition of Na_2MoO_4 and KIO_3 in NaCl solution for Cu–Fe causes shifts of the corrosion potential (Ecorr) to more positive values and increases the values of corrosion current density (Icorr), while the presence of high $K_2Cr_2O_7$ shifts the corrosion potential of the working electrodes to more negative values. Immersion of Cu–Fe–Al in the test solution shifts the corrosion potential, Ecorr, to more positive values and increases the values of Icorr.

From impedance measurements, Nyquist curves may be regarded as semi-circle and arc (semi-circle) with elapse of time. Increase or decrease of the diameter of the semi-circle with time that proportionally direct with passive film resistance was recorded, indicating an increase or

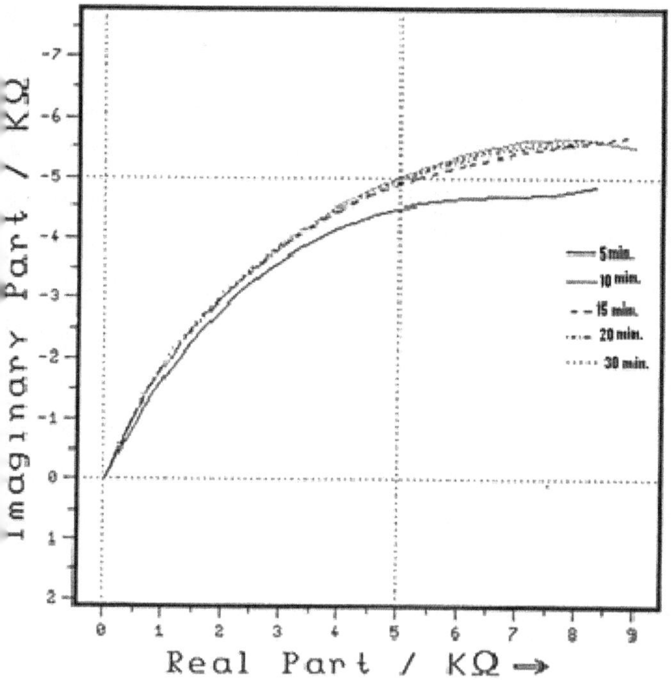

igure 9b. Nyquist plot for Cu–Fe–Al alloy in 3.5 NaCl +0.01 M $_2Cr_2O_7$ solution as a function of immersion time at 20℃.

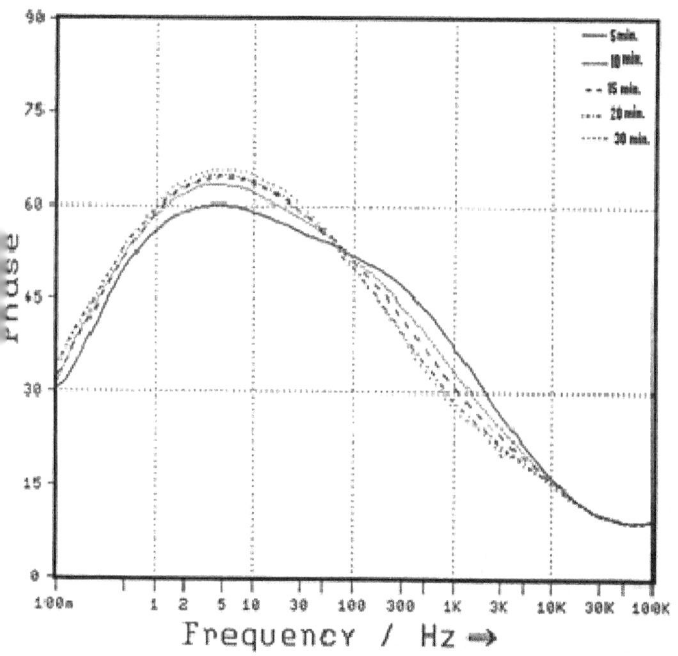

igure 9c. Phase plot for Cu–Fe–Al alloy in 3.5 NaCl + 0.01 M $_2Cr_2O_7$ solution as a function of immersion time at 20℃.

decrease of the protective properties of the surface layer. Bode plots indicated that at both high and low frequency limit the phase angle (θ°) assumes a value of about θ, corresponding to the resistive behavior of solution resis-

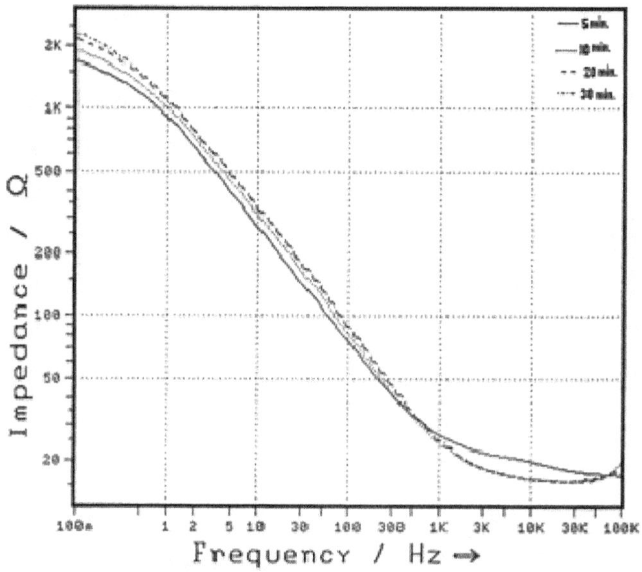

Figure 10a. Bode plot for Cu–Fe–Al alloy in 3.5 NaCl + 0.01 M KIO_3 solution as a function of immersion time at 20℃.

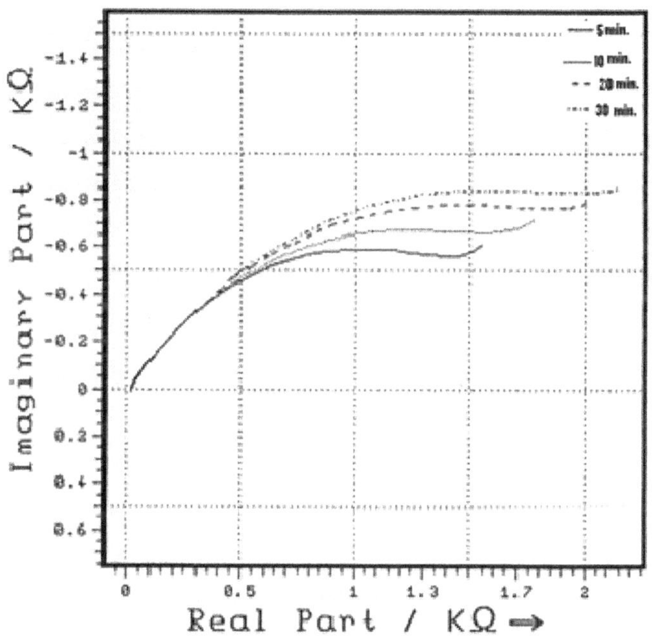

Figure 10b. Nyquist plot for Cu–Fe–Al alloy in 3.5 NaCl +0.01 M KIO_3 solution as a function of immersion time at 20℃.

tance, R_p and (R_s+R_p), respectively. The dialectic constant of the oxide film is a proof for the capacitive behavior of the system at intermediate frequency. The value tend to be more than 65°, the capacitive time constant is attributed to the formation of the oxide layer or the oxide layer itself, that is, the oxide film can be considered to be a parallel circuit of a resistor (due to the

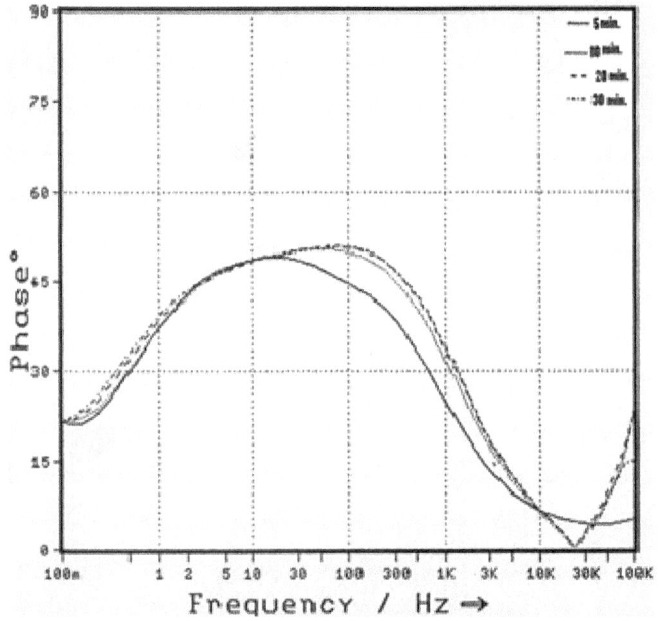

Figure 10c. Phase plots for Cu–Fe–Al alloy in 3.5 NaCl + 0.01 M KIO_3 solution as a function of immersion time at 20°C.

ionic conduction in the oxide) and a capacitance (due to the dielectric properties of this oxide). The impedance diagrams obtained were found to best fit for the simple Randle circuit: where R_p is the polarization resistance and C_{dl} is the capacitance of double layer. Nyquest plots tend to be semi-circle. The depression angles of these circles are correlated to the microscopic roughness of the metal/solution interface and with modification in the morphology of the electrode surface (Levie, 1965). Depressed capacitive semi-circles are often referred to as surface inhomogeneity (Metikos et al., 1994). The main parameters that can be deduced from the analysis of the Nyquist diagram are the charge transfer of the corrosion reaction, R_{ct} and the capacity of the double layer, C_{dl}. R_p values are calculated from the difference in impedance at lower and higher frequencies (Mansfeld et al., 1981; Tsuru et al., 1978). The double-layer capacitance, C_{dl}, can be calculated from the following relationship (Bentiss et al., 2000):

Cdl = π Rp f(zim,max)

Where f(zim,max) is the frequency at which the imaginary part of the impedance is maximum. The calculated values of R_p and C_{dl} are listed in Tables 3a and b, for the alloys (I, II), respectively, in NaCl solutions containing different concentrations of the investigated additives.

Conclusion

The investigated inhibitors behave as anodic ones, where they increase the anode polarization and hence move the corrosion potential, E_{corr}, in the positive direction, some of these inhibitors and under definite conditions behave as cathodic inhibitors, where in the corresponding manner displaces the corrosion potential, E_{corr}, in the negative direction. Also, these inhibitors greatly retard the localized attack on the electrodes surface.

REFERENCES

Ateya BG, Ashour EA, Sayed M (1994). (Corrosion of α-Al-Bronze in Saline Water) Electrochem. J. Soc. 141.

Barcia OO, Mattos NP´eb`ere B Tribollet J (1993). Electrochem. Soc. 140: 2825.

Bengough G, Jones GR, Pirret RJ (1920). Inst. Metals. 23, 65.

Bianchi G Fiori, F Mazza (1978). Corrosion. 334: 396

Bonfiglio C, Albaya H, Cobo O, (1973). Corrosion. Sci. 713, 717.

Brossard L (1984). Corrosion, 40: 420.

Bentiss F, Traisnel M, Lagrenee M (2000). Corrosion. 335, 315.

Blundy R, Pryor M (1972). (The potential dependence of reaction product composition on copper-nickel alloys). Corrosion Sci. 12: 65.

Cairs JH (1990). Ferrous sulfate treatment to prevent corrosion of condenser and exchange tubes, Technical Bulletin TB/7, Yorkshire Imperial Metals Ltd, Leeds.

Cartledge GH (1955). Phys. Chem. J. 59: 979.

De chialov MGR, Savarezza RC, Vazquez D, Arvia AJ (1985). Electrochim. Acta. 30: 501.

Faita G, Fiori G, Salvadore D (1975). (Copper behaviour in acid and alkaline brines—I kinetics of anodic dissolution in 0.5M NaCl and free-corrosion rates in the presence of oxygen).Corros. Sci. 15, 383.

El-Feky HE, Helal NH Negem MR (2009). (Electrochemical behavior of some copper alloys in sulfuric acid solutions). Materials and corrosion 60: 9999.

Gudas HJP (1979). Corrosion. 35: 259.

Lee HP, Nobe K (1984). Electrochem J Soc. 131: 1236. 25

Lee H, Nobe K, Pearlstein A (1985). Electrochem J Soc. 10: 1031

Lee H, Nobe K, Pearlstein A (1955). Corrosion. 11: 335.

Lee H, Nobe K, Pearlstein A (1966). J. Electrochem. Soc. 113, 328.

Rozenfeld IL, Maksimchuk VP (1961). Dokl. Akad. Nauk. SSSR. 119: 986.

Mansfeld F, Kending MW, Tsai S (1981). Corrosion 337: 301

Mayre EO (1960). Symposium Europeau Sur less Inhibiteurs de Corrosion, Ferrara, Italy, 273.

Metikos-Hukovic M, Bobic R, Gwabac Z, Brinic S (1994). Electrochem. Appl. J. 24: 772.

Nomoto N, Chingping T, Ohta M, Yamakawa K (1999). 62-Hitachi Cable Review.

Ohsawa M, Suetaka W, (1979). Corros Sci. 19: 709.

Rozenfeld IL (1981). Corrosion Inhibitors. McGraw – Hill International Book Company. New York. pp. 134-297

Sury P (1970). Dissertation philosophical Faculty of the Univ. Zurich.

Purification of an industrial aluminum alloy by melt stirring during ohno continuous casting process

Simbarashe Fashu*, Li Jianguo and Hu Qiao Dan

School of Material Science and Engineering, Shanghai Jiao Tong University, Shanghai, Dongchuan Road 800, 200240, China.

With the purpose of upgrading the purity of a dilute industrial aluminum alloy during ohno continuous casting (OCC) process through the use of forced melt stirring; the critical stirring intensity required under a casting speed of 2 mm/min was numerically determined. The determined critical stirring intensity induces convective velocity magnitudes into the melt which are effective in transporting solutes from the solute boundary layer into the bulk melt. It was found that, with adequate melt stirring, production of substantially long and high purity ingots during OCC is possible before their purity diminishes to lower than the initial alloy purity.

Key words: Segregation, impurities, purification, stirring.

INTRODUCTION

The mode of redistribution of excess solutes accumulating at the solid-liquid interface during ohno continuous casting (OCC) is a key factor determining the final purity of the produced ingots (Hashimoto et al., 1995). In absence of adequate melt convection, the rejected solutes forms a thick enriched solute boundary layer at the solid-liquid interface upon which solute transport is solely through diffusion (Wanqi, 2001). Thus, purification of the ingot occurs only during the transient stage before the accretion of the solute boundary layer at the stationary solid-liquid interface and after that, the solid will maintain the initial composition of the alloy due to a high solute concentration in the liquid at the interface (Tiller et al., 1995). In order to obtain high purity ingots during OCC process, a sufficiently strong convection towards the solidification front, reduces the thickness of the boundary layer (Willers et al., 2008), thus lowering the solute concentration in the solid. Purification of metals during solidification by taking advantage of segregation and melt convection is now an established process and is more effective if the partition coefficients of solutes are much less than unity. Mullin and Hulme (1958) equipped the zone refining process with an alternating magnetic field and obtained an enhanced purification rate in zone refining of semiconductors due to forced melt mixing. The emphasis of this article is placed upon using computer simulation to determine the adequate mechanical critical melt stirring intensity, required under a casting speed of 2 mm/min during OCC of aluminum rods to produce ingots of high purity. The lower casting speed (solidification rate) is used to increase the size of the mushy zone and the time for the solute rich liquid in the mushy zone to flow.

*Corresponding author. E-mail: sfashu04@gmail.com.

OCC process

Ohno continuous casting (OCC) process is a recently developed heated mold unidirectional continuous casting process used for producing superior, unidirectional rods and wires of substantial length (Ohno, 1986). The process is reliable in producing high purity wires and rods required mostly for the telecommunications industry and modern science technology, for example, video and audio cables (Motoyasu *et al.*, 1995). As shown schematically in Figure 1, the OCC process involves the continuous introduction of molten metal into an externally heated mould which is held slightly above the melting

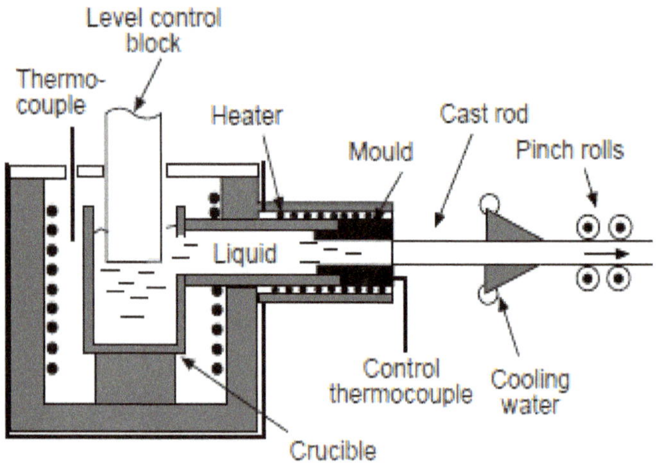

Figure 1. Schematic diagram of the horizontal OCC process.

point of the metal being cast; heat is extracted from the cast product by means of a cooling device located some distance from the mould exit. The solidified ingot is pulled with the pinch rollers placed some distance from the water cooling region.

MATHEMATICAL MODEL

Mechanical stirring OCC configuration

The OCC domain considered in our simulation is shown in Figure 2 and the regions of interest are the liquid in the heated mold and solid ingot, surrounding the solid-liquid interface where solidification and mass transfer occurs. Melt stirring was induced into the melt through mechanical rotation of a stirrer positioned at 30 mm from the mold exit as shown in Figure 3 to produce an effective axial flow field required for solute transport.

Governing equations

A macro-transfer model, based on the continuum model of Bennon and Incropera (1987), and reassessed by Prescott and Incropera (1987), was used for predicting the solute distribution during OCC of a multicomponent aluminum alloy.

Assumptions

1. All of the properties of the mixture can be obtained from the properties of its components in each phase.
2. All transport properties of each phase, such as thermal and electrical conductivity or viscosity, are constants.
3. The density is constant in each phase and the densities of the two phases are similar.
4. The mushy region is modeled by means of using the mixture viscosity.
5. Local equilibrium is assumed to exist at the solid-liquid interface, that is, the phase diagram is applied; $C_{i,s} = K_i C_{i,l}$.
6. The flow of the liquid phase is assumed to be incompressible, laminar and Newtonian.
7. Solutes diffusion coefficients are anisotropic and constant in each phase.

Based on the afore mentioned assumptions, the set of governing equations of heat, momentum and solute transport are:

Mass conservation equation:

$$\frac{\partial \rho}{\partial t} + \nabla.\left(\rho \vec{u}\right) = 0 \tag{1}$$

Where t is the time, ρ is the density, $\vec{u}$ is the average velocity of the mixture, which can be defined as $\vec{u} = f_s \vec{u_s} + f_l \vec{u_l}$, f_s and f_l are solid and liquid fractions, respectively.

Momentum conservation equation:

$$\frac{\partial\left(\rho \vec{u}\right)}{\partial t} + \nabla.\left(\rho \vec{u}\vec{u}\right) = -\nabla p + \nabla.(\mu_l \nabla \vec{u}) - \frac{(1-f_l)^2}{(f_l+\varepsilon)^3}(\vec{u}-\vec{u_s})A_{mush} + S_m \tag{2}$$

where p is the pressure, μ_l is the dynamic viscosity of the melt, K is the permeability constant of the mushy zone, ε is a small number (0.001) to prevent division by zero, u_s is the solid velocity due to ingot pulling from the mushy zone, S_m is the momentum source which in our simulation is only due to mechanical stirring.

The enthalpy-porosity technique was used to treat the mushy zone as a pseudo porous medium, with the porosity of each cell being set to its liquid fraction. The variation of permeability is approximated as function of liquid mass fraction f_l using the Carman-Kozeny relation:

$$K = K_0 \frac{f_l^{\,3}}{(1-f_l)^2} \tag{3}$$

Where K_o is a permeability constant depending on the morphology of the two-phase mushy region. $A_{mush} = \frac{\mu_l}{K_0}$ is the mushy zone constant and depends on the material. The value of K_o is usually approximated as $K_0 = \frac{d^2}{180}$, where d denotes the dendrite arm spacing.

Energy conservation equation:

$$\frac{\partial(\rho H)}{\partial t} + \nabla.(\rho H \vec{u}) = \nabla.(k \nabla T) + S_h \tag{4}$$

Where H is the enthalpy, k is the thermal conductivity; T is the temperature and S_h the source term, is written as:

$$S_h = -\nabla.[\rho(H_l - H)(\vec{u}-\vec{u_s})]$$

The first term on the right hand side of Equation 4 represent the net

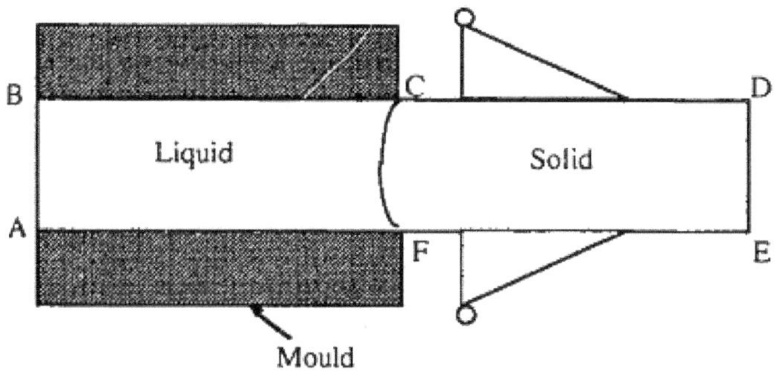

Figure 2. OCC domain considered in simulation.

Fourier diffusion flux. The enthalpy was computed as the sum of the sensible enthalpy, h, and the latent heat, ΔH :

$$H = h + \Delta H \tag{5}$$

Where:

$$h = h_{ref} + \int_{T_{ref}}^{T} C_p dT \tag{6}$$

Where h_{ref} = reference enthalpy, T_{ref} = reference temperature, C_p = specific heat at constant pressure. The latent heat content written in terms of the latent heat of the material, L is:

$$\Delta H = f_l L \tag{7}$$

The latent heat content varies between zero (for a solid) and L (for a liquid) and f_l is obtained from Equation 8. The liquid fraction is assumed to be linearly (Lever rule) proportional to temperature as:

$$fi = \begin{cases} 1 & when \quad T \geq Tl \\ \dfrac{T - Ts}{Tl - Ts} & when \quad Tl \geq T \geq Ts \\ 0 & when \quad T \leq Ts \end{cases} \tag{8}$$

Assuming that the solidus and liquidus lines are straight lines, the solidus and liquidus temperatures of solutes mixture are expressed as:

$$T_l = T_{melt} + \sum_{i=1}^{n} m_i C_i \tag{9}$$

$$T_s = T_{melt} + \sum_{i=1}^{n} \frac{m_i C_i}{k_i} \tag{10}$$

Where T_m is the fusion temperature, m_i is the slope of the liquidus line of binary phase diagram of solute i and aluminum and k_i is the partition coefficient of solute i in aluminum.

The solution for temperature is an iteration between the energy of Equation 4 and the liquid fraction of Equation 7, using the method suggested by Voller and Swaminathan (1991) to update the liquid fraction.

Solute conservation equation:

$$\frac{\partial(\rho C_i)}{\partial t}+\nabla.(\rho f_l \vec{u}_l C_{i,l}+(1-f_l)\rho \vec{u}_s C_{i,s}]=\nabla.[\rho f_l D_{i,ml}\nabla C_{i,l}+(1-f_l)\rho D_{i,ms}\nabla C_{i,s}]+S_c \tag{11}$$

Where, C_i is the total mass fraction of species i, $C_i = f_s C_s^i + f_l C_l^i$, D is the solute mass diffusion coefficient, $S_c = -\nabla.[\rho(C_l^i - C_i)(\vec{u} - \vec{u}_s)]$ is the advection source term in the mushy zone due to the relative phase motion. The subscripts i, l, c, m and s represents the species number, liquid, casting, mixture and solid respectively. The term on the right hand side of Equation 11 represents the net Fickian diffusion flux of solutes associated with the solid and liquid phases. The liquid velocity, u_l is found from the average velocity as follows:

$$\vec{u}_l = \frac{\left(\vec{u} - \vec{u}_s f_s\right)}{f_l} \tag{12}$$

The mixture quantities are defined in the following manner:

$f_l + f_s = 1$, $H = H_l f_l + H_s f_s$, $D = D_l f_l + D_s f_s$, $k = k_l f_l + k_s f_s$

The energy and species conservation equations are closed, using the equilibrium phase diagram and the liquid fraction in the whole domain, f_l–T relationship is defined by the Lever rule as shown in Equation 7. The assumption of local equilibrium at the interface does not preclude the existence of non-equilibrium conditions on a macroscopic scale, which are considered in the macroscopic Equation 9.

Numerical solution

The generalized format of the governing equations contains the transient terms, convective terms, diffusion terms and source terms, hence, the numerical procedure of solving them only requires slight modifications for the format of source terms as suggested by Patankar (1980).The solidification model was introduced and iteratively solved in CFD Fluent 6.3.26 software with the aid of User Defined Subroutines and convenient source terms of governing equations were introduced through User Defined Functions (UDFs). A fully implicit control volume based formulation was used for the time dependent terms, and the combined convection/diffusion coefficients were discretized using an upwind scheme. In order to resolve the pressure-velocity coupling in the momentum equations, the SIMPLE algorithm was used and pressure was discretized by the PRESTO! method. A momentum source term due to mechanical stirring was introduced into the melt through Multiple Reference Frame (MRF) model. A full, two dimensional single domain, fixed uniform grid of Figure 2 was used for the numerical calculation. Because of the high nonlinearity and full coupled nature of the governing equations, under-relaxation factors were reduced accordingly to obtain converged solutions. Double-diffusive convection in the melt was intentionally neglected in order to test theoretical models for diffusive regime and compare them with simulation results of solute redistribution in absence of melt

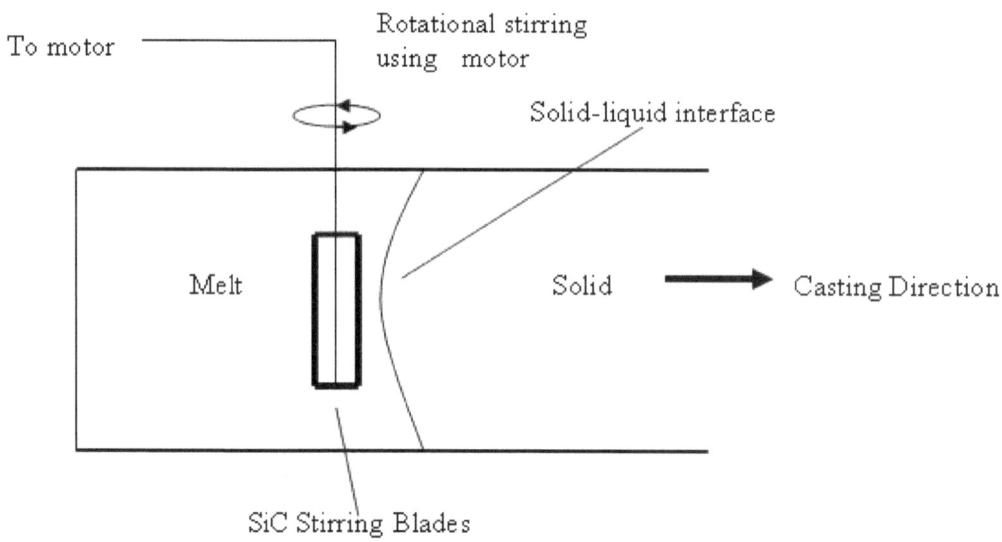

Figure 3. Positioning of a stirring blade in the melt near the interface for stirring.

Table 1. Thermo physical properties of Al 0.12wt%Cu 0.11wt%Si alloy.

Property	Units	Va values
Density of liquid /solid	kg/m^3	2400/2600
Specific heat	J/kgK	1080
Thermal conductivity of liquid	w/mK	90
Thermal conductivity of solid	w/mK	180
Diffusion coefficient of Cu in liquid	m^2/s	6e-9
Diffusion coefficient of Si in liquid	m^2/s	4e-9
Diffusion coefficient of Cu in solid	m^2/s	5e-13
Diffusion coefficient of Si in solid	m^2/s	3e-13
Partition coefficient of Si/Cu in Al	-	0.13/0.14
Melting temperature of Al	K	933
Slope of liquidus line Al-Cu/Al-Si	-	-3.43/-6
Viscosity of melt	Pa*s	0.0013
Melting heat	kJ	390

Table 2. Initial and boundary conditions.

Property	Units	Value
Melt inlet temperature	K	938
Melt inlet velocity	mm/min	2
Ingot pulling speed	mm/min	2
Heat transfer coefficient of Al in air	w/m^{2}K	414
Heat transfer coefficient of Al in water spray	w/m^{2}K	10000-50000
Temperature of heated mold	K	938

convection. Furthermore, the alloy is dilute, such that, solutal diffusion is negligible, and the use of a heated mold in horizontal OCC with gravity vector being perpendicular to thermal gradient makes thermal convection effects to be insignificant (Verhoeven, 1968).

The industrial aluminum alloy under consideration is composed of copper and silicon as major impurities. Thermo-physical properties of the industrial dilute aluminum alloy and the initial and boundary conditions used in this simulation are shown in Tables 1 and 2, respectively. The solid-liquid interface position was controlled to be just near the stirring blade through adjustment of cooling in the water spray region.

SIMULATION RESULTS AND DISCUSSION

Validation of solidification results with published literature

In order to validate the solidification model which was used in our computations, the model was compared against theory and experimental results from literature. Lee et al. (2004) used Bridgman growth to grow ingots of 0.4 and 5.5 mm diameter in absence of external melt convection using Al-4%wtCu at a casting speed of 0.4 µm/s and produced 40 mm ingots during transient state before the solute boundary layer had sufficient time to build. The Electron Microprobe Analysis (EMPA) composition profile of short solid samples and liquid was done as shown in Figure 4.

Simulation to show the capability of the solidification model was done using the same conditions as those used by Lee et al. (2004). Figure 5 shows the simulated copper solute redistribution after OCC of 5.5 mm ingots during the transient state. This result is in good agreement with experimental results obtained by Lee et al. (2004) as shown in Figure 4, showing that the model is capable of computing solute redistribution at the casting speeds considered.

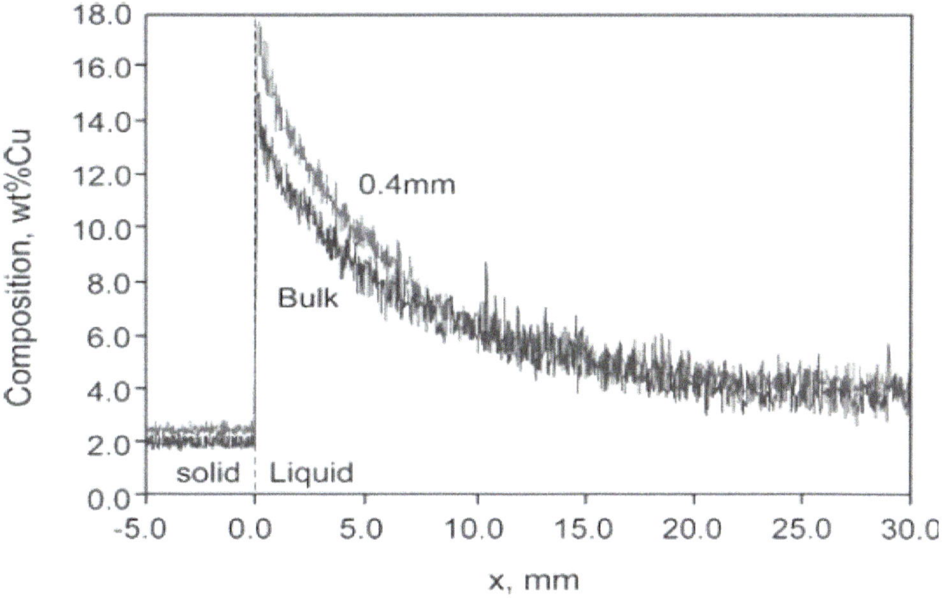

Figure 4. Line scan EPMA result of the composition profiles in 0.4 and 5.5 mm (bulk) diameter and 40 mm long ingots of Al-4%wtCu alloy unidirectionally solidified with Bridgman furnace at 0.4 μm/s (Lee et al., 2004).

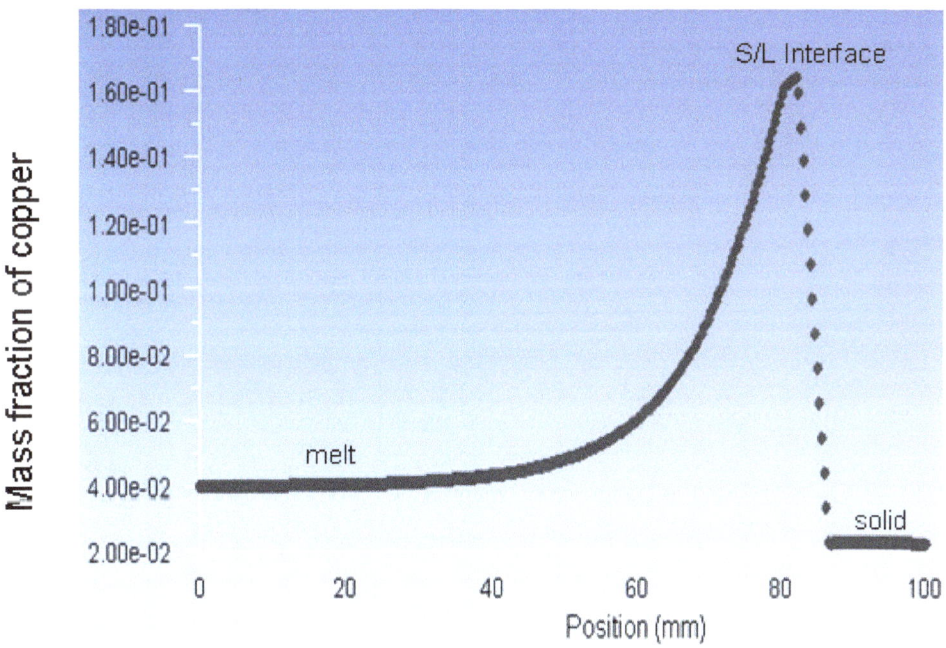

Figure 5. Solute redistribution obtained by simulation using experimental conditions from Lee et al. (2004).

Solidification and mechanical stirring results

Figure 6 shows the temperature distribution profile during steady OCC process. The solidus and liquidus fronts are stationary and solidification starts at the centre of the ingot, thus avoiding casting defects. Figure 7 shows the aluminum redistribution after the OCC process has reached steady state in composition when there is no fluid flow and the solute distribution is caused only by the diffusion mechanism. This is in agreement with theory of Tiller et al. (1995), which states that the solute redistribution is constant, once steady state condition is

Figure 6. Temperature distribution profile.

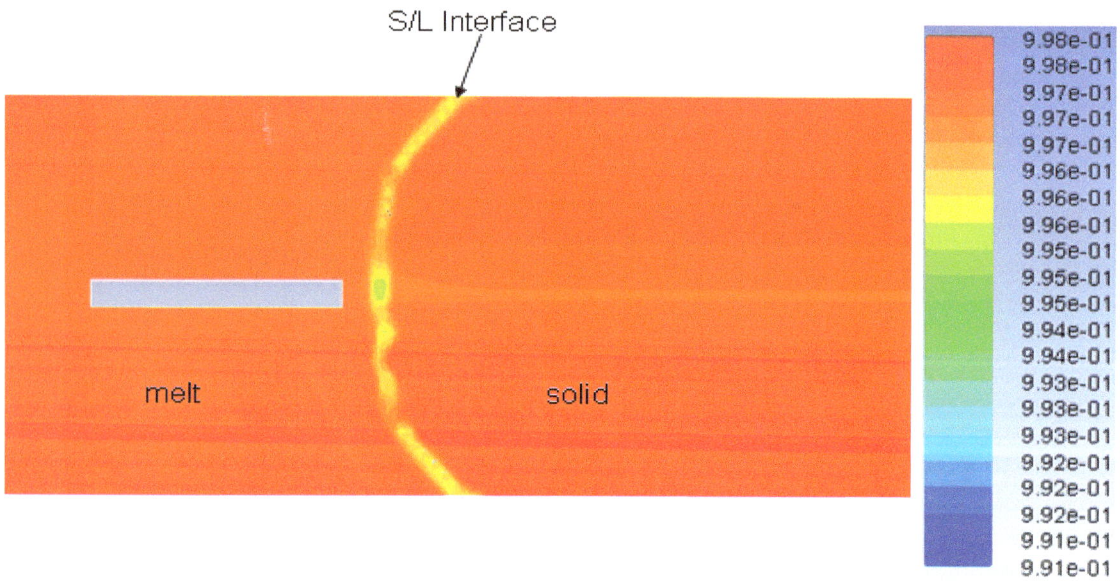

Figure 7. Aluminum redistribution at steady state in absence of melt stirring.

attained. Figure 8 shows the corresponding axial aluminum redistribution near the center of the ingot. Purification of the ingot occurs slightly during the transient process before the solute boundary layer is established. Using unsteady state simulation, it was found that about 500 mm ingots were cast to reach that state in composition. After attaining steady state composition, the accumulated solutes prevent solute redistribution and the solute concentration in the entire casting remains equal to the initial value and remains constant during OCC process because solute diffusion is a very slow process.

Thus, the solute diffusion coefficient, D_l, of the macroscopic solute transport in Equation 11 will be almost zero, and although the solute will be microscopically rejected at the solid-liquid interface, there will be no mechanism to transport the solute. This same solute redistribution behavior was also shown by Diao and Tsai (1993) during directional solidification of an Al-4.1%wtCu alloy in absence of melt convection. The critical stirring intensity required to produce ingots of highest purity during OCC process was obtained from rigorous trial and error using different stirring rates. Figures 9 and 10 show the solute distribution in ingot and

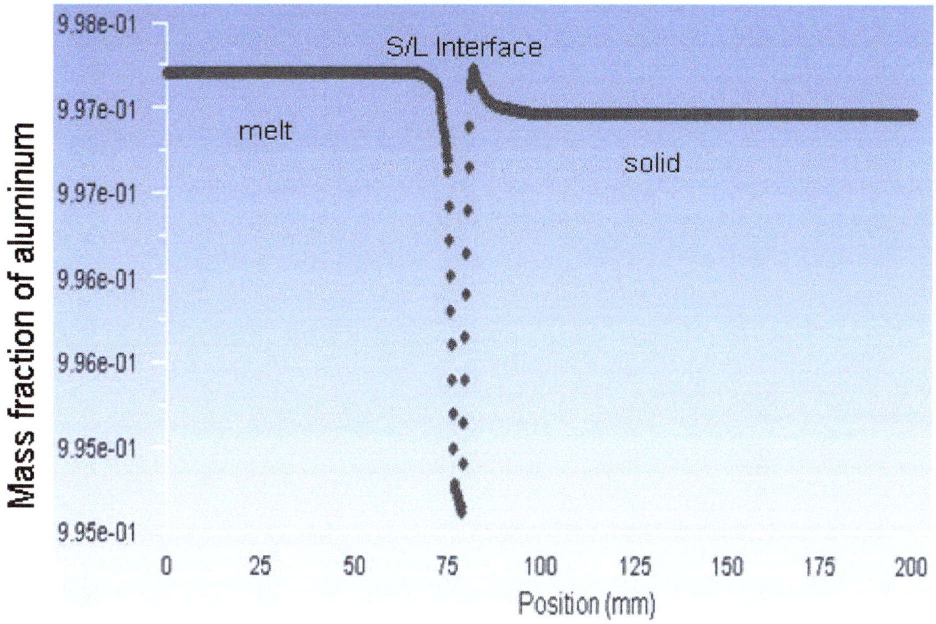

Figure 8. Axial distribution of aluminum in absence of melt stirring.

Figure 9. Aluminum redistribution in presence of optimum melt stirring of 10 rpm, after casting 2 m ingot.

its corresponding axial distribution, in presence of optimum stirring intensity, (Figure 11) for casting of 2 m ingot. Stirring significantly increased the ingot purity level from 99.77 to above 99.925%Al at the center of the ingot where stirring is more effective. The sides of the ingots are a bit less pure due to the poor design of the stirrer used. Since the fresh melt will continuously move into the heated mold during OCC process, it will dilute the bulk melt and maintain it at a low solute composition for a long time. When the solute level of the ingots becomes higher than the initial alloy being used, then the process should be stopped and the melt discarded for other uses.

Conclusions

The critical stirring intensity required under an optimum casting speed to produce aluminum ingots of high purity

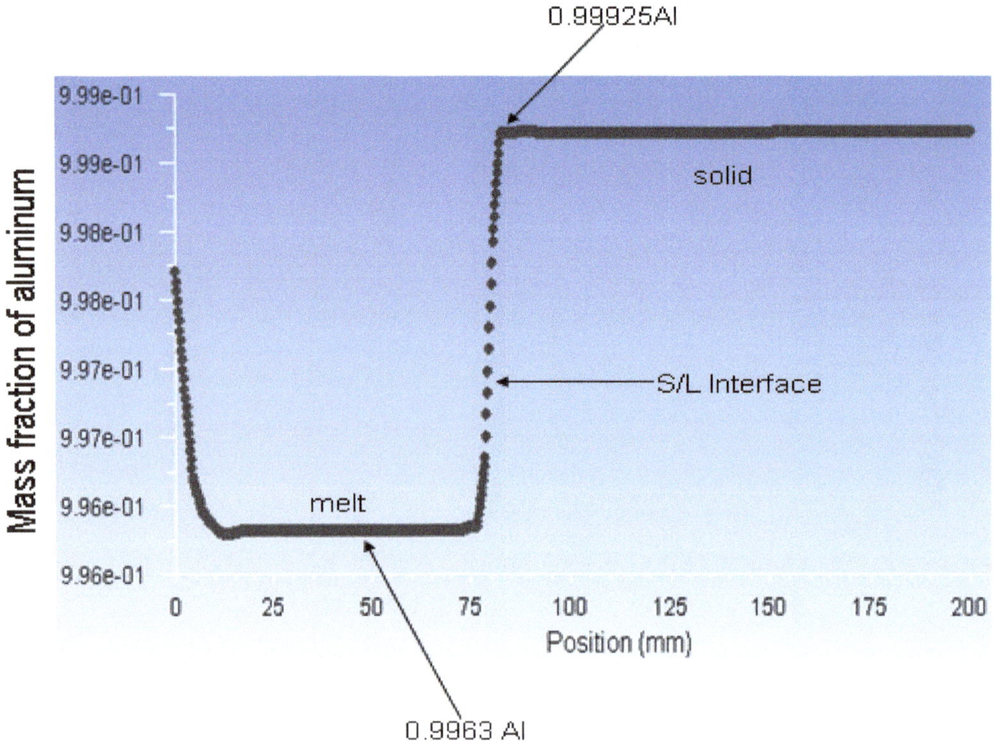

Figure 10. Axial aluminum redistribution at in presence of optimum melt stirring of 10 rpm, after casting 2 m ingot.

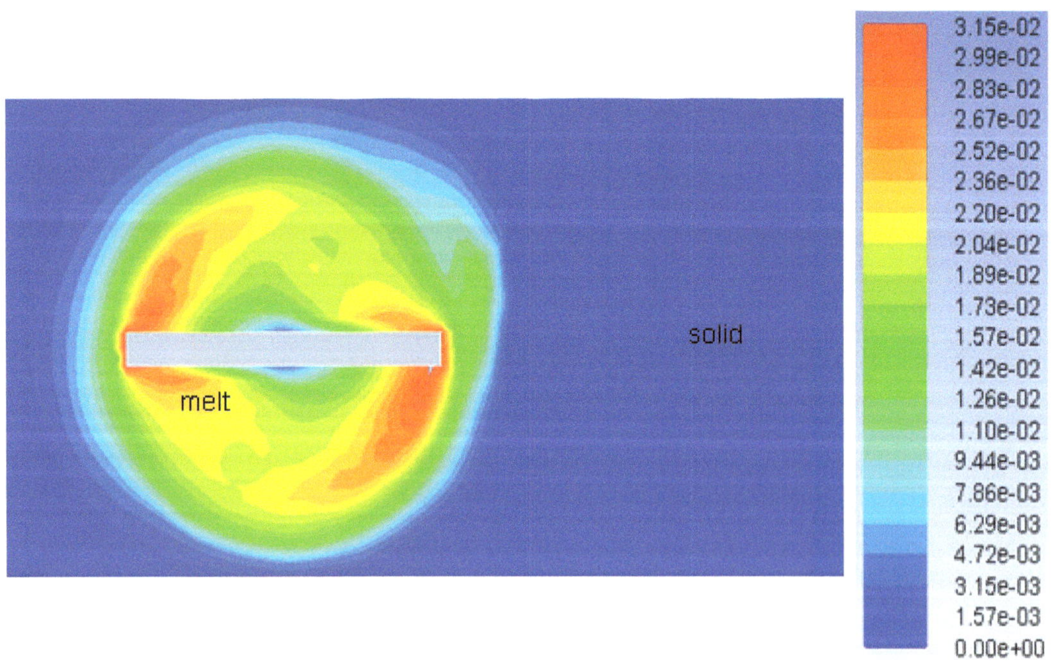

Figure 11. Velocity profile in m/s for optimum mechanical melt stirring of 10 rpm.

during OCC was determined by numerical simulation. This was done through melt stirring near the phase change region to produce a special axial rotating melt flow field of adequate magnitude and direction. The industrial aluminum alloy is very dilute, and continuous flow into the heated mold makes it possible to produce long and pure ingots before their purity level falls below the initial alloy composition. Thus, with optimization of casting speed and stirring intensity, high purity ingots at high production rates are possible to achieve during OCC process. This, of course should be accomplished with a good stirrer design, effective in mechanically stirring near the whole interface region.

ACKNOWLEDGEMENTS

This work was supported by the National Natural Science Foundation of China (50804030), China Post-Doctoral Science Foundation through the First Class Funding (20080430084) and the Special Funding (200902230) programs, and Shanghai Post-Doctoral Scientific Program (08R214135). We also acknowledge Instrumental Analysis Centre, SJTU for help in microstructure characterization. Thanks to Professor Lu Hao, for helpful discussion, and provision of CFD Fluent 6.3.26 Software.

REFERENCES

Bennon WD, Incropera FP (1987). Continuum model for momentum, heat and species transport in binary solid-liquid phase change systems – I. J. Heat Mass Transf., 30(20): 2166-2170.

Diao QZ, Tsai HL (1993). Modeling of solute redistribution in the mushy zone during solidification of aluminum-copper alloys. Metall. Trans., 24A: 963-973.

Hashimoto E, Ueda Y, Kondo T (1995). Purification of Ultra-High Purity Aluminum. J. Phys., 5: 153-157.

Lee JH, Shan L, Miyahara H, Trivedi R (2004). Diffusion coefficient measurement in liquid metallic alloys. Metall. Trans., 35B: 909-917.

Motoyasu G, Soda H, McLean A, Wang Z (1995). Pilot scale casting of single crystals copper wires by the Ohno Continuous Casting Process. J. Mater. Sci., 30(21): 5438-5448.

Mullin JB, Hulme KF (1958). The use of electromagnetic stirring in zone refining. J. Electron. Control., 14: 170-174.

Ohno A (1986). Continuous casting of single crystal ingot by the OCC process. J. Metals., 38: 14-16.

Patankar SV (1980). Numerical Heat Transfer and Fluid Flow. Hemisphere, Washington DC, Hemisphere Publishing Corporation, pp. 126-134.

Prescott PJ, Incropera FP (1991). Modeling of dendritic solidification systems: Reassessment of the continuum momentum equation. Int. J. Heat Mass Transf., 34: 2351–2359.

Tiller WA, Jackson KA, Rutter JW, Chalmers B. (1953). The redistribution of solute atoms during the solidification of metals. Acta Metall., 1: 428-437.

Verhoeven JD. (1968). The growth of off-eutectic composites from stirred melts. Trans., AIME, 242: 1937-1942.

Voller VR, Swaminathan CR. (1991). General source-based method for solidification phase change. Numer. Heat Transf., 19B(2): 175-189.

Wanqi J (2001). Solute redistribution and segregation in solidification processes. Sci. Technol. Adv. Mater., 2: 29-35.

Willers B, Eckert S, Nikrityuk PA, Räbiger D, Dong J, Eckert K, Gerberth G. (2008). Efficient melt stirring using pulse sequences of a rotating magnetic field Part II. Metall. Met. Transf., 39B: 304-316.

Preparation and characterization of exfoliated nano-composite of polyvinyl acetate and montmorillonite

Hui-Wang Cui[1,2,3]* and Guan-Ben Du[1,2]

[1]College of Wood Science and Technology, Nanjing Forestry University, Nanjing 210037, Jiangsu, China.
[2]College of Wood Science and Technology, Southwest Forestry University, Kunming 650224, Yunnan, China.
[3]Sino-Swedish Microsystem Integration Technology Center, Shanghai University, Shanghai 200072, China.

To improve the chemical, physical, and mechanical properties of polyvinyl acetate (PVAc), montmorillonite (MMT) was added in the synthesis of PVAc in conventional lab conditions. The final product was studied mainly by X-ray diffraction (XRD), Fourier transform infrared spectroscopy (FTIR), transmission electron microscopy (TEM), static tensile, and the rheology was also investigated by the power-law function equation and the Cross-Williamson model viscous equation. The exfoliated nano-composite of PVAc-MMT was obtained. Linear macromolecular chains of PVAc were formed in MMT layers. MMT was exfoliated into layers or sheets of nano-particles, and randomly dispersed in PVAc matrix. The particle diameter of MMT ranged from 50 to 100 nm, while PVAc's ranged from 250 to 500 nm, the smaller MMT particles were adsorbed around the larger PVAc particles to form the "Strawberry" structure. In addition, PVAc and PVAc-MMT were pseudo-plastic Non-Newtonian fluids, they all possessed the normal stress effect (or Weissenberg effect), that was the pole-climbing phenomenon.

Key words: Polyvinyl acetate (PVAc), montmorillonite (MMT), chemical structure, dispersion, particles, rheology, static tensile.

INTRODUCTION

Generally, polyvinyl acetate (PVAc) is synthesized by the monomer of vinyl acetate (VAc) in the mixture of polyvinyl alcohol (PVA) as protective colloid, non-ion emulsifier, initiator, and water. It has many characteristics, being non-poisonous, non-harmful, easily produced, low price, convenient in application, economizing resources, etc.. Because of these reasons, PVAc is widely applied in bonding of many porous materials, such as wood processing, furniture packaging, building decoration, texture bonding, and print bonding. Montmorillonite (MMT) is a natural nano-mineral. Its crystal structure is a 2:1 type layered silicate formed by a layer of aluminum (or magnesium) octahedral inserted in the middle of two layers of silicon oxygen tetrahedron. The structural units are joined by the molecular strength between layers, so it is very loose. Water or other organic molecules can enter into the layers which can lead to expansion through absorbing water, high dispersion and absorption, and can also cause it to easily make mud, be activated, organized and modified.

Now, MMT has been applied in many polymers for many years, such as polyacrylate ester (Ray and Bhowmick, 2001; Ye et al., 2004), poly(methyl methacrylate) (Essawy et al., 2004; Metin and Yelda, 2007), polyurethane (Deng et al., 2007; Esposito et al., 2008), epoxy (Kommann et al., 2001; Cui et al., 2007), polyethylene (Zhao et al., 2004; 2005), and polypropylene (Qin et al., 2005; Chen et al., 2006), etc.. Ding et al. (2002, 2004a, 2004b, 2005) also reported the application in PVAc, but they mainly focused on the preparation of PVAc-MMT initiated by γ-ray radiation intercalation polymerization and the co-blending of PVAc-MMT with high density polyethylene. In the paper, based on our previous work (Cui and Du, 2011), and under the conventional conditions in the lab, with no ultrasonic dispersion and no radiation, PVAc-MMT was prepared by VAc and MMT. The obtained product was investigated mainly by X-ray diffraction (XRD), Fourier transform infrared spectroscopy (FTIR), transmission electron microscopy (TEM) and static tensile. Its rheology was also investigated by the power-law function equation

*Corresponding author. E-mail: cuihuiwang@hotmail.com.

and the Cross-Williamson model viscous equation.

MATERIALS AND METHODS

Samples

1.00 g of MMT was immersed into 25.00 g of VAc for 24 h. It was then mixed with 70.00 g of 10% PVA solution, 0.50 g of alkylphenol polyoxyethylene (10) ether, 3.75 g of 10% ammonium persulfate solution, 6.25 g of sodium lauryl sulfate and 250 g of water whilst stirring vigorously for 8 h. When the mixture became a homogeneous emulsion, the temperature rose to 70℃. Whilst stirring vigorously, 3.75 g of 10% ammonium persulfate solution and 25.00 g of VAc were gradually added into the homogeneous emulsion in 6 h for polymerization.

Subsequently, the temperature rose from 85 to 90℃ the emulsion further polymerized for 0.5 to 1 h. After polymerization, the temperature dropped to below 50℃, 4.00 g of ethanol, 3.00 g of water, 0.30 g of sodium benzoate, 0.18 g of sodium bicarbonate, and 6.00 g of Di-n-butyl phthalate were added into the emulsion. Finally, PVAc-MMT was obtained. PVAc was also prepared by the same process but without MMT.

Apparent viscosity

The samples were tested by NDJ-1 rotary viscometer; the shear rate (γ) was 0.63, 1.26, 3.14 and 6.28 s^{-1} respectively. The apparent viscosity (η, mPa·s) under different shear rate was obtained.

Solid content

The solid content (O, %) is calculated by:

$$O=\frac{m_1-m_2}{m_1}\times 100\%$$

Where m_1 (g) is the original weight of samples, and m_2 (g) is the weight of samples dried for 1 h at 105℃.

XRD

The samples were tested by DX-2000 of XRD under Cu Kα of radiating, 40 kV of tube voltage, 30 mA of tube current, scanning from 0.5 to 15° at a rate of 0.02°·s^{-1}, and λ=1.54184 Å of wavelength. The large value of $d_{(001)}$ for MMT is calculated by Bragg law of $\lambda=2d\sin\theta$, where λ is the wavelength representing the intensity of X-ray, d is the distance between layers of MMT, and θ is the diffraction angle.

FTIR

The samples mixed with kalium bromatum powder were pressed into plates, and then tested by the NICOLET 380 of FTIR.

TEM

The samples were observed by JEM-3100F TEM, the amplification was 20000 times.

Dispersion

3 to 5 drops of PVAc or PVAc-MMT were added to 20 ml of water in a glass dish (diameter 90 mm). When they became homogeneous, the dispersion was observed with water as reference.

Static tensile

PVAc and PVAc-MMT were made into films of 50×10×0.5 mm. After drying them to constant weight at room temperature, they were tested by SANS CMT 5000 computer-controlled electronic universal testing machine under a tension rate of 10 mm·min^{-1} and a test temperature of 25℃. The test should be finished in 10 min.

Storage time

The glue was stored away from light at room temperature till it appeared as gel and/or delamination. The cycle was the storage time (S, days).

RESULTS AND DISCUSSION

Chemical structure

The diffraction peak representing $d_{(001)}$ of layers for MMT appeared in XRD patterns (Figure 1), its 2θ was 7.00°, its $d_{(001)}$ was 1.263 nm (Cui and Du, 2009). However, there was no diffraction peak from 1 to 12° of 2θ for PVAc-MMT. This phenomenon showed that the linear macromolecular chains of PVAc were formed in MMT layers, and MMT was greatly exfoliated into layers or sheets of nano-particles, randomly dispersing in PVAc matrix. The exfoliated nano-composite of PVAc-MMT was obtained. PVAc and PVAc-MMT were also tested by FTIR (Figure 2). No chemical bond was found between MMT and PVAc, but physical effect. The absorption bands of PVAc-MMT were stacked by those of MMT and PVAc. Neither new absorption bands formed, nor did existing absorption bands disappear. They also showed that MMT was greatly exfoliated into layers or sheets of nano-particles, and randomly dispersed in the matrix of PVAc. The results coincided with XRD results.

Dispersion and particles

From the visual, PVAc and PVAc-MMT were the same. They were viscous, milk-white, homogeneous, fine emulsions. All had no coarse particles, no foreign bodies and no delamination. However, when some dropped in water, their dispersion was a little different (Figure 3). Some coarse particles were found out in the dispersion of PVAc referred with water, but not PVAc-MMT. The dispersion of PVAc-MMT was much better.

PVAc and PVAc-MMT were further observed by TEM to investigate their particles (Figure 4). From the TEM comparison between PVAc and PVAc-MMT, the

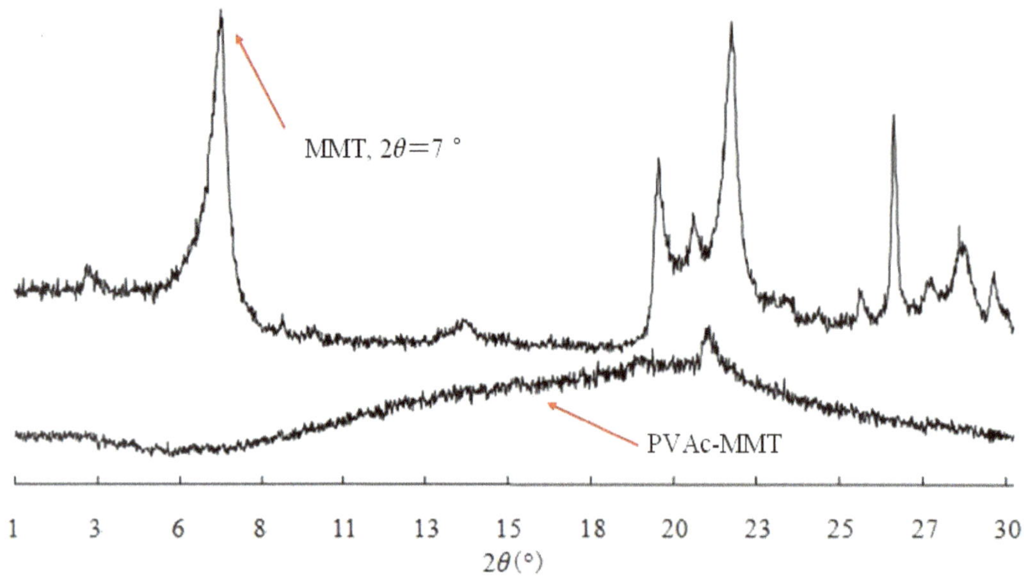

Figure 1. XRD patterns.

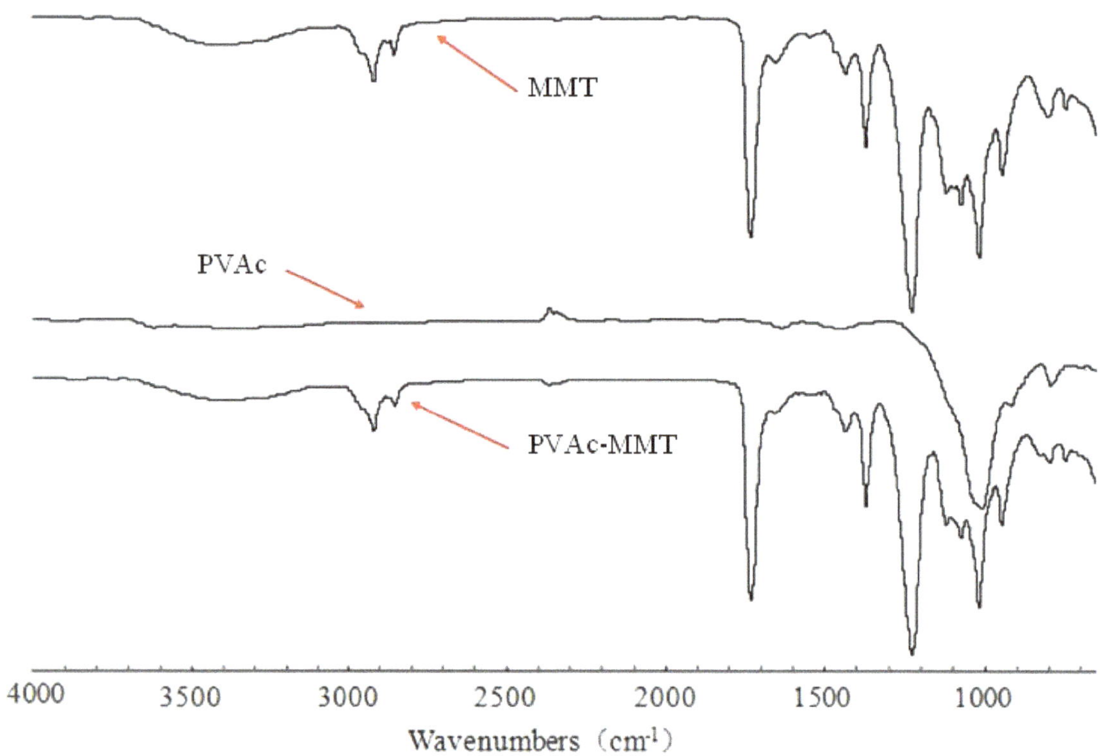

Figure 2. FTIR spectra.

differences were clearly apparent. The larger particles were PVAc, their diameter was from 250 to 500 nm, and the smaller particles were MMT, their diameter was from 50 to 100 nm. TEM also showed that MMT particles randomly dispersed with PVAc particles, this phenomenon coincided with the results of XRD and FTIR.

In addition, a "strawberry" structure was also found in the dispersion. This structure was only mentioned by Suprakas and Mosto (2005). As observed by TEM, the smaller MMT particles were adsorbed around the larger PVAc particles, and they formed the "strawberry" structure.

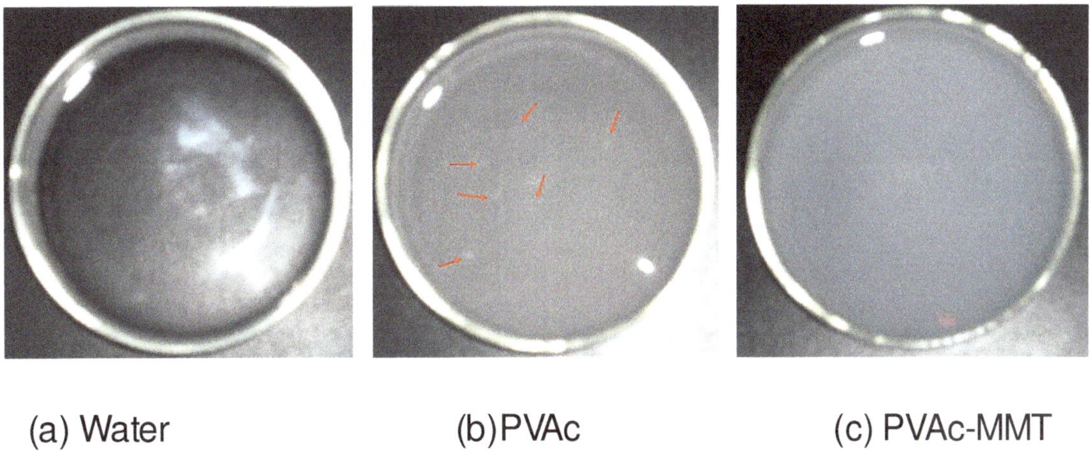

Figure 3. Dispersion.

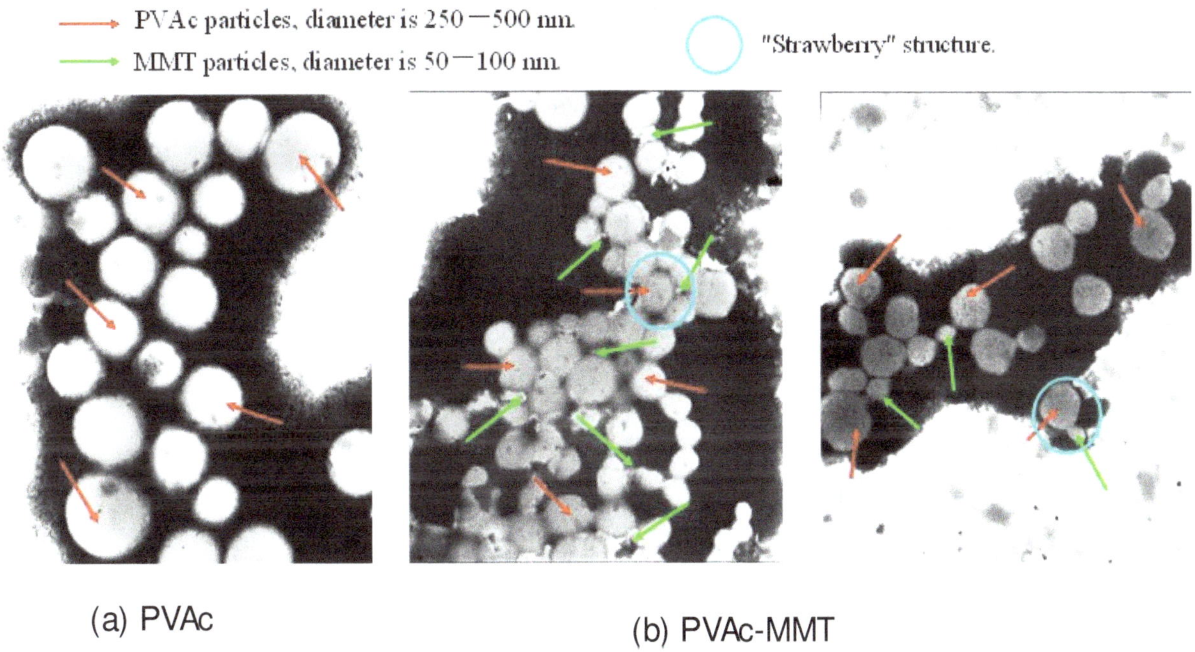

Figure 4. TEM images.

Table 1. Apparent viscosity.

η (mPa·s)	γ (s^{-1})			
	0.63	**1.26**	**3.14**	**6.28**
PVAc	15	14	13	13
PVAc-MMT	14	13	13	12

Rheology

The apparent viscosity (Table 1) under different shear rate decreased with the increase of the shear rate. This phenomenon indicated that both PVAc and PVAc-MMT were pseudo-plastic Non-Newtonian fluids. As we know, a pseudo-plastic fluid is one of the most common Non-Newtonian fluids. Rubber, most polymers, and their plastic melt and concentrated solution all are pseudo-plastic fluids. Their characteristic is that the apparent viscosity decreases with the increase of the shear rate. The orientation of the slender molecular chains of polymer in the flow direction causes the decrease of viscosity, so it is often called as the shear thinning fluid. The rheology of pseudo-plastic Non-Newtonian fluid is always described by power-law function equation:

Table 2. Rheological results.

	i	Ï (mPa·s)	ζ (s)	η_0 (mPa·s)	η_∞ (mPa·s)	M_w	M_n
PVAc	0.93	14	4.6×10^{12}	1.2×10^2	0	1.0×10^2	1.0×10^2
PVAc-MMT	0.93	14	4.8×10^3	1.4×10^2	0	1.1×10^2	1.1×10^2

$$\tau = Ï\gamma^i$$

Where τ is the shear stress, Ï (Pa·s) is the fluid consistency, γ (s^{-1}) is the shear rate, and i is the flow index, also called as the Non-Newtonian index.
The bigger Ï is, the more viscous the fluid is. i is used to judge the difference between fluid and Newtonian fluid. If i is farther away from 1, the Non-Newtonian behavior is more obvious. For the Newtonian fluid $i=1$, then Ï is equivalent to Newtonian viscosity (μ); for the pseudo plastic fluid, $i<1$.

Compare the power-law function equation with the Newtonian fluid flow equation:

$$\tau = \mu\gamma$$

Combine them into the equation:

$$\tau = (Ï\gamma^{i-1})\gamma$$

And make:

$$\eta = Ï\gamma^{i-1}$$

The power-law function equation can be written as:

$$\tau = \eta\gamma$$

Where η is the apparent viscosity of Non-Newtonian fluid that is the tested viscosity (Pa·s or mPa·s).

For Newtonian fluid and pseudo-plastic fluid, the relationship among their zero shear viscosity (η_0), η, and limit viscosity (η_∞) under lower γ is $\eta_0>\eta>\eta_\infty$, and $\eta_\infty=0$. According to the Cross-Williamson model viscous equation:

$$\eta = \frac{\eta_0}{1+|\zeta\gamma|^{1-i}}$$

η_0 is:

$$\eta_0 = \eta(1+|\zeta\gamma|^{1-i})$$

Where ζ (s) is the characteristic time of materials.

The polymer has a critical molecular weight (M_c), when the weight average molecular weight (M_w) $<M_c$, η_0 is proportional to M_w; when $M_w>M_c$, η_0 has a 3.4 exponential relationship with the increase of M_w. They illustrate that a large number of entanglement zones are formed among polymer chains. The substantial increase in viscosity is due to the contribution of the number of entangled points in entanglement zones and the maximum molecular weight. That is related to M_w. Their relationship can be expressed as:

$$\eta_0 = Ж_1 M_w \ (M_w<M_c)$$

$$\eta_0 = Ж_2 M_w^{3.4} \ (M_w>M_c)$$

Where $Ж_1$ and $Ж_2$ are the empirical constants, $Ж_1$ is about 1 to 1.6, and $Ж_2$ is about 2.5 to 5.0. For the linear and narrow width molecular weight polymer, the relationship between M_w and number average molecular weight (M_n) is $M_w \approx M_n$ when $M_w<M_c$; the relationship is $M_w/M_n<2$ when $M_w>M_c$.

From the above calculations, the rheological results of PVAc-MMT were obtained (Table 2). i was all the same regardless of the addition of MMT, it was 0.93 and less than 1, so PVAc and PVAc-MMT were pseudo-plastic Non-Newtonian fluids. PVAc-MMT's Ï, ζ, η_0, M_w, and M_n respectively were 14 mPa·s, 1.4×10^2 mPa·s, 1.1×10^2, and 1.1×10^2. ζ is a characteristics constant of materials, it is always expressed by time, usually measured in seconds. PVAc's ζ was 4.6×10^{12} s, and PVAc-MMT's ζ was 4.8×10^3 s. As previously set, η_∞ was 0 mPa·s all the time whether for PVAc or PVAc-MMT. In other words, MMT could improve the rheology of PVAc. As with other polymers, both PVAc and PVAc-MMT also possessed the normal stress effect that was the pole-climbing phenomenon, it is also called as the Weissenberg effect.

Static tensile

In the static tensile test, the obtained results only were the maximum break force of PVAc and PVAc-MMT films. Many other static tensile data would be calculated from the maximum break force through equations.

The static tensile strength (or static tensile stress) represented by σ (MPa=N·mm^{-2}) is calculated by:

$$\sigma = F/S$$

Where F (N) is the maximum break force and S (mm^2) is

Table 3. Static tensile.

	σ (MPa)	ε (ε_x) (mm·mm^{-1})	ε_y (mm·mm^{-1})	εt (%)	E' (MPa)	ı	O (%)	S (days)
PVAc	2.07	1.92	0.30	192	1.08	0.16	13.35	≥180
PVAc-MMT	3.49	2.03	0.20	203	1.72	0.10	15.01	≥180

Note: Each data in this table was averaged from 20 sets of results.

the cross sectional area of samples.

The break elongation represented by εt (%) and the break strain represented by ε (mm·mm^{-1}) are calculated by:

$$\varepsilon t = (l - l_0)/l_0 \times 100\%$$

$$\varepsilon = (l - l_0)/l_0$$

Where l_0 (mm) is the original length of samples and l (mm) is the break length of samples.

The elastic modulus is represented by E′ (MPa). It is calculated by Hooke's law :

$$E' = \sigma/\varepsilon$$

Where σ (N·mm^{-2}) is the stress and ε (mm·mm^{-1}) is the strain.

When a material is under tensile or compression, it experiences not only the vertical deformation along longitudinal direction, but also simultaneously the horizontal deformation as it is reduced or increased along a transverse direction. In the elastic deformation range, the horizontal strain (ε_y) is proportional to the vertical strain (ε_x), the ratio is known as Poisson's ratio (ı) of material. It is generally calculated by the following equation:

$$\imath = \left|\frac{\varepsilon_y}{\varepsilon_x}\right|$$

Where $\varepsilon_x = \varepsilon$, $\varepsilon_y = (b - b_0)/b_0$, b_0 (mm) is the original width of samples, and b (mm) is the break width of samples.

The results of static tensile were calculated (Table 3). σ of PVAc-MMT was 3.49 MPa, 1.69 times that of PVAc's 2.07 MPa, and tts ε was 2.03 mm·mm^{-1}, 1.06 times that of PVAc's 1.92 mm·mm^{-1}. Therefore, E' of PVAc-MMT was still higher, it was 1.72 MPa, 1.59 times that of PVAc's 1.08 MPa. εt of PVAc-MMT was 203%, 1.06 times that of PVAc's 192%. Their ε_x were different as well as ε, their ε_y was almost the same, one was 0.30 mm·mm^{-1}, and the other one was 0.20 mm·mm^{-1}, so their ı must be greatly different. ı of PVAc-MMT was 0.10, and PVAc's ı was 0.16. All indicated that the static tensile properties of PVAc-MMT were much better than PVAc. This is because:

(1) MMT was greatly exfoliated into layers or sheets of nano particles, randomly dispersing in PVAc matrix,
(2) The smaller MMT particles were adsorbed around the larger PVAc particles to form the "strawberry" structure,
(3) MMT possessed the small size effect as a natural nano-mineral.

In a word, the static tensile properties of PVAc were greatly improved because of the addition of MMT in polymerization.

Moreover, the solid content of PVAc-MMT was 15.01% (Table 3), a little higher than PVAc. The storage time of PVAc and PVAc-MMT was good. They all were more than 180 days. This showed their storage stability was very great.

Conclusions

In the synthesis, MMT was added and polymerized with VAc. From the analysis of XRD, FTIR, TEM, and static tensile, and the rheology investigated by the power-law function equation and the Cross-Williamson model viscous equation, the following conclusions were obtained:

(i) The exfoliated nano-composite of PVAc-MMT was obtained. The linear macromolecular chains of PVAc were formed in MMT layers. There was no chemical bond between MMT and PVAc, but physical effect. MMT was exfoliated into layers or sheets of nano-particles, randomly dispersing in PVAc matrix.
(ii) The dispersion of PVAc-MMT was better than PVAc. The particle diameter of PVAc was from 250 to 500 nm and MMT's was from 50 to 100 nm. They randomly dispersed together, and the smaller MMT particles were adsorbed around the larger PVAc particles to form the "strawberry" structure.
(iii) MMT could improve the rheology of PVAc. PVAc and PVAc-MMT were pseudo-plastic Non-Newtonian fluids, their i was 0.93, and their storage stability was very good. Both PVAc and PVAc-MMT possessed the normal stress effect (or Weissenberg effect) that was the pole-climbing phenomenon.
(iv) The static tensile properties of PVAc-MMT were much better. Its E' was 1.69 times that of PVAc, its εt was 1.06 times that of PVAc, its ı was 0.10, and PVAc's ı was

0.16.

ACKNOWLEDGEMENTS

The authors are grateful for the financial support of the Program for New Century Excellent Talents in University of the Ministry of Education of China (NCET-06-0825).

REFERENCES

Chen K, Yang RC, Cheng SW (2006). Research on tribologicalbehavior of OMMT/polypropylene composites. Tribol., 26: 561-565.

Cui HW, Du GB, (2009). Influences of long alkane chain quaternary ammonium salts on the structure of layers in OMMT. Non-Met. Min., 32: 17-20.

Cui HW, Du GB (2011). Preparation and characterization of PVAc-MMT-DOAB. High Perform. Polym., 23: 40-48.

Cui J, Yu L, Zhang C, Mi X, Fu T, Na H (2007). Studies on epoxy resin/montmorillonite nanocomposites cured with 4,4'-diamino diphenyl ether diphenyl ketone. Chem. J. Chinese U., 28: 592-595.

Deng X, Liu F, Luo Y, Chen Y, Jia D (2007). Preparation, structure and properties of comb-branched waterborne polyurethane/OMMT nanocomposites. Prog. Org. Coat., 60: 11-16.

Ding Y, Zhang Z, Liu Y, Shi T (2002). Preparation and characterization of PVAc/montmorillonite nanocomposites initiated by radiation polymerization. Acta. Polym. Sin., 18: 504-508.

Ding Y, Zhang Z (2004a). Study on the blend of HDPE and PVAc-montmorillonite nanocoposite prepared via -ray irradiation. J. Radiat. Res. Radiat. Process., 22: 193-196.

Ding Y, Zhang Z (2004b). Studies on the blend composed of HDPE, PA6 and PVAc-montmorillonite nanocomposite prepared via γ-ray irradiation. Chem. J. Chinese U., 25: 1351-1354.

Ding Y, Zhang Z (2005). Effects of MMT nanoparticles on the crystallinity of HDPE in PVAc-MMT/PA-6/HDPE blends. Polym. Mater. Sci. Eng., 21: 256-259.

Esposito CC, Prinari P, Cannoletta D, Mensitieri G, Maffezzoli A (2008). Synthesis and characterization of clay-nanocomposite solvent-based polyurethane adhesives. Int. J. Adhes. Adhes., 28: 91-100.

Essawy H, Badran A, Youssef A, Abd El-Hakim AE (2004). Synthesis of poly(methyl methacrylate)/montmorillonite nanocomposites via in situ intercalative suspension and emulsion polymerization. Polym. Bull., 53: 9-17.

Kommann X, Lindberg H, Berglund LA (2001). Synthesis of Epoxy-clay Nanocomposites: influence of the nature of the clay on structure. Polym., 42: 1303-1310.

Metin Tanoğlu, Yelda Ergün (2007). Porous nanocomposites prepared from layered clay and PMMA[poly(methyl methacrylate)]. Compos., Part A 38: 318-322.

Qin H, Shi-min, Chungui Z, Hu G, Yang M (2005). Flame retardant mec Zhang hanism of polymer/clay nanocomposites based on polypropylene. Polym., 46: 8386-8395.

Ray S, Bhowmick AE (2001). Synthesis characterization and properties of montmorillonite clay-polyaerylate hybrid material and its effect on the properties of engagechy hybrid composite. Rubber Chem. Technol., 74: 835-845.

Suprakas SR, Mosto B (2005). Biodegradable polymers and their layered silicate nanocomposites: In greenng the 21st century materials world. Prog. Mater. Sci., 50: 962-1079.

Ye L, Zhong A, Chen D, Yang F, An B (2004). Preparation and rheologic behavior of MMT laminating adhesive. Chem. Res. Appl., 16: 219-221.

Zhao C, Qin H, Gong F, Feng M, Zhang S, Yang M (2005). Mechanical, thermal and flammability properties of polyethylene/clay nanocomposites. Polym. Degrad. Stab., 87: 183-189.

Zhao H, Yang F, Zhang X, Li G, Cai H, Chen B, Feng Z, Huang B (2004). Preparation and characterization of polyethylene/montmorillonite nanocomposites via *in situ* polymerization. Polym. Mater. Sci. Eng., 20: 63-65.

Biosorption of copper(II) and lead(II) ions from aqueous solutions by modified loquat (*Eriobotrya japonica*) leaves (MLL)

Akl M. Awwad[1]* and Nidà M. Salem[2]

[1]Royal Scientific Society, P. O. Box 1438 Al-Jubaiha 11941 Amman, Jordan.
[2]Faculty of Agriculture, Jordan University, Amman, Jordan

A modified loquat leaves (MLL) for the removal of copper(II) and lead(II) ions from aqueous solutions has been investigated. The biosorption of copper(II) and lead(II) ions was found to be dependent on solution pH, initial metal ion concentrations, biosorbent dose, contact time, and temperature. The experimental equilibrium biosorption data were analyzed by three widely used two-parameters; Langmuir, Freundlich, and Temkin isotherm models. Langmuir isotherm model provided a better fit with the experimental data than other models by high correlation coefficients R^2. Kinetic studies showed that pseudo-second order described the biosorption experimental data better than the pseudo-first order kinetic model.

Key words: Biosorption, loquat leaves, copper, lead, adsorption isotherms, kinetic.

INTRODUCTION

Copper(II) and lead(II) ions have been released to the environment through electrical wiring, air conditioning tubing, plumbing, electroplating and pigments. There is a growing demand to find relatively efficient, low cost and easily available adsorbents for the adsorption of toxic heavy metals. Different agro-wastes have been investigated for the biosorption of Cu(II) and Pb(II) ions such as sugar beet pulp treated with NaOH and citric acid (Altundogan et al., 2007), *pinus pinaster* bark (Vàzquez et al., 1994) apple residues (Lee et al., 1998) agricultural waste sugar beet pulp (Aksu and İşoğlu, 2005), black carrot residues (Güzel et al., 2008), acid treated sugar cane bagasse (Sousa et al., 2009), dehydrated wheat bran (Őzer et al., 2004), wheat shell (Basici et al., 2004), sour orange residue (Khormaei et al., 2007), bagasse pith based sulphurised activated carbon (Krishnan and Anirudhan, 2008), chemically modified orange peel (Ning-chan et al., 2009), lemon peel (Bhatanagar et al., 2010), pine cone powder (Ofomaja et al., 2010), biomatrix from rice husk (Krishnani et al., 2008).

In the present work, We have studied the potential of copper(II) and lead(II) biosorption on a agro material which is a modified loquat leaves (MLL) coming from loquat fruit tree waste. Results from this study can be used to assess the utility of MLL for copper(II) and lead(II) removal from water and industrial wastewaters.

*Corresponding author. E-mail: aklm@rss.gov.jo;amawwad2000@yahoo.com.

MATERIALS AND METHODS

Adsorbent

The raw loquat leaves (LL) was collected from a local plantation, Amman, Jordan. This agricultural waste was thoroughly rinsed with water to remove dust and soluble material. Then it was allowed to dry at room temperature. The dried waste was grounded to a fine powder in a grinding mill (Retsch RM 100) and sieved to get size fraction <44 μm. The fine powder was then treated with sodium hydroxide (0.2 mol/L) to improve the biosorption capacities to copper and lead ions. For this purpose, 100 gram of dried LL soaked in solution 500 mL NaOH (0.2 mol/L) for 24 h. By filtration, the sodium hydroxide modified loquat leaves (MLL) was washed with de-ionized water until the pH value of the solution reached 7.0, and then dried in an oven at 60°C for 24 h.

Materials

All the chemicals used were of analytical reagent (AR) grade. Stock solutions of 1000 mg/L of copper(II) and lead(II) was prepared from copper nitrate [Cu $(NO_3)_2$. $3H_2O$] and lead nitrate [Pb $(NO_3)_2$], respectively using double distilled water. Desired test solutions of Cu (II) and Pb (II) ions were prepared using appropriate subsequent dilutions of the stock solution. The range of concentrations of Cu (II) and Pb (II) ions prepared from standard solution varies between 10 and 100 mg/L. Before mixing the adsorbent, the pH of each test solution was adjusted to the required value with 0.1 M NaOH or 0.1 M HCl.

Analysis

The concentrations of Cu(II) and Pb(II) ions in the solutions before and after equilibrium were determined by AAS6300 Atomic absorption spectrometer (Shimadzu, Japan). The pH of the solution was measured with a WTW pH meter using a combined glass electrode. Fourier Transform Infrared Spectroscopy, FTIR (IR Prestige-21, Shimadzu, Japan) was used to identify the different chemical functional groups present in the modified loquat leaves (MLL). FTIR analyses also used to determine the functional groups which are responsible for the metal binding with MLL. The analysis was carried out using KBr and the spectral range varying from 4000 to 400 cm^{-1}.

Biosorption experiments

Batch biosorption experiments were conducted by mixing biosorbent with Cu(II) or Pb(II) ion solutions with desired concentration in 250 ml glass flask. The glass flasks were stoppered during the equilibration period and placed on a temperature controlled shaker at a speed of 120 r/min. The effect of pH on the equilibrium biosorption of Cu(II) and Pb(II) was investigated by mixing The amount of biosorption was calculated based on the difference between the initial (C_o, mg/L) and final concentration (C_e, mg/L) in every flask, as follows:

$$q_e = \frac{C_o - C_e}{M} xV \quad (1)$$

Where q_e is the metal uptake capacity (mg/g), *V* the volume of the metal solution in the flask (L) and *M* is the dry mass of biosorbent (g).

RESULTS AND DISCUSSION

FTIR analysis

To investigate the functional groups of MLL and metal loaded LL, a FT-IR study was carried out and the spectra are shown in Figure 1a and b. The modified loquat leaves (MLL) displays a number of absorption peaks, reflecting the complex nature of the loquat leaves. A strong and broad peak at 3410 cm^{-1} results due to the stretching of the N–H bond of amino groups and indicative of bonded hydroxyl group. A change in peak position to 3360 cm^{-1} Figure 1b in the spectrum of the metal loaded loquat leaves, indicates the binding of copper or lead with amino and hydroxyl groups (Nakbanpote et al., 2007; Munagapati et al., 2010). The strong absorption peak at 2927 cm^{-1} could be assigned to –CH stretching vibrations of $-CH_3$ and $-CH_2$ functional groups. The peak at 1621 cm^{-1} indicates the fingerprint region of CO, C–O and O–H groups, which exists as functional groups of MLL. Shifting of this peak to 1651 cm^{-1}, indicated involvement of these groups in metal binding. The region between 1520 and 1000 cm^{-1} is the fingerprint region, OH, and C-H bending vibration and C-O stretching vibration absorption bands The absorption peaks at 1442, 1377 and 1319 cm^{-1} could be attributed to the presence of C–O stretching,. These peaks were shifted to 1519, 1450 and 1373 cm^{-1}, respectively. The intense band at 1060 cm^{-1} can be assigned to the C-O of alcohols and carboxylic acids. The shift of the peak from 1060 to 1035 cm^{-1} also suggests the involvement of C–O group in binding Cu (II) (Nakbanpote et al., 2007; Munagapati et al., 2010). The shifts in the absorption peaks generally observed indicate the existence of a metal binding process taking place on the surface of the loquat bark.

Effect of pH

To find the optimal pH for the effective biosorption of Cu(II) and Pb(II) onto MLL It was found the uptake of metal ions increased with the increase in pH from 2.0 to 6.0. At lower pH values Cu(II) and Pb(II) removal was inhibited, possibly as a result of the competition between hydrogen and metal ions on the sorption sites, with an apparent preponderance of hydrogen ions, which restricts the approach of metal cations as in consequence of the repulsive force. As the pH increased, the active sites in MLL would be exposed, increasing the negative charge density on the MLL surface, increasing the attraction of metal ions with positive charge and allowing the biosorption onto the MLL surface. In this study, the Cu(II) and Pb (II) ions at pH 6.0 would be expected to interact more strongly with the negatively charged binding sites in the biosorbent, Figure 2. The decrease in Cu(II) and Pb(II) biosorption above pH 6.0 was probably due to the precipitation of Cu(II) and Pb(II) ions as copper and lead hydroxides and not due to biosorption. As a result, the optimum pH for copper and lead ions biosorption was found as 6.0 and the other biosorption experiments were performed at this pH value.

Effect of contact time

The data obtained from the biosorption of Cu(II) and Pb(II) ions onto MLL powder showed that a contact time of 60 min was sufficient to achieve equilibrium, Figure 3 and the biosorption did not change significantly with further increase in contact time. Therefore, the uptake, q_e (mg/g) and the adsorbed metal concentrations C_e, (mg/L) at the end of 60 min are given as the equilibrium values

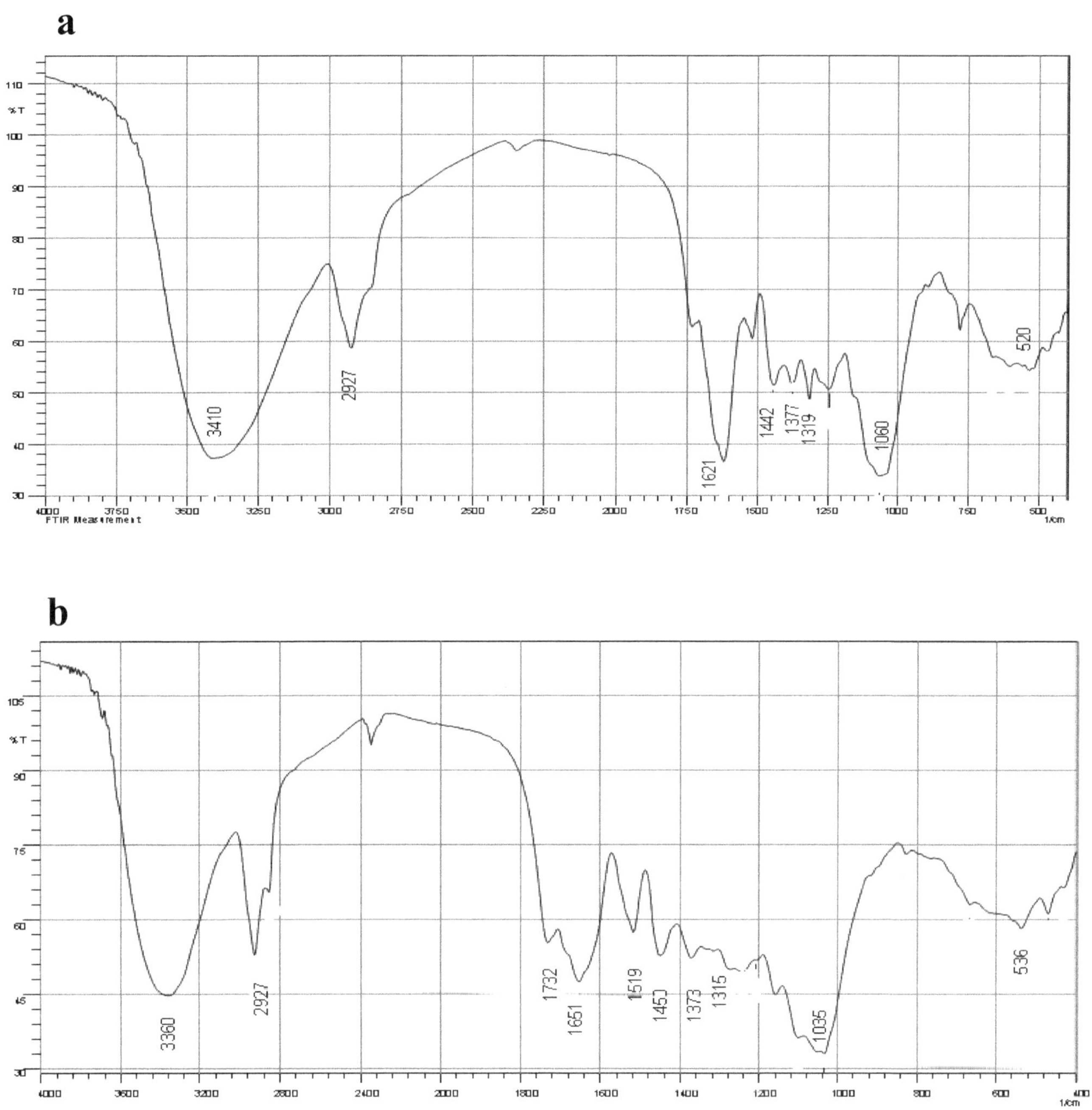

Figure 1. FTIR of (a) MLL and (b) Cu(II) loaded MLL.

and the other biosorption experiments were conducted at this contact time of 60 min.

It was found up to 400 mg/L initial metal concentrations, the amount of adsorbed Cu(II) and Pb(II) ions per unit mass of the MLL increased. These data is reasonable and indicating the presence of many active sites on MLL surface available to bind with Cu(II) and Pb(II) ions at low concentration. Beyond the initial metal ion concentrations 400 mg/L, the adsorption capacity of MLL biosorbent remained relatively constant. This is probably due to saturation of metal ion binding sites at MLL surface at concentration higher than 300 mg/L.

Effect of adsorbent concentration

The adsorbent concentration effect on the percent removal at equilibrium conditions was investigated. It was observed that the amount of Cu(II) and Pb(II) biosorbed varied with varying adsorbent concentration. The amount

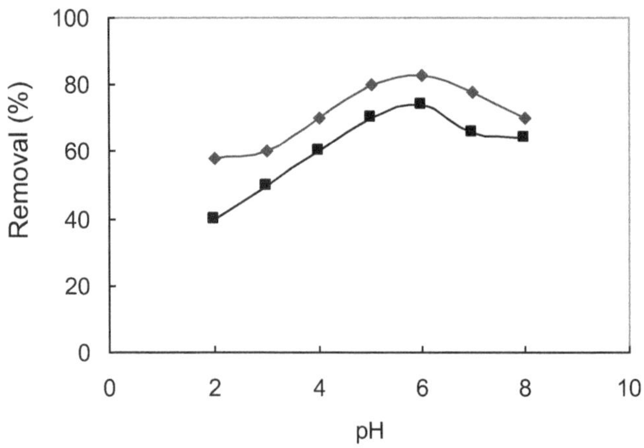

Figure 2. Effect of pH on biosorption of ♦, Cu (II) and ■, Pb (II) at 303 K, initial metal concentration 40 mg/L, agitation time 120 rpm, MLL concentration 0.5 g/L.

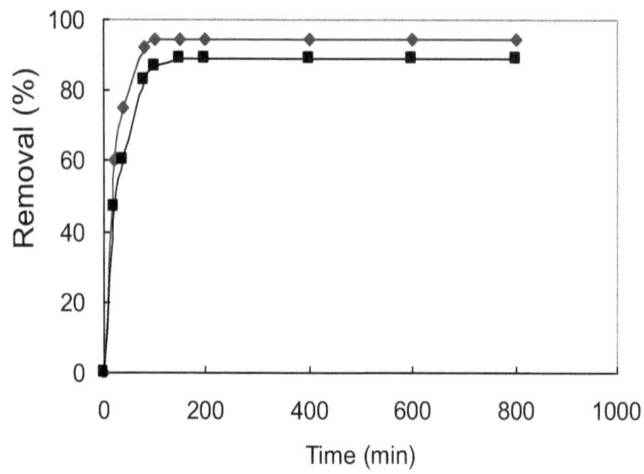

Figure 3. Effect o0f contact time on ♦, Cu (II) and ■, Pb (II) biosorption onto MLL at 303 K. C_o = 40 mg/L, pH = 6.0, MLL dose 0.5 g/L. Effect of metal ion concentration.

of Cu(II) and Pb(II) adsorbed increases with an increase in adsorbent concentration from 0.1 to 0.5 g. The percentage of metal removal was increased from 60.32 to 98.88% for an increase in MLL concentration from 0.1 to 0.5 g at initial concentration of 40 mg/L. The increase in the adsorption of the amount of solute is obvious due to increasing MLL surface area.

Biosorption isotherms

An adsorption isotherm describes the fraction of sorbate molecules that are partitioned between liquid and solid phases at equilibrium. Adsorption of Cu(II) and Pb(II) ions onto MLL particles was modeled using four adsorption isotherms:

Freundlich isotherm

The Freundlich isotherm model is the well known earliest relationship describing the adsorption process. This model applies to adsorption on heterogeneous surfaces with the interaction between adsorbed molecules and the application of the Freundlich equation also suggests that sorption energy exponentially decreases on completion of the sorptional centers of an adsorbent. This isotherm is an empirical equation and can be employed to describe heterogeneous systems and is expressed as follows in linear form (Freundlich, 1939).

$$\log q_e = \log K_F + \frac{1}{n}\log C_e \quad (2)$$

Where K_F is the Freundlich constant related to the bonding energy. $1/n$ is the heterogeneity factor and n (g/L) is a measure of the deviation from linearity of adsorption.

Freundlich equilibrium constants were determined from the plot of log q_e versus log C_e , Figures 4 and 5 on the basis of the linear of Freundlich Equation (2). The n value indicates the degree of non-linearity between solution concentration and adsorption as follows: if $n = 1$, then adsorption is linear; if $n < 1$, then adsorption is a chemical process; if $n > 1$, then adsorption is a physical process. The n value in Freundlich equation was found to be 2.26-3.49 for MLL, Table 1. Since n lie between 1 and 10, this indicate the physical biosorption of copper(II) and lead(II) onto MLL. The values of regression coefficients R^2 are regarded as a measure of goodness of fit of the experimental data to the isotherm models.

Langmuir isotherm

The Langmuir isotherm assumes monolayer adsorption on a uniform surface with a finite number of adsorption sites (Langmuir, 1918). Once a site is filled, no further sorption can take place at that site. As such the surface will eventually reach a saturation point where the maximum adsorption of the surface will be achieved. The linear form of the Langmuir isotherm model is described as:

$$\frac{C_e}{q_e} = \frac{1}{K_L q_{\max}} + \frac{1}{q_{\max}} Ce \quad (3)$$

Where K_L is the Langmuir constant related to the energy of adsorption and q_{max} is the maximum adsorption capacity (mg/g). Values of Langmuir parameters q_{max} and K_L were calculated from the slope and intercept of the

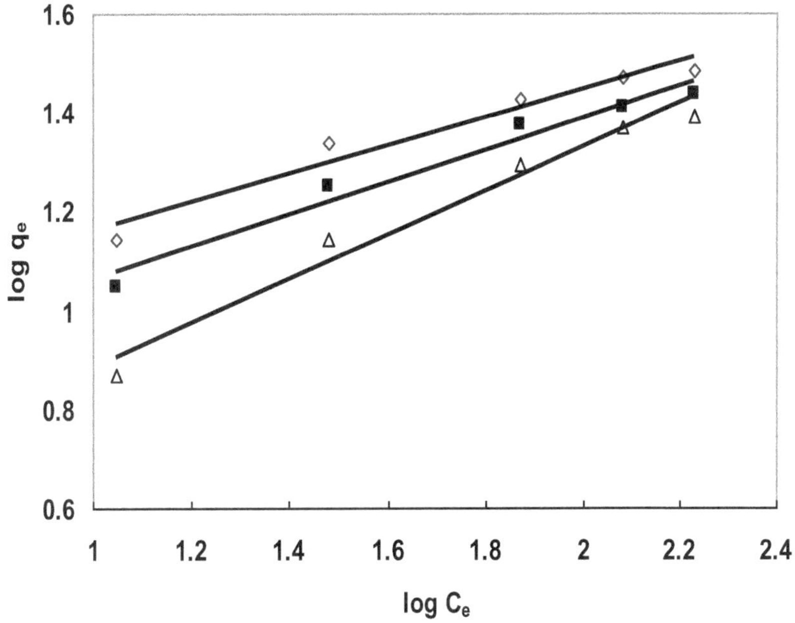

Figure 4. Freundlich adsorption isotherm of Cu(II) onto MLL at ◊, 293 K; ϒ, 303 K, Δ, 313 K.

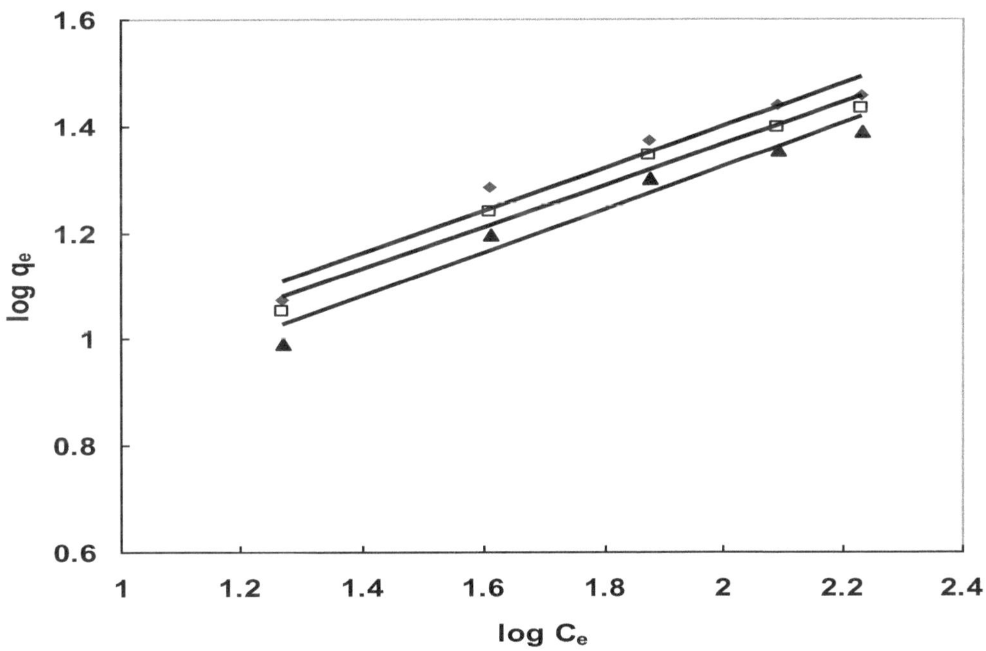

Figure 5. Freundlich adsorption isotherm of Pb (II) onto MLL at ♦, 293 K; ϒ, 303 K, ▲, 313 K.

linear plot of C_e/q_e versus C_e as shown in Figures 6 and 7.

Values of q_{max}, K_L and regression coefficient R^2 are listed in Table 1. These values for MLL biosorbent indicated that Langmuir theory describes the biosorption phenomena favorable.

The essential characteristics of the Langmuir isotherm parameters can be used to predict the affinity between the sorbate and sorbent using separation factor or dimensionless equilibrium parameter, R_L expressed as in

Table 1. Langmuir, Freundlich, and Temkin constants for biosorption of Cu(II) onto MLL at different temperatures.

T(°C)	Q_m (mg/g)	*b* (mg/g)	R^2
		Langmuir	
20	33.33	0.062	0.9994
30	29.67	0.051	0.9996
40	28.33	0.032	0.9997
	n	K_f	R^2
		Freundlich	
20	3.49	7.49	0.9501
30	3.06	5.53	0.9639
40	2.26	2.78	0.9713
	b_T	K_T	R^2
		Temkin	
20	396.43	1.0	0.9838
30	434.26	1.0	0.9924
40	402.13	0.99	0.9981

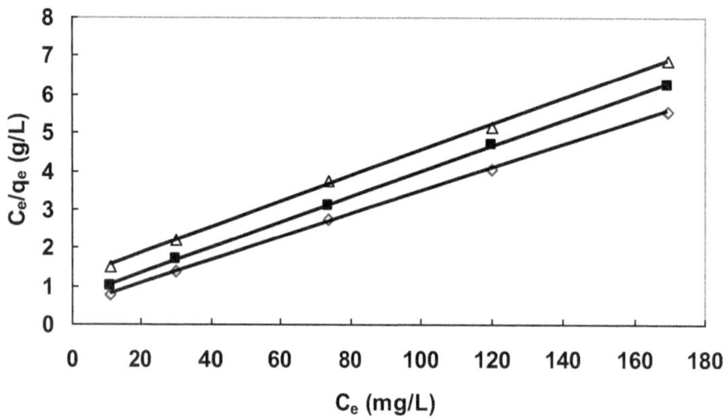

Figure 6. Langmuir adsorption isotherms for Cu (II) ions onto MLL at ◊, 293 K; ■, 303 K, Δ, 313 K.

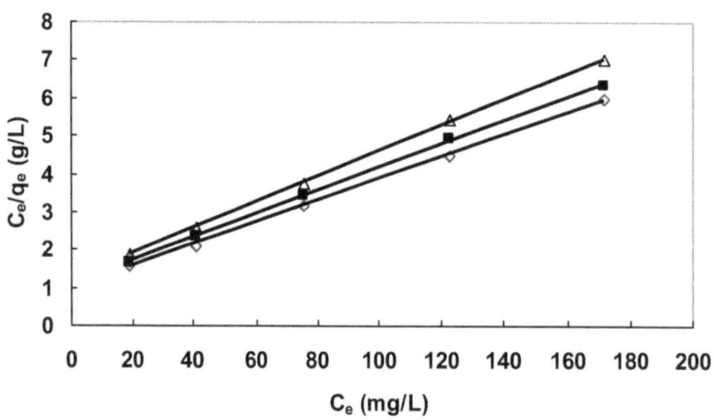

Figure 7. Langmuir adsorption isotherms for Pb (II) ions onto MLL at ◊, 293 K; ■, 303 K, Δ, 313 K.

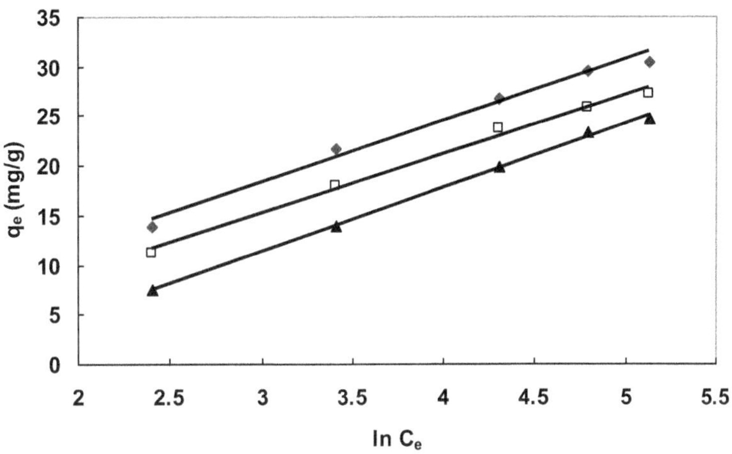

Figure 8. Temkin adsorption isotherms for Cu (II) ions onto MLL at ♦, 293 K;ϒ, 303 K, ▲, 313 K.

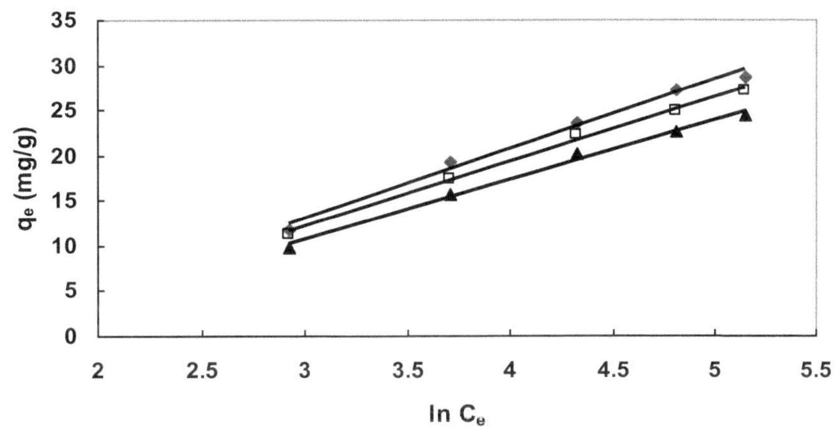

Figure 9a. Temkin adsorption isotherms for Pb (II) ions onto MLL at ♦, 293 K;ϒ, 303 K, ▲, 313 K.

the following equation (Őzer et al., 2004):

$$R_L = \frac{1}{(1 + K_L C_o)} \quad (4)$$

Where K_L is the Langmuir constant and C_o is the initial concentration of nickel(II) ions. The value of separation parameter R_L provides important information about the nature of adsorption. The value of R_L indicated the type of Langmuir isotherm to be irreversible ($R_L = 0$), favorable ($0 < R_L < 1$), linear ($R_L = 1$) or unfavorable ($R_L > 1$).

The R_L was found to be 0.25 to 0.57 for concentration of 10 to 100 mg/L of Cu(II) and Pb(II). They are in the range of 0 to 1 which indicates the favorable biosorption.

Temkin isotherm

The Temkin isotherm model assumes the adsorption energy decreases linearly with the surface coverage due to adsorbent-adsorbate interactions. The linear form of Temkin isotherm model (Aharoni and Ungarish 1977; Boparai et al., 2011) is defined by:

$$q_e = RT \ln K_T + \frac{RT}{b_T} \ln C_e \quad (5)$$

Where b_T is the Temkin constant related to heat of sorption *(J/mol)* and K_T is the Temkin isotherm constant (L/g). These constants were obtained from plotting q_e versus ln C_e, Figures 8 and 9a. Values of b_T and K_T are listed in Table 1.

Adsorption kinetic

Parameters from two kinetic models, pseudo first-order

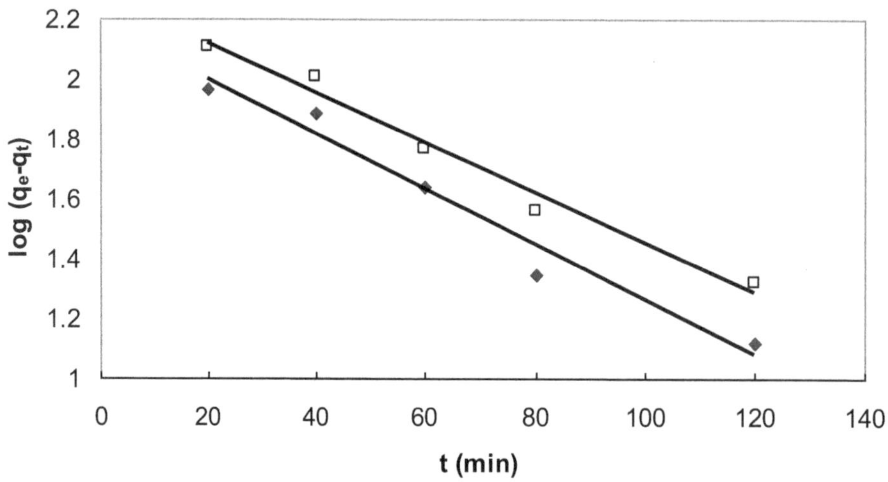

Figure 9b. Pseudo-first order kinetic model for □,Cu(II) and ♦, Pb(II) biosorption onto MLL at 303 K.

and pseudo second-order were fit to experimental data to examine the adsorption kinetics of cadmium uptake by MLL.

Pseudo first-order kinetics

The pseudo-first order equation of Lagergren is generally expressed as follows:

$$\frac{dq_t}{dt} = k_1(q_e - q_t) \quad (6)$$

Where q_e and q_t are the sorption capacities at equilibrium and at time t, respectively (mg/g) and k_1 is the rate constant of pseudo-first order sorption, (1/min). After integration and applying boundary conditions, $q_t = 0$ to $q_t = q_t$ at $t = 0$ to $t = t$; the integrated form of Equation (6) becomes:

$$\log(q_e - q_t) = \log q_e - \frac{k_1}{2.303}t \quad (7)$$

The equation applicable to experimental results generally differs from a true first order equation in two ways: The parameter $k_1(q_e - q_t)$ does not represent the number of available sites; and the parameter log q_e is an adjustable parameter which is often not found equal to the intercept of a plot of log ($q_e - q_t$) against t, whereas in a true first order sorption reaction log q_e should be equal to the intercept of log($q_e - q_t$) against t. In order to fit Equation (7) to experimental data, the equilibrium sorption capacity, q_e must be known. In many cases is unknown and as chemisorption tends to become unmeasurably slow, the amount sorbed is still significantly smaller than the equilibrium amount. In most cases in the literature, the pseudo-first order equation of Lagergren does not fit well for the whole range of contact time and is generally applicable over the initial 20 to 30 min of the sorption process. Furthermore, one has to find some means of extrapolating the experimental data to $t = \infty$, on treating q_e as an adjustable parameter to be determined by trial and error. For this reason, it is therefore necessary to use trial and error to obtain the equilibrium sorption capacity, in order to analyze the pseudo-first order model kinetics.

The pseudo first order rate constant can be obtained from the slope of plot between log ($q_e - q_t$) against time, t. Figure 9b shows the Lagergren pseudo-first order kinetic plot for the adsorption of copper onto MLL powder. The pseudo first order rate constant values were calculated from the slope of Figure 9b. The calculated values and their corresponding linear regression correlation coefficient values are shown in table 3. The linear regression correlation coefficient value found 0.9747, which shows that this model cannot be applied to predict the adsorption kinetic model.

Pseudo-second order kinetics

The pseudo second-order rate expression, which has been applied for analyzing chemisorption kinetics rate (Ho, 2006) is expressed as:

$$\frac{dq_t}{dt} = k(q_e - q_t)^2 \quad (8)$$

Where q_e and q_t are the sorption capacity at equilibrium and at time t, respectively (mg/g) and k is the rate constant of pseudo-second order sorption, (g/mg min).

Table 2. Langmuir, Freundlich, and Temkinm constants for biosorption of Pb (II) onto MLL at different temperatures.

T(°C)	Q_m (mg/g)	*b* (mg/g)	R^2
		Langmuir	
20	34.60	0.029	0.9993
30	32.48	0.028	0.9996
40	29.47	0.027	0.9997
	n	K_f	R^2
		Freundlich	
20	2.52	4.02	0.9521
30	2.55	3.82	0.9713
40	2.47	3.24	0.962
	b_T	K_T	R^2
		Temkin	
20	318.06	0.996	0.9892
30	353.7	0.996	0.9968
40	395.88	0.997	0.9943

For the boundary conditions to $q_t = 0$ to $q_t = q_t$ at $t = 0$ to $t = t$; the integrated form of Equation (8) becomes:

$$\frac{1}{(q_e - q_t)} = \frac{1}{q_e} + kt \qquad (9)$$

Equation (9) can be rearranged to obtain:

$$q_t = \frac{t}{\frac{1}{kq_e^2} + \frac{t}{q_e}} \qquad (10)$$

The linear form of Equation (10)

$$\frac{t}{q_e} = \frac{1}{kq_e^2} + \frac{1}{q_e}t \qquad (11)$$

Where *t* is the contact time (min), q_e (mg/g) and q_t (mg/g) are the amount of the solute adsorbed at equilibrium and at any time, *t*.

If pseudo-second order kinetics is applicable, the plot of t/q_t versus *t* of Equation (11) should give a linear relationship, from which q_e and *k* can be determined from the slope and intercept of the plot, Figure 10. The pseudo-second order rate constant *k,* the calculated q_e value and the corresponding linear regression correlation coefficient value are given in Table 3. At all initial metal concentrations, the linear regression correlation coefficient R^2 values were higher. The higher values confirm that the adsorption data are well represented by pseudo-second order kinetics.

Desorption and regeneration studies

In order to investigate biosorption on Cu(II) and Pb(II) ions from metal- loaded MLL, the metal–looaded biosorbent was treated with HCl. Desorption studies were performed with different hydrochloric acids concentration. Complete elution of the loaded Cu(II) and Pb(II) 99.24% could be achieved by using 0.1N HCl. To keep the biosorption process cost down, our results indicate that MLL biosorbent could be used repeatedly in Cu(II) and Pb(II) biosorption.

Conclusion

The potential of modified loquat leaves (MLL) for the removal of Cu (II) and Pb (II) ions from aqueous solutions and electroplating wastewater was dependent on biosorption process such as pH, initial metal ions concentrations, biosorbent dose, contact time, and temperature. The equilibrium data have been analyzed using Langmuir, Freundlich and Temkin isotherms. The characteristic parameters for each isotherm and related correlation coefficients, R^2 were determined. The Langmuir biosorption isotherms were demonstrated to provide the best correlation for the biosorption of Cu(II) and Pb(II) ions onto MLL. The maximum monolayer adsorption capacity of MLL was found to be 33.33 mg of

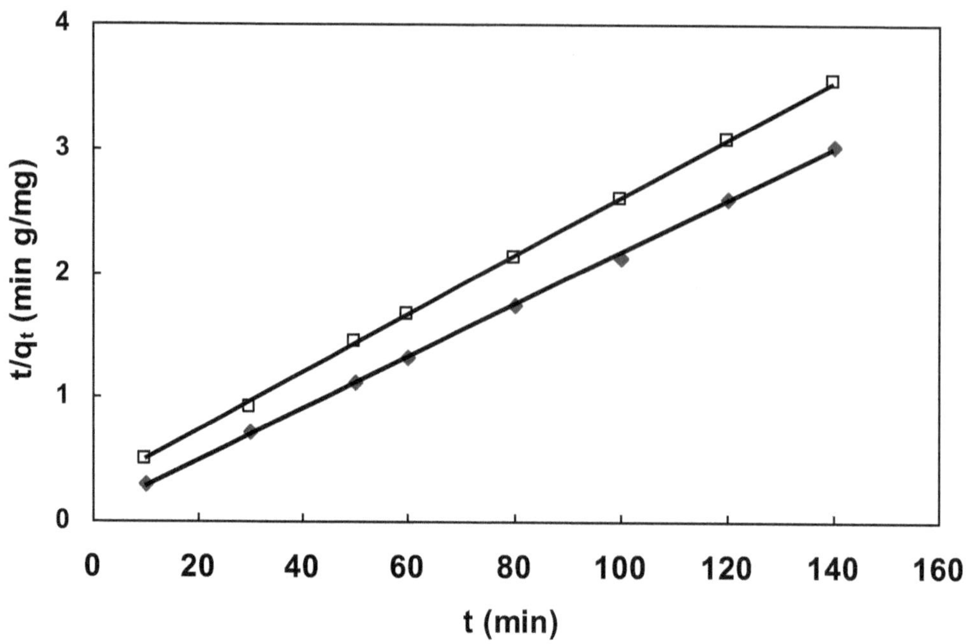

Figure 10. Pseudo-second order kinetic model for □, Cu (II) and ♦,Pb (II) biosorption onto MLL at 303 K.

Table 3. Biosorption kinetic model parameters for Cu (II) and Pb (II) by MLL.

Pseudo-first order		k_1	q_e	R^2
	Cu (II)	0.021	151.64	0.9658
	Pb (II)	0.019	185.35	0.9802
Pseudo-second order		k_2	q_e	R^2
	Cu (II)	0.002	47.85	0.9995
	Pb (II)	0.002	42.74	0.9997

Cu(II)/g of MLL And 34.6 mg of Pb(II)/g of MLL. The kinetic results provided the best correlation of the experimental data of biosorption of Cu(II) and Pb(II) onto MLL by pseudo second-order equation. it can be concluded that since the MLL is an easily, locally available, low-cost adsorbent and has a considerable high biosorption capacity, it may be treated as an alternative adsorbent for treatment of wastewater containing copper(II) and lead(II) ions.

ACKNOWLEDGEMENT

Authors are thankful to Abdul Hameed Shoman Fund for Scientific Research, Jordan for financial support and Royal Scientific Society, Jordan for having given feasibilities to carry out the research work.

REFERENCES

Aharoni A, Ungarish M (1977). Kinetics of activated chemisorption-part 2. Theoretical models . J Chem Soc Faraday Trans, 73: 456-464.

Aksu Z, İşoğlu A (2005). Removal of copper (II) ions from aqueous solution by biosorption onto agricultural waste sugar beet pulp. Process Biochem., 40: 3031-3044.

Altundogan HS, Arslan NE, Tumen F (2007). Copper removal from aqueous solution by sugar beet pulp treated by NaOH and citric acid. J. Hazard. Mater., 149: 432-439.

Basici N, Kocadagistan E, Kocadagistan B (2004). Biosorption of copper (II) from aqueous solutions by wheat shell. Desalination, 164: 135-140.

Bhatanagar A, Minocha AK, Sillanpaa M (2010). Adsorptive removal of cobalt from aqueous solution by utilizing lemon peel as biosorbent. Biochem. Eng. J., 48: 181-186.

Boparai HK, Joseph M, O'Carroll DM (2011). Kinetics and thermodynamics of Cadmium ion removal by adsorption onto nano zerovalent iron particles. J. Hazard. Mater., 186: 458-465.

Freundlich H (1939). Adsorption in solution, J. Amer. Chem. Soc., 61: 2-28.

Güzel F, Yakut H, Topal G (2008). Determination of kinetic and equilibrium parameters of the batch adsorption of Mn(II), Co(II), Ni(II), and Cu(II) from aqueous solution by black carrot (*Daucua carota L.)* residues. J. Hazard. Mater., 153: 1275-1287.

Ho YS (2006). Review of second-order models for adsorption systems. J. Hazard. Mater., 136: 681-689.

Khormaei M, Nasernejad B, Edrisi M, Eslamzadeh T (2007). Copper biosorption from aqueous solution by sour orange residue. J. Hazard. Mater., 149: 269-274

Krishnan KA, Anirudhan TS (2008). Kinetic and equilibrium modelling of cobalt(II) adsorption onto bagasse pith based sulphurised activated carbon. Chem. Eng. J., 137: 257-264.

Krishnani KK, Meng X, Christodoulatos C, Boddu VM (2008). Biosorption mechanism of nine different heavy metals onto biomatrix from rice husk. J. Hazard. Mater., 153: 1222-1234.

Langmuir I (1918). The adsorption of gases on plane surface of glass, mica, and platinum. J. Amer. Chem. Soc., 40: 1361-1403

Lee SH, Jung CH, Chung H, Lee MY, Yang JW (1998). Removal of heavy metals from aqueous solution by apple residues. Process. Biochem., 33: 205-211.

Munagapati VS, Yarramuthi V, Adavala SK, Alla SR, Abburi (2010). Biosorption of Cu(II), Cd(II) and Pb(II) by Acacia leucocephala bark powder: Kinetics and thermodynamics. Chem. Eng. J., 157: 357-365.

Nakbanpote W, Goodman BA, Thiravetyan P (2007). Copper adsorption on rice husk derived materials studied by EPR and FTIR. Colloids and surfaces A: Physicochem. Eng. Aspects, 304: 7-13.

Ning-chan F, Xue-yi G, Sha L (2009). Kinetic and thermodynamic studies on biosorption of Cu(II) by chemically modified orange peel. Trans. Nonferrous Met. Soc. China, 19: 1365-1370.

Ofomaja AE, Naidoo EB, Modise SJ (2010). Dynamic studies and pseudo-second order modeling of copper(II) biosorption onto pine cone powder. Desalination, 251: 112-122.

Őzer A, Őzer D, Őzer A (2004). The adsorption of copper(II) ions on to dehydrated wheat bran (DWB): determination of the equilibrium and thermodynamic parameters. Process Biochem., 39: 2183-2191.

Sousa FW, Sousa MJ, Isadora RN, Oliveira IRN, Oliveira AG, Cavalcante RM, Fechine PBA, Neto VOS, de Keukeleire D, Nascimento RF (2009). Evaluation of a low-cost adsorbent for removal of toxic metal ions from wastewater of an electroplating factory. J. Environ. Manage., 90: 3340-3344

Vàzquez G, Antorrena G, Gonzàlez J, Doval MD (1994). Adsorption of heavy metal ions by chemically modified Pinus pinaster Bark. Bioresour. Technol., 48: 251-255.

Characterization of stainless steel SS-1672

Farah Noori[1]* and Solveig Melin[2]

[1]Physics Science, Baghdad University, Baghdad, Iraq.
[2]Division of Mechanics, Lund University, Lund, Sweden.

This research concerns characterization of stainless steel SS-1672 with 0.46 wt.% carbon concentration, at room temperature with respect to mechanical properties and morphology. Specimens were divided into two groups according to heat treatment, and were hardened at three different temperatures (250, 450 and 650°C) to obtain different martensitic content. Hardness results are compared with those of the normalized condition and simulated by using the finite element method. The results show that hardness decrease with increasing martensitic content. Fatigue tests were performed and simulated by using the finite element method and fatigue life analyses by using the fatigue life prediction method in Ansys were carried out. One of the experiments survived 10^6 cycles at applied stress 241 MPa. This indicates the endurance limit, σ_e, to be slightly above 240 MPa. The microstructure observed by optical microscopic and scanning electron microscope (SEM) showed pearlite and martinsite phases clearly distinguishable. Fractography studies also indicated the nature of typical fracture surfaces at low and high cycle fatigue.

Key words: Fatigue [(Low Cycle Fatigue (LCF) and high cycle fatigue (HCF)], stainless steel, morphology, toughness of metallic alloys.

INTRODUCTION

The mechanical resistance of engineering components to avoid failure in case of repeated loads is dependent on the material together with the type of loading and load level. Repeated stresses might lead to serious consequences in terms of failure due to fatigue. Mechanical fatigue is the dominant source of failure, causing about 90% of collapses in engineering components. The collapses due to fatigue occur under stress levels below collapses under static loading, therefore fatigue testing is a very important mechanical test to pay attention to. A fatigue related failure might occur suddenly without prior warning and the fatigue behavior interpreted by a traditional cycle diagram is useful (Robert, 2005; Yung et al., 2005).

In general, it has been observed that fatigue crack growth can be explained by growth or nucleation of small stage I fatigue cracks in slip bands or in grain boundaries. During repeated loading such cracks extends and crosses nearby grain-boundaries to eventually line up perpendicular to the applied stress to become a stage II crack following the Pairs law (Yung et al., 2005).

Many means for raising the capability of fatigue resistance have been developed, that is, cold rolling or different heat treatments such as annealing, quenching and tempering, depending on the type of application. Medium carbon steels can be used for manufacturing mechanical engineering parts, industrial, medical or civil (Eric, 2006). Case hardening steels are particularly suitable.

Research that address the failure of metals, with focus on the nature of the fatigue mechanisms of fracture, starting from crack initiation and the behavior during growth is thus of utter most importance. Here, the stainless steel SS-1672, which contains 46% carbon steel will be investigated. Two paramters were controlling fatigue failar of steel are chemical composition and morphology. It is possible used of cold-rolled or subjected to heat treatment (annealing, quenching and tempering), depending on the application for which it is intended. It is commonly used for manufacturing

*Corresponding author. E-mail: farah.noori@scbaghdad.edu.iq.

Table 1. Chemical composition of pearlite SS-1672 alloy.

Elements of alloy	C	Si	Mn	P	S
Wt.%	0.46	0.25	0.80	Max. 0.035	Max. 0.030

mechanical parts, such as clutches, springs, saws, valves, measuring tape. Case hardening steels are particularly suitable.

The suitable way to follow the fatigue test when designing is to understand the Wohler's curve of the material used. The Wohler's curve is known as the S-N curve which related between the stress (S) and number of cycles to failure (N). Two different regions are noticeable the high cycle fatigue in the right, has low stresses and elastic deformation corresponding to a high fatigue life and low cycle fatigue in the left, has high stress and plastic deformation and low fatigue life.

EXPERIMENTAL WORK

Material

Characterization of steel SS-1672 is performed with respect to tensile properties, fatigue properties, hardness and microstructure. The chemical composition of the material which examine with the optical emission spectrometer device is seen in Table 1 (Fagersta and Fagersta, 1962).

Heat treatments

Hardness test specimens were manufactured with cylindrical shape with 1 cm diameter and 2 cm height. The specimens were heat treated before the hardness tests using a Linn Electro Them furnace, able to supply a maximum temperature of 1000°C.

For these experiments, the specimens were first heated to a temperature of 850°C and held there for a period of 15 min, where after they were quenched in cold water in order to obtain a transformation to a martinsite phase (Jaykant, 2009).

For tempering, the hardness specimens were heated for periods of 30 min at three different anneal temperatures; 250, 450 and 650°C, and in between the different anneal treatments, the specimens were cooled in air before tested. The tempering was done to remove internal stresses and to get different microscopic phases for the purpose of studying the hardness after each anneal treatment (Jaykant, 2009).

Microstructure examination

Samples for microstructure examination with optical and scanning electron microscope were prepared. The specimens were cut in cylindrical shape with 7 mm diameter and 5 mm height, and to minimize the surface roughness, the samples were grinded using different grades of wet SiC papers; 120, 320, 500, 1200 and 2400 particles/inch, respectively, and finally polished on a rotating disc using a high performance diamond paste, DP-suspension, with a particle diameter of 3 μm. Grinding and polishing were performed using a Steruers Rotopol-2. It works by spreading a grind over the rotating disc and pressing the sample onto the spinning disc. Distilled water and alcohol were used in-between the grinding operations to clean the samples. Etching with Nital (2% HNO3) immersion was carried out during a few seconds to be able to reveal different phases, followed by cleaning with water and alcohol.

Mechanical properties tests

Tensile test

One tensile test was performed. The specimen was manufactured according to the ASTM E8 recommendations (ASM handbook, 1996). The diameter of the specimen was 7 mm and nominal measuring length 10 cm. The specimen was after machining grained with SiC papers with 500 and 1200 particles/inch. The tensile test was executed at room temperature using an universal electronic digital Instron 8500 machine at an elongation speed of 2 mm/min. The Instron machine was calibrated before the test and the calibration curve is seen in Figure 1.

For the calibration, an extensometer was mounted on to a calibration specimen with a gap of 25 mm and the specimen was loaded by a force that was increased in steps and kept constant at each stress level to ensure that the stress level kept constant with time. Initially, the machine was unloaded for 12 s to ensure that no disturbances in the surrounding created a false signal. Thereafter, the following sequence of force levels was applied: 0.449, 1.000 and 1.495 kN during 8.5 s each. As seen from Figure 1, the drift is very slight.

Fatigue testing

Six fatigue experiments were performed and the specimens were manufactured according to the ASTM E1823 recommendations (ASM handbook, 1996). The diameters of all specimens were 7 mm and nominal measuring length 3.5 cm. The specimens were prepared by grinding the surfaces with SiC papers with 500 and 1200 particles/inch. For the fatigue tests a MTS machine was used, applying sinusoidal loading with stress ratio $R= \sigma_{min}/\sigma_{max} = -1$, with σ denoting the amplitude of applied load as shown in Figure 2. All six tests were performed at room temperature at a frequency of 15 Hz This choice of frequency ensured that no thermal softening of the material occurred.

Hardness tests

Rockwell hardness tests conducted with two shapes of indenter for two groups of SS-1672 alloy specimens were performed: Test B (HRB) with a steel sphere with diameter 1/16 inch, indenting at a load of 100 kgf, and test C (HRC) with a 120° diamond cone at a load of 150 kgf. The hardness of each sample was read directly from the testing device.

RESULTS AND DISCUSSION

Tensile test

The tensile test was performed to characterize the tensile

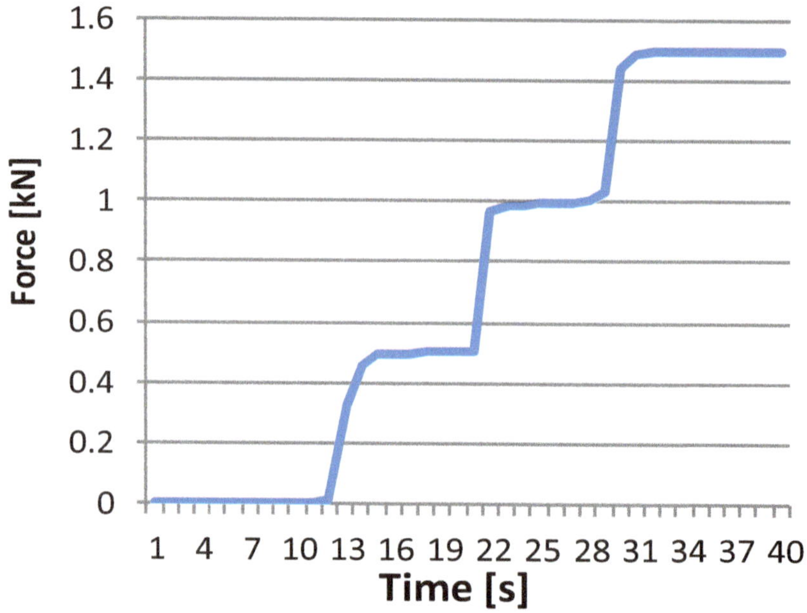

Figure 1. Calibration curve for Instron 8500.

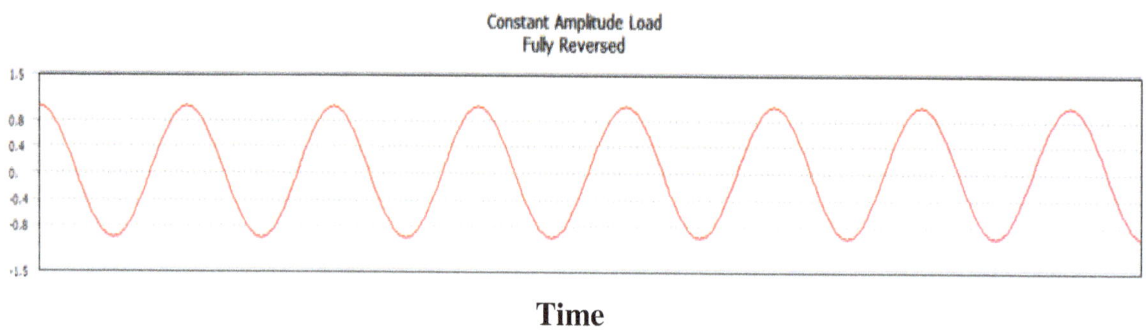

Figure 2. Normalized applied stress as function of time.

behavior of the SS-1672 alloy at room temperature, in addition to extract data from the elastic portion of the stress-strain curve to find proper load levels for the fatigue experiments. The stress-strain curve can be seen in Figure 3. From the tensile curve the yield stress, σ_y, the ultimate strength, σ_T, and the Young's modulus, E, were determined. Further, the relative elongation was determined as the difference between the specimen length before and after rupture, normalized by the original length. The contraction was determined as the ratio between the neck area after rupture and the initial specimen cross section area. In Figure 4, the specimen before and after rupture can be seen. The toughness was determined from the area under the stress-strain curve in Figure 4 as:

$$w = \int_0^{\varepsilon_f} \sigma d\varepsilon \qquad (1)$$

with w denoting the toughness and ε_f the strain upon failure. The result is $w = 18.9$ Nm/m^3.

The values resulting from the tensile test are shown in Table 2. It was observed from Table 2 that the measured value of the mechanical properties as regards yield stress, ultimate strength, Young's modulus, relative elongation and contraction values of the test sample compared well with literature, cf. [7], material number 141650.

Hardness tests

The results of the Rockwell HRB and HRC hardness tests are displayed in Table 3. The values in Table 3 refer to HRB hardness measures of the material before heat treatment and to HRC hardness measures of specimens after hardening heat treatment at 850°C to obtain a

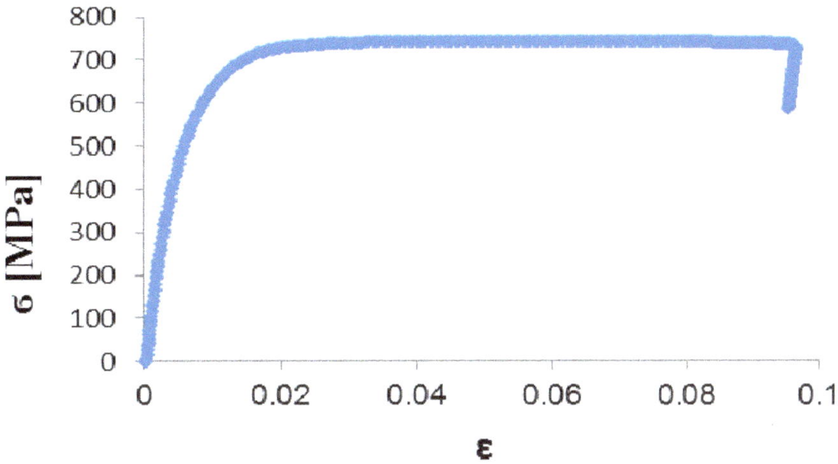

Figure 3. Stress as function of strain for SS-1672.

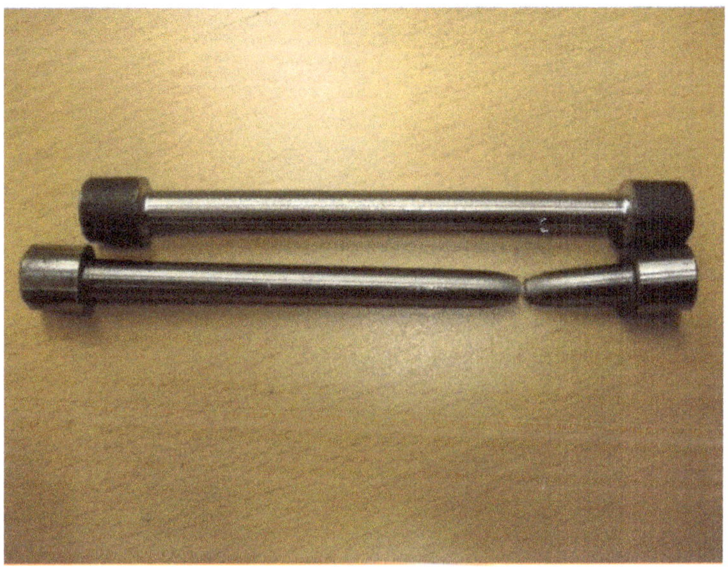

Figure 4. The tensile specimen before and after rupture.

Table 2. Tensile test results.

Mechanical properties	Literature [7]	Experimental
σ_y MPa	860	742
σ_T MPa	>550	594
E GPa	206	206
Relative elongation	9%	9%
Contraction	52%	53%

martinsite phase. Martinsite is a hard and brittle phase, while pearlite is a very soft and ductile phase. Therefore, it is logical to expect an increase in hardness in dual phase martinstic steel as the amount of the softer ferrite matrix is reduced.

Figure 5 shows the variation of HRC hardness of SS-1672 with tempering which consists of heating the hardened steel at different anneal temperatures below hardening temperature. The three measured values were connected by a fitting line using the least square method.

Table 3. The hardness test results.

Pearlite		**HRB**		**Average**
SS-1672	103.6	104.2	103.4	103.73
Martensite		**HRC**		**Average**
SS-1672	61	61.3	61.2	61.15

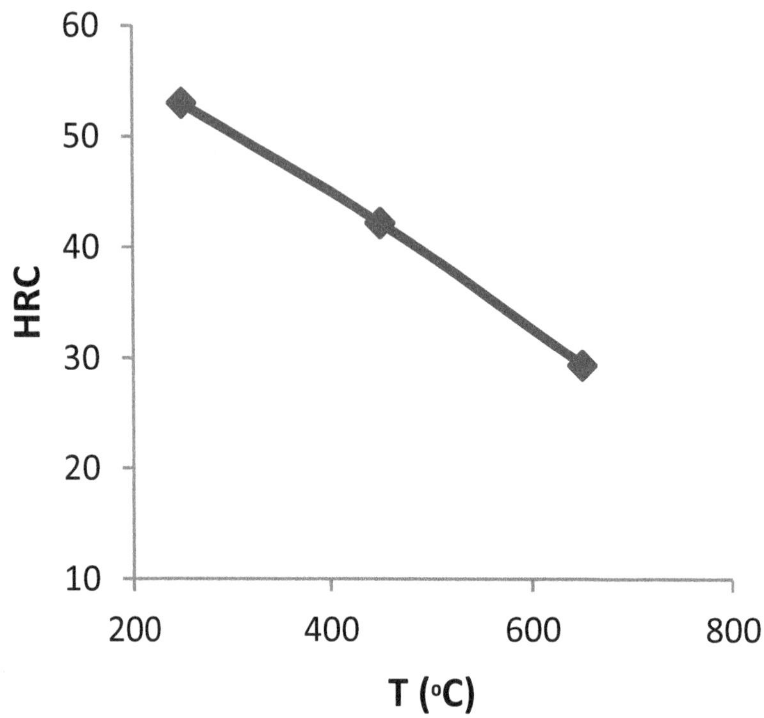

Figure 5. HRC as a function of anneal tempering temperature.

Table 4. Maximum force applied maximum depth and residual depth.

	Experimental (HRC)	**Simulation**
Fmax	1470 N	1451 N
hmax	0.49 mm	0.50 mm
hc	0.47 mm	0.47 mm

It is observed that the hardness decreases with increasing tempering temperature due to the increasing ductility of the martinsite phase induced by tempering. Tempering increases the resistance of the material to fracture and relieves the internal stresses.

The applied loads for the HRB test was 980 N and for the HRC test 1470 N. The maximum depth h_{max} and residual depth h_c after unloading was measured using a Reichert Austria with an accuracy of 0.01 mm. The results of the tests can be seen in Table 4.

Hardness simulation

A simulation with the finite element program Abaqus is used to determine the relationship between the applied force at a diamond cone indenter and the displacement depth in the specimen together with the corresponding stress distribution. A von Mises (Ottosen and Ristinmaa, 2005) material model is used with material parameters taken from Table 2. The model is plane axisysmetric and the element type is 4 node linear square elements and in

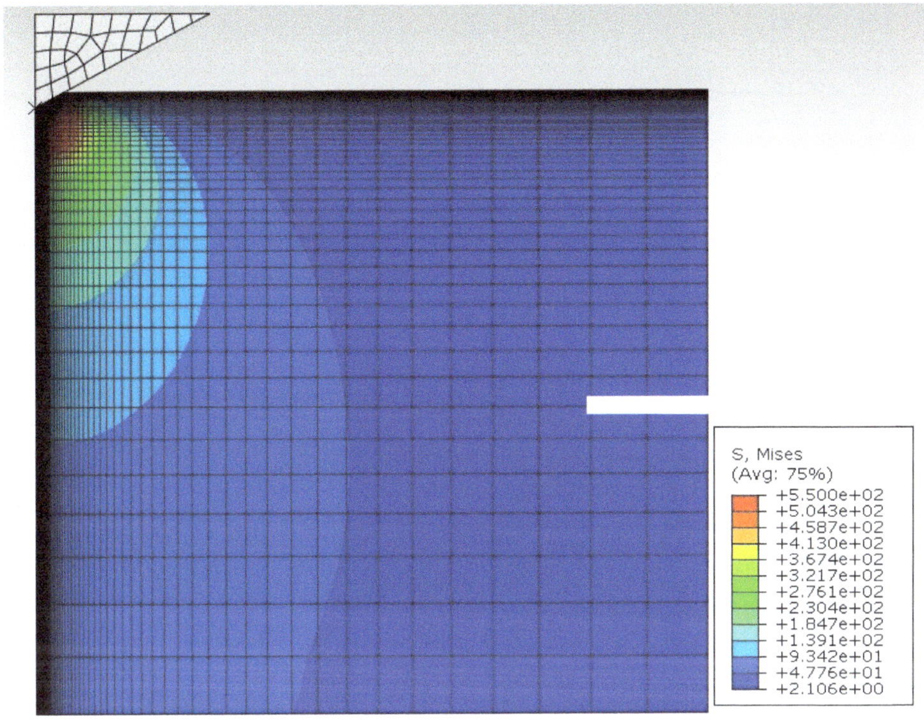

Figure 6. Stress distribution in the hardness specimen from FEM simulation.

Table 5. The stress levels and the number of cycles to failure.

Force level kN	Stress MPa	Number of cycles
21.9	640	543
19.6	546	1060
16.6	431	23049
14.6	348	94482
11.6	301	303084
8.9	241	>10E+6

all 5184 elements were used. The simulated test piece thickness should be at least 10 times the indentation depth (ISO, 2002), as empirically determined.

The exact limits of influence of the boundary conditions on the simulation depend on the geometry of the indenter used and the material properties of the specimen. In this case, the experimental value of h_{max} = 0.5 mm, which means that the model height should be at least 5 mm. Here, the height 5 mm was chosen.

A stiff cone was pressed vertically into the sample in the simulation by applying a force along the symmetry line so that simulated values of h_{max} and h_c came to agree with the experimental results. The experimental and simulation data for maximum force F_{max}, maximum displacement depth h_{max} and residual depth h_c after removal of the applied load are shown in Table 4 for a HRC test for the material as received. It is seen that the agreement is very good.

Figure 6 shows a plot of the simulated stress distribution in the specimen. The yield stress is about 740 MPa according to Table 2, so the simulation predicts the effective plastic zone to extend 0.94 mm along the symmetry line if the yield stress is set equal to the von Mises stress. Thus the plasticity is well contained within the model.

Fatigue tests

The fatigue tests were performed under six constant applied stress amplitudes with R = -1 with test specimens from material as received. The stress levels and the number of cycles to failure at each test are seen in Table 5. One of the experiments survived 10^6 cycles at applied stress 241 MPa. This indicates the endurance limit, σ_e, to be around or slightly above 240 MPa. This agrees exactly

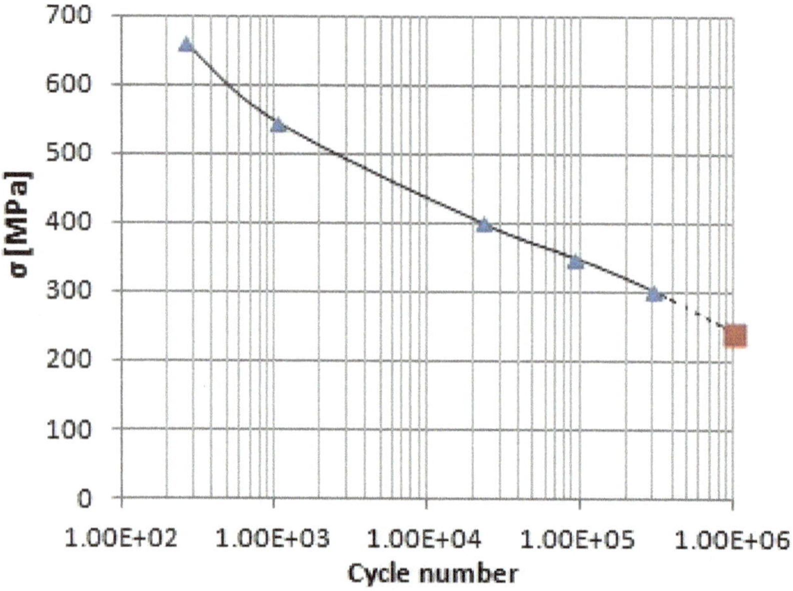

Figure 7. Stress levels as a function to number of cycles, ■ indicates the run-out test.

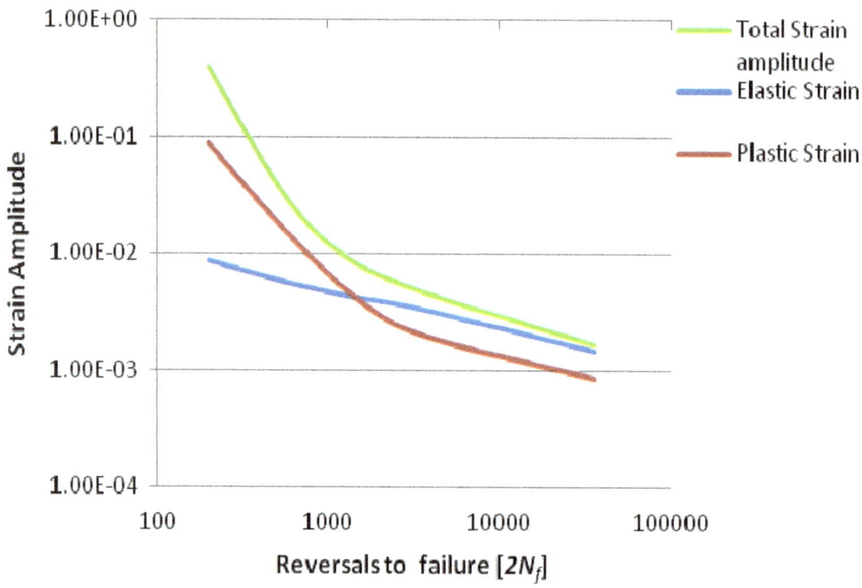

Figure 8. Strain amplitudes as a function of reversals to failure.

with σ_e taken from literature (Handbook in Solid Mechanics, 2008). The stress to number of cycles to failure is shown in Figure 7.

The process of failure is due to the progressively propagation of flaws and initiation of micro cracks under cyclic loading. The process is enhanced by the stress concentrations induced by at such flaws and cracks (Balbi et al., 2009).

Analysis of the fatigue data

The total strain as function of reversals to failure was approximated from the tensile test curve and agrees with the simulated curve from Ansys (FEM model). The total strain amplitude, $\Delta\varepsilon/2$, as well as its elastic, $\Delta\varepsilon_e /2$, and plastic, $\Delta\varepsilon_p /2$, parts versus number of cycles to failure are shown in Figure 8. The elastic and plastic strain

Table 6. The marrow parameters.

Fatigue strength coefficient MPa (σf)	Fatigue ductility coefficient (έf)	Fatigue strength exponent (b)	Fatigue ductility exponent (c)
1154	0.18	-0.06	-0.53

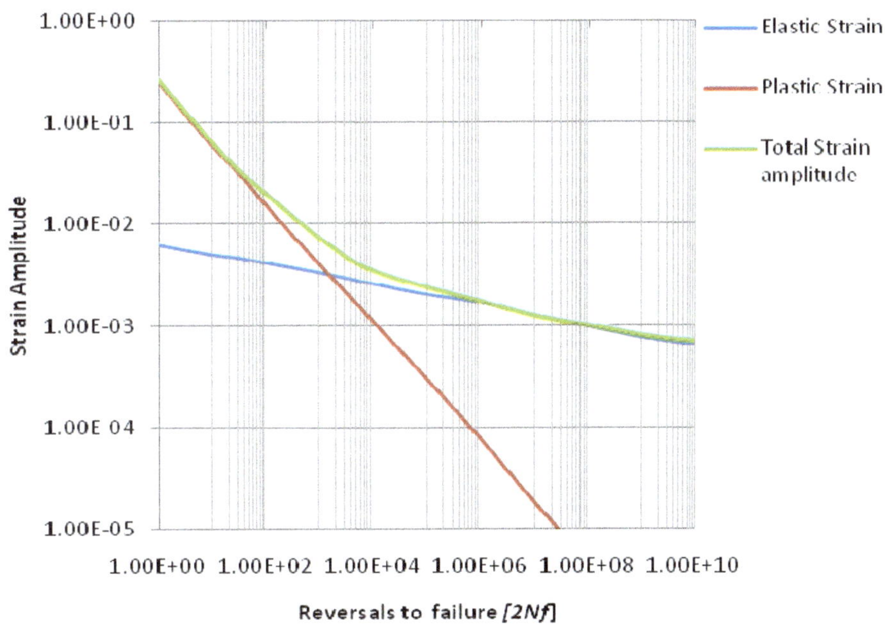

Figure 9. Total, elastic and plastic strain amplitudes as functions of reversals to failure with Excel.

amplitudes are determined from the total strain amplitude by using Equations (2) and (3) (Suresh, 2001; Chiu, 2005):

$$\Delta\varepsilon_e = \frac{\Delta\sigma}{E} \qquad (2)$$

$$\Delta\varepsilon_p = \Delta\varepsilon - \Delta\varepsilon_e \qquad (3)$$

Fatigue simulation

To determine the regions of high and low cycle fatigue, respectively, the transition point can be estimated from the Morrow relation:

$$\frac{\Delta\epsilon}{2} = \frac{\sigma_f'}{E}(2N_f)^b + \epsilon_f'(2N_f)^c \qquad (4)$$

where $2N_f$ is the number of reversals to failure, σ_f' and ϵ_f' are the fatigue strength and the fatigue ductility coefficients, respectively, approximately equal to the true fracture strength and the true fracture strain. The constants *b* and *c* are the fatigue strength and fatigue ductility exponents, respectively. Their values are seen in Table 6 (Kim et al., 2002).

The transition life is obtained from (4) by putting the elastic and plastic parts of the strain amplitudes equal. The result is given by Equation (5):

$$2N_t = \left(\frac{\epsilon_f'}{\sigma_f'} E\right)^{\frac{1}{b-c}} \qquad (5)$$

From the values of Table 6 one obtains:

$2N_t = 1565$

Taking the total strain curve from Figure 8 and approximating the slops at low and high numbers of cycles to failure, the transition life taken from the crossing of the lines is estimated to $2N_t = 1300$. This is shown in Figure 9 and 10 is the result of a finite element simulation of fatigue using the finite element program Ansys as

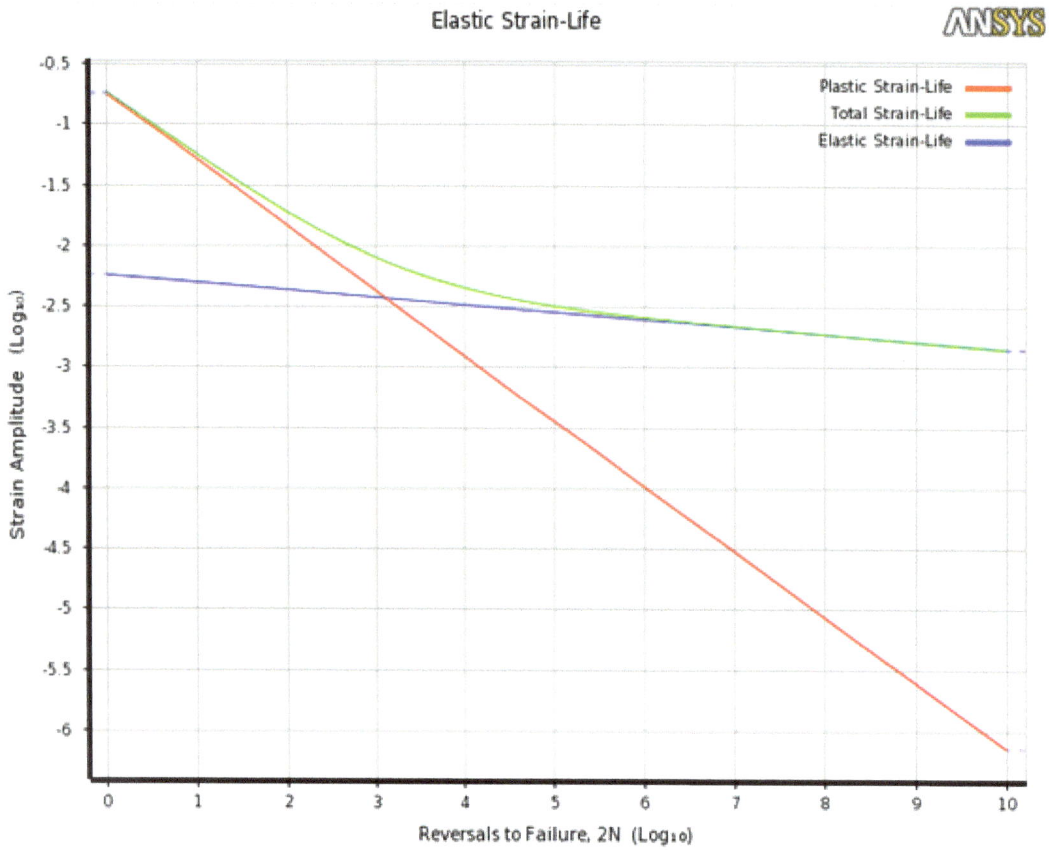

Figure 10. Total, elastic and plastic strain amplitudes as functions of reversals to failure from Ansys analysis.

described subsequently. The estimate from Figure 10 gives $2N_f$ = 1270. These estimates compose well. The value from Equation (5) is somewhat higher, probably due to the somewhat imprecise values of the constants in Table 6, adopted from steel with the same carbon content as SS-1672.

FEM model

Numerical simulation of the dynamic behavior of the sample is better step to predict the building model and understanding the objectives components is used in service during fatigue is performed with the finite element program Ansys and fatigue life analyses by using the fatigue life prediction method in Ansys (Smolnicki et al., 2007). The material constitutive model is the same as for the indentation simulation with values from Table 2. The element type is 8 node linear hexagonal and tetrahedron elements, and 5280 elements were used. The model is seen in Figure 11.

Figure 11 shows the fatigue life distribution in the specimen under maximum tension with a load of 21 kN. The image show that small regions within the measuring length close to the specimen ends connections have slightly lower fatigue life than the specimen center. This also compases to the actual fracture site at Low Cycle Fatigue (LCF). From the simulation, a minimum life of approximately 570 cycles is obtained. This also compases to the actual number of cycles to failure of approximately 550 cycles at a load level of 21.9 kN (Table 5).

Microstructure analysis

The micrographs in Figures 12 and 13 are taken in a SEM with magnifications 876 and 2000 X, respectively. Before heat treatment it is observed from Figure 12 that the phases which are present are ferrite (α - iron) and pearlite ($Fe3C$ + α - iron). In Figure 13, taken after hardening heat treatment at 850°C, the martinstic phase is clearly distinguished.

Martinsite is a distorted body-centered γ-iron with an additional carbon atom in its middle, this causes internal stresses that make this structure very hard but brittle and by heat treatments with the temperatures as 250, 450, 650°C other structures similar to Martensite produced,

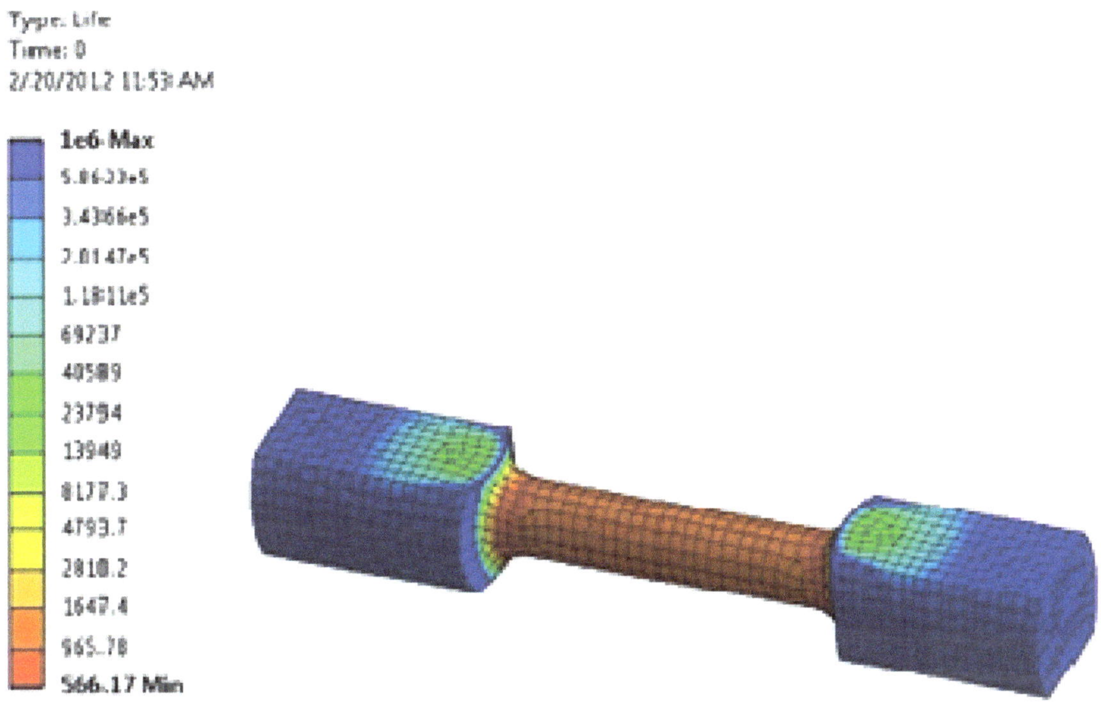

Figure 11. Life distribution in the fatigue specimen from FEM simulation.

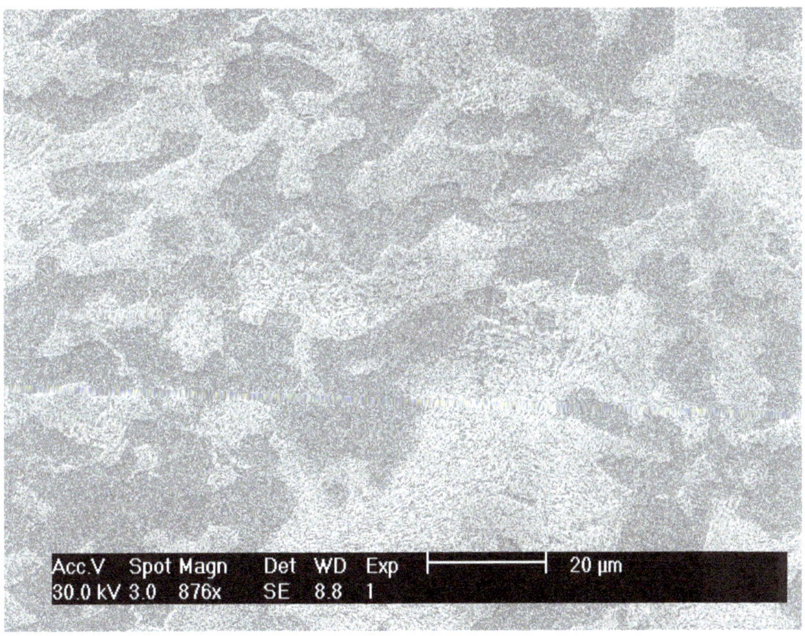

Figure 12. Pearlite (Fe3C+α-iron) microstructure at 876 X magnification factor.

such as Trostite and Sorbite, respectively were shown in Figure 14, that are not as hard, but more ductile at grains with plastic formation and composition of striation. Figure 14 Show the tempered Martensite with 250 and 650°C, respectively.

Fractography

Fractography studies of specimens failed under both Low Cycle Fatigue (LCF) and high cycle fatigue (HCF) conditions were performed. Pictures of the fracture

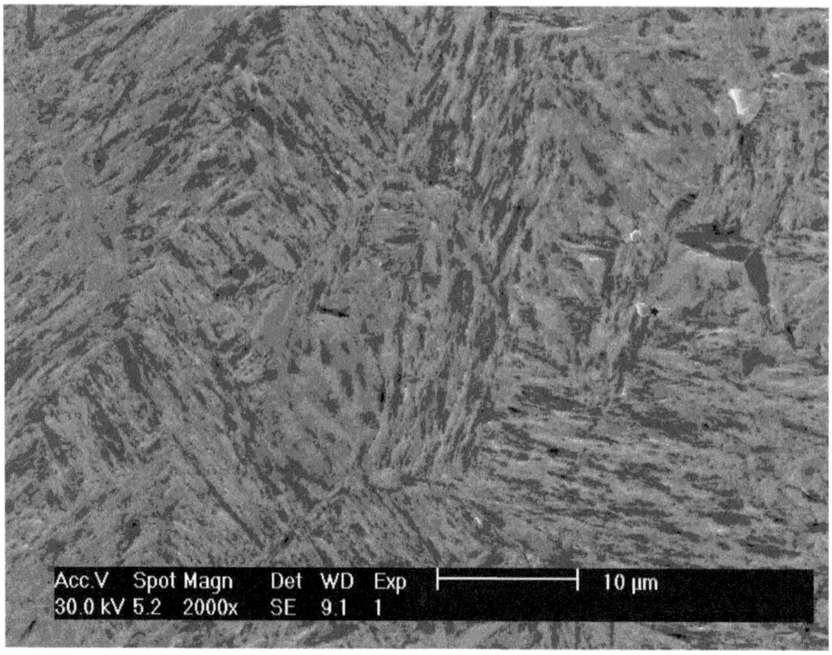

Figure 13. Martinsite microstructure at 2000X magnification factor.

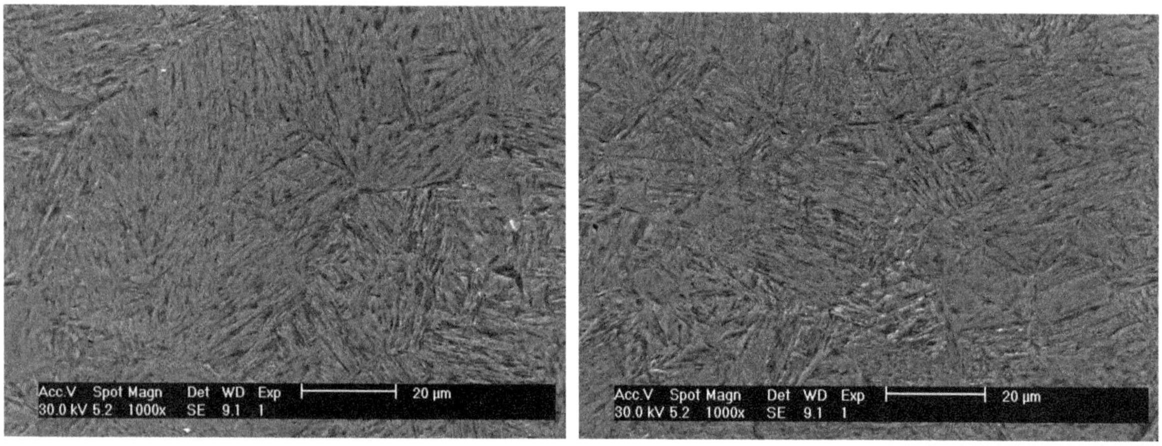

Figure 14. Tempered Martinsite (Trostite and Sorbite) microstructure at 2000 X magnification factor at 250 and 650°C, respectively.

surfaces were taken in a SEM at magnifications 1600, 3200 and 3206 X. The results are seen in Figure 15 (LCF) and 17 (HCF). The specimen in Figure 15 was loaded by 11.6 MPa and the specimen in Figure 16 by 19.6 MPa (Table 5). From the graphs in Figure 15, in the LCF region at magnifications 1600 and 3206 X, ductile dimples and a number of micro voids are observed.

From the graphs in Figure 15, in the HCF region at magnifications 1600 and 3200 X, brittle cleavage fracture is seen as well as bright or white lines in the facets, which show how the grains have split by cleavage during the fatigue fracture.

Conclusion

An investigation of the characteristics of steel SS-1672 has been performed. For the martinsite phase, it is observed that the hardness decreases with increasing tempering temperature due to the increasing ductility of the tempering martinsite phase induced by the tempering, increasing the resistance of the material to fracture and relieving the internal stresses. An Abaqus simulation predicts the effective plastic zone to extend 0.94 mm along the symmetry line when the yield stress is set equal to the von Mises stress.

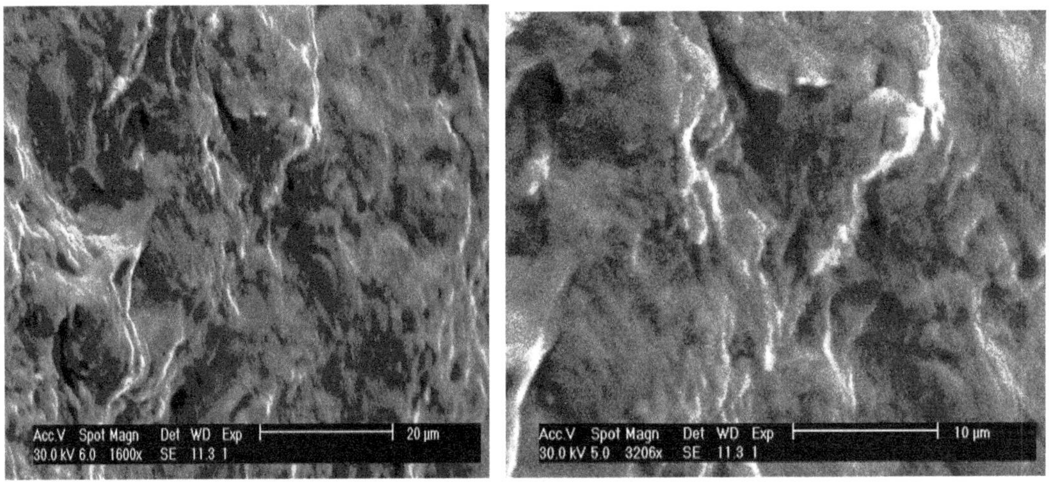

Figure 15. Fracture surface microstructure of a LCF region sample with magnifications 1600 and 3200 X.

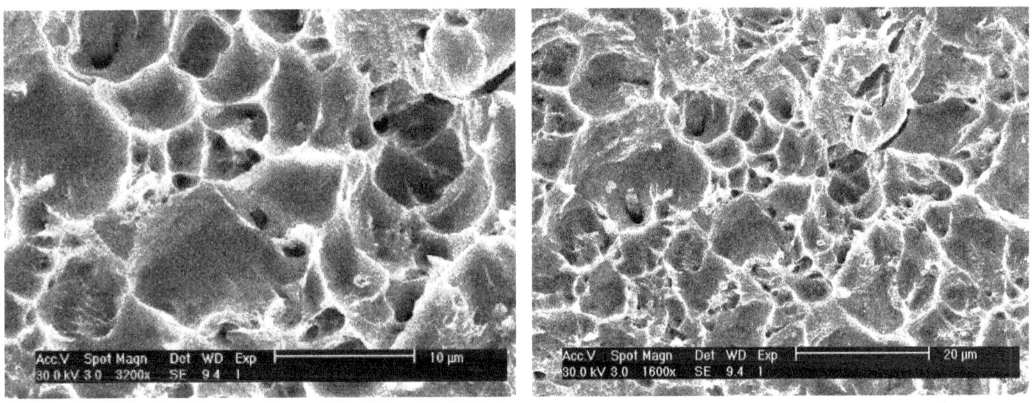

Figure 16. Fracture surface microstructure of a HCF region sample with magnifications 3200 and 1600 X.

From the simulation a minimum life of approximately 570 cycle is obtained. This also compasses to the actual number of cycles to failure of approximately 550 cycles at a load level of 21.9 kN. The transition life estimated from the crossing of the lines of slops at high and low number of cycles to failure is lower than when using the Morrow relation, probably due to the lack of some parameter influences such as the ductility exponent in the Morrow relation.

With the finite element program Ansys, it was shown that small regions close to the ends of the fatigue specimens have slightly lower fatigue life than in the specimen center. The stress in these regions increases due to the shape of the structure.

The specimens failed under both LCF and HCF conditions. The LCF region showed ductile dimples and a number of micro voids and for the HCF region brittle cleavage fractures and bright or white lines in the facets were observed, indicating how the grains have undergone cleavage.

ACKNOWLEDGEMENTS

The authors would like to thank everyone who so kindly responded to my request for fatigue and morphology tests throughout the research. Special thanks go to Zivorad Zivkovic, Ling Chen and Behrouz Afzali Far. I greatly acknowledge the support from the Divisions of Material Science and Mechanics at Lund University, Sweden.

REFERENCES

ASM Handbook (1996). Fatigue and fracture. ASM International, 19: 1057, ISBN: 978-0-97170-385-9.

Balbi M, Avalos M, El Bartali A, Armas I A (2009). Microcrack growth and fatigue behavior of a duplex stainless steel. Int. J. Fatigue 31:2006-2013.

Chiu PK, Weng KL, Wang SH, Yang JR, Huang YS, Fang J (2005). Low- cycle fatigue-induced martensitic transformation in SAF 2205 duplex stainless steel. Mater. Sci. Eng. A 398:349-359.

Eric PA (2006). Stainless Steel in the Food & Beverage Industry. First edition, Materials and applications series, 7, Euro Inox. Fagersta datablad for steel, Fagersta Bruks AB, Fagersta, Sweden, (1962).

Handbook in Solid Mechanics (2008). Department of Solid Mechanics, KTH, Stockholm, Sweden, 3 ed., pp:372-373

ISO (2002). Metallic materials - instrumented indentation test for hardness and material parameters. Part 1: test method, Geneva, Switzerland, 14577-1.

Jaykant GU (2009). "Mechanical and Wear Properties of Carburizes Mild Steel Samples. Master thesis Roll No. 207ME 207, Department of Mechanical Engineering, national Institute of Technology, Rourkela, Orissa, India.

Kim KS, Chen X, Han C, Lee HW (2002). Estimation methods for fatigue properties of steels under axial and tensional loading. Int. J. Fatigue, 24:783-793.

Ottosen N, Ristinmaa M (2005). The mechanics of constitutive modeling. First edition, ISBN: 0-008-044606-X, pp. 279-285.

Robert LN (2005). Mechanic Design: An Integrated Approach. 3ed, Prentice- Hall.

Smolnicki T, Rusiński E, Karliński J (2007). FEM modelling of fatigue loaded bolted flange joints J. Achiev. Mater. Manuf. Eng. 22(1):69-71.

Suresh S (2001). Fatigue of Materials. Second edition Cambridge University press,

Yung LI, Jwo PA, Richard BH, Mark EB (2005). Fatigue Testing and Analysis. Theory and fatigue practice, Elsevier's Science & Technology rights Department in Oxford, UK.

28

Comparison of unconstrained search methods for correlation of binary vapor liquid equilibrium (VLE) by five variants of Wilson model

A. Aravind Kumar[1]* and A. Ravi Prasad[2]

[1]Department of Chemical Engineering, University of Petroleum and Energy Studies, Bidholi, Through Prem Nagar, Dehradun-248007, India.
[2]Department of Chemical Engineering, Andhra University College of Engineering, Visakhapatnam-530003, India.

A comparative evaluation of five variants of Wilson model and eight unconstrained optimization methods for correlating the binary VLE data of 2-butanol-Tetrachloroethene system at 101.08 kPa pressure is presented. The data reduction and correlation for vapor/liquid equilibrium of the above system is presented. The objective of this paper was to find the parameters of the five variants of the Wilson models for the 2-butanol-tetrachloroethene, using optimization methods. The application of this optimization method for thermodynamic calculations is versatile due to its reliability, and efficiency is tested using phase equilibrium and parameter estimation problems. The obtained results indicated that proposed methodology is generally robust for the minimization of the objective functions involved in flash calculations using Gibbs energy minimization and in the calculation of homogeneous in non-reactive mixtures.

Key words: Binary VLE, 2-butanol, tetrachloroethene, unconstrained search methods, Wilson model, phase equilibria.

INTRODUCTION

The correlation of binary vapor liquid equilibrium data is very important as it provides parameters in describing the G^E/activity coefficient models which can in turn be used for the design and simulation of distillation equipment, and also for testing theories of liquid mixtures. Most of the liquid mixtures encountered in commercial operations are non ideal. The non ideality in a liquid mixture can be concisely represented by the Excess Gibbs free energy $\left(G^E/RT = g^E\right)$. There are several empirical and semi-theoretical models proposed in literature (Ravi, 2004). The g^E models can be broadly classified as classical models and models based on local composition concept. Often, they are classified as enthalpic (h^E) models, entropic (s^E) models and complete g^E models $\left(g^E = h^E - Ts^E\right)$ (Ravi Prasad, 2004). The latest G^E models offer the advantage that the multi-component VLE data can be predicted using the parameters of the constituent binary pairs, only without any experimental data on the multi-component mixture. The correlation of binary VLE (P-T-x-y) data consists in the determination of the constants in the G^E model by minimizing the objective function $\sum\left(g^E_{exp} - g^E_{cal}\right)^2$. Minimization of other objective functions is also possible. The constants can be obtained by several methods based on either solution of sets of non linear equations or optimization.

In Chemical Engineering, several thermodynamic calculations can be formulated as optimization problems with or without restrictions. As indicated by Henderson et al. (2004), the formulation of thermodynamic calculations

*Corresponding author. E-mail: aravind991@gmail.com.

for optimization problems offers some advantages: (a) the use of a robust optimization method, (b) the possibility of using a direct optimization method which requires only calculations of the objective function and (c) the use of an iterative procedure whose convergence is almost independent of the initial guesses. Some examples of these calculations are phase stability analysis, phase equilibrium problems, parameter estimation in thermodynamic models, calculation of critical points, among others. In this paper, as part of a continuing study of comparative evaluation of G^E/γ models and parameter estimation methods, eight methods of unconstrained optimization, namely Hooke-Jeeves, Powell, Rosenbrock, Nelder-Mead simplex, Cyclic coordinate, Random search, Simulated annealing and Genetic algorithm (Ashok and Tirupati, 1999) are applied to obtain parameters in five variants of Wilson model namely Wilson (1964), Enthalpic Wilson (Bruin, 1976), Effective Wilson (Sabarattinam et al., 1977), Modified Wilson (Huang and Lee, 1996) and T. K. Wilson (Tsuboka and Katayama, 1975) to describe the vapor liquid equilibrium data of 2-butanol-tetrachloroethene at 101.08 kPa (Venkateswara Rao and Ravi Prasad, 1984).

MATERIALS AND METHODS

Procedure for parameter estimation

1. The properties needed for processing and the experimental P, T, x, y data is read.
2. The g^E_{exp} is calculated using $g^E_{exp} = x_1 \ln \gamma_1 + x_2 \ln \gamma_2$ with $\gamma_i = y_i P / x_i P_i^{sat}$. The vapor phase is assumed to be ideal.
3. The g^E_{cal} values are calculated using the selected model equation and using the assumed set of parameters.
4. The expression $\left(g^E_{exp} - g^E_{cal}\right)^2$ is minimized using the selected optimization method.
5. With the optimized parameters vapor composition is calculated and absolute average deviation in experimental vapor composition and calculated vapor composition is evaluated

Variants of Wilson G^E models

The five variants of Wilson G^E models used in this study are listed below.

Wilson (1964)

$$\Lambda_{12} = \frac{V_2}{V_1} EXP\left(-\frac{A_{12}}{RT}\right); \qquad \Lambda_{21} = \frac{V_1}{V_2} EXP\left(-\frac{A_{21}}{RT}\right); \qquad \text{and}$$

$$\left(\frac{G^E}{RT}\right)_{CAL} = -X_1 \ln(X_1 + \Lambda_{12} X_2) - X_2 \ln(X_2 + \Lambda_{21} X_1)$$

Enthalpic Wilson (Bruin, 1976)

$$\left(\frac{G^E}{RT}\right)_{CAL} = -\frac{X_1 X_2 \ln\left[EXP\left(-\frac{A_{12}}{RT}\right) EXP\left(-\frac{A_{21}}{RT}\right)\right]}{\left[X_1 + X_2\left(\frac{V_2}{V_1}\right) EXP\left(-\frac{A_{12}}{RT}\right)\right]\left[X_2 + X_1\left(\frac{V_1}{V_2}\right) EXP\left(-\frac{A_{21}}{RT}\right)\right]}$$

Effective Wilson (Sabarathinam et al., 1977)

$$\Lambda_{12} = \frac{V_2}{V_1} EXP\left(-\frac{A_{12}}{RT}\right); \qquad \Lambda_{21} = \frac{V_1}{V_2} EXP\left(-\frac{A_{21}}{RT}\right); \qquad \text{and}$$

$$\left(\frac{G^E}{RT}\right)_{CAL} = X_1 \ln(X_1 + X_2 \Lambda_{21}) + X_2 \ln(X_2 + X_1 \Lambda_{12})$$

Modified Wilson (Huang-Lee) (Huang and Lee, 1996)

$$\tau_{12} = EXP\left(-\frac{A_{12}}{RT}\right); \quad \tau_{21} = EXP\left(-\frac{A_{21}}{RT}\right); \quad R_1' = \left(\frac{V_1}{V_2}\right)^{1/3},$$

$$R_2' = \left(\frac{V_2}{V_1}\right)^{1/3}; \quad \Phi_1' = \frac{X_1 R_1'}{X_1 R_1' + X_2 R_2'}; \quad \Phi_2' = \frac{X_2 R_2'}{X_1 R_1' + X_2 R_2'};$$

and

$$\left(\frac{G^E}{RT}\right)_{CAL} = -X_1 R_1' \ln(\Phi_1' + \Phi_2' \tau_{12}) - X_2 R_2' \ln(\Phi_2' + \Phi_1' \tau_{21}) + X_1 \ln\left(\frac{\Phi_1'}{X_1}\right) + X_2 \ln\left(\frac{\Phi_2'}{X_2}\right)$$

T. K. Wilson (Tsuboka and Katayama, 1975)

$$\tau_{12} = \frac{V_2}{V_1} EXP\left(-\frac{A_{12}}{RT}\right); \quad \tau_{21} = \frac{V_1}{V_2} EXP\left(-\frac{A_{21}}{RT}\right);$$

and

$$\left(\frac{G^E}{RT}\right)_{CAL} = X_1 \ln\left(\frac{X_1 + \left(\frac{V_2}{V_1}\right) X_2}{X_1 + X_2 \tau_{12}}\right) + X_2 \ln\left(\frac{X_2 + \left(\frac{V_1}{V_2}\right) X_1}{X_2 + X_1 \tau_{21}}\right)$$

Optimization methods

The brief details of the eight optimization methods used in this study are given below. All the optimization methods used are search methods and they do not require any derivative information during the calculation procedure.

Hooke-Jeeves method (HOOK)

In this method, an initial step size is chosen and the search is initiated from a given point. The method involves steps of exploration and pattern search. The step size *s* may be chosen in the range 0.05 to 1. Values beyond 1 can be tried; exploration about a point y: Let x = y. If e_i is unit vector along the coordinate direction x, then the function is evaluated at x + se_i. If the function reduces, then x is updated to be x + se_i. If function does not reduce, then the function is evaluated at x-se_i. If the function reduces, then x is updated to be x-se_i. If both fail, the original x is retained. The searches are performed with i ranging from 1 to n. At this stage, the initial point is at y and the new point is at x. Exploration is said to be successful if the function value at x is lower than the value at y by a predetermined amount.

Powell's method (POWE)

Powell developed an idea of constructing the conjugate directions without using derivatives. After searching along all conjugate directions, a spacer step is introduced where a search is made from the current point along the coordinate directions. This is to ensure convergence. If a stage starts from point x^1 and ends at point x^2, then the convergence is said to be achieved when the norm $|x^2 - x^1|$ is less than a small parameter ε or by checking the function improvement from x^1 to x^2.

Rosenbrock's method (ROSE)

In Rosenbrocks method, the search is carried out in n orthogonal directions at any stage. New orthogonal directions are established for the next stage. The orthogonal setting makes this method robust and efficient. In this algorithm, the steps are taken in cyclic manner and repeated until there is at least one success and one failure in each direction.

Nelder-Mead simplex method (NELD)

In this simplex method, in an n-dimensional space n + 1 points form a simplex. A triangle is an example of a simplex in two dimensions. In three dimensions, a tetrahedron (four points) forms a simplex. An initial simplex in n dimensions is easily created by choosing the origin as one corner and n points, each marked at a distance from the origin along the coordinate axes. During the computation the operations of reflection and contraction (expansion) are performed on the simplex till the optimum is reached.

Cyclic coordinate search (CYCL)

In this method, the search is conducted along each of the coordinate directions for finding the minimum. If e_i is the unit vector along the coordinate direction i, we determine the value α_i minimizing $f(\alpha) = f(x + \alpha e_i)$, where α is a real number. A move is made to the new point $x + \alpha_i e_i$ at the end of the search along the direction i. In an n-dimensional problem, we define the search along all the directions as one stage. The function value at the end of the stage is compared to the value at the beginning of the stage in establishing the convergence. The length of the move at each stage is another parameter for convergence consideration. The search for the minimum in each direction follows the steps of the establishment of a three-point pattern and the search for the minimum in the interval established.

Random search (RAND)

Random Search through probability algorithm (RSPA) uses probabilities to guide search for an optimal solution. The role of left digit is more important than the role of right digit for evaluating objective function. We calculated probabilities for searching correct values of digits from left digits to right digits of every variable. The complexity of RSPA of a problem is not based on type of expressions in objective function or constraints (linear or nonlinear), but on the relation of decided variables in the formula of object function or constraints; therefore if there are k dependent variables, we select k variables to change the value of variables for every iteration. We cannot calculate exactly a number of iterations for searching an optimal solution because RSPA is a random algorithm; therefore we use unfixed number of iterations which has capability to find an optimal solution.

Simulated annealing (SIMA)

Generally we seek to update a point when the function has a lower value in the conventional methods of minimization and this strategy may lead to a local minimum. Simulated annealing is a method seeking global minimum by adopting a strategy where a higher value of a function is acceptable under some conditions. The analogy with annealing process where stresses are relived from a previously hardened body has been observed by Metropolis that the probability of higher energy is larger at higher temperatures and there is some chance of a high energy as the temperature drops. Energy in the annealing process some times increases even while the trend is a net decrease. This property applied to optimization problems is referred to as the Metropolis algorithm.

Genetic algorithm (GENE)

The genetic algorithm is a technique that draws its analogy from nature, and it simulates the evaluation process. This algorithm revolves around the genetic reproduction process and survival of the fittest strategies. The various steps in applying algorithm are: creation of initial population, evaluation, Creation of a meeting pool, cross over operation, mutation and evaluation.

RESULTS AND DISCUSSION

The binary system 2-Butanol-Tetrachloroethene at 101.08 kPa is chosen as an example system. The experimental data for this system show positive deviations from Roult's law and exhibit a minimum boiling azeotrope. The results of processing the isobaric binary vapor liquid equilibrium data of the above system to make a comparative study of five variants of Wilson G^E models and eight methods of optimization for estimating the parameters of the models are presented in Table 1. While processing the data with the various G^E models, the vapor phase is taken to be ideal. The molar volumes of the components are calculated using the modified Racket (1970) equation. The absolute average deviation between the experimental and calculated G^E/RT values and vapor compositions for each model and for each method of optimization are shown. The parameters of the selected variants of the Wilson model are also presented.

As can be observed from the values of Δy, all the variants of Wilson model selected for this study are satisfactorily describing the experimentally observed data. The lowest deviation (Δy) is observed with the Effective Wilson model. The calculated vapor composition by the Effective Wilson model using the eight

Table 1. Absolute average deviation in G^E/RT, y and parameter values for each model, system: 2-butanol-tetracholoroethene pressure at 101.08 kPa.

Model	Optimization Method	$\Delta G^E/RT$	Δy	A_{12} (j/mol)	A_{21} (j/mol)
Wilson	Hooke and Jeeves	0.004249	0.004143	1426.757	3395.993
	Powell	0.004236	0.004139	1458.596	3425.231
	Rosenbrock	0.004239	0.004138	1455.541	3427.840
	Nelder Mead simplex	0.004223	0.004146	1472.358	3404.644
	Cyclic Coordinate	0.004239	0.004134	1456.724	3426.736
	Random search	0.004240	0.004134	1455.908	3427.725
	Simulated Annealing	0.004236	0.004126	1460.929	3421.765
	Genetic algorithm	0.004240	0.004134	1455.975	3427.643
Enthalpic Wilson	Hooke and Jeeves	0.004116	0.004514	666.4464	1996.071
	Powell	0.004114	0.004577	660.3227	2002.525
	Rosenbrock	0.004122	0.004529	659.5076	2003.082
	Nelder Mead simplex	0.004147	0.004288	660.0463	2002.989
	Cyclic Coordinate	0.004115	0.004521	663.8833	1998.722
	Random search	0.004120	0.004527	660.6419	2001.926
	Simulated Annealing	0.004122	0.004508	661.3513	2001.672
	Genetic algorithm	0.004120	0.004525	661.1623	2001.425
Effective Wilson	Hooke and Jeeves	0.004072	0.004100	-1896.929	-1538.071
	Powell	0.004102	0.004052	-1921.426	-1515.730
	Rosen brock	0.004071	0.004101	-1896.496	-1538.460
	Nelder Mead simplex	0.004101	0.004196	-1900.522	-1532.090
	Cyclic Coordinate	0.004062	0.004103	-1891.247	-1543.402
	Random search	0.004071	0.004096	-1896.718	-1538.452
	Simulated Annealing	0.004058	0.004069	-1893.628	-1542.497
	Genetic algorithm	0.004071	0.004096	-1896.779	-1538.383
Modified Wilson	Hooke and Jeeves	0.004212	0.004228	1962.571	2920.429
	Powell	0.004212	0.004275	1955.128	2928.496
	Rosenbrock	0.004219	0.004226	1955.581	2929.923
	Nelder Mead simplex	0.004220	0.004192	1960.406	2926.663
	Cyclic Coordinate	0.004215	0.004237	1958.415	2925.172
	Random search	0.004215	0.004237	1957.929	2925.783
	Simulated Annealing	0.004212	0.004261	1957.702	2923.799
	Genetic algorithm	0.004215	0.004240	1957.943	2925.479
T. K. Wilson	Hooke and Jeeves	0.004195	0.004270	2408.562	2417.438
	Powell	0.004185	0.004250	2428.834	2393.785
	Rosenbrock	0.004187	0.004252	2419.698	2404.217
	Nelder Mead simplex	0.004207	0.004232	2402.560	2427.461
	Cyclic Coordinate	0.004188	0.004255	2417.793	2406.445
	Random search	0.004187	0.004252	2419.629	2404.306
	Simulated Annealing	0.004209	0.004055	2414.197	2410.280
	Genetic algorithm	0.004187	0.004252	2419.634	2404.300

optimization methods are presented in Table 2. As can be observed from the results, all the optimization methods selected for this study are equally good for describing the VLE data of the example system. The graphical representation of the deviation between the experimental and the calculated vapor composition of the

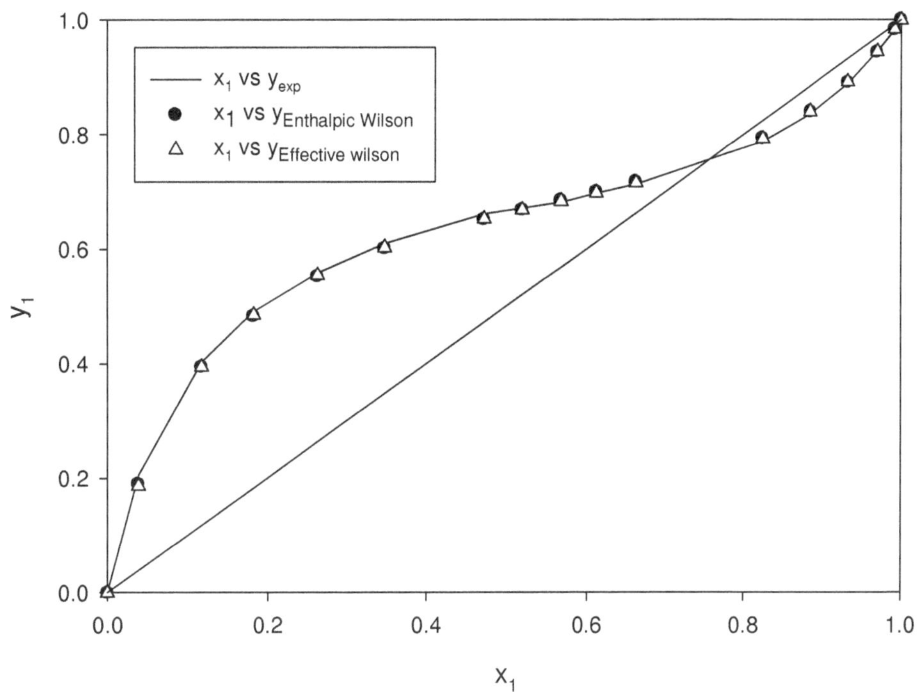

Figure 1. Comparison of vapour composition using enthalpic and effective Wilson thermodynamic models for the system: 2-butanol-tetrachloroethene at 101.08 kPa.

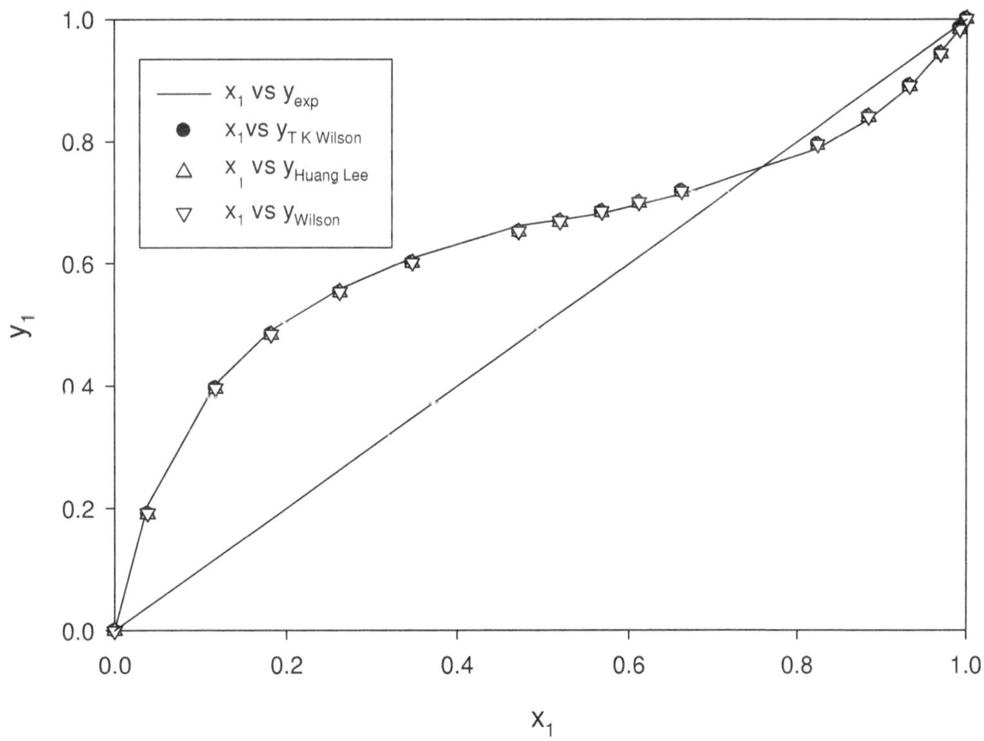

Figure 2. Comparison of vapour composition using Wilson, T. K. Wilson and Modified Wilson thermodynamic models for the system 2- butanol-tetrachloroethene at 101.08 kPa.

Enthalpic and Effective Wison model is shown in Figure 1. The optimization method chosen is Hook and Jeeves for the study. Figure 2 is representing the deviations of vapor composition using Wilson, Modified Wilson and T.

Table 2. Calculated vapor composition by Effective Wilson equation, system: 2-butanol-tetracholoroethene at 101.08 kPa.

T	x_1	y_1	y_1 Calculated			
			HOOK	POWE	ROSE	NELD
394.250	0.000	0.000	0.000000	0.000000	0.000000	0.000000
387.650	0.038	0.205	0.191899	0.192327	0.191901	0.191915
380.250	0.117	0.403	0.396964	0.397372	0.396974	0.396996
376.650	0.182	0.491	0.485309	0.485537	0.485326	0.485347
374.050	0.263	0.559	0.553854	0.553869	0.553876	0.553894
372.550	0.348	0.610	0.601642	0.601490	0.601665	0.601680
371.350	0.472	0.662	0.651066	0.650777	0.651086	0.651094
370.950	0.520	0.672	0.667153	0.666842	0.667170	0.667175
370.550	0.568	0.682	0.682698	0.682383	0.682712	0.682714
370.350	0.612	0.697	0.697293	0.696989	0.697303	0.697302
370.250	0.662	0.713	0.714944	0.714669	0.714948	0.714943
370.250	0.824	0.788	0.792467	0.792402	0.792450	0.792434
370.750	0.885	0.836	0.839820	0.839855	0.839797	0.839778
371.350	0.933	0.889	0.891231	0.891321	0.891209	0.891192
371.650	0.970	0.946	0.944104	0.944185	0.944089	0.944078
372.450	0.992	0.984	0.983756	0.983787	0.983751	0.983748
372.650	1.000	1.000	1.000000	1.000000	1.000000	1.000000

T	x_1	y_1	y_1 Calculated			
			CYCL	RAND	SIMA	GENE
394.250	0.000	0.000	0.000000	0.000000	0.000000	0.000000
387.650	0.038	0.205	0.191711	0.191903	0.192007	0.191904
380.250	0.117	0.403	0.396819	0.396972	0.397111	0.396972
376.650	0.182	0.491	0.485270	0.485320	0.485443	0.485319
374.050	0.263	0.559	0.553931	0.553866	0.553962	0.553865
372.550	0.348	0.610	0.601801	0.601653	0.601719	0.601652
371.350	0.472	0.662	0.651276	0.651075	0.651097	0.651073
370.950	0.520	0.672	0.667363	0.667160	0.667166	0.667158
370.550	0.568	0.682	0.682897	0.682703	0.682692	0.682702
370.350	0.612	0.697	0.697471	0.697297	0.697271	0.697295
370.250	0.662	0.713	0.715087	0.714944	0.714901	0.714943
370.250	0.824	0.788	0.792431	0.792458	0.792369	0.792459
370.750	0.885	0.836	0.839713	0.839807	0.839715	0.839809
371.350	0.933	0.889	0.891100	0.891220	0.891140	0.891221
371.650	0.970	0.946	0.944006	0.944096	0.944047	0.944097
372.450	0.992	0.984	0.983722	0.983754	0.983738	0.983754
372.650	1.000	1.000	1.000000	1.000000	1.000000	1.000000

K .Wilson. From the figures, it is clearly evident that both the models are correlating and the experimental data vary accurately.

Conclusion

The comparative evaluation of five variants of Wilson model and eight optimization methods for describing the VLE of 2-butanol-tetrachloroethene is presented. Effective Wilson model is found to produce the best agreement with the experimental data for the selected system.

REFERENCES

Ashok DB, Tirupathi, RC (1999). Optimization concepts and applications in Engineering, Pearson Education Asia, Singapore.

Bruin S (1976). Activity Coefficient Relations in Miscible and Partially Miscble Multi Component Systems. I&EC Fundam. 9:305-314.

Henderson N, Freitas L, Platt GM (2004). Prediction of critical points: A new methodology using global optimization. AIChE J. 50:1300.

Huang J, Lee L (1996). Simultanious Estimation of Excess Enthalpy, Excess Gibbs Energy and Vapor-Liquid Equilibrium using Modified Wilson Model. Fl. Ph. Equalib. 121:27-43.
Rackett HG (1970). Equation of State for Saturated Liquids. J. Chem. Eng. Data 15:514-517.
Ravi PA (2004). VLE data generation, correlation and prediction. Ph. D. Thesis, Andhra University, Viskhapatnam, India.
Sabarathinam PL, Andiappan AN, Lakshmanan SM (1977). Application of Effective Local Composition Wilson Equation to Vapor-Liquid Equilibrium of systems with and Without Salts. Ind. J. Technol. 15:150-153.
Tsuboka T, Katayama T (1975). Modified Wilson Equation for Vapor Liquid and Liquid Liquid Equilibria. J. Chem. Eng. Jpn. 8:181-187.
Venkateswara Rao K, Ravi Prasad A (1984). Measurement, Correlation and Prediction of Binary Vapor-Liquid Equilibrium for Alcohol Tetrachloroethene Systems. Can. J. Chem. Eng. 62:142-148.
Wilson GM (1964). Vapor-Liquid Equilibrium XI. A New Expression for Excess free Energy of Mixing. J. Am. Chem. Soc. 86:127-130.

Solar photocatalytic degradation of phenol using nanosized ZnO and α-e_2O_3

Hamza A., Fatuase J.T., Waziri, S. M. and Ajayi, O. A.

Department of Chemical Engineering, Ahmadu Bello University, Zaria, Nigeria.

This work focuses on the solar photocatalytic degradation of phenol using nanosized ZnO and α-Fe_2O_3 synthesized via precipitation route. The synthesized ZnO and α-Fe_2O_3 were characterized using XRD, UV/Vis and surface area analysis. α-Fe_2O_3 was found to exhibit better solar photocatalytic activity than ZnO under identical experimental conditions due to the larger surface area and low band gap of α-Fe_2O_3. The photocatalytic degradation of phenol follows pseudo-first order kinetics on both ZnO and α-Fe_2O_3.

Key words: Photocatalysis, solar, phenol, ZnO, α-Fe_2O_3.

INTRODUCTION

Photocatalysis involves the use of semiconductors and electromagnetic radiation especially in the visible and ultraviolet regions to drive various photochemical processes (destruction of a wide variety of organic and inorganic pollutants and microorganisms, generation of hydrogen gas from water, etc). Some of the major advantages of photocatalysis include: ambient operating temperature and pressure conditions, low cost of photocatalysts, complete destruction of water pollutants and pathogens without generation of secondary pollutants, possibility of using solar energy to drive the process, etc. (Chong et al., 2010; Fujishima et al., 2007).

The photocatalytic activities of various semiconducting materials (TiO_2, ZnO, Fe_2O_3, CdS, GaP, ZnS, etc) and their composites to mineralize a wide variety of water pollutants have been investigated by many researchers. TiO_2 and ZnO are the most widely used photocatalyst due to their good photocatalytic activity, non-toxicity and low cost. ZnO was found to be more photocatalytically active than TiO_2 under solar irradiation (Hayata et al., 2011). The major disadvantage of TiO_2 and ZnO is their large band gap of about 3.2 eV (~388 nm) which makes them effective only in the presence of UV light with wavelengths shorter than 380 nm (Casbeer et al., 2012). On the other hand, α-Fe_2O_3 as a visible light active photocatalyst with a bandgap of about 2.2 eV has been applied for photocatalytic degradation of a wide range of pollutants (Jaafar et al., 2012; Bakardjieva et al., 2007). Solar energy has the highest potential of all the renewable energies especially in tropical Africa (Blanco et al., 2009). Most developing countries have problems of wastewater management. Fortunately, there are high levels of solar radiation in most of these countries. Hence, there is the need to develop materials that can harness solar energy and use it to solve the problems of wastewater treatment. Phenol and its derivatives are found in the effluents of various industries such as petrochemical, coal, pharmaceutical, wood, paint, pulp and paper industries (Ahmaruzzaman, 2008). Phenol is easily absorbed through human skin causing skin and eye burns upon contact. Phenol affects the liver, kidneys, lungs, and vascular system. Excessive exposure to phenol may cause comas, convulsions, cyanosis and even death (Ahmaruzzaman, 2008).

Photocatalytic degradation of phenolic compounds by various photocatalysts has been reported by several researchers (Hayata et al., 2011; Chiou et al., 2008; Valenzuela et al., 2002). For example, Valenzuela et al. (2002) investigated photocatalytic degradation of phenol and found ZnO to be more active than α-Fe_2O_3 under artificial ultraviolet illumination. The aim of this work was to synthesize nanosized ZnO and α-Fe_2O_3 using precipitation route and compare their effectiveness for solar photocatalytic degradation of phenol under identical experimental conditions.

MATERIALS AND METHODS

Synthesis and characterization of ZnO and α-Fe_2O_3

Analytical grade reagents purchased from a local chemical store in Zaria, Kaduna State, Nigeria were used throughout the work. 13.75 g of zinc nitrate was dissolved in 250 ml of deionized water. 13.85 ml of ammonium hydroxide was added at a constant rate to the solution with vigorous stirring at 50°C and a pH of 8. The formed precipitate was aged for 24 h, washed with deionized water and then filtered under vacuum. The product was dried overnight in an oven at 110°C. The dried product was further calcined in a furnace at 1000°C for 4 h to yield ZnO.

38.45 g of iron nitrate was dissolved in 250 ml of deionized water. 21 ml of ammonium hydroxide was then added. The mixture was stirred at 50°C and a pH of 8.0 for 3 h. The formed precipitate was aged for 24 h, washed with deionized water and then filtered under vacuum. The product was dried overnight in an oven at 110°C. The dried product was further calcined in a furnace at 1000°C for 4 h to yield α-Fe_2O_3.

X-ray diffraction (XRD) patterns of the synthesized ZnO and α-Fe_2O_3 were recorded using a powder X-ray diffractometer (Shimadzu, model 6000) employing Cu Kα radiation source (λ=0.154nm). Absorption spectra of the synthesized ZnO and α-Fe_2O_3 were recorded using a UV/Vis spectrophotometer (Jenway 6405). The specific surface area of the photocatalyst was obtained using a modified BET method which is based on the adsorption of water molecules on the surface of solid samples (Adefila et al., 2003).

Photocatalytic experiments

The concentration of phenol solution used in all the experiments was 500 mg/L. The pH of the solution is 4.0. 100 ml of phenol solution was mixed with 0.1 g of the photocatalyst (ZnO or α-Fe_2O_3) in the photoreactor and magnetically stirred for 2 h in the dark in order to establish adsorption-desorption equilibrium of phenol on the photocatalyst. Thereafter, the mixture was exposed to sunlight under continuous stirring at room temperature (28±2°C) for 6 h. After every hour, samples were taken for analysis after separating the photocatalyst particles. These samples were analyzed for residual concentration of phenol at 270 nm using a UV/Vis spectrophotometer (Jenway, 6405). The experiments were performed between 10 am to 4 pm in September, 2012. Experiments were repeated in duplicate for getting the reproducibility of results. The percentage photodegradation of phenol was calculated using Equation 1:

$$Degradation = \frac{(C_0 - C_t)}{C_0} \times 100\% \quad (1)$$

Where Co and C_t are the initial concentration of phenol and the concentration of phenol after irradiation time (t) respectively.

RESULTS AND DISCUSSION

Characterization of the synthesized ZnO and α-Fe_2O_3

The XRD patterns of the synthesized ZnO and α-Fe_2O_3 are displayed in Figures 1 and 2, respectively. In Figure 1, diffraction peaks were observed at 2θ = 24.3°, 33.3°, 35.7°, 40.9°, 49.5°, 54.1°, 62.5° and 64.0°. These peaks are similar to the typical peaks of the hematite phase of Fe_2O_3 with rhombohedral symmetry (JCPDS file No 33-0664). In Figure 2, diffraction peaks were observed at 2θ = 31.75°, 34.42°, 36.24°, 47.53°, 56.58°, 62.84° and 67.91°. These peaks are similar to the typical peaks of the wurtzite phase of ZnO (JCPDS file No 33-0664).

The crystallite sizes of ZnO and α-Fe_2O_3 were determined using the Debye–Scherrer equation (Equation 2).

$$\tau = \frac{k\lambda}{\beta \cos\theta} \quad (2)$$

Where τ is the particle size, λ is the wavelength of the X-ray radiation (Cu Kα = 0.1542 nm), k is the shape factor (k = 0.94), β is the full line width at half maximum height and θ is the angular position of the peak maximum. The estimated crystallite sizes of ZnO and α-Fe_2O_3 were 58.4 nm and 58.2 nm, respectively. Hence, the synthesized ZnO and α-Fe_2O_3 are nanoparticles because their crystallite sizes are less than 100 nm. The UV-Vis absorption spectra of the catalysts are shown in Figure 3. ZnO absorption is exclusively in the ultraviolet region. Whereas, α-Fe_2O_3 exhibit broad absorption band that extends to the visible region (up to about 550 nm). The specific surface areas of ZnO and α-Fe_2O_3 were 39.5 m^2/g and 162.3 m^2/g, respectively, as determined by the modified BET method.

Photocatalytic degradation of phenol over ZnO and α-Fe_2O_3

Control experiments were performed by mixing 0.1 g of the synthesized photocatalyst (α-Fe_2O_3 or ZnO) with 100 ml of phenol solution in the dark. The results show that 6 h of adsorption of phenol in the dark on both ZnO and α-Fe_2O_3 resulted in the removal of less than 12 % of phenol. In the absence of photocatalyst (photolysis), the percentage degradation of phenol was only 21% in an irradiation time period of 6 h. The effect of irradiation time on the percentage degradation of phenol over ZnO and α-Fe_2O_3 as well as in the absence of photocatalyst is depicted in Figure 4. The percentage degradation of phenol increased with increase in the irradiation time, and

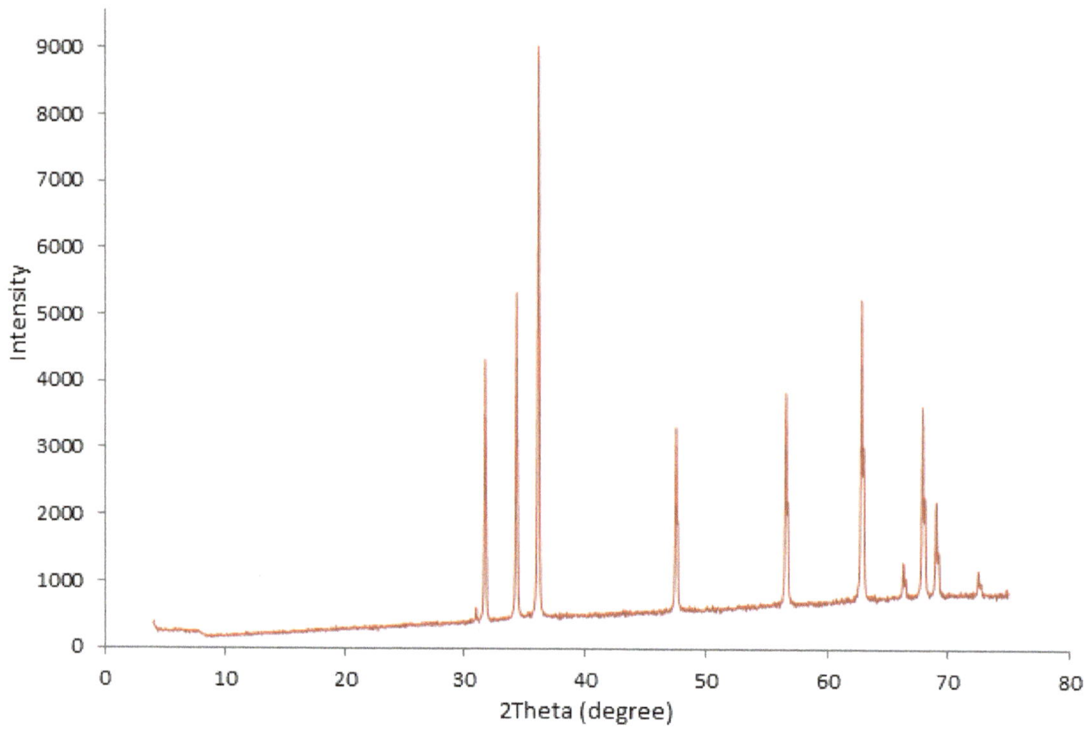

Figure 1. XRD pattern of the synthesized ZnO.

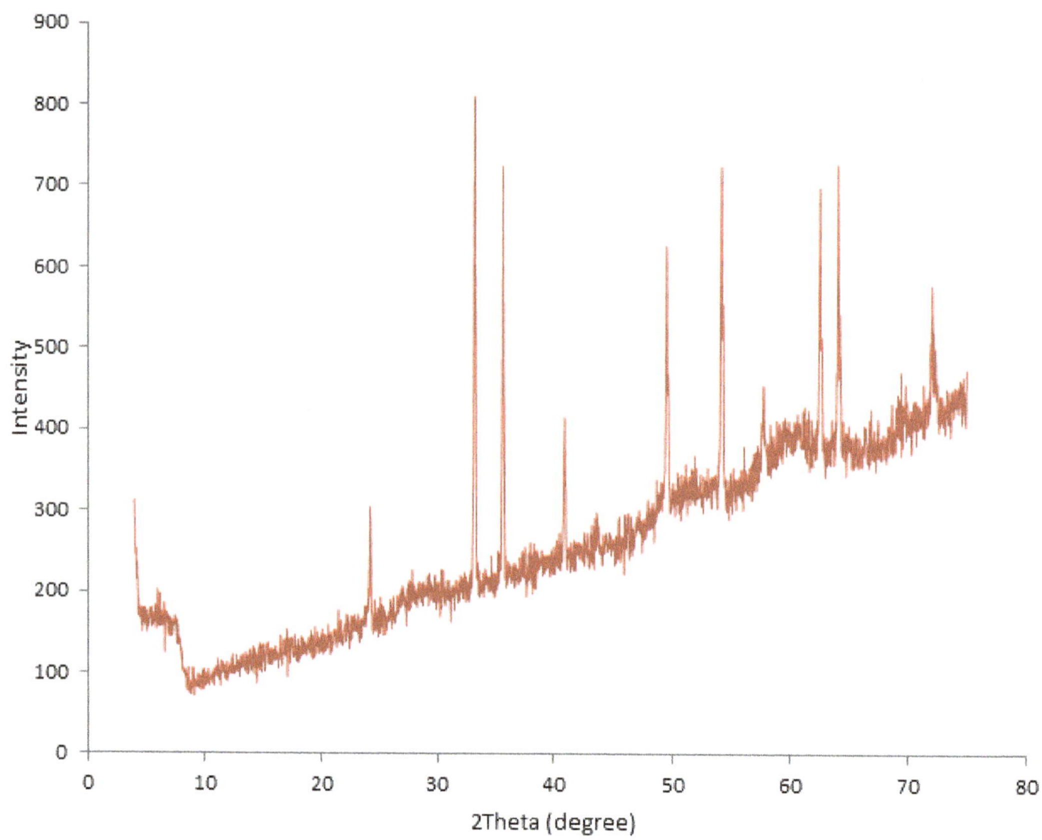

Figure 2. XRD pattern of the synthesized α-Fe_2O_3.

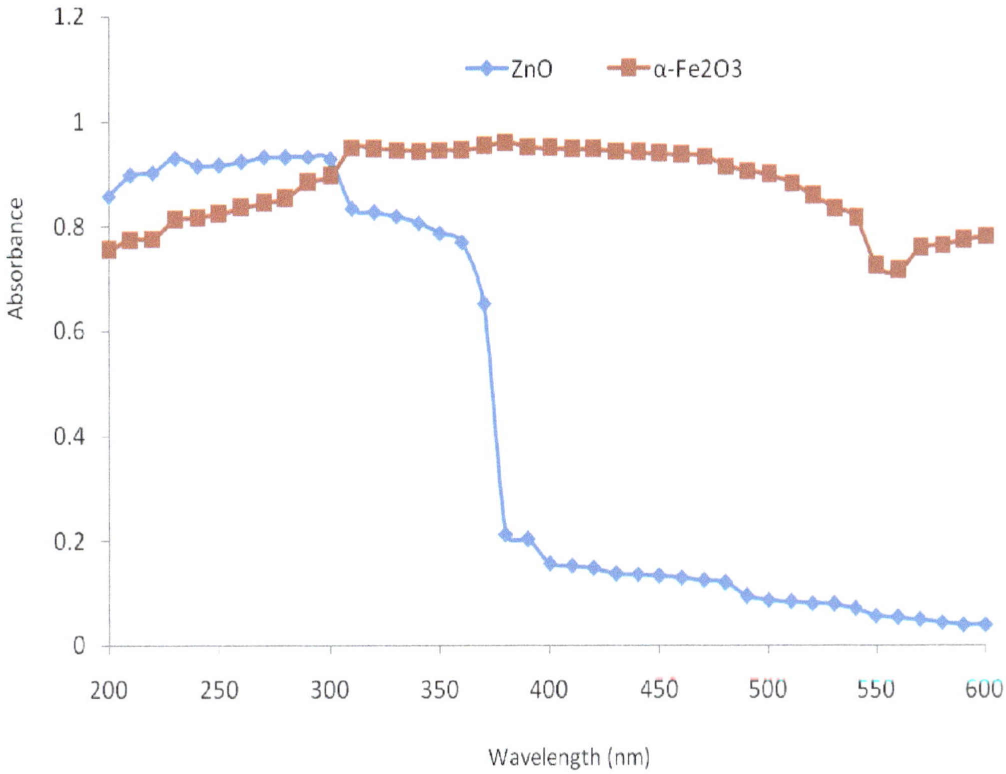

Figure 3. UV-Vis absorption spectra of the synthesized ZnO and α-Fe_2O_3.

reached 52% and 86% for ZnO and α-Fe_2O_3, respectively, within 6 h. This result indicates that the synthesized α-Fe_2O_3 has a better photocatalytic activity for the degradation of phenol than ZnO.

The pseudo first order kinetics equation (3) is often used to describe the kinetics of photocatalytic processes (Hamza et al., 2011).

$$-\ln\frac{C_0}{C} = k_{app}t \qquad (3)$$

Where k_{app} and t are the apparent rate constant of photocatalytic reactions and the irradiation time, respectively. The apparent rate constant is often chosen as the basic kinetic parameter for comparing different photocatalysts because it enables the determination of photocatalytic activity independent of the previous adsorption period in the dark and the concentration of solute remaining in the solution (Hamza et al., 2011). As seen in Figure 5, a linear relationship (with $R^2 > 0.90$) exists between $-\ln(C_0/C)$ and irradiation times, t, for both ZnO and α-Fe_2O_3. Hence, the photocatalytic degradation of phenol on both ZnO and α-Fe_2O_3 follows pseudo-first order kinetics. The apparent rate constants (k_{app}) were found to be 0.139 hr^{-1} and 0.310hr^{-1} for ZnO and α-Fe_2O_3, respectively.

The higher phtocatalytic activity of α-Fe_2O_3 can be attributed to the low band gap of α-Fe_2O_3. The band gap α-Fe_2O_3 is 2.1 eV (590 nm) which falls in the visible light region of the electromagnetic spectrum (400 - 800 nm). Visible light constitutes about 46% of the solar spectrum (Casbeer et al., 2012). Hence, α-Fe_2O_3 is a good absorber of solar light and utilizes the absorbed solar light photons to drive photochemical reactions under visible light irradiation better than ZnO. The band gap ZnO is 3.2 eV which falls in the ultra violet light region of the electromagnetic spectrum (200 -400 nm). This makes ZnO less active photocatalyst under solar irradiation. It should be noted Valenzuela et al. (2002) found ZnO to be more active than α-Fe_2O_3 under ultraviolet illumination (Valenzuela et al., 2002). Ultraviolet light constitutes about 5% of the solar spectrum (Casbeer et al., 2012). The larger specific surface area of α-Fe_2O_3 (162.3 m^2/g) may also be responsible for the observed higher photocatalytic activity of α-Fe_2O_3 when compared with ZnO (39.5 m^2/g). Generally, the efficiency of heterogeneous catalytic processes increases with increase in the specific surface area of the catalyst.

Conclusions

ZnO and α-Fe_2O_3 nanoparticles were synthesized and characterized using XRD, UV/Vis and modified BET techniques. α-Fe_2O_3 was found to exhibit better photocatalytic activity than ZnO under solar light irradiation due to the narrower band gap and larger

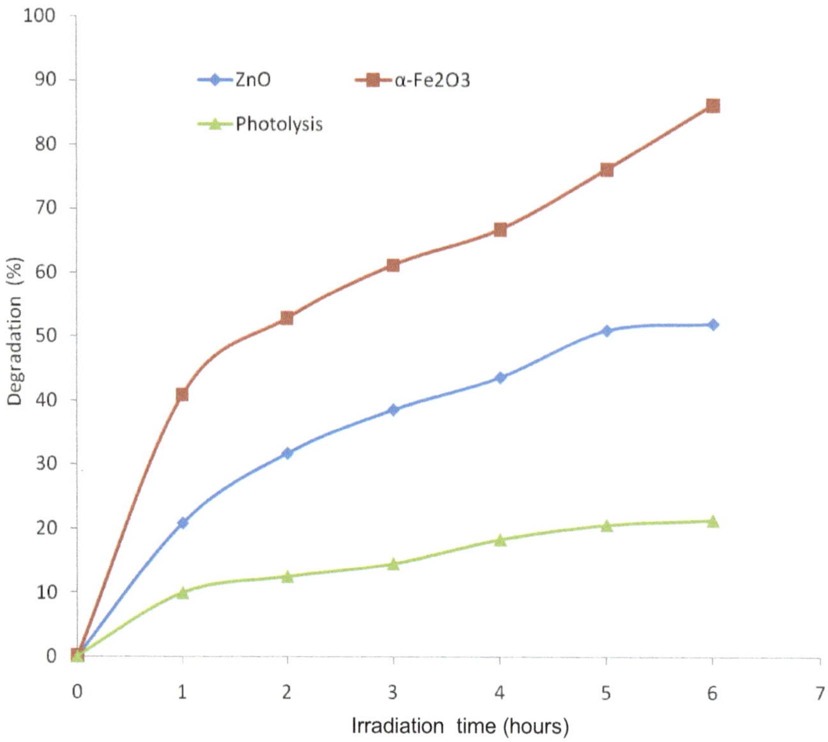

Figure 4. Effect of irradiation time on the photolytic and photocatalytic degradation of phenol using ZnO and α-Fe_2O_3.

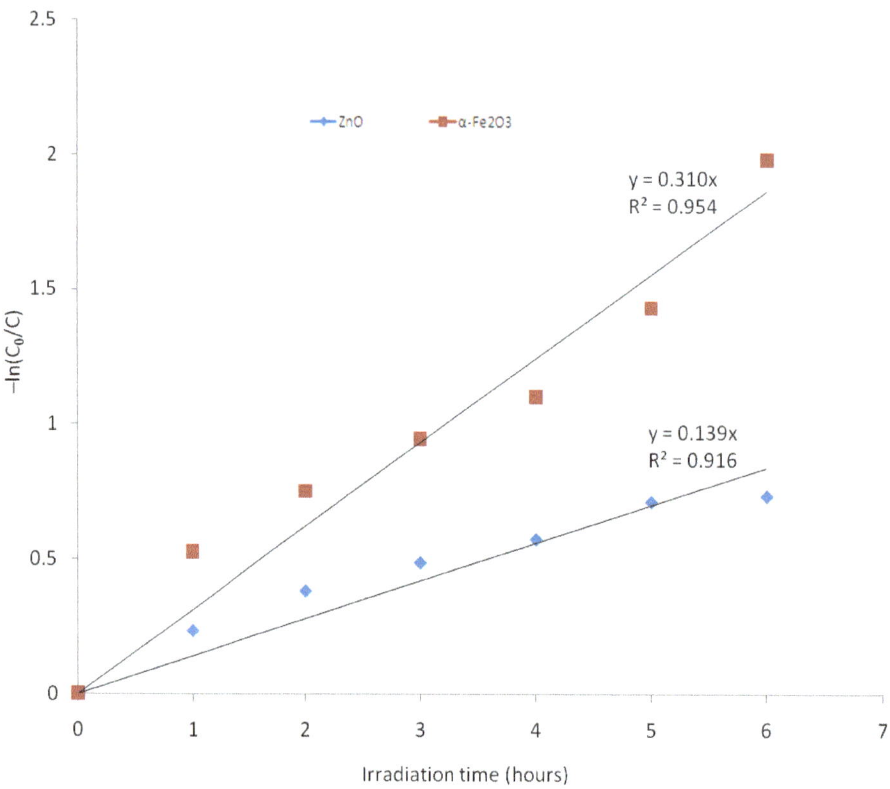

Figure 5. Pseudo first order kinetic plot for the photocatalytic degradation of phenol using ZnO and α-Fe_2O_3.

specific surface area of α-Fe_2O_3 when compared with ZnO. The photocatalytic degradation of phenol follows pseudo-first order kinetics on both ZnO and α-Fe_2O_3.

REFERENCES

Adefila SS, Aderemi BO, Ajayi OA, Baderin DA (2003). Comparative surface area determination using water adsorption method. Nig. J. Eng. 11:88-95.

Ahmaruzzaman M (2008). Adsorption of phenolic compounds on low-cost adsorbents: A review. Adv. Coll. Interf. Sci. 143:48–67.

Bakardjieva S, Stengla V, Subrta J, Houskovaa V, Kalend V (2007). Photocatalytic efficiency of iron oxides: Degradation of 4-chlorophenol. J. Phys. Chem. Solids 68.

Blanco J, Malato S, Fernandez P, Gernjak W, Maldonado MI (2009). Review of feasible solar energy applications to water processes. Renew. Sustain. Energy Rev. 13:1437–1445.

Casbeer E, Sharma, VK, Li XI (2012). Synthesis and photocatalytic activity of ferrites under visible light: A review . Separ. Purif. Technol. 87:1–14.

Chiou CH, Wu CY Juang RS (2008). Influence of operating parameters on photocatalytic degradation of phenol in UV/TiO_2 process. Chem. Eng. J. 139:322–329.

Chong MN, Jin B, Chow C WK, Saint C (2010). Recent developments in photocatalytic water treatment technology: A review. Water Res. 44:2997-3027.

Fujishima AZ, Tryk DA (2007). Heterogeneous photocatalysis: From water photolysis to applications in environmental cleanup Int. J. Hydrogen Energy 32:2664–2672.

Hamza A, Abdallah NM, Aliyu A (2011). Photocatalytic activity of Nigerian ilmenite. Nig. J. Eng. 18:111-114.

Hayata K, Gondal MA, Khaled MM, Ahmed S, Shemsi AM (2011). Nano ZnO synthesis by modified sol gel method and its application in heterogeneous photocatalytic removal of phenol from water. Appl. Catal. A: General 393:122–129.

Jaafar N, Abdul Jalil A, Triwahyono S, Muhid M, Sapawe N, Satar M, Asaari, H (2013). Photodecolorization of methyl orange over α-Fe_2O_3-supported HY catalysts: The effects of catalyst preparation and dealumination. Chem. Eng. J. 191:112-122.

Valenzuela MA, Bosch P, Jimenez-Becerrill J, Quiroz O, Paez AI (2002) Preparation, characterization and photocatalytic activity of ZnO, Fe_2O_3 and $ZnFe_2O_4$. J. Photochem. Photobiol. A. 148:177–182.

Time dependent mathematical model of air pollutants emitted from time-dependent elevated line source into a stable atmospheric boundary layer

Moka Shekhu[1] and C. Sulochana[2]

[1]Department of Science, Government Polytechnic, Chennasandra, Bangalore-560067, India.
[2]Department of Mathematics, Gulbarga University, Gulbarga, Karnataka, India.

A time dependent atmospheric model is represented for chemically reactive primary pollutants emitted from an elevated line source into a stable atmospheric boundary layer, over rough surface terrains. The time dependent model was obtained through an analytical solution from a multiple Inverse Laplace transform of the atmospheric diffusion equation with the parabolic eddy-diffusion coefficient (exchange coefficient) and the wind velocity as functions of vertical height. The pollutants were considered to be of chemically reactive primary pollutants emitted from a time-dependent line source of, step-Function type. In order to facilitate the application of the model, the results for the general situation that includes chemical reaction rate and time dependent line source are incorporated in the model. In this model, the effect of step-function type elevated line source studied near the source. The results obtained in this model are compared with that of the continuous line sources in the previous work. The results obtained in this model are of good agreement with that of the continuous line source.

Key words: Stable-atmospheric boundary layer (SBL), Step-function type line sources, parabolic eddy-diffusion coefficient.

INTRODUCTION

The study of the atmospheric dispersion of air pollutants from different types of sources received a great deal of attention during the last few decades. The effects of various sources of chemically reactive pollutants are been investigated in the recent studies. In addition, the depletion of pollutants plumes will effect the pollutants concentration especially when the deposition occurs over a long distance (Robson, 1983). An analytical model for air pollutant transport and dispersion from a point source is studied (Donald, 1976). This paper addresses an analytical model of air pollutant dispersion and removal from a step-function type elevated line sources. The removal (washed out) of pollutant depends on the nature of gaseous effluents emitted from the sources and the percentage of water vapors present in the atmosphere. The step-function type of line sources were assumed the series of industries or highways on during certain period of time and not continuous. The results were validated by comparing that of Robson (1987) by chemically non-reactive and time independent case. It is interesting to note that when there is no chemical reaction of primary pollutants, the results of the present model coincide with the results of the model presented by Robson (1987). This paper analyzed unsteady state dispersion of pollutants from a line source into stable atmosphere boundary layer. The quadratic diffusion coefficient and step-function type line source with variable wind profile like constant, constant shear and parabolic wind are incorporated. We adopted multiple Laplace transform and Green's Function technique for the effective solution. In most of the previous models, numerical methods used either fully or partly. In model where the Laplace transform is generally inverted numerically (Robson, 1987). Whereas in this model, the multiple Laplace transform inverted analytically this gives an exact solution. The solutions of this model verified with that of (Robson, 1987) by assuming no delayed removal (chemical reaction rate) and time independent case.

METHODOLOGY

We consider the whole atmospheric boundary layer as subdivided into two layers viz: (i) the surface layer (SL) that is, $0 \le z \le z_o$ below the source height and (ii) the layer above SL and up to the inversion layer that is, protected zone $z_o \le z \le H$ above source height. Here, H is the height of SL and z_o is the source height. The situation analyzed is that the chemically reactive pollutant was emitted from an elevated line source, which is of step-function type. It is assumed that the emission of contaminant are in the gaseous state from a time dependent elevated line source of infinite length into a stable ABL, the variable wind velocity depending on the nature of the surface terrains viz. flat, buildup areas and mountains like elevated regions. The diffusion coefficient (exchange coefficient) was assumed to be of quadratic in nature, the boundary layer assumed to be of stable atmospheric boundary layer, where the mixing of pollutants is limited. To study the variation in the effect of pollutants, the various wind profiles like constant, constant shear and parabolic type are considered. The present work focused on the step-function type of source, where certain industries work during the certain allotted period and then switched off. The pollutant transport governed by the atmospheric advection-diffusion equation with delayed removal is as follows:

$$\frac{\partial C}{\partial t}+u\frac{\partial C}{\partial x}+v\frac{\partial C}{\partial y}+w\frac{\partial C}{\partial z}=\frac{\partial}{\partial x}\left(K_x\frac{\partial C}{\partial x}\right)+\frac{\partial}{\partial y}\left(K_y\frac{\partial}{\partial y}\right)+\frac{\partial}{\partial z}\left(K_z\frac{\partial C}{\partial z}\right)-K''C+S \tag{1}$$

Where, C is the concentration of pollutants in the atmosphere at any location (x, y, z) at time t; u, v, w are the components of velocity; K_x, K_y, K_z are the coefficients of eddy diffusivity (exchange coefficient) in the x, y, z directions respectively; K'' is the first order delayed removal rate, and S is the source of the effluents. The height of the atmospheric boundary layer is assumed to be z = H, from the surface.

The basic Equation (1) is simplified by assuming the following assumptions:

1. Model is of time dependent and advection dominates over horizontal diffusion

$$u\frac{\partial C}{\partial x} >> \frac{\partial}{\partial x}\left(K_x\frac{\partial C}{\partial x}\right)$$

2. The line source is in the direction of y-axis, which leads the concentration gradient and flux gradient becomes zero along y-direction.

$$v\frac{\partial C}{\partial y}=0 \quad \& \quad \frac{\partial}{\partial y}\left(K_y\frac{\partial C}{\partial y}\right)=0$$

3. The vertical diffusion dominates over advection

$$\frac{\partial}{\partial z}\left(K_z\frac{\partial C}{\partial z}\right) >> w\frac{\partial C}{\partial z}$$

4. The source (S) is of time dependent line source and is assumed to be

$$S = QW(t)\delta(x-x_0)\delta(z-z_0)$$

5. The diffusion coefficient (exchange coefficient) is of parabolic form $(1-z)^2$, suited for stable atmospheric boundary layer, where the penetration of pollutants through the inversion layer does not takes place.

Based on the aforementioned assumptions, the transport and diffusion Equation (1) becomes:

$$\frac{\partial C}{\partial t}+u\frac{\partial C}{\partial x}=\frac{\partial}{\partial z}\left(K_z\frac{\partial C}{\partial z}\right)-K''C+QW(t)\delta(x-x_0)\delta(z-z_0) \tag{2}$$

Initial and boundary conditions are:

$$C(x,y,z)=0\,, \quad at\; t=0, \qquad \forall\, x, z \in R^+ \tag{3a}$$

$$C(x,y,z)=0\,, \quad at\; x=0, \qquad t \ge 0, \quad z \ge 0 \tag{3b}$$

$$K_z\frac{\partial C(x,y,z)}{\partial z}=0, \quad z=1 \qquad \forall\; x,t \ge 0 \tag{3c}$$

$$K_z\frac{\partial C(x,y,z)}{\partial z}=0, \quad z=0 \qquad \forall\; x,t \ge 0 \tag{3d}$$

W (t) = $w_o U(t-t_o) = \begin{cases} w_o\,, & 0 < t \le t_o \\ 0\,, & t > t_o \end{cases}$ ------ step-function type. (4)

Introducing the non-dimensional parameters

$$Z=\frac{z}{H},\quad \frac{K_z}{K_o}=(1-Z)^2,\quad X=\frac{K_o}{H^2 u_o}x,\quad \frac{u(z)}{u_o}=\left[\frac{z}{H}\right]^n=Z^n,\quad V_D=\frac{K_o}{H^2}v_d$$

$$T=\frac{K_o}{H^2}t,\quad C=\frac{H<u>}{W_o Q}c,\quad \text{where } <u>=\frac{u_o}{n+1},\quad W(T)=\frac{W(t)}{w_o},$$

And $u_o = \frac{4.7u_* H}{kL}, K_o = \frac{k}{4.7}u_* L$ (Robson, 1987)

where: u_* = is friction velocity; L=the Monin-Obukovlenght; k=Vankarman constant; H=the boundary layer height and n=the constant decided on the stability of boundary layer.

In addition, non-dimensional source term is as follows:

$$W(T) = \frac{W(t)}{W_o} = U(T - T_o)$$

Incorporating all the aforementioned non-dimensional parameters in the Equation (2) and dropping out the capitals, we get:

$$\frac{\partial C}{\partial t} + z^n \frac{\partial C}{\partial x} = \frac{\partial}{\partial z}\left((1-z)^2 \frac{\partial C}{\partial z}\right) - \alpha C + \frac{U(t-t_o)}{n+1}\delta(x-x_0)\delta(z-z_0) \quad (5)$$

Where, $\alpha = \frac{K''H^2}{K_0}$ (6)

α is the non-dimensional chemical reaction.

Initial and boundary conditions are:

$$C(x,z,t) = 0, \quad at\ x = 0, \quad t \geq 0, z \geq 0 \quad (7a)$$

$$C(x,z,t) = 0, \quad at\ t = 0, \quad x \geq 0, z \geq 0 \quad (7b)$$

$$(1\text{-}z)^2 \frac{\partial C(x,z,t)}{\partial z} = 0, \quad at \quad z = 1 \quad (7c)$$

$$(1\text{-}z)^2 \frac{\partial C(x,z,t)}{\partial z} = 0, \quad at \quad z = 0 \quad (7d)$$

Solution

Taking Laplace transforms along t and using initial conditions, we get:

$$s\bar{C}(x,z,s) - C(x,z,0) + z^n \frac{\partial \bar{C}}{\partial x} = \frac{\partial}{\partial z}\left((1\text{-}z)^2 \frac{\partial \bar{C}}{\partial z}\right) - \alpha \bar{C} + \frac{\overline{W(s)}}{n+1}\delta(x-x_0)\delta(z-z_0) \quad (8)$$

Initial and Boundary conditions:

$$C(x,z,t) = 0 \quad at\ t = 0, \quad x \geq 0, z \geq 0 \quad (9a)$$

$$(1\text{-}z)^2 \frac{\partial \overline{C}(x,z,s)}{\partial z} = 0, \quad at \quad z = 1 \quad (9b)$$

$$(1\text{-}z)^2 \frac{\partial \overline{C}(x,z,s)}{\partial z} = 0, \quad at \quad z = 0 \quad (9c)$$

Where,

$$\bar{C}(x,z,s) = \int_0^\infty e^{-st} C(x,z,t)dt \quad (10)$$

Again, taking Laplace transform over (8) along x-axis, which assumed of semi-infinite region, then it becomes

$$s\hat{C}(p,z,s) + z^n\left(p\hat{C}(p,z,s) - \bar{C}(0,z,s)\right) = \frac{d}{dz}\left((1\text{-}z)^2 \frac{d\hat{C}}{dz}\right) - \alpha\hat{C} + \frac{\overline{W(s)}}{n+1}e^{-sx_o}\delta(z-z_o) \quad (11)$$

Where

$$\hat{C}(p,z,s) = \int_0^\infty e^{-st}\,\overline{C}(x,z,s)dt \quad (12)$$

Differential Equation (11) becomes

$$\frac{d}{dz}\left((1\text{-}z)^2 \frac{d\hat{C}}{dz}\right) - (\alpha + s + pz^n)\hat{C} = -\frac{\overline{W(s)}}{n+1}e^{-sx_o}\delta(z-z_0) \quad (13)$$

boundary conditions:

$$(1\text{-}z)^2 \frac{d\hat{C}(p,z,s)}{dz} = 0, \quad at \quad z = 1 \quad (14a)$$

$$(1\text{-}z)^2 \frac{d\hat{C}(p,z,s)}{dz} = 0, \quad at \quad z = 0 \quad (14b)$$

Case 1

When the wind profile is of constant (n = 0)

The movements of air near the earth's surface retarded by frictional effect proportional to the surface roughness. Thus, the nature of the terrain, the location and density of trees, the location and size of lakes, rivers, hills and building produces different wind velocity gradients in the vertical direction (Warn and Walker, 1967). The wind of constant type may be experienced in case of flat open country, lakes and seas, then the Equation (13) becomes

$$\frac{d}{dz}\left((1\text{-}z)^2 \frac{d\hat{c}}{dz}\right) - (\alpha + s + p)\hat{c} = -\overline{W(s)}e^{-px_o}\delta(z-z_0) \quad (15)$$

Under the boundary conditions:

$$(1\text{-}z)^2 \frac{d\hat{C}(p,z,s)}{dz} = 0, \quad at \quad z = 1 \quad (16a)$$

$$(1\text{-}z)^2 \frac{d\hat{C}(p,z,s)}{dz} = 0, \quad at \quad z = 0 \quad (16b)$$

The complete solution of equation 16a and b, through Green's function technique Beck et al. (1992) assumed to be of the following form:

$$\hat{C}(p,z,s) = \begin{cases} G_1(z/z_0), & z<z_0 \\ G_2(z/z_0), & z>z_0 \end{cases} \tag{17}$$

$$\hat{C}(p,z,s) = \begin{cases} -\dfrac{\overline{W(s)}\, e^{-px_0}\, \phi_a(z)\phi_b(z_0)}{A_1}, & z<z_0 \\[2ex] -\dfrac{\overline{W(s)}\, e^{-px_0}\, \phi_a(z_0)\phi_b(z)}{A_1}, & z>z_0 \end{cases} \tag{18}$$

Where, $\phi_a(z)$ and $\phi_b(z)$ are the two independent solutions of homogeneous linear ordinary differential equation,

$$\frac{d}{dz}\left((1\text{-}z)^2 \frac{d\hat{\phi}}{dz}\right) - (\alpha + s + p)\phi = 0 \tag{19}$$

Under the boundary conditions:

$$(1\text{-}z)^2 \frac{d\phi}{dz} = 0, \qquad at \quad z = 1 \tag{20a}$$

$$(1\text{-}z)^2 \frac{d\phi}{dz} = 0, \qquad at \quad z = 0 \tag{20b}$$

And are found to be

$$\chi_a = (1-z)^{\frac{-(\beta+1)}{2}} + R_1(1-z)^{\frac{(\beta-1)}{2}} \tag{21a}$$

$$\chi_b = (1-z)^{\frac{(\beta-1)}{2}} \tag{21b}$$

In addition, A_1 is obtained through Wronskian of

$$W[\phi_a, \phi_b]_{z=z_0} = \phi_a(z)\phi_b'(z) - \phi_a'(z)\phi_b(z) = \frac{A_1}{(1-z_o)^2} \tag{22}$$

By Wronskian A_1 found to be $A_1 = -\beta$
Where,

$$\beta = \sqrt{1 + 4(\alpha + s + p)} \tag{23}$$

and

$$R_1 = \frac{\beta+1}{\beta-1} \tag{24}$$

The arbitrary constant R_1 determined as shown in the Equation (24) by using the boundary conditions on the solution (20a). There of the solution $\varphi_a(z)$ and $\varphi_b(z)$ satisfies the b.c's., at z = 0 and z=1 respectively. Taking inverse Laplace transform using standard tables (Erdelyi et al., 1954) the solution (18) becomes:

$$C_1(x,z,t) = \delta(t-x)\frac{e^{-(\alpha+\frac{1}{4})x}}{2\sqrt{(1-z)(1-z_0)}}\left\{\frac{e^{-\frac{g_1^2}{4x}} + e^{-\frac{g_2^2}{4x}}}{\sqrt{\pi x}} + e^{\frac{-g_2+x^2}{2}} Erfc\left(\frac{g_2}{2\sqrt{x}} - \frac{\sqrt{x}}{2}\right)\right\}, \quad z<z_0 \tag{25}$$

Where,

$$g_1 = \log\left(\frac{1-z}{1-z_o}\right) \tag{26}$$

$$g_2 = \log\left(\frac{1}{(1-z)(1-z_o)}\right) \tag{27}$$

The complete solution for $z > z_o$ is obtained by interchanging z and z_o in the above solution.

Case 2

When the wind profile is of constant shear (n=1)

The wind profile of constant shear $u(z) = z$ is experienced in the metro cities of highly buildup areas. The velocity varies linearly with vertical height then differential equation (13) is reduced to the form for quadratic diffusion coefficient.

$$\frac{d}{dz}\left((1\text{-}z)^2 \frac{d\hat{C}}{dz}\right) - (\alpha + s + pz)\hat{C} = -\frac{\overline{W(s)}}{2} e^{-px_o}\delta(z - z_o) \tag{28}$$

Following boundary conditions,

$$(1\text{-}z)^2 \frac{d\hat{C}}{dz} = 0, \qquad at \quad z = 1 \tag{29a}$$

$$(1\text{-}z)^2 \frac{d\hat{C}}{dz} = 0, \qquad at \quad z = 0 \tag{29b}$$

The solution of Equation 29 through Green's function technique is assumed to be of the following form:

$$\hat{C}(p,z,s)=\begin{cases} -\dfrac{W(s)e^{-px_o}\psi_a(z)\psi_b(z_o)}{2A_2}, & z<z_0 \\ -\dfrac{W(s)e^{-px_o}\psi_a(z_o)\psi_b(z)}{2A_2}, & z>z_0 \end{cases} \quad (30)$$

Where A_2 is obtained through Wronskian of $\psi_a(z)$ & $\psi_b(z)$

$$W[\psi_a,\psi_b]_{z=z_o}=\psi_a\psi_b'-\psi_a'\psi_b=\frac{A_2}{(1-z_o)^2} \quad (31)$$

$\psi_a(z)$ and $\psi_b(z)$ are the two linearly independent solutions of the homogeneous differential equation:

$$\frac{d}{dz}\left((1\text{-}z)^2\frac{d\psi}{dz}\right)-(\alpha+s+pz)\psi=0 \quad (32)$$

Under the boundary conditions:

$$(1\text{-}z)^2\frac{d\psi}{dz}=0, \qquad at\ z=1 \quad (33a)$$

$$(1\text{-}z)^2\frac{d\psi}{dz}=0, \qquad at\ z=0 \quad (33b)$$

Let $(1-z)=\eta$

$$\psi''+\frac{2}{\eta}\psi'+\left(\frac{p}{\eta}-\frac{\alpha+p+s}{\eta^2}\right)\psi=0 \quad (34)$$

This is the Bessel's differential equation of first kind since the order ($\nu=\sqrt{1+4(\alpha+s+p)}$) of Equation 34 is assumed to be of non-integer when it is compared with the generalized Bessel's equation

$$R''+\left(\frac{1-2m}{x}-2\alpha\right)R'+\left[p^2a^2x^{2p-1}+\alpha^2+\frac{\alpha(2m-1)}{x}+\frac{m^2-p^2\upsilon^2}{x^2}\right]R=0 \quad (35)$$

(Sherwood and Reed, 1939), and the solutions are found to be

$$\psi_a(z)=\frac{1}{\sqrt{\eta}}\left[J_{-\nu}(2\sqrt{p\eta})+R_2J_\nu(2\sqrt{p\eta})\right] \quad (36)$$

ν is not an integer

$$\psi_b(z)=\frac{J_\nu(2\sqrt{p\eta})}{\sqrt{\eta}} \quad (37)$$

Where,

$$\nu=\sqrt{1+4(\alpha+s+p)} \quad (38)$$

An arbitrary constant R_2 is determined by imposing the boundary condition at z = 0 on Equation (35)

$$R_2=\frac{1}{p^\nu}\frac{\Gamma(1+\nu)}{\Gamma(1-\nu)}\left(\frac{\nu+1}{\nu\text{-}1}\right) \quad (39)$$

Wronskian determines A_2 with help of the above solutions we get:

$$A_2=-\frac{\sin\nu\pi}{\pi} \quad (40)$$

The solutions of Equation 32 followed by the boundary conditions, which is defined in two regions are follows:

$$\hat{C}(p,z,s)=\begin{cases} \dfrac{\pi\overline{W(s)}}{2}\dfrac{1}{\sqrt{\eta}}\left[J_{-\nu}(2\sqrt{p\eta})+R_2J_\nu(2\sqrt{p\eta})\right]\dfrac{J_\nu(2\sqrt{p\eta_o})}{\sin\nu\pi}, & z<z_0 \\ \dfrac{\pi\overline{W(s)}}{2}\dfrac{1}{\sqrt{\eta}}\left[J_{-\nu}(2\sqrt{p\eta_o})+R_2J_\nu(2\sqrt{p\eta_o})\right]\dfrac{J_\nu(2\sqrt{p\eta})}{\sin\nu\pi}, & z>z_0 \end{cases} \quad (41)$$

As limiting case, where $p\to 0$ that is, $x>>1$, for large values of x, $J_\nu(2\sqrt{p\eta})$ and $J_{-\nu}(2\sqrt{p\eta})$ are approximated (Beck et al., 1992) as follows:

$$J_\nu(2\sqrt{p\eta})\cong\frac{\left(\sqrt{p\eta}\right)^\nu}{\Gamma(1+\nu)} \quad (42)$$

$$J_{-\nu}(2\sqrt{p\eta})\cong\frac{\left(\sqrt{p\eta}\right)^{-\nu}}{\Gamma(1-\nu)}, \qquad \nu\notin \text{integers} \quad (43)$$

Then the solution for $z<z_0$ solution is

$$\psi_1(x,z,t) \cong U(t-x-t_0)\frac{e^{-(\alpha+\frac{1}{4})x}}{2\sqrt{(1-z)(1-z_0)}}\left\{F(x)+\frac{e^{\frac{-2h_2+x}{8}}}{2}Erfc\left(\frac{h_2}{2\sqrt{x}}-\frac{\sqrt{x}}{2}\right)\right\} \quad (44)$$

Where,

$$F(x)=\frac{e^{-\frac{h_1^2}{4x}}+e^{-\frac{h_2^2}{4x}}}{2\sqrt{\pi x}} \quad (45)$$

$$h_1=\log\left(\frac{1-z}{1-z_o}\right) \quad (46)$$

$$h_2=\log\left(\frac{1}{(1-z)(1-z_o)}\right) \quad (47)$$

This is the complete solution of the Equation (32) followed by the boundary conditions 33(a)- (b) for $z<z_o$. Similarly, the solution for $z>z_o$ can be obtained by interchanging z and z_o in the above solution. However, we dealt only with the case below the source height $z<z_o$, that is, at the ground level.(z=0).

Case 3

When the wind profile is of parabolic type (n = 2)

The wind profile of parabolic type $u(z)=z^2$ (when n = 2) experienced within the study area located near the deep valley and oceans, then the differential Equation (4.3.6) reduced to the form with diffusion coefficient (exchange coefficient) is of quadratic in nature.

$$\frac{d}{dz}\left((1\text{-}z)^2\frac{d\hat{C}}{dz}\right)-(\alpha+s+pz^2)\hat{C} = -\frac{\overline{W(s)}}{3}e^{-px_o}\delta(z-z_0) \quad (48)$$

Boundary conditions:

$$(1\text{-}z)^2\frac{d\hat{C}}{dz}=0, \qquad at\ \ z=1 \quad (49a)$$

$$(1\text{-}z)^2\frac{d\hat{C}}{dz}=0, \qquad at\ \ z=0 \quad (49b)$$

Solutions of Equation (48) through Green's function are assumed to be of the form:

$$\hat{C}(p,z,s)=-\overline{W(s)}e^{-px_o}\begin{cases}\dfrac{\xi_a(z)\xi_b(z_o)}{3A_3}, & z<z_0\\[2ex] \dfrac{\xi_a(z_0)\xi_b(z)}{3A_3}, & z>z_0\end{cases} \quad (50)$$

Where A_3 is determined by the Wronskian

$$W[\xi_a,\xi_b]_{z=z_0}=\xi_a\xi_b'-\xi_a'\xi_b=\frac{A_3}{(1-z_o)^2} \quad (51)$$

In addition, $\xi_a(z)$ and $\xi_b(z)$ are the linearly independent solutions of the differential equation.

$$\frac{d}{dz}\left((1\text{-}z)^2\frac{d\xi}{dz}\right)-(\alpha+s_1+pz)\xi=0 \quad (52)$$

With boundary conditions

$$(1\text{-}z)^2\frac{d\xi}{dz}=0, \qquad at\ \ z=1 \quad (53a)$$

$$(1\text{-}z)^2\frac{d\xi}{dz}=0, \qquad at\ \ z=0 \quad (53b)$$

Taking,

$$\mu=2\sqrt{p}(1-z) \quad (54)$$

in the above differential equation then it reduces to the form

$$\mu^2\frac{d^2\xi}{d\mu^2}+2\mu\frac{d\xi}{d\mu}+\left\{\frac{-\mu^2}{4}+\lambda_2\mu-\lambda_1\right\}\xi=0 \quad (55)$$

Taking the transformation on the independent variable

$$\xi(\mu)=\frac{1}{\mu}W(\mu) \quad (56)$$

in Equation (55), we get,

$$\frac{d^2W}{d\mu^2}+\left\{\frac{-1}{4}+\frac{\lambda_2}{\mu}-\frac{\lambda_1}{\mu^2}\right\}W=0 \quad (57)$$

This is a Whittaker's differential equation, where $\lambda_1=\alpha+s+p$

and $\lambda_2 = \sqrt{p}$ the two independent solutions of Equation (52) are the Whittaker's Confluent Hyper geometric function as follows:

$$W_a(\mu) = W_{k,-m}(\mu) + R_3 W_{k,m}(\mu) \tag{58}$$

$$W_b(\mu) = W_{k,m}(\mu) \tag{59}$$

Where,

$$W_{k,m}(\mu) = \mu^{\frac{1}{2}+m} e^{-\frac{\eta}{2}} F(\frac{1}{2} - k + m, 1 + 2m, \mu) \tag{60}$$

$$W_{k,-m}(\mu) = \mu^{\frac{1}{2}-m} e^{-\frac{\eta}{2}} F(\frac{1}{2} - k - m, 1 - 2m, \mu) \tag{61}$$

and also Inter-relation between Whittaker's to Bessel's function is derived as follows (Carslaw and Jeager)

$$F(\alpha, \beta, \frac{a^2}{4t}) = e^{\frac{a^2}{4t}} (\alpha - \beta)!(\beta - 1)!\left(\frac{a}{2}\right)^{1-\beta} .t^{\beta-\alpha} .p^{\frac{\beta}{2}-\alpha+\frac{1}{2}} J_{\beta-1}\left(ap^{\frac{1}{2}}\right) \tag{62}$$

Where,

$$p^n = \frac{t^n}{(-n)!} \tag{63}$$

Where, $F(\alpha, \beta, x)$ is the confluent Hyper geometric functions and is defined as

$$F(\alpha, \beta, x) = \sum_{k=0}^{\infty} \frac{(\alpha)_k}{k!(\beta)!} x^k \tag{64}$$

By the transformations (62), the solution (60) and (61) becomes

$$W_a(\mu) = \sqrt{\mu}\, e^{\frac{\mu}{2}} \left[J_{-2m}(2\sqrt{\frac{\mu}{\pi}}) + R_3 J_{2m}(2\sqrt{\frac{\mu}{\pi}}) \right] \tag{65}$$

$$W_b(\mu) = \sqrt{\mu}\, e^{\frac{\mu}{2}} \left[J_{2m}(2\sqrt{\frac{\mu}{\pi}}) \right] \tag{66}$$

Then complete solution of the Equation (55) becomes

$$\xi_a(z) = \frac{e^{\frac{\mu}{2}}}{\sqrt{\mu}} \left[J_{-2m}(2\sqrt{\frac{\mu}{\pi}}) + R_3 J_{2m}(2\sqrt{\frac{\mu}{\pi}}) \right] \tag{67}$$

$$\xi_a(\mu\mu = \frac{e^{\frac{\mu}{2}}}{\sqrt{\mu}} \left[J_{2m}(2\sqrt{\frac{\mu}{\pi}}) \right] \tag{68}$$

The solution (49) using standard tables, Erdyle et al. (1954), Schaum Series (1950) Beck et al. (1992), and the inverse Laplace transform can be carried out for the solutions assuming $Xo = 0$ we get,

$$\hat{C}(p,z,s) = -\frac{\overline{W(s)} e^{\left(\frac{\mu+\mu_o}{2}\right)}}{3\sqrt{\mu\mu o}} \begin{cases} \dfrac{\left[J_{-2m}(2\sqrt{\frac{\mu}{\pi}}) + R_3 J_{2m}(2\sqrt{\frac{\mu}{\pi}}) \right] J_{2m}(2\sqrt{\frac{\mu o}{\pi}})}{A_3}, & z<z_0 \\ \dfrac{\left[J_{-2m}(2\sqrt{\frac{\mu_o}{\pi}}) + R_3 J_{2m}(2\sqrt{\frac{\mu_o}{\pi}}) \right] J_{2m}(2\sqrt{\frac{\mu}{\pi}})}{A_3}, & z>z0 \end{cases} \tag{69}$$

Where R_3 can be determined through on of the boundary conditions at z= 0, and is found to be

$$R_3 = -\left(\frac{\pi^2}{4p}\right)^m \left[\frac{2\sqrt{p} - 2m - 1}{2\sqrt{p} + 2m - 1}\right] \tag{70}$$

and A_3 is determined though Wronskian and is found to be

$$A_3 = -m \frac{e^{2\sqrt{p}(1-zo)}}{\sqrt{p}} \tag{71}$$

Then Equation (69) becomes by assuming that the concentration is far from the source, that is, distance $x >> 1$ or $p \to 0$. Then the Bessel's function is approximated

$$\hat{C}(p,z,s) = -\frac{\overline{W(s)}}{3 - m\frac{e^{2\sqrt{p}(1-zo)}}{\sqrt{p}}} \begin{cases} \dfrac{e^{\mu/2}}{\sqrt{\mu}} \left[\left(\frac{\mu}{\pi}\right)^{-m} + R_3 \left(\frac{\mu}{\pi}\right)^m \right] \dfrac{e^{\mu_o/2}}{\sqrt{\mu_o}} \left(\frac{\mu_o}{\pi}\right)^m, & z<z_0 \\ \dfrac{e^{\mu_o/2}}{\sqrt{\mu_o}} \left[\left(\frac{\mu_o}{\pi}\right)^{-m} + R_3 \left(\frac{\mu_o}{\pi}\right)^m \right] \dfrac{e^{\mu/2}}{\sqrt{\mu}} \left(\frac{\mu}{\pi}\right)^m, & z>z_0 \end{cases} \tag{72}$$

$$\hat{C}_1(p,z,s) = \overline{W(s)} \frac{e^{\sqrt{p}(z_o - z)}}{6\sqrt{(1-z)(1-z_o)}} \left[\frac{e^{-mh_1}}{m} - \left(\frac{\sqrt{p} - m - \frac{1}{2}}{\sqrt{p} + m - \frac{1}{2}} \right) \frac{e^{-mh_2}}{m} \right], \quad z < z_0 \tag{73}$$

Where,

$$h_1 = \log\left(\frac{1-z}{1-z_o}\right) \tag{74}$$

$$h_2 = \log\left(\frac{1}{(1-z)(1-z_o)}\right) \tag{75}$$

$$m=\sqrt{\frac{1}{4}+p+s+\alpha} \tag{76}$$

Taking inverse Laplace transform along s, we get

$$\bar{c}_1(p,z,s)=\frac{e^{\sqrt{p}(z_o-z)}}{6\sqrt{(1-z)(1-z_o)}}L^{-1}\left\{\overline{W(s)}\left[\frac{e^{-mh_1}}{m}-\left(\frac{\sqrt{p}-m-\frac{1}{2}}{\sqrt{p}+m-\frac{1}{2}}\right)\frac{e^{-mh_2}}{m}\right]\right\} \tag{77}$$

$$L^{-1}\{\overline{W(s)}\}=W(t)=U(t-t_o) \tag{78}$$

$$L^{-1}\left\{\frac{e^{-mh_1}}{m}-\left(\frac{\sqrt{p}-m-\frac{1}{2}}{\sqrt{p}+m-\frac{1}{2}}\right)\frac{e^{-mh_2}}{m}\right\}=e^{-(\frac{1}{4}+\alpha+p)t}L^{-1}\left\{\frac{e^{-h_1\sqrt{s}}}{\sqrt{s}}-\left(\frac{\sqrt{p}-\sqrt{s}-\frac{1}{2}}{\sqrt{p}+\sqrt{s}-\frac{1}{2}}\right)\frac{e^{-h_2\sqrt{s}}}{\sqrt{s}}\right\}$$

$$=e^{-(\frac{1}{4}+\alpha+p)t}\left\{\frac{e^{-\frac{h_1^2}{4t}}+e^{-\frac{h_2^2}{4t}}}{\sqrt{\pi t}}-2\left(-\frac{1}{2}+\sqrt{p}\right)Erfc\left(\frac{h_1}{2\sqrt{t}}\right)\right\} \tag{79}$$

$$L^{-1}\left\{\overline{W(s)}\left[\frac{e^{-mh_1}}{m}-\left(\frac{\sqrt{p}-m-\frac{1}{2}}{\sqrt{p}+m-\frac{1}{2}}\right)\frac{e^{-mh_2}}{m}\right]\right\}$$

$$=\frac{e^{\frac{-z_0^2}{4x}}}{6\sqrt{\pi x^3}(1-z_0)}\left\{\left[K_1F_1(t)-K_2F_2(t)-\frac{F_3(t)}{2(\alpha+\frac{1}{4})}\right]G_1(x,z_o)-z_o\left(K_3F_4(t)+K_4F_5(t)\right)\right\} \tag{80}$$

Where,

$$G_1(x,z_o)=zo\left(1+\frac{1}{x}\right)+2 \tag{81}$$

$$K_1(t)=\frac{e^{h_1\sqrt{\left(\frac{1}{4}+\alpha\right)}}}{4\alpha+1} \tag{82}$$

$$K_2(t)=\frac{e^{-h_1\sqrt{\left(\frac{1}{4}+\alpha\right)}}}{4\alpha+1} \tag{83}$$

$$K_3(t)=\frac{e^{h_1\sqrt{\left(\frac{1}{4}+\alpha\right)}}}{\sqrt{\alpha+\frac{1}{4}}} \tag{84}$$

$$K_4(t)=\frac{e^{-h_1\sqrt{\left(\frac{1}{4}+\alpha\right)}}}{\sqrt{\alpha+\frac{1}{4}}} \tag{85}$$

$$F_1(t)=Erfc\left(\frac{h_1}{2\sqrt{t}}+\sqrt{\left(\frac{1}{4}+\alpha\right)}\right)-Erfc\left(\frac{h_1}{2\sqrt{t-t_o}}+\sqrt{\left(\frac{1}{4}+\alpha\right)(t-t_o)}\right) \tag{86}$$

$$F_2(t)=Erfc\left(\frac{h_1}{2\sqrt{t}}-\sqrt{\left(\frac{1}{4}+\alpha\right)}\right)-Erfc\left(\frac{h_1}{2\sqrt{t-t_o}}-\sqrt{\left(\frac{1}{4}+\alpha\right)(t-t_o)}\right) \tag{87}$$

$$F_3(t)=e^{-(1/4+\alpha)t}Erfc\left(\frac{h_1}{2\sqrt{t}}\right)-e^{-(1/4+\alpha)(t-t_o)}Erfc\left(\frac{h_1}{2\sqrt{t-t_o}}\right) \tag{88}$$

$$F_4(t)=Erf\left(\frac{h_1}{2\sqrt{t}}+\sqrt{\left(\frac{1}{4}+\alpha\right)}\right)-Erf\left(\frac{h_1}{2\sqrt{t-t_o}}+\sqrt{\left(\frac{1}{4}+\alpha\right)(t-t_o)}\right) \tag{89}$$

$$F_4(t)=Erf\left(\frac{h_1}{2\sqrt{t}}\sqrt{\left(\frac{1}{4}+\alpha\right)}-\frac{h_1}{2\sqrt{t}}\right)-Erf\left(\sqrt{\left(\frac{1}{4}+\alpha\right)(t-t_o)}-\frac{h_1}{2\sqrt{t-t_o}}\right) \tag{90}$$

RESULTS AND DISCUSSION

This paper analyzed unsteady state dispersion of pollutants from a line source into stable atmosphere boundary layer. The quadratic diffusion coefficient and step-function type line source with variable wind profiles like constant, constant shear and parabolic wind are incorporated. We adopted multiple Laplace transform and Green's Function technique for the effective solution. In most of the previous models, numerical methods used either fully or partly. In model where the Laplace transform is generally inverted numerically (Robson, 1987). Whereas in this model, the multiple Laplace transform inverted analytically this gives an exact solution. The solutions of this model verified with that of (Robson, 1987) by assuming no delayed removal (chemical reaction rate) and time independent case. It is interesting to note that the effect of chemical reaction rte has virtually no impact t the short travel time but has significant effect t large travel time. It is interesting to note that when there is no chemical reaction of pollutant; the results of the present model coincide with the results of the model presented by Sulochana and Moka Shekhu (2009).

The results obtained in this model are illustrated graphically in the Figures 1 to 8. Concentration profiles of

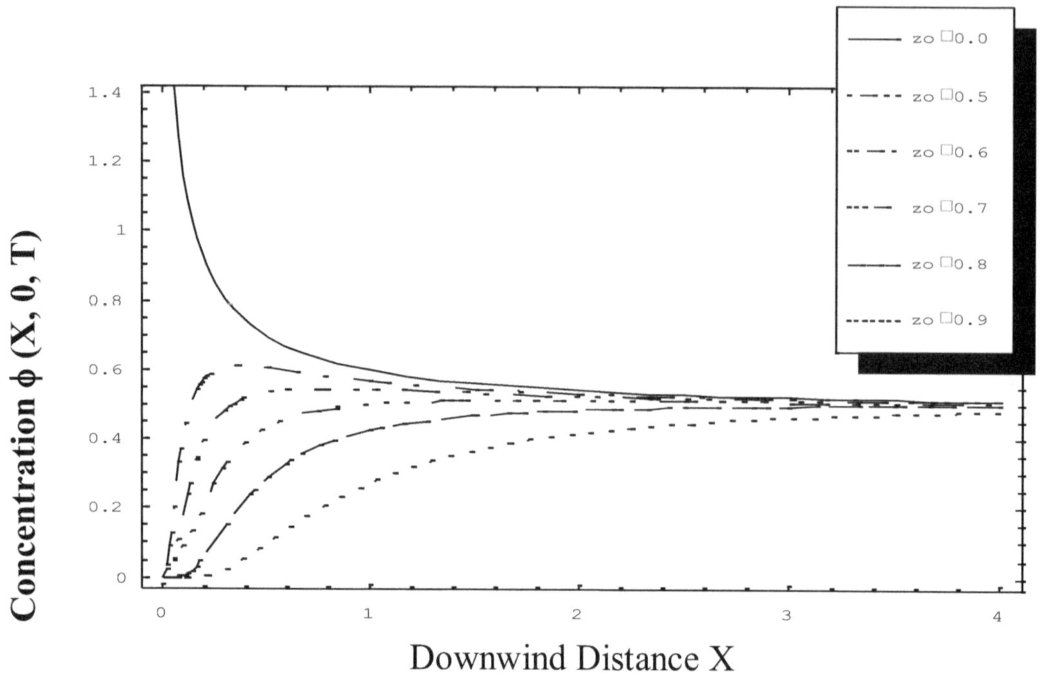

Figure 1. Ground level concentration for various source heights $z_o = 0, z_o = 0.5, z_o = 0.6, z_o = 0.7, z_o = 0.8, z_o = 0.9$ corresponding to the quadratic diffusion and Wind coefficient of constant shear. Comparing the results with Robson (1987) as limiting case where chemical reaction rate alpha=0.

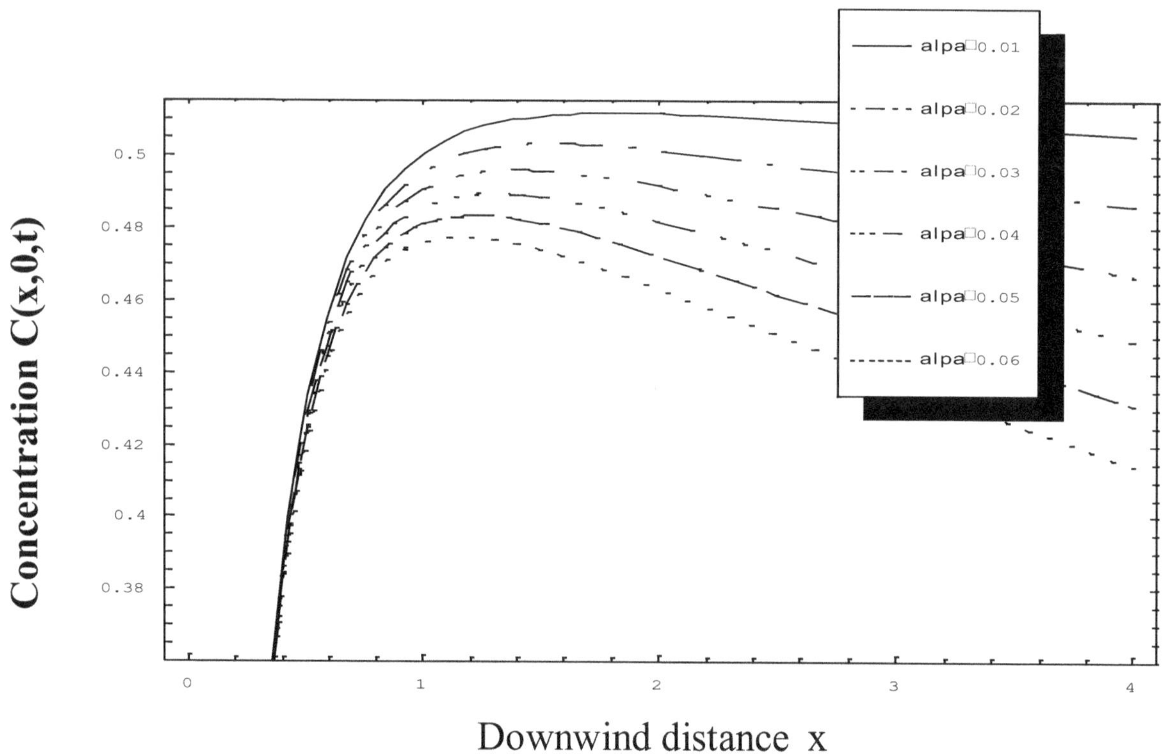

Figure 2. Ground level concentration for variable chemical reaction rate corresponding to the quadratic diffusion and wind coefficient of constant shear, $z_o = 0.0$.

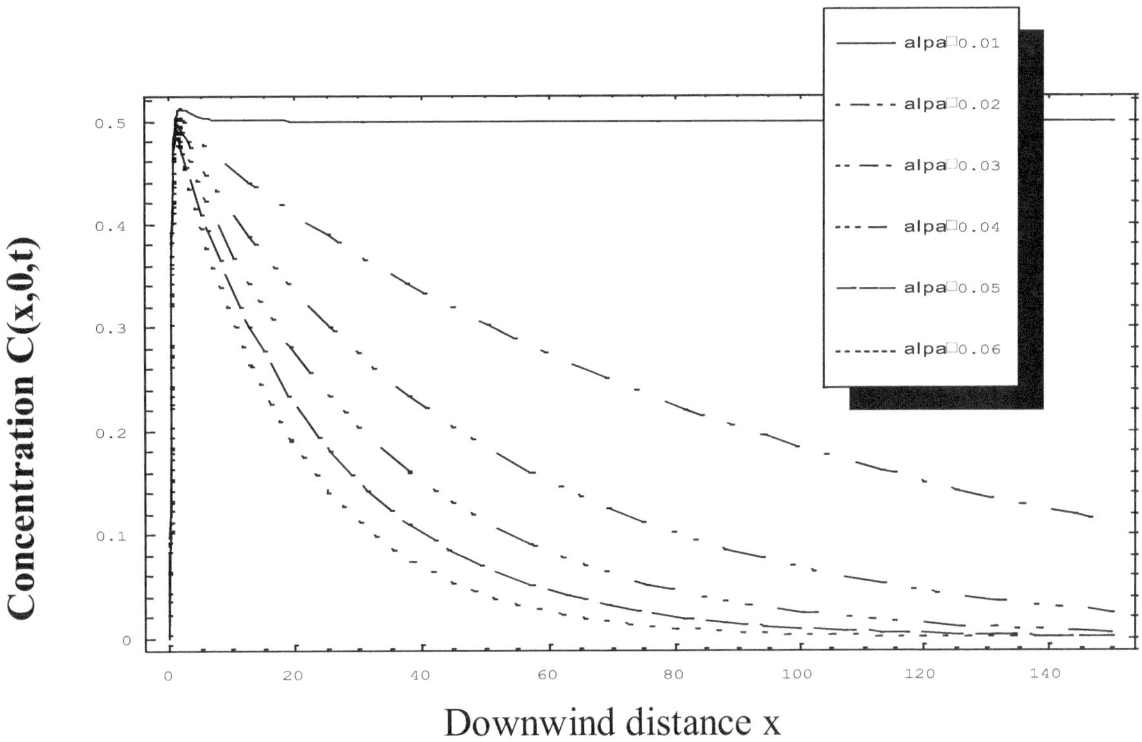

Figure 3. Ground level concentration for variable chemical reaction rate corresponding to the quadratic diffusion and wind coefficient of constant shear, $z_o = 0.7$.

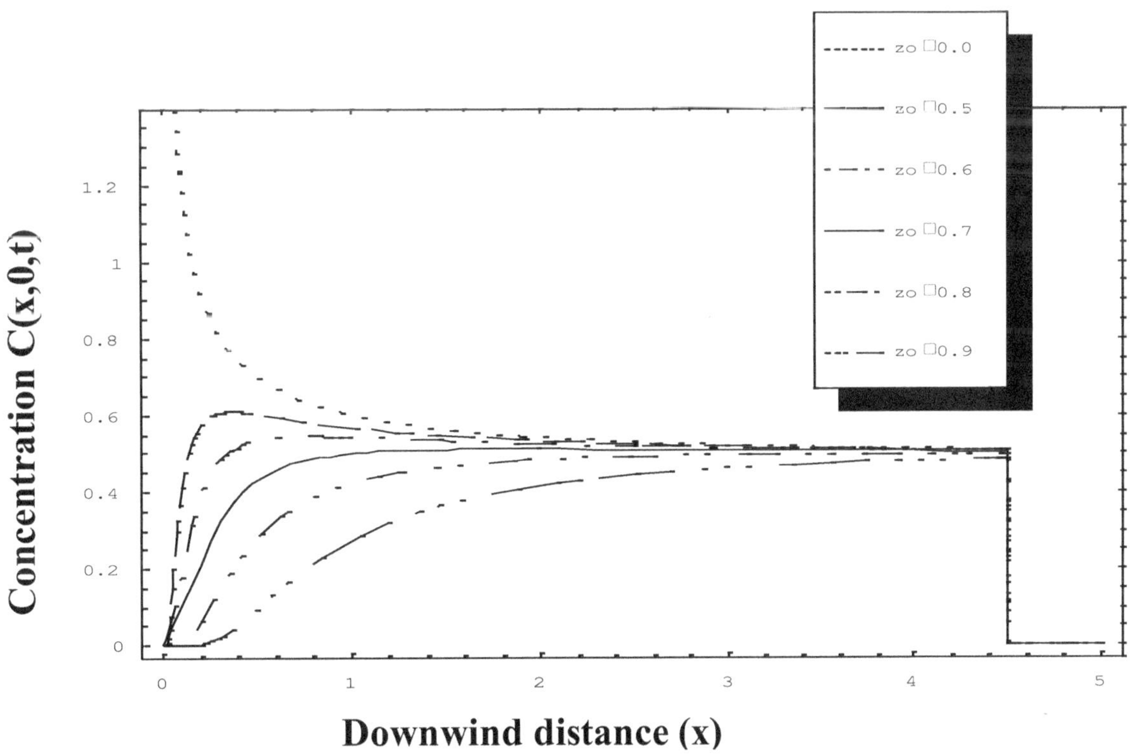

Figure 4. Ground level concentration vs. source heights corresponding to the quadratic diffusion and constant shear wind coefficient, chemical reaction rate $\alpha = 0.0$ at time $t = 5.0,\quad t_o = 0.5$.

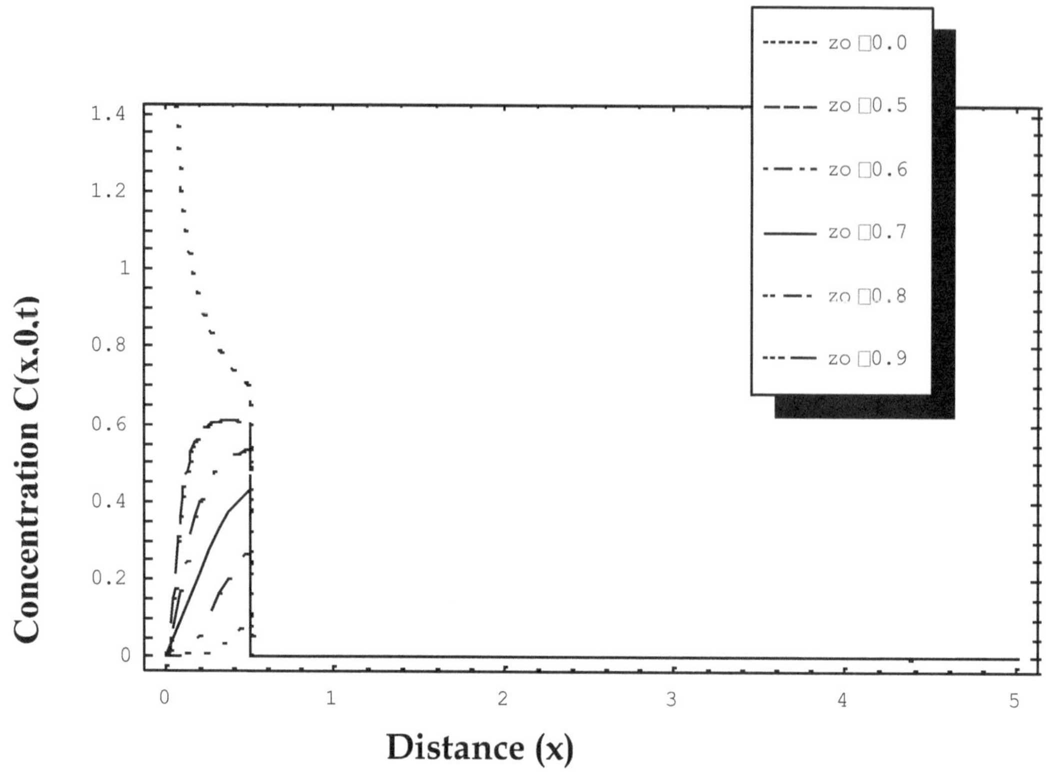

Figure 5. Ground level concentration versus source heights corresponding to the quadratic diffusion and constant shear wind coefficient, Chemical reaction rate $\alpha = 0.0$ at time $t = 1, \quad t_o = 0.5$

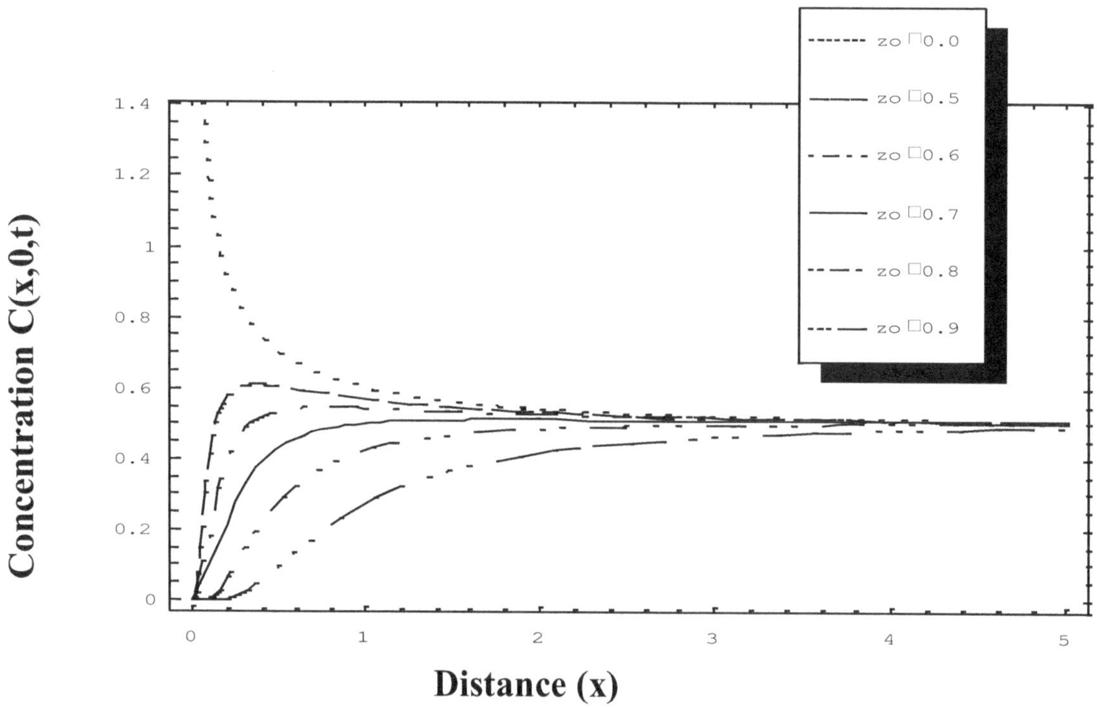

Figure 6. Ground level concentration versus source heights corresponding to the quadratic diffusion and constant shear wind coefficient, Chemical reaction rate $\alpha = 0.0$ at time $t = 15, \quad t_o = 0.5$

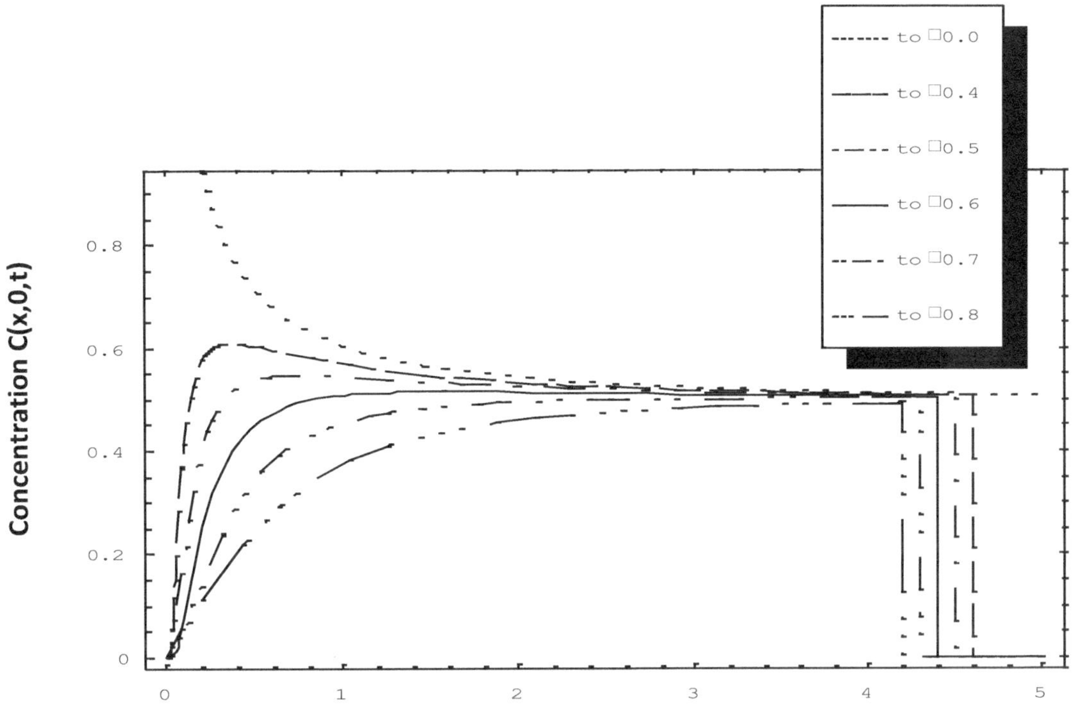

Figure 7. Ground level concentration versus downwind distance corresponding to the quadratic diffusion and constant shear wind coefficient, settling velocity for various time t_o Chemical reaction rate $\alpha = 0.0$ at time $t = 5, \quad z_o = 0.5$.

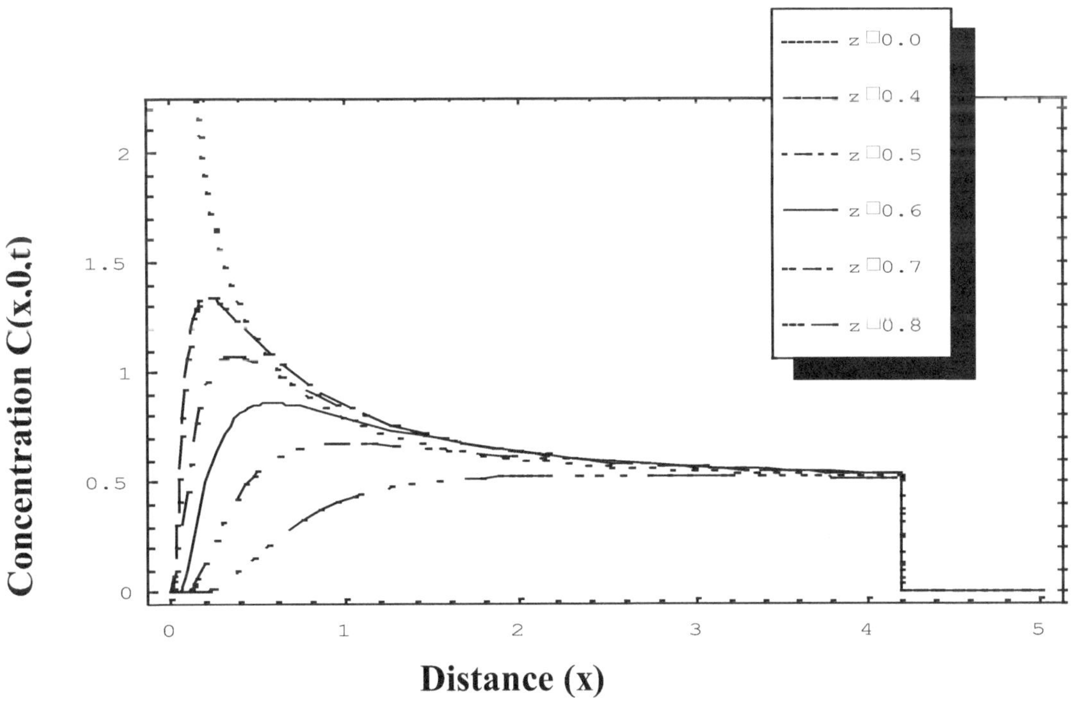

Figure 8. Ground level concentration versus distance for various heights on the surface to the quadratic diffusion and constant shear wind coefficient, chemical reaction rate $\alpha = 0.0$ at time $t = 5, \quad t_o = 0.5$

chemical reactions at large travel time are shown in Figures 2 to 8.

Nomenclatures

C (x, y, z) = Pollutant concentration (ppm)

H= height of the inversion layer (m)

z_o = height of the source (m)

(x_o, z_o) = the location of the source in xz -plan

K_z = Exchange coefficient along vertical direction

Q=Source strength at (x_o, z_o)

δ(.)= Dirac Delta function

K''=First order delayed removal

t = time in seconds

W (t) = time dependent source of step-function type and is of the form:

α = is the non-dimensional chemical reaction.

REFERENCES

Beck JV, Cole KD, Haji-Sheikh, Litkouhi (1992). Heat conduction Using Green's functions'. Hemisphere Publishing Corporation, Washington, DC.

Donald LE (1976). An analytical model for air pollutant transport and deposition from point source'. Atmos. Environ. 11:231-237.

Erdelyi A, Magnus W, Oberhettinger FG (1984). Tables of Integral Transforms, Vol. 1, McGraw-Hill, New York pp. 321-328.

Nieuwstadt, FTM (1984). Some aspects of the turbulent stable boundary layer. Bound. Layer Meteorol. 30:31-55.

Nieuwstadt FTM (1984). The turbulent structure of the stable, nocturnal boundary layer. J. Atmos. Sci. 41:2202–2216.

Robson RE (1987). Turbulent Dispersion in a Stable Layer with a Quadratic exchange Coefficient. Bound. Layer Meteorol. 39:207-218.

Robson RE(1983). On the theory of plume trapping by elevated inversion. Atmos. Environ. 17:1923-1930.

Sherwood TK, Reed CE (1939). 'Applied Mathematics in chemical Engineering.' Mc.Graw-Hill, New York.

Sulochana C, Moka S (2009). Dispersion of Chemically Reactive Pollutants emitted from an Instantaneous Line source into a Stable Atmospheric Boundary Layer. Pacific-Asian J. Math. 3(1-2):103-117.

Sulochana C, Moka S (2009). Mathematical Model of Dispersion and Diffusion of Air pollutant from a Step-function type line source into a stable atmospheric boundary layer. Indian J. Math. Math. Sci. 5(2):125-143.

Uma A, Sunitha G, Gupta RS (1991). Removal model suitable for air pollutants emitted from an elevated source. Appl. Math. Model. 15:386-389.

Inter-provincial cooperative control of air pollution in China

Jian Xue[1,2] and Changmin Li[1]

[1]School of Management, Shanghai University, 200444, Shanghai, P. R. China.
[2]Training Center of China Post Group Corporation, 050021, Hebei, Shijia Zhuang, P. R. China.

Based on the status of the air pollution in China, three basic assumptions are proposed. The China's inter-provincial cooperative control of air pollution (ICCAP) model is established by the combination of national optimization model and allocation model of cooperation benefit. A case study for the pollutant SO_2 control in China shows that the proposed model is superior to the current command-order regulation. Our proposed model not only solves the problem of conflicts over air pollution across regional boundaries but also efficiently utilizes the resources in the whole of China.

Key words: Air pollution, bubble policy, cooperative control, modified MCRS method.

INTRODUCTION

China's economy has skyrocketed, but at a price. Power plants, factories and heavy industries are all belching out black, dirty air, at the cost of our health and our environment. Major pollutants produced by human activity include: Sulfur oxides (SOx) especially sulfur dioxide are emitted from burning of coal and oil, Nitrogen oxides (NOx) especially nitrogen dioxide are emitted from high temperature combustion and so on. Nowadays, a kind of compound pollutant composed by Ozone and atmospheric haze is emerging. Smog-filled cities, in Pearl River Delta, Yangtze Delta, and Beijing-Tianjin-Hebei area and so, are ringed with heavy industry, metal smelters, and coal-fired power plants, all critical to keeping the fast growing economy going even as they spew tons of carbon, gases, and soot into the air (Report of the Chinese Academy of Environmental Sciences). China has become the severest polluted country in the world (Tao, 2011). With an increasing number of air pollution episodes and low visibility days reported by the media (Wang, 2011), much attention has been paid to reducing air pollutant emissions and to improving air quality across cities, municipalities, and provinces in China. Air pollution will become the biggest health threat in China unless the government takes greater steps to monitor and publicize the dangers of smog and take measure to reduce the pollution. China's air pollution control urgently calls for sustainable strategy.

Since 1970, air pollution control has attracted many researchers and become an especially hot topic for aspiring researchers invested in environmental program. Walter et al. (1976) applied linear programming method to the nationally emission control problems. Duan (1991), Hu (2000) and Pan et al. (2004) extended it to large scale cities to comprehensively improve their environmental quality. Zeng (1996) studied the natural cybernetics theory to facilitate the solution of the problems "environment and development". Halkos (1994) deals with the transboundary nature of the acid rain problem and presented general forms of cooperative and non-cooperative equilibria in explicit and implicit set-up of the model under the assumptions of complete and incomplete information (Halkos, 1993, 1996). Becker and Easter (1997) applied cooperative game theory for allocating water from the Great Lakes. By forming different partial coalitions, they concluded that the non-cooperative coalitions could be induced to cooperate ones and the grand coalition was shown to be the best solution for all players. Chander and Tulkens (1994) showed that, under certain conditions, there exists a very simple cost-sharing rule that not only lies in the core but that also sustains full cooperation as an equilibrium, whether countries are symmetric or asymmetric. Jorgensen and Zaccour (2001) derived a rule to share surplus gained when two countries or regions agree to coordinate their policies to reduce downstream pollution. International environmental agreements involving substantial emission reduction efforts are unlikely to be reached without provisions for international transfers. The reason is that, although there is generally a substantial surplus to be gained from cooperation, there are most often countries for which the abatement effort required by the world optimum is so large that they end up worse off under this world optimum compared to the non-cooperative laissez-faire situation. Germain et al. (2003) provided a transfer scheme for which a core property is proved analytically in a dynamic (closed-loop) game theoretic context. Grabisch and Lange (2007) and Hsiao and Raghavan (1993) extended cooperative games in characteristic function form to multichoice cooperative games. Hsial (1992) analyzed the necessity of allocating the optimal treatment cost fairly and reasonably among each polluter in total emission control was analyzed and the possibility of applying game theory to the issue of the optimal treatment cost allocation is discussed in this paper. Then the multiple-person cooperation game model of the optimal treatment cost allocation and the solution approaches including the method based on Shapley value, core method, CGA method, MCRS (Minimum Costs Remaining Savings) method and so on are delineated. Tijs and Driessen (1986) introduced a relatively fair Surplus distribution principle to investigate the cooperative control of transboundary water pollution in China's river basins. Chen and Hou (1999) investigated the optimization models of the coal-consumption cost, reserve cost and SO_2 emission cost of generators. Then, the Shapley model, as a means of solving multiplayer cooperative game, was applied to allocate profits between the generators and power supply companies.

Regarding the development pattern of China's air pollution control, we have experienced the evolvement of the sole pursuit of economic growth to coordinated, comprehensive and sustainable development. In the past decade, there has been an explosion of interest in pollution control field among researchers and policy-makers. There are three major channels for pollution reduction in China: command-order regulation, governmental macro-regulation (such as emission charges), and market regulation (such as emission trading). Tan et al. (2011) pointed out that none of these is perfect in practice. The current pollution control strategy in China is non-cooperative command-order regulation (NCCOR), under which each province separately complete his task after the central government gives his total emission target. There are two main questions in the process of using the NCCOR. Since any province has its peculiar industry structure, corporate ownership composition and etc, one questions that it does not fully take consideration of costs differentiation for various provinces; the other relates to integrating social resource. According to the non-cooperative means, it forbids pollution transfer. Some provinces may have stronger productive and pollution treatment ability, while others may insist it implies that it misses the opportunity for the optimal allocation of resources.

Based on the above analysis, this paper attempts to build the Inter-provincial Cooperative Control of Air Pollution (ICCAP). In view of the nonlinear optimization theory and cooperative game theory, we give full consideration to the significant difference in the cost functions for different provinces. And we finally take SO_2 pollutant control as an example to show the validity of our results.

INTER-PROVINCIAL COOPERATIVE CONTROL OF AIR POLLUTION IN CHINA

Basic assumptions

Assumption 1: China is in a bubble and all their pollution is trapped inside their world. "Bubble policy" is first established by American government in 1979. In this bubble, pollutant can be transferred freely among all provinces no matter how far they are.

Assumption 2: There is no environmental damage if the level of the pollutant in a region is below the maximum environmental carrying capacity, which is a percentage of the national annual pollutant reduction quota for that pollutant. We further assume that there is no environmental damage if the level of the pollutant in each province, is below the national standard for that pollutant.

Assumption 3: Pollutant reduction or transfer within the lake basin is controlled by the provinces themselves. Each province is capable of controlling the distribution of pollutants, including quantities retained by itself and transferred to other provinces.

Notations

The basic variables used in this paper are described as follows:

P_i: The annual pollutant reduction by province $i = 1, 2, \cdots, 31$

P_i^*: The optimal annual pollutant reduction by

province $i=1,2,\cdots,31$

P_{0i}: The annual initial pollutant quantity produced by province $i=1,2,\cdots,31$.

P_{1i} : Annual air pollutant industry quantity produced by province $i=1,2,\cdots,31$

P_{ei}: The maximum environmental carrying capacity of province $i=1,2,\cdots,31$ based on the national pollutant quality standard.

Sh_{ij}: The fee the province i pays province j for transferred pollutant cost, $i\neq j$

π_i: The environmental cost for province $i=1,2,\cdots,31$

π_q: The total environmental cost

AC_i: The reduction cost for province $i=1,2,\cdots,31$

W_i: Exhaust volume for province $i=1,2,\cdots,31$

α_i: The coefficients of the upper limits of annual pollutant quality produced by province $i=1,2,\cdots,31$

β_i: The coefficients of the lower limits of annual pollutant industry quality produced by province $i=1,2,\cdots,31$

γ_i: The coefficients of the upper limits of annual pollutant industry quality produced by province $i=1,2,\cdots,31$.

DESCRIPTION OF CHINA'S ICCAP OF AIR POLLUTION PROBLEM

Here, we develop the ICCAP model consisting of 31 provinces or decision-makers. There are two main parts for our model: the national optimization model in which 31 provinces try to minimize their total environmental costs under the total volume of waste gas emission cap constraint and the national standard for that pollutant associated with the "Bubble Policy"; as well as the allocation model of cooperation benefit by the modified MCRS method.

National optimization model

For province $i=1,2,\cdots,31$, the environmental cost is defined as follows:

$$\pi_i(P_i)=AC_i(P_i)+EC_i+Sh_{ij} \quad (1)$$

Wang (2011) and Wu (2007) explored the characterization of the reduction cost function. They showed that the reduction cost function can be defined as follows:

$$AC_i=\theta\cdot W_i^{\sigma}\cdot P_i^{\mu} \quad (2)$$

where θ,σ,μ are constants, W_i and P_i are the volume of waste gas emission and the volume of reduction waste gas for province $i=1,2,\cdots,31$.

In view of Assumption 2, there is no environmental damage if the level of the pollutant in each province $i=1,2,\cdots,31$, is below the national standard for that pollutant, that is

$$EC_i=0.$$

Thereby, for any province its environmental cost

$$\pi_i(P_i)=AC_i(P_i)+\sum_{\substack{j=1\\ j\neq i}}^{31}Sh_{ij},\quad i=1,2,\cdots,31. \quad (3)$$

Since the waste gas can flow among all 31 provinces in the national bubble and Sh_{ij} represents the rewards or the punishment due to the waste gas was reduced by itself or the others, the aggregation of the fee between any two provinces is zero, that is,

$$\sum_{i=1}^{31}\sum_{\substack{j=1\\ j\neq i}}^{31}Sh_{ij}=0$$

Combining (3), the national total reduction cost is

$$\pi_q=\sum_{i=1}^{31}\pi_i=\sum_{i=1}^{31}AC_i(P_i) \quad (4)$$

According to Assumption 3, assumed that the air treatment facilities of province i can handle the quality of pollutant no more than α_i times of maximum environmental carrying capacity,

$$P_{0i}-P_i\leq\alpha_i\cdot P_{ei} \quad (5)$$

where $\alpha_i\geq 1$.

Furthermore, province $i\in I$ minimizes its environmental cost (π_i) by adjusting its pollutant reduction (r_i) under the unified transferrable pollutant price, synchronously subject to the following industry quantity produced:

$$\beta_i\cdot P_{1i}\leq P_i\leq\gamma_i\cdot P_{1i}\quad i=1,2\cdots31 \quad (6)$$

where β_i and γ_i represent the coefficients of the lower and upper limits of annual pollutant industry quality produced by province $i=1,2,\cdots,31$.

There is no environmental damage if the level of the pollutant in all regions is below the total national standard for that pollutant, that is,

$$\sum_{i=1}^{31} P_{0i} - \sum_{i=1}^{31} P_i \le \sum_{i=1}^{31} P_{ei} \,. \qquad (7)$$

Meanwhile,

$$\sum_{i=1}^{31} P_i = \sum_{i=1}^{31} P_{0i} - \sum_{i=1}^{31} P_{ei} \qquad (8)$$

From the above analysis, we can obtain the optimal reduction $P^{(*)} = (P_1^{(*)}, P_2^{(*)}, \cdots, P_{31}^{(*)})^T$ by minimizing the objective function (4) with the constraints (5) to (8).

Compared with the command-control strategy, the reduction costs for some provinces are decreasing under the proposed reduction $P^{(*)} = (P_1^{(*)}, P_2^{(*)}, \cdots, P_{31}^{(*)})^T$ at the price of others' reduction costs are inevitably increasing due to the cooperative principle. This policy cannot be carried out well and put into practice efficiently since it is contrary to the individual rationality. Therefore, the key point is to design a popular tactics taking the individual rationality and the national optimization into account.

Allocation model of cooperation benefit by the modified MCRS method

Cooperative cost distribution model has been a subject of substantial researches, a number of solution methods have been proposed, such as Shapley value, CGA (Cost Gap Allocation), MCRS and core methods (Tijs and Driessen, 1986). Considering the computation complexity, we modify the MCRS method to solve the benefit distribution problem due to the large scale decision-makers.

The MCRS method are described as follows (Tijs and Driessen, 1986): The benefit for any $k \in K$ is

$$\bar{Z}_k = Z_{k\min} + \frac{Z_{k\max} - Z_{k\min}}{\sum_{k\in K}(Z_{k\max} - Z_{k\min})}\left[C(K) - \sum_{k\in K} Z_{k\min}\right], \quad \forall k \in K, \qquad (9)$$

where K is a set, $C(K)$ is the total cost computed before cooperation, $Z_{k\max}$ and $Z_{k\min}$ are constants, and can be computed as follows (Chen and Hou, 1999):

$$Z_{k\min} = C(K) - C(K - \{k\}), \qquad Z_{k\max} = X_k, \qquad \forall k \in K$$
(10)

In view of the above analysis, (9) and (10) are the way to solve cooperative cost distribution. In order to deal with the cooperative benefit allocation problem in our problem, we are now in the position to propose the modified MCRS method. For any $i = 1, 2, \cdots, 31$, we modified (9) and (10), and let

$$Z_i^* = Z_{i\max} + \frac{Z_{i\max}}{\sum_{i=1}^{31} Z_{i\max}}\left[\pi_q - \sum_{i=1}^{31} Z_{i\max}\right],$$

where

$Z_{i\max} = \pi_q - C(\{1, 2, \cdots, 31\} - \{i\})$, and $C(\{1, 2, \cdots, 31\} - \{i\})$ represents all provinces become an alliance except province *i*.

CASE STUDY FOR CHINA'S SO_2 REDUCTION

According to seven years (2003 to 2009) of statistical data from China Environment Yearbook, the estimation of the parameters of the SO_2 reduction cost functions are obtained by means of SPSS software. Table 1 gives the acceptable results according to *t*-test, p-value and R^2-test.

Taking the date of the volume of waste gas emission $W_i, i = 1, 2, \cdots, 30$ for each province in 2009 into Table 1, then we can obtain the reduction cost function for each province and the total reduction cost function. Drawing on the experience of pollutant SO_2 control in Zhejiang province, for $i = 1, 2, \cdots, 30$, we adopt $\alpha_i = 1.3,\ \beta_i = 0.4,\ \gamma_i = 0.9$. Recalling (4) to (8), we obtain the national optimization model as follows:

$$\min_{P_i} \quad \pi_q = 219.23 \cdot P_1^{1.835} + 26.03 \cdot P_2^{2.323} + \cdots + 5571.33 \cdot P_{30}^{-0.253}$$

$$\begin{aligned} s.t. \quad & 5104.3 - \sum_{i=1}^{30} P_i \le 2307.24, \\ & \sum_{i=1}^{30} P_i = 2929.9, \\ & 2.4 \le P_1, \qquad 6.9 \le P_1 \le 15.5, \\ & 17.5 \le P_2, \qquad 17.2 \le P_2 \le 38.7, \\ & 78 \le P_3, \qquad 91.2 \le P_3 \le 205.2, \\ & \qquad \cdots\cdots \\ & 0 \le P_{30}, \qquad 21.8 \le P_{30} \le 49. \end{aligned}$$

Using Matlab software, the optimal reduction for each province is computed as listed in Table 2.

According to Table 3, under the national optimization model, the total reduction cost is $3534678.1 thousand in 2009. Compared with the current NCCOR, the national optimization model reduces the total environmental cost by $1132385.3 thousand (24.26% of the current measures).

Table 3 shows that, after cooperation, all most provinces' reduction cost (costs for Heilongjiang province

Table 1. Reduction cost functions for each province.

Province	Reduction cost
Beijing	$AC_1 = 1398404633 \cdot W_1^{-1.867} \cdot P_1^{1.835}$
Tianjin	$AC_2 = 7.939 \cdot 10^{11} \cdot W_2^{-2.776} \cdot P_2^{2.323}$
Hebei	$AC_3 = 0.000309019 \cdot W_3^{-1.616} \cdot P_3^{0.529}$
Shanxi	$AC_4 = 9.684 \cdot 10^{-27} \cdot W_4^{7.841} \cdot P_4^{-1.499}$
Inner Mongolia	$AC_5 = 276.79671 \cdot W_5^{-0.181} \cdot P_5^{1.52}$
Liaoning	$AC_6 = 0.1276909 \cdot W_6^{0.073} \cdot P_6^{2.784}$
Jilin	$AC_7 = 0.100893 \cdot W_7^{0.911} \cdot P_7^{1.102}$
Heilongjiang	$AC_8 = 0.000186941 \cdot W_8^{1.638} \cdot P_8^{1.044}$
Shanghai	$AC_9 = 7.656 \cdot 10^{-11} \cdot W_9^{3.342} \cdot P_9^{0.995}$
Jiangsu	$AC_{10} = 282.8921 \cdot W_{10}^{-0.194} \cdot P_{10}^{1.66}$
Zhejiang	$AC_{11} = 2.463 \cdot 10^{-15} \cdot W_{11}^{4.907} \cdot P_{11}^{-0.523}$
Anhui	$AC_{12} = 50.097014 \cdot W_{12}^{0.321} \cdot P_{12}^{0.830}$
Fujian	$AC_{13} = 327.72937 \cdot W_{13}^{-0.009} \cdot P_{13}^{1.415}$
Jiangxi	$AC_{14} = 1.075229584 \cdot W_{14}^{0.502} \cdot P_{14}^{1.322}$
Shandong	$AC_{15} = 0.005465597 \cdot W_{15}^{1.143} \cdot P_{15}^{1.056}$
Henan	$AC_{16} = 2.397 \cdot 10^{9} \cdot W_{16}^{-2.198} \cdot P_{16}^{2.463}$
Hubei	$AC_{17} = 11602766505 \cdot W_{17}^{-3.212} \cdot P_{17}^{4.175}$
Hunan	$AC_{18} = 0.0068699 \cdot W_{18}^{1.908} \cdot P_{18}^{-0.368}$
Guangdong	$AC_{19} = 11675312 \cdot W_{19}^{-1.286} \cdot P_{19}^{1.802}$
Guangxi	$AC_{20} = 2.653468703 \cdot W_{20}^{0.09} \cdot P_{20}^{2,134}$
Hainan	$AC_{21} = 3.132 \cdot 10^{-6} \cdot W_{21}^{2.843} \cdot P_{21}^{0.697}$
Chongqing	$AC_{22} = 33.13797558 \cdot W_{22}^{0.715} \cdot P_{22}^{0.295}$
Sichuan	$AC_{23} = 329.0780414 \cdot W_{23}^{-0.252} \cdot P_{23}^{1.89}$
Guizhou	$AC_{24} = 8.205853697 \cdot W_{24}^{0.352} \cdot P_{24}^{1.246}$
Yunnan	$AC_{25} = 1.24228833 \cdot W_{25}^{0.682} \cdot P_{25}^{1.036}$
Shaanxi	$AC_{26} = 2.879 \cdot 10^{-5} \cdot W_{26}^{2.202} \cdot P_{26}^{0.19}$
Gansu	$AC_{27} = 2.525 \cdot 10^{-7} \cdot W_{27}^{2.608} \cdot P_{27}^{0.877}$
Qinghai	$AC_{28} = 261465811 \cdot W_{28}^{-1.541} \cdot P_{28}^{2.818}$
Ningxia	$AC_{29} = 2.031 \cdot 10^{-8} \cdot W_{29}^{3.115} \cdot P_{29}^{0.38}$
Xinjiang	$AC_{30} = 10.618345 \cdot W_{30}^{0.708} \cdot P_{30}^{-0.253}$

Tibet province is removed from this table since its actual reduction is near to zero.

Table 2. SO_2 reduction results for China in 2009 (10^4 t).

Province	Actual reduction	Optimal reduction
Beijing	11.3	6.9
Tianjin	25.7	17.5
Hebei	123.8	181
Shanxi	143.4	219.9
Inner Mongolia	164.6	121.1
Liaoning	102.2	77.7
Jilin	17.1	38.5
Heilongjiang	19.7	19.7
Shanghai	38.5	25
Jiangsu	185.6	140.2
Zhejiang	130.6	178.4
Anhui	161.9	189.5
Fujian	35.5	30.1
Jiangxi	143.0	124.2
Shandong	273.3	213.7
Henan	141.4	103.6
Hubei	88.5	65.5
Hunan	84.8	134.7
Guangdong	187.4	146.4
Guangxi	87.8	68.5
Hainan	6.0	5.4
Chongqing	78.0	123
Sichuan	83.8	71.4
Guizhou	131.8	94
Yunnan	141.7	126
Shaanxi	74.5	133.8
Gansu	191.3	168.2
Qinghai	5.6	5.6
Ningxia	29.3	51.4
Xinjiang	21.8	49

and Qinghai province do not change) become lower than before. It implies that the cooperative pollution control is reasonable and acceptable. In comparison with the non-cooperative command-order regulation, the inter-provincial air pollution cooperatively control model not only effectively solves the environmental and social resource allocation, but also gain a "win-win" effect. It can well make a compromise between the economic development and the environment protection.

Conclusion

Recently, China is vigorously promoting energy conservation to improve air quality. Our research is just considered in such a macro background that the Twelfth Five-Year Plan of China clearly points out the target of energy conservation. Under the national discharge pollutant standard, we proposed the China's inter-provincial cooperative control of air pollution model. After computing the optimal reduction volume, we design the allocation

Table 3. Comparison of reduction costs in 2009 ($\$10^4$).

Province	SO_2 reduction cost in NCCOR (A)	Benefit allocation by modified MCRS method (B)	Reduction cost after benefit allocation in (C)=(A)-(B)	ICCAP Operative Optimal SO_2 reduction cost before benefit allocation (D)	Money pay to/(receive from) other provinces (E)=(C)-(D)
Beijing	2759.25	1288.33	1470.92	1116.04	354.87
Tianjin	7214.84	3602.99	3611.85	2954.85	657.00
Hebei	23438.46	430.41	23008.04	28654.41	-5646.36
Shanxi	16940.50	15227.38	1713.13	8924.85	-7211.72
Inner Mongolia	15284.31	1476.85	13807.46	9586.32	4221.13
Liaoning	15458.85	6134.49	9324.37	7207.79	2116.57
Jilin	1093.77	171.62	922.14	2675.15	-1753.00
Heilongjiang	2185.03	0.00	2185.03	2185.00	0.00
Shanghai	10109.24	2620.90	7488.34	6578.68	909.66
Jiangsu	33274.46	8025.93	25248.53	20886.76	4361.76
Zhejiang	27073.03	8342.06	18730.97	22998.53	-4267.56
Anhui	11025.41	740.74	10284.67	12564.26	-2279.60
Fujian	6909.36	994.62	5914.74	5470.59	444.15
Jiangxi	10359.93	143.41	10216.51	8598.68	1617.84
Shandong	46986.25	4945.30	42040.96	36236.76	5804.19
Henan	19585.97	6900.85	12685.12	9103.68	3581.44
Hubei	15795.24	9382.45	6412.79	4496.18	1916.61
Hunan	10072.18	6026.25	4045.93	8495.00	-4449.07
Guangdong	53364.22	15313.63	38050.59	34200.00	3850.59
Guangxi	12870.63	3745.35	9125.28	7577.79	1547.48
Hainan	1281.42	34.52	1246.91	1190.69	56.21
Chongqing	15000.21	1783.74	13216.46	17157.35	-3940.89
Sichuan	19100.74	4229.81	14870.91	14112.65	758.27
Guizhou	12365.47	638.71	11726.76	8115.44	3611.32
Yunnan	17246.07	1913.13	15332.94	14177.06	1155.88
Shaanxi	7630.35	4452.61	3177.74	8528.24	-5350.50
Gansu	30383.51	1235.66	29147.87	27141.18	2006.69
Qinghai	18556.18	0.00	18556.18	18555.88	0.00
Ningxia	2965.80	1110.94	1854.87	3671.91	-1817.05
Xinjiang	375.68	2325.87	-1950.19	306.07	-2256.26
National total	466706.33	113238.53	353467.80	353467.81	0.00

model of cooperation benefit interest coordination in regional planning by the proposed modified MCRS method. Finally, we take the pollutant SO_2 as an example. Compared with the current NCCOR, the results shows that our method can reduce the total control cost sharply. It can result in substantial environmental and public health benefits. It has potential to guide administration (such as the Environmental Protection Ministry) to effectively complete the air pollution reduction targets in the future. These indicate that our results might be successfully extended to cope with more transboundary environmental problems in the Twelfth Five-Year period.

ACKNOWLEDGEMENTS

This study was supported by grants from the National Natural Science Foundation of China (project No. 90924030), and Program for New Century Excellent Talents in University (project No. NCET-10-0938) in China, the Chinese Ministry of Education, Humanities and Social Sciences (No. 13YJC630196 and 13YJC630072) and Shanghai Philosophy and Social (2013EGL010).

REFERENCES

Becker N, Easter KW (1997). Water diversion from the Great Lakes: is a cooperative approach possible? Water Resour. Dev. 13(1):53-65.
Cao D, Song CY, Wang JN (2009). Pollutant reduction cost function of the establishment of joint and Empirical Analysis. Res. Environ. Sci. 22(3):371-376 (in Chinese).
Chander P, Tulkens H (1994). A core-theoretic solution for the design of cooperative agreements on transfrontier pollution. Int. Tax Public Finan. 2:279-293.

Chen WY, Hou D (1999) Optimal treatment cost allocation based on multiple person cooperation game in total emission control. Acta Scientiae Circumstantiae. 19(1):57-62 (In Chinese).
Du N, Cao D, Yang HF (2007). Research on Air Pollution Abatement Cost Function of Industrial Enterprise. Sci. Technol. Eng. 17(6):1116-1118 (In Chinese).
Duan (1991). Comprehensive urban environmental improvement optimal planning method. Beijing: China Environmental Science Press.
Germain M, Toint P, Tulkens H, De Zeeuw A (2003). Transfers to sustain in dynamic core theoretic cooperation in international stock pollutant control. J. Econ. Dyn. Control 28:79-99.
Grabisch M, Lange F (2007). Games on lattices, multichoice games and the Shapley value: A new approach. Math. Meth. Oper. Res. 65(1):153-167.
Halkos GE (1993). Sulphur abatement policy: Implications of cost differentials. Energy Pol. 1:1035-1043.
Halkos GE (1994). Optimal abatement of sulphur in Europe. Environ. Resour. Econ. 4:127-150.
Halkos GE (1996). Incomplete information in the acid rain game. Empirica. 23:129-148.
Hsiao CR Raghavan TES (1993). Shapley value for multichoice cooperative games, I. Games Econ. Behav. 5(2):240-256.
Hsial C (1992). Multichoice cooperative games. Games Theory Econ. Appl. 38:170-188.
Hu BQ (2000). The discrete planning in total control. Beijing: China Environmental Science Press.
Jorgensen S, Zaccour G (2001). Time consistent side payments in a dynamic game of downstream pollution. J. Econ. Dyn. Control 25:1973-1987.
Pan XZ, Wang QY, Han WM, Shen XD. Hua RH (2004). Atmospheric environment capacity calculation method and application of practice. The Atmospheric Environment Science and Technology Research Progress-11th National Atmospheric Environment Conference Proceedings, Jinan pp. 110-120.
Tan ZF, Yu C, Li L, Jiang HY, Cai CK (2011). Benefits distribution optimization models of energy-saving and SO_2 emission-reducing based on power generation side and supply side's cooperation. Syst. Eng.-Theor. Pract. 31(6):1182-1190. (in Chinese).
Tao HY (2011). Haze days, are man-made fault of "bad"!. Worker's Daily. December 2, 2011, 6th edition.
Tijs SN, Driessen TSH (1986). Game theory and cost allocation problems. Manag. Sci. 8:1015-1028.
Walter S, Clifford SR, Robert AK (1976). Environmental Quality Management Application to the Lower Delacoare Valley. Washington DC: Resources for the Future.
Wang S (2011). The Twelfth Five-year Plan reduction program announced: Index distribution linked to the environment. Jinghua Times. October 27, 5th edition. (in Chinese).
Wang S (2011). Determine the PM2.5 monitoring from next year in Beijing-Tianjin-Hebei. Jinghua Times. December 22, 9th edition.
Wu L (2007). Theoretical basis of the emissions trading system research. Chin. Agric. Sci. Bull. 23(4):497-501.
Zeng QC (1996). Natural cybernetics. Clim. Environ. Res. 28(1):16-21.
Zhao LJ, Cao W (2011). Study of regulating model of cooperation and reallocation of Benefits on transboundary pollution of lake basin. J. Syst. Eng. 26(3):367-372. (in Chinese).

Extreme pressure property of Carbon Nano Tubes (CNT) based nanolubricant

Shubrajit Bhaumik, S.Prabhu, VarunKarpurapu and Nihit Uppala

School of Mechanical Engineering, SRM University, Chennai - 603203 India.

The paper predicts the exceptional extreme pressure properties of multi walled carbon nano tube as compared with traditional mineral oil which also contained some additives and graphite. The extreme pressure property was predicted by four ball tester. In the test the load bearing capacity of the newly developed nano oil was predicted to have a better value than the commercial available mineral oil. The weld load of the nanolubricant was found to be 200 kgf as compared with the base mineral oil which got welded at 160 kgf. Thus, the finding would be helpful in developing new nanolubricants which would increase the life of the mating parts, reducing the chances of failure of machine and increasing the productivity.

Key words: Mineral oil, multi walled carbon nano tube (MWCNT), pass load, weld load, load bearing capacity, Carbon Nano Tubes (CNT).

INTRODUCTION

With the advancement of new technologies, new critical applications have also taken their birth in various fields. The birth of critical applications has also made the researchers to think regarding the protection of the parent material. Lubrication is an important phenomenon in any process industries. Initially, mineral oil was commonly used. With the advancement of the modern era of mechanisms, the researchers are trying to extend the life of the mating parts. For that, various types of additives have been formulated and added to the base mineral oil. For many years phosphorus, sulfur, nitrogen and chlorine compounds are used as successful AW and EP additives. The use of these compounds decreased because of increase in price of sulfur and nitrogen compounds. These compounds also pose environmental and toxicological hazards. So current research is focused on development of eco-friendly additives. In recent years graphite and molybdenum disulfide have attracted the attention because of their excellent emergency lubricating properties which plays an important role when lubricant film begins to break. The nanoscale materials have drawn ample attention of researchers from distinct research fields in recent years because of its propitious physical and chemical properties. In tribology nanomaterials have been proved to have greater potential as lubricating additives. Carbon nanotubes are one such nanomaterial whose existence was observed in 1990s. Due to superior physical and mechanical properties, carbon nano tubes has found its place in many critical applications. Due to its tubular structures and higher mechanical properties it can be predicted that carbon nano tubes may work as rolling bearings without getting broken between two mating parts which may reduce the frictional coefficient and increase the load bearing capacity of the CNT based lubricant.

Curasu et al. (2012) predicted the frictional reduction capability and excellent AW/EP properties of SWCNTs and obtained the optimum concentration as 0.5 wt%. Hong et al. (2010) concluded that CNT greases distinguish themselves from more common graphitic materials as solid grease additives. The performance of CNT grease could be much better with the improvement of nanotube quality and purity. Marquis and Chiabnte (2005) concluded that CNT is found to considerably increase the thermal conductivity of many heat transfer fluids such as mineral and synthetic oils, water, water/ethylene glycol mixtures and other commercial heat transfer fluids such as antifreeze. Wang et al. (2007) concluded that when CaF_2 nanocrystals are used as lubricant additive in lithium grease, it improves the wear resistance, load carrying capacity and anti friction with the optimum concentration 1 wt%. Xianbing et al. (2011) concluded that when $CaCo_3$ nanoparticles as an additive in lithium grease significantly improve its anti-wear performance, friction reduction property, load carrying capacity, and extreme pressure property with 0.5%wt optimum concentration.

In this paper, the extreme pressure property of MWNTs has been predicted. MWCNTs have been used as solid additives to base mineral oil in different percentages (0.1, 0.5 and 0.6% wt). The extreme pressure properties of the specimens were compared to that of the base mineral oil without additives and the graphite as an additive. A Four Ball Tester is used to evaluate the tribological properties. The properties of the specimens are listed in Table 1 as follows:

EXPERIMENTAL DETAILS

Base oil: Mineral oil ISO VG 250 is chosen because 250 cSt oils are used in heavy duty industrial applications like gear boxes of many process plants.

Additives: Multi-walled carbon nanotubes are used as an additive for comparing the tribological properties with classical additives (Graphite).

Preparation of nano lubricant

CNT have good dispersibility in mineral oil. When CNTs are directly added to the base mineral oil, they are not stable. Agglomeration of CNT occurs within no time. In order to get a stable lubricant, we can either use a surfactant or mix CNT to the mineral oil using ultrasonication. In the present work, stable lubricant for a longer time has been produced using ultra-sonication. No surfactant has been used. In ultra-sonication the length of the CNT are broken to increase the ease of dispersion of CNT in oil and hence maintains stability for a longer time. Specimens with various concentrations of MWNTs dispersed in mineral oil were stable up to 156 h.

Extreme pressure test

Four ball tester

A Four Ball Tester is used to evaluate the Extreme Pressure (EP) property of lubricant. The tribosystem consists of three stationary balls fixed in a ball pot which are pressed against the fourth ball fixed to the collet at required pressure. The top ball rotates at defined speed according to the test standards. The load is uniformly distributed over the three points where the three balls touch the fourth ball. The pot is filled with the lubricant. Rotation of the driving spindle causes a frictional torque that produces a scar on the lower balls. The test balls used had a 12.7 mm diameter and a roughness of R_a= 0.016 mm and were made of AISI E 52100 steel. The test standards were performed according to the ASTM D-2783 standard.

The test is performed by adding weights to the apparatus and will run the apparatus for specific time according to the test standards. After the completion of test with respective weight the balls will be removed from the apparatus and are observed. If the scar diameter is above 4 mm or if the ball gets welded then the respective load is called weld load of the specimen and the previous load where the balls do not get welded or the scar diameter is below 4 mm is called pass load.

SEM analysis

A scanning electron microscope (SEM) is a type of electron microscope that produces images of a sample by scanning it with a focused beam of electrons. The electrons interact with electrons in the sample, producing various signals that can be detected and that contain information about the sample's surface topography and composition. The morphology of the rubbed surfaces was investigated using an FEI Inspect scanning electron microscope with a maximum resolution of 3 nm.

RESULTS AND DISCUSSION

Four ball test results

To study the behavior of MWNTs under the above mentioned conditions, five specimens were compared under same conditions: Base mineral oil, Base mineral oil with CNT as an additive in three different concentrations (0.1, 0.5 and 0.6 wt%) and Base mineral oil with graphite as an additive. The Four Ball Test results in Table 2 show the load bearing capacity of each specimen.

As observed from Figure 1 and Table 2 the pass load and weld load of the base mineral oil is very less when compared to that of the other specimens. From Figure 1 it can be predicted that pass load and weld load values of CNT specimens and graphite specimens are same but is higher than the pure base oil which is being used in industries. This may be because both additives are allotropes of carbon. Since the pass load and weld load values are same it does not mean that all the specimens have same efficiency. The most efficient lubricant among the additive added specimens can be obtained by comparing the scar diameter at their pass Loads. The lower the scar diameter, the higher will be the load bearing capacity. The scar diameters of the four additive added specimens are shown in Table 3.

Though the graphite and CNT are allotropes of carbon, CNT have better physical properties compared to that of graphite. As the elastic modulus of CNT is much higher than graphite, the CNT can withstand more loads than graphite without breaking. So graphite added specimen has high scar diameter as compared to CNT specimens. Among the CNT added specimens there exists variation in scar diameters because of different concentrations of CNT. In 0.1 wt% CNT specimen the amount of CNT

Table 1. Physical properties of the lubricants.

Property	Mineral oil + 0.1% CNT	Mineral oil + 0.5% CNT	Mineral oil + 0.6% CNT	Mineral oil + 0.5% Graphite
Viscosity	249.70	251.85	242.56	249
Flash point	220	218	222	218
Fire point	232	228	234	226

Table 2. Pass Load and Weld Load values from Four Ball Test.

Specimen	Pass load (kg)	Weld load (kg)
Base mineral oil	126	160
Base mineral oil + 0.1 wt% CNT	160	200
Base mineral oil + 0.5 wt% CNT	160	200
Base mineral oil + 0.6 wt% CNT	160	200
Base mineral oil + Graphite	160	200

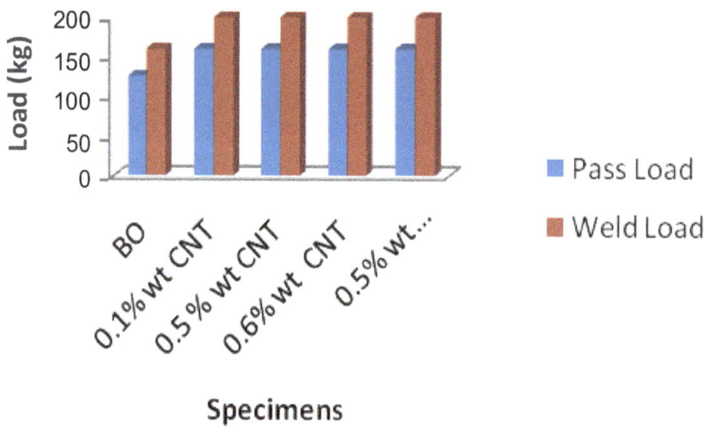

Figure 1. Pass Load and Weld Load values from the Four Ball Test.

Table 3. Scar diameters of different specimens.

Specimen	Scar diameter at pass load (mm)
Base mineral oil + 0.1 wt% CNT	2.431
Base mineral oil + 0.5 wt% CNT	2.225
Base mineral oil + 0.6 wt% CNT	2.313
Base mineral oil + Graphite	3.136

present is not sufficient to make it a better efficient lubricant whereas in 0.6 wt% CNT specimen the amount of CNT present became excessive because the oil reached its solubility state and excess CNT cannot be dissolved in the oil which leads to agglomeration of the nanotubes. This may also increase the fluid friction between tribopairs. Thus, an increasing trend has been observed after 0.5% wt as shown in Figure 2. Hence, 0.5 wt% CNT is the optimum concentration.

SEM analysis

SEM images predict the microstructure of the balls. It can be predicted that the heat generated in case of the mineral oil and mineral oil + 0.5%wt graphite sample is higher due to which white layers are being observed in Figure 3(a) and (c). Thus, it can be predicted that the lubricant film is not stable in case of mineral oil and mineral oil + 0.5%wt graphite. On the contrary, the image

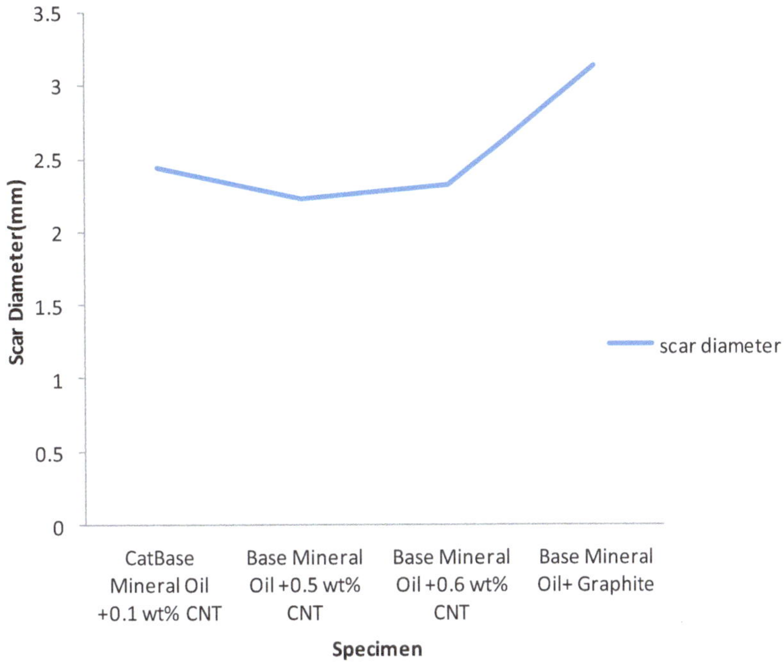

Figure 2. Scar diameters on balls.

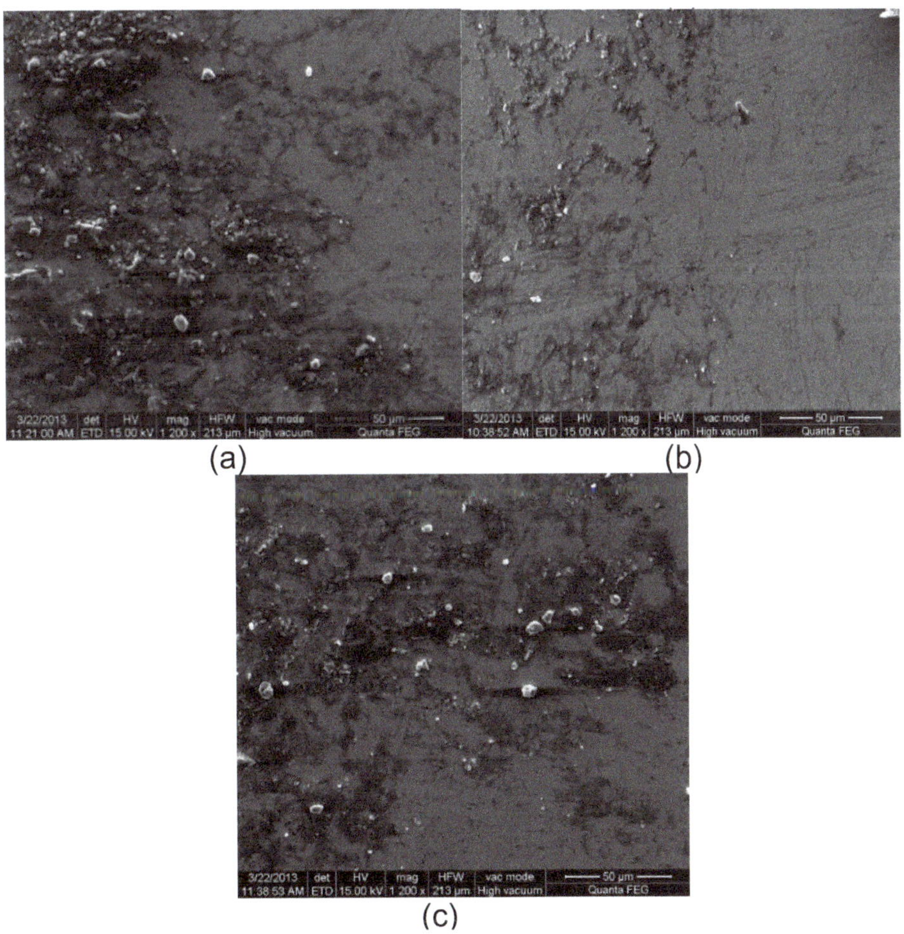

Figure 3. SEM image of metal ball used in four ball tester for pass load of (a) mineral oil (b) Mineral oil +0.5% wt CNT (c) Mineral oil + 0.5% Graphite.

of mineral oil + 0.5%wt CNT predicts lesser formation of heat effected zone. Thus, CNT lubricant film is predicted to be more stable than the mineral oil and mineral oil + 0.5% graphite sample.

Conclusion

From the results predicted in the present work we can conclude that MWCNTs are much more efficient additives than commonly used graphite. The load carrying capacity, which is an important parameter for the lubricant has been predicted to be higher in case of nano lubricants than the conventional lubricant. The present result would be beneficial in developing new nano fluids.

REFERENCES

Curasu DL, Andronescu C, Pirva C, Ripeanu R (2012). The efficiency of Co based SWCNT as an AW/EP additive for mineral base oil. Wear 290-291:133-139.

Hong H, Thomas D, Waynick A, Yu W, Smith P, Roy W (2010). Carbon Nanotube grease with enhanced thermal and electrical conductivities. J. Nanoparticle Res. 12:529-535

Marquis FDS, Chiabnte LPF (2005). Improving the Heat Transfer of Nanofluids and Nanolubricants with Carbon Nanotubes. JOM 57:32-43.

Wang L, Wang B, Wang X, Liu W (2007). Tribological Investigation of CaF_2 nanocrystals as grease additives. Tribol. Int. 40:1179-1185

Xianbing J, Yinxia C, Galqlng Z, Xiaobo W, Weimin L (2011). Tribological properties of CaCO3 nanoparticles as an additive in lithium grease. Tribol. Lett. 41:113–119.

Permissions

All chapters in this book were first published in JCEMS, by Academic Journals; hereby published with permission under the Creative Commons Attribution License or equivalent. Every chapter published in this book has been scrutinized by our experts. Their significance has been extensively debated. The topics covered herein carry significant findings which will fuel the growth of the discipline. They may even be implemented as practical applications or may be referred to as a beginning point for another development.

The contributors of this book come from diverse backgrounds, making this book a truly international effort. This book will bring forth new frontiers with its revolutionizing research information and detailed analysis of the nascent developments around the world.

We would like to thank all the contributing authors for lending their expertise to make the book truly unique. They have played a crucial role in the development of this book. Without their invaluable contributions this book wouldn't have been possible. They have made vital efforts to compile up to date information on the varied aspects of this subject to make this book a valuable addition to the collection of many professionals and students.

This book was conceptualized with the vision of imparting up-to-date information and advanced data in this field. To ensure the same, a matchless editorial board was set up. Every individual on the board went through rigorous rounds of assessment to prove their worth. After which they invested a large part of their time researching and compiling the most relevant data for our readers.

The editorial board has been involved in producing this book since its inception. They have spent rigorous hours researching and exploring the diverse topics which have resulted in the successful publishing of this book. They have passed on their knowledge of decades through this book. To expedite this challenging task, the publisher supported the team at every step. A small team of assistant editors was also appointed to further simplify the editing procedure and attain best results for the readers.

Apart from the editorial board, the designing team has also invested a significant amount of their time in understanding the subject and creating the most relevant covers. They scrutinized every image to scout for the most suitable representation of the subject and create an appropriate cover for the book.

The publishing team has been an ardent support to the editorial, designing and production team. Their endless efforts to recruit the best for this project, has resulted in the accomplishment of this book. They are a veteran in the field of academics and their pool of knowledge is as vast as their experience in printing. Their expertise and guidance has proved useful at every step. Their uncompromising quality standards have made this book an exceptional effort. Their encouragement from time to time has been an inspiration for everyone.

The publisher and the editorial board hope that this book will prove to be a valuable piece of knowledge for researchers, students, practitioners and scholars across the globe.

List of Contributors

E. Chinedu Ekuma
Department of Physics, University of Port Harcourt, P. M. B. 5323, Choba, Port Harcourt, Rivers State, Nigeria

E. Ndubuisi Idenyi
Department of Industrial Physics, Ebonyi State University, P. M. B. 053, Abakaliki, Ebonyi State, Nigeria

O. Israel Owate
Department of Physics, University of Port Harcourt, P. M. B. 5323, Choba, Port Harcourt, Rivers State, Nigeria

Nadia H. Helal
Chemistry Department, Faculty of Science, Fayoum University - Fayoum, Egypt

M. Kavanloui
Department of Materials Science and Engineering, School of Engineering, Shiraz University, Shiraz, Iran

B. Hashemi
Department of Materials Science and Engineering, School of Engineering, Shiraz University, Shiraz, Iran

E. Nourafkan
Department of Chemical and Petroleum Engineering, School of Engineering, Shiraz University, Shiraz, Iran

Mortaza Gholizadeh
Department of Chemical Engineering, Islamic Azad University-Ahar Branch, Ahar, I.R Iran

Mosen Momayyeza
Department of Chemical Engineering, Islamic Azad University-Ahar Branch, Ahar, I.R Iran

S. Missaoui
Electronics Laboratory, Department of Physics, Faculty of Science, Tunis El Manar, Tunisia

A. Gharbi
Electronics Laboratory, Department of Physics, Faculty of Science, Tunis El Manar, Tunisia

M. Kaddour
Sfax Faculty of Science, Tunisia

Lutvo Kurić
Bosnia and Herzegovina, Novi Travnik, Kalinska 7

L. S. Neiva
Academic Unit of Engineering of Materials, Federal University of Campina Grande - 58109-970, Avenue Aprígio Veloso, 882 - Bodocongó - Campina Grande, PB - Brazil

H. M. C. Andrade
Academic Unit of Engineering of Materials, Federal University of Campina Grande - 58109-970, Avenue Aprígio Veloso, 882 - Bodocongó - Campina Grande, PB - Brazil

L. Gama
Academic Unit of Engineering of Materials, Federal University of Campina Grande - 58109-970, Avenue Aprígio Veloso, 882 - Bodocongó - Campina Grande, PB - Brazil

C. Koroneos
Laboratory of Heat Transfer and Environment, Aristotle, University of Thessaloniki, Box 483, 54124 Thessaloniki, Greece

S. Lykidou
Department of Chemistry, Laboratory of Organic Chemistry Technology, Aristotle University of Thessaloniki, 54124 Thessaloniki, Greece

Joel Hemanth
Rajiv Gandhi Institute of Technology, R. T. Nagar Post, Hebbal, Bangalore-560032, Karnataka, India

Hamid Reza Goshayeshi
Islamic Azad University-Mashhad Branch (IAUM), Mashhad, Iran

Mohammad Reza Safaiy
Young Researchers Club, Islamic Azad University-Mashhad Branch, Iran

Ranjana Sharma
School of Applied Sciences, Netaji Subhas Institute of Technology, Sector- 3, Dwarka, New Delhi - 110078

Himani Joshi
School of Applied Sciences, Netaji Subhas Institute of Technology, Sector- 3, Dwarka, New Delhi - 110078

Purnima Jain
School of Applied Sciences, Netaji Subhas Institute of Technology, Sector- 3, Dwarka, New Delhi - 110078

Anjali Goel
Department of Chemistry, KGM, Gurukul Kangri University, Post - Jwalapur, Hardwar-249407 (UK) India

Ruchi Sharma
Department of Chemistry, KGM, Gurukul Kangri University, Post - Jwalapur, Hardwar-249407 (UK) India

K. R. Genwa
Department of Chemistry, Jai Narain Vyas University, Jodhpur - 342005, India

C. P. Sagar
Department of Chemistry, Jai Narain Vyas University, Jodhpur - 342005, India

M. A. Moharram
Spectroscopy Department, Physics Division, National Research Center, El Behooth Street, 12622 Doki, Giza, Egypt

F. H. Abd El-kader
Faculty of Science, Physics Department, Cairo University, Egypt

Kh. I. Grais
Spectroscopy Department, Physics Division, National Research Center, El Behooth Street, 12622 Doki, Giza, Egypt

M. G. Khafagi
Spectroscopy Department, Physics Division, National Research Center, El Behooth Street, 12622 Doki, Giza, Egypt

Fathia Mamdouh
Spectroscopy Department, Physics Division, National Research Center, El Behooth Street, 12622 Doki, Giza, Egypt

Stephen A. Adegbite
School of Chemical Engineering, University of Birmingham, Edgbaston, Birmingham B15 2TT, United Kingdom

Alpaben K. Patel
Directorate of Economics and Statistics, Sector-18, Gandhinagar-382009, India

Jyoti Divecha
Department of Statistics, Sardar Patel University, Vallabh Vidyanagar, 388120, India

Sarmila Dutta
Department of Physics, IMPS College of Engineering and Technology Malda, India

Shibabrata Basak
School of Applied Sciences, Haldia Institute of Technology, Haldia, India

Pijus Kanti Samanta
Department of Physics, Ghatal Rabindra Satabarsiki Mahavidyalaya, Ghatal, Paschim Medinipur, India

I. Yusufu, Mohammed
Department of Food Science and Technology, Federal Polytechnic, Idah- Nigeria

C. C. Ariahu
Department of Food Science and Technology, University of Agriculture, Makurdi, Nigeria

N. N. Nkpa
Department of Chemistry, University of Agriculture, Makurdi, Nigeria

B. D. Igbabul
Department of Food Science and Technology, University of Agriculture, Makurdi, Nigeria

Rashid Mahmood
Division of Science and Technology, University of Education, Township Lahore-54000, Pakistan

Muhammad Javed Iqbal
Centre of Nano-Science and Technology, Preston University, H-8 Islamabad

Mohammadine El Haddad
Equipe de Chimie Analytique and Environnement, Faculté Poly-disciplinaire, Université Cadi Ayyad, BP 4162, 46000 Safi, Maroc

H. V. Rajeshwari
Department of Studies in Chemistry, Karnatak University, Pavate Nagar, Dharwad 580003, India

K. S. Byadagi
Department of Studies in Chemistry, Karnatak University, Pavate Nagar, Dharwad 580003, India

S. T. Nandibewoor
Department of Studies in Chemistry, Karnatak University, Pavate Nagar, Dharwad 580003, India

S. A. Chimatadar
Department of Studies in Chemistry, Karnatak University, Pavate Nagar, Dharwad 580003, India

Nadejda Taneva
University of Architecture, Civil Engineering and Geodesy, "Christo Smirnenski"blvd, Sofia 1046, Bulgaria

H. E. El-Feky
Department of Chemistry, Faculty of Science, University of Fayoum, Fayoum, Egypt

N. H. Helal
Department of Chemistry, Faculty of Science, University of Fayoum, Fayoum, Egypt

M. R. Negem
Department of Chemistry, Faculty of Science, University of Fayoum, Fayoum, Egypt

Simbarashe Fashu
School of Material Science and Engineering, Shanghai Jiao Tong University, Shanghai, Dongchuan Road 800, 200240, China

Li Jianguo
School of Material Science and Engineering, Shanghai Jiao Tong University, Shanghai, Dongchuan Road 800, 200240, China

Hu Qiao Dan
School of Material Science and Engineering, Shanghai Jiao Tong University, Shanghai, Dongchuan Road 800, 200240, China

Hui-Wang Cui
College of Wood Science and Technology, Nanjing Forestry University, Nanjing 210037, Jiangsu, China
College of Wood Science and Technology, Southwest Forestry University, Kunming 650224, Yunnan, China
Sino-Swedish Microsystem Integration Technology Center, Shanghai University, Shanghai 200072, China

Guan-Ben Du
College of Wood Science and Technology, Nanjing Forestry University, Nanjing 210037, Jiangsu, China
College of Wood Science and Technology, Southwest Forestry University, Kunming 650224, Yunnan, China

Akl M. Awwad
Royal Scientific Society, P. O. Box 1438 Al-Jubaiha 11941 Amman, Jordan

Nidà M. Salem
Faculty of Agriculture, Jordan University, Amman, Jordan

Farah Noori
Physics Science, Baghdad University, Baghdad, Iraq

Solveig Melin
Division of Mechanics, Lund University, Lund, Sweden

A. Aravind Kumar
Department of Chemical Engineering, University of Petroleum and Energy Studies, Bidholi, Through Prem Nagar, Dehradun-248007, India

A. Ravi Prasad
Department of Chemical Engineering, Andhra University College of Engineering, Visakhapatnam-530003, India

A. Hamza
Department of Chemical Engineering, Ahmadu Bello University, Zaria, Nigeria

J.T. Fatuase
Department of Chemical Engineering, Ahmadu Bello University, Zaria, Nigeria

S. M. Waziri
Department of Chemical Engineering, Ahmadu Bello University, Zaria, Nigeria

O. A. Ajayi
Department of Chemical Engineering, Ahmadu Bello University, Zaria, Nigeria

Moka Shekhu
Department of Science, Government Polytechnic, Chennasandra, Bangalore-560067, India

C. Sulochana
Department of Mathematics, Gulbarga University, Gulbarga, Karnataka, India

Jian Xue
School of Management, Shanghai University, 200444, Shanghai, P. R. China
Training Center of China Post Group Corporation, 050021, Hebei, Shijia Zhuang, P. R. China

Changmin Li
School of Management, Shanghai University, 200444, Shanghai, P. R. China

Shubrajit Bhaumik
School of Mechanical Engineering, SRM University, Chennai - 603203 India

S.Prabhu
School of Mechanical Engineering, SRM University, Chennai - 603203 India

VarunKarpurapu and Nihit Uppala
School of Mechanical Engineering, SRM University, Chennai - 603203 India